Lecture Notes in Networks and Systems

Volume 348

The series "Lecture Notes in Networks and Systems" publishes the latest developments in Networks and Systems—quickly, informally and with high quality. Original research reported in proceedings and post-proceedings represents the core of LNNS.

Volumes published in LNNS embrace all aspects and subfields of, as well as new challenges in, Networks and Systems.

The series contains proceedings and edited volumes in systems and networks, spanning the areas of Cyber-Physical Systems, Autonomous Systems, Sensor Networks, Control Systems, Energy Systems, Automotive Systems, Biological Systems, Vehicular Networking and Connected Vehicles, Aerospace Systems, Automation, Manufacturing, Smart Grids, Nonlinear Systems, Power Systems, Robotics, Social Systems, Economic Systems and other. Of particular value to both the contributors and the readership are the short publication timeframe and the world-wide distribution and exposure which enable both a wide and rapid dissemination of research output.

The series covers the theory, applications, and perspectives on the state of the art and future developments relevant to systems and networks, decision making, control, complex processes and related areas, as embedded in the fields of interdisciplinary and applied sciences, engineering, computer science, physics, economics, social, and life sciences, as well as the paradigms and methodologies behind them.

Indexed by SCOPUS, INSPEC, WTI Frankfurt eG, zbMATH, SCImago.

All books published in the series are submitted for consideration in Web of Science.

For proposals from Asia please contact Aninda Bose (aninda.bose@springer.com).

More information about this series at https://link.springer.com/bookseries/15179

M. Shamim Kaiser · Kanad Ray ·
Anirban Bandyopadhyay · Kavikumar Jacob ·
Kek Sie Long
Editors

Proceedings of the Third International Conference on Trends in Computational and Cognitive Engineering

TCCE 2021

Springer

Editors
M. Shamim Kaiser
Jahangirnagar University
Dhaka, Bangladesh

Kanad Ray
Amity University Rajasthan
Jaipur, India

Anirban Bandyopadhyay
Surface Characterization Group
National Institute for Materials Science
Tsukuba, Japan

Kavikumar Jacob
Faculty of Applied Sciences
and Technology
Tun Hussein Onn University of Malaysia
Parit Raja, Johor, Malaysia

Kek Sie Long
Faculty of Applied Sciences
and Technology (FAST)
Universiti Tun Hussein Onn Malaysia
Johor, Malaysia

ISSN 2367-3370 ISSN 2367-3389 (electronic)
Lecture Notes in Networks and Systems
ISBN 978-981-16-7596-6 ISBN 978-981-16-7597-3 (eBook)
https://doi.org/10.1007/978-981-16-7597-3

This Springer imprint is published by the registered company Springer Nature Singapore Pte Ltd.
The registered company address is: 152 Beach Road, #21-01/04 Gateway East, Singapore 189721, Singapore

Organization

Chief Patron

Prof. Datuk Ts. Dr. Wahid Bin Razzaly, Vice-chancellor, UTHM, Malaysia

Conference Co-chairs

Dr. Chi-Sang Poon, Massachusetts Institute of Technology, USA
Dr. Anirban Bandyopadhyay, National Institute for Materials Science, Japan
Dr. Kanad Ray, Amity University, Rajasthan, India

Steering Committee

Anirban-Bandyopadhyay, National Institute For Materials Science, Japan
Anirban Dutta, The State University of New York at Buffalo, USA
Chi-Sang Poon, Massachusetts Institute of Technology, USA
J. E. Lugo, University of Montreal, Canada
Jocelyn Faubert, University of Montreal, Canada
Kanad Ray, Amity University Rajasthan, India
Luigi M. Caligiuri, The University of Calabria, Italy
Mufti Mahmud, Nottingham Trent University, UK
M. Shamim Kaiser, Jahangirnagar University, Bangladesh
Subrata Ghosh, Northeast Institute of Science & Technology, India
Shamim Al Mamun, Jahangirnagar University, Bangladesh

Advisory Committee

Alamgir Hossain, Teesside University, UK
Hashim Bin Saim, UTHM, Malaysia
Jalil Bin Ali, UTM, Malaysia
Joarder Kamruzzaman, Federation University Australia
Nafarizal Bin Nayan, UTHM, Malaysia
Kazi M. Ahmed, UAP, Bangladesh
Kok Lay Teo, Sunway University, Malaysia
Rosli Bin Omar, UTHM, Malaysia
Md. Abu Taher, BUP, Bangladesh
Md Abdur Razzaque, DU, Bangladesh
Md. Atiqur Rahman Ahad, DU, Bangladesh
Md. Saiful Islam, BUET, Bangladesh
Md. Obaidur Rahman, DUET, Bangladesh
Md. Hanif Ali, Jahangirnagar University, Bangladesh
Md Roshidul Hasan, BSMRAU, Bangladesh
Mohamad Zaky Bin Noh, UTHM, Malaysia
Muhammad H. Rashid, University Of West Florida, USA
Sarwar Morshed, AUST, Bangladesh
Satya Prashad Mazumder, BUET, Bangladesh
Subrata Kumar Aditya, SHU, Bangladesh

Organizing Committee

Mohd Helmy Abd Wahab, UTHM, Malaysia (Chair)
Syed Zuhaib Haider Rizvi, UTHM, Malaysia (Co-Chair)
M. Shamim Kaiser, JU, Bangladesh (Co-Chair)
Muhammad Sufi Bin Roslan, UTHM, Malaysia (Secretary)
Kek Sie Long, UTHM, Malaysia (Registration Chair)
Umar Abu Bakar, UTHM, Malaysia (Treasurer)
Maizatulazrina Binti Yaakob, UTHM, Malaysia
Syed Zulkarnain Syed Idrus, UTHM, Malaysia
Haryana Mohd Hairi, UTHM, Malaysia
Saiful Najmee Bin Mohamad, UTM, Malaysia
Iliana Md. Ali, TATIUC, Malaysia
Shamim Al Mamun, JU, BD
Syed Asim Shah, UTHM, Malaysia

Technical Program Committee Chairs

Anirban-Bandyopadhyay (Chair), National Institute For Materials Science, Japan
Sie Kek, UTHM, Malaysia (Co-Chair)
M. Arif Jalil, UTM, Malaysia (Co-Chair)

Technical Program Committee

Kavi Kumar, UTM, Malaysia
Nabihah Ahmad, UTM, Malaysia
S. P. Tiwari, IIT, Dhanbad, India
Su Rong, NTU, Singapore
Said Broumi, University Of New Mexico, USA
Pal Madhumangal, Vidyasagar University, India
F. Suresh Singh, University Of Kerala, India
Sabariah Saharan, UTHM, Malaysia
Arif Jalil, UTM, Malaysia
Muhammad Arifur Rahman, Jahangirngar University, Bangladesh
Sajjad Waheed, MBSTU, Bangladesh
Md. Zahidur Rahman, GUB, Bangladesh
Muhammad Golam Kibria, ULAB, Bangladesh
Md. Majharul Haque, Deputy Director, Bangladesh Bank
Samsul Arefin, CUET, Bangladesh
Md. Obaidur Rahman, DUET, Bangladesh
Mustafa Habib Chowdhury, IUB, Bangladesh
Marzia Hoque-Tania, Oxford University, UK
Antesar Shabut, CSE, Leeds Trinity University, UK
Md. Khalilur Rhaman, BRAC University, Bangladesh
Md. Hanif Seddiqui, University of Chittagong, Bangladesh
M. M. A. Hashem, KUET, Bangladesh
Tomonori Hashiyama, The University of Electro-Communications, Japan
Wladyslaw Homenda, Warsaw University of Technology, Poland
M. Moshiul Hoque, CUET, Bangladesh
A. B. M. Aowlad Hossain, KUET, Bangladesh
Sheikh Md. Rabiul Islam, KUET, Bangladesh
Manohar Das, Oakland University, USA
Kaushik Deb, CUET, Bangladesh
Carl James Debono, University of Malta, Malta
M. Ali Akber Dewan, Athabasca University, Canada
Belayat Hossain, Loughborough University, UK
Khoo Bee Ee, Universiti Sains Malaysia, Malaysia
Ashik Eftakhar, Nikon Corporation, Japan

Md. Tajuddin Sikder, Jahangirnagar University, Bangladesh
Mrs. Shayla Islam, UCSI, Malaysia
Antony Lam, Mercari Inc. Japan
Ryote Suzuki, Saitama University, Japan
Hishato Fukuda, Saitama University, Japan
Md. Golam Rashed, Rajshahi University, Bangladesh
Md. Sheikh Sadi, KUET, Bangladesh
Tushar kanti Shaha, JKKNIU, Bangladesh
M. Shazzad Hosain, NSU, Bangladesh
M. Mostafizur Rahman, AIUB, Bangladesh
Tabin hassan, AIUB, Bangladesh
Aye Su Phyo, Computer University Kalay, Myanmer
Md. Shahedur Rahman, Jahangirnagar University
Lu Cao, Saitama University, Japan
Nihad Adnan, Jahangirnagar University
Mohammad Firoz Ahmed, Jahangirnagar University
A. S. M. Sanwar Hosen, JNU, South Korea
Mahabub Hossain, ECE, HSTU, Bangladesh
Md. Sarwar Ali, Rajshahi University, Bangladesh
Risala T. Khan, Jahangirnagar University, Bangladesh
Mohammad Shahidul Islam, Jahangirnagar University, Bangladesh
Manan Binte Taj Noor, Jahangirnagar University, Bangladesh
Md. Abu Yousuf, Jahangirnagar University, Bangladesh
Md. Sazzadur Rahman, Jahangirnagar University, Bangladesh
Rashed Mazumder, Jahangirnagar University, Bangladesh
Md. Abu Layek, Jagannath University, Bangladesh
Saiful Azad, Universiti Malaysia Pahang, Malaysia
Mostofa Kamal Nasir, MBSTU, Bangladesh
Mufti Mahmud, NTU, UK
A. K. M. Mahbubur Rahman, IUB, Bangladesh
Al Mamun, Jahangirnagar University, Bangladesh
Al-Zadid Sultan Bin Habib, KUET, Bangladesh
Anup Majumder, Jahangirnagar University, Bangladesh
Atik Mahabub, Concordia University, Canada
Bikash Kumar Paul, MBSTU, Bangladesh
Md. Obaidur Rahman, DUET, Bangladesh
Nazrul Islam, MIST, Bangladesh
Ezharul Islam, Jahangirnagar University, Bangladesh
Farah Deeba, DUET, Bangladesh
Md. Manowarul Islam, Jagannath University, Bangladesh
Md. Waliur Rahman Miah, DUET, Bangladesh
Rubaiyat Yasmin, Rajshahi University, Bangladesh
Sarwar Ali, Rajshahi University, Bangladesh
Rabiul Islam, Kulliyyah of ICT, Malaysia
Dejan C. Gope, Jahangirnagar University, Bangladesh

Sk. Md. Masudul Ahsan, KUET, Bangladesh
Mohammad Shahriar Rahman, ULAB, Bangladesh
Golam Dastoger Bashar, Boise State University, USA
Md. Hossam-E-Haider, MIST, Bangladesh
H. Liu, Wayne State University, USA
Imtiaz Mahmud, Kyungpook National University, Korea
Kawsar Ahmed, MBSTU, Bangladesh
Kazi Abu Taher, BUP, Bangladesh
Linta Islam, Jagannath University, Bangladesh
Md. Musfique Anwar, Jahangirnagar University, Bangladesh
Md. Sanaul Haque, University of Oulu, Finland
Md. Ahsan Habib, MBSTU, Bangladesh
Md. Habibur Rahman, MBSTU, Bangladesh
M. A. F. M. Rashidul Hasan, Rajshahi University, Bangladesh
Md. Badrul Alam Miah, UPM, Malaysia
Mohammad Ashraful Islam, MBSTU, Bangladesh
Mokammel Haque, CUET, Bangladesh
Muhammad Ahmed, ANU, Australia
Nazia Hameed, University of Nottingham, UK
Partha Chakraborty, CoU. Bangladesh
Kandrapa Kumar Sarma, Gauhati University, India
Vaskar Deka, Gauhati University, India
K. M. Azharul Islam, KUET, Bangladesh
Tushar Sarkar, RUET, Bangladesh
Surapong Uttama, Mae Fah Luang University, Thailand
Sharafat Hossain, KUET, Bangladesh
Shaikh Akib Shahriyar, KUET, Bangladesh
A. S. M Sanwar Hosen, Jeonbuk National University, Korea

Publication Committee

Kanad Ray, Amity University Rajasthan, India
Kavikumar S/O Jacob, UTHM, Malaysia
Mohd Helmy Bin Abd Wahab, UTHM, Malaysia
Kek Sie Long, UTHM, Malaysia
Mufti Mahmud, NTU, UK

Preface

The Third International Conference on Trends in Computational and Cognitive Engineering, TCCE 2021, was held from October 21 to 22, 2021. The conference is hosted by Universiti Tun Hussein Onn Malaysia (UTHM), 86400 Parit Raja, Batu Pahat, Johor, Malaysia. The inaugural and second events of this series took place at Central University of Haryana in Mahendergarh, India, on November 28–30, 2019 and at Jahangirnagar University in Dhaka, Bangladesh, on December 17–18, 2020. The Center for Computational and Cognitive Engineering (TCCE) is dedicated to the experimental, theoretical, and applied aspects of computational and cognitive engineering. Computational and cognitive engineering analyzes diseases and behavioral disorders using computer and mathematical methods that are widely used in all fields of science, engineering, technology, and industry. The conference's objective is to bring together researchers, educators, and business professionals involved in related fields of research and development. This volume compiles the peer-reviewed papers presented at the meeting.

The conference on TCCE 2021 attracted 127 full papers from 9 countries in 6 tracks. These tracks include—Artificial Intelligence and Soft Computing, Cognitive Science and Computational Biology, IoT and Data Analytics, Network and Security, and Computer Vision. The submitted papers underwent a single-blind review process, soliciting expert opinion from at least two experts: at least two independent reviewers, the track co-chair, and the respective track chair. Following rigorous review reports from the reviewers and the track chairs, the technical program committee has selected 52 high-quality full papers from 09 countries that were accepted for presentation at the conference. Consequently, this volume of the TCCE 2021 conference proceedings contains those 45 full papers presented on October 21 and 22, 2021. Due to the COVID-19 pandemic, the Organizing Committee decided to host the event virtually. However, the research community reacted amazingly in this challenging time.

The book series will be insightful and fascinating for those interested in learning about computational intelligence and cognitive engineering that explores the dynamics of exponentially increasing knowledge in core and related fields. We are thankful to the authors who have made a significant contribution to the conference and have developed relevant research and literature in computation and cognitive engineering.

We would like to express our gratitude to the Organizing Committee and the Technical Committee members for their unconditional support, particularly the Chair, the Co-Chair, and the Reviewers. Special thanks to the advisors Prof. Dr. Jalil Ali and A. P. Dr. Nafarizal bin Nayan for their thorough support. TCCE 2021 could not have taken place without the tremendous work of the team and the gracious assistance. We would like to thank IEEE UTHM Student Branch. We are grateful to Mr. Aninda Bose, Ms. Sharmila Mary Panner Selvam, and other team members of Springer Nature for their continuous support in coordinating this volume publication. We would also like to thank Mr. Md. Mahfuzur Rahman of DIU and Milon Biswas of BUBT for continuous support. Last but not the least, we thank all of our contributors and volunteers for their support during this challenging time to make TCCE 2021 a success.

Dhaka, Bangladesh
Jaipur, India
Tsukuba, Japan
Parit Raja, Malaysia
Johor, Malaysia
October 2021

M. Shamim Kaiser
Kanad Ray
Anirban Bandyopadhyay
Kavikumar Jacob
Kek Sie Long

Contents

Network and Security

Signal Processing, Computer Vision and Rhythm Engineering

Editors and Contributors

About the Editors

Dr. M. Shamim Kaiser is currently working as Professor at the Institute of Information Technology of Jahangirnagar University, Savar, Dhaka-1342, Bangladesh. He received his Bachelor's and Master's degrees in Applied Physics Electronics and Communication Engineering from the University of Dhaka, Bangladesh, in 2002 and 2004, respectively, and the Ph.D. degree in Telecommunication Engineering from the Asian Institute of Technology, Thailand, in 2010. His current research interests include data analytics, machine learning, wireless network and signal processing, cognitive radio network, big data and cyber security, renewable energy. He has authored more than 100 papers in different peer-reviewed journals and conferences. He is Associate Editor of the IEEE Access Journal, Guest Editor of Brain Informatics Journal, and Cognitive Computation Journal. He is Life Member of Bangladesh Electronic Society; Bangladesh Physical Society. He is also a senior member of IEEE, USA, and IEICE, Japan, and active volunteer of the IEEE Bangladesh Section. He is founding Chapter Chair of the IEEE Bangladesh Section Computer Society Chapter. He organized various international conferences such as ICEEICT 2015-2018, IEEE HTC 2017, IEEE ICREST 2018, and BI2020.

Kanad Ray (Senior Member, IEEE) received the M.Sc. degree in physics from Calcutta University and the Ph.D. degree in physics from Jadavpur University, West Bengal, India. He has been Professor of Physics and Electronics and Communication and is presently working as Head of the Department of Physics, Amity School of Applied Sciences, Amity University Rajasthan (AUR), Jaipur, India. His current research areas of interest include cognition, communication, electromagnetic field theory, antenna and wave propagation, microwave, computational biology, and applied physics. He has been serving as Editor for various Springer book series. He was Associate Editor of the Journal of Integrative Neuroscience (The Netherlands: IOS Press). He has been visiting Professor to UTM & UTeM, Malyasia and visiting Scientist to NIMS, Japan. He has established MOU with UTeM

Malaysia, NIMS Japan, and University of Montreal, Canada. He has visited several countries such as Netherlands, Turkey, China, Czechoslovakia, Russia, Portugal, Finland, Belgium, South Africa, Japan, Singapore, Thailand, and Malaysia for various academic missions. He has organized various conferences such as SoCPROS, SoCTA, ICOEVCI, and TCCE as General Chair and Steering Committee Member.

Anirban Bandyopadhyay is Senior Scientist in the National Institute for Materials Science (NIMS), Tsukuba, Japan. He possesses a Masters of Science in Condensed Matter Physics, Computer, Numerical Analysis, and Astrophysics from North Bengal University and Doctor of Philosophy in Physics from Jadavpur University. He received his Ph.D. from the Indian Association for the Cultivation of Science (IACS), Kolkata. 2004–2005, where he worked on supramolecular electronics and multi-level switching. He has developed a resonance chain based complete human brain model that is fundamentally different than Turing tape essentially developing an alternate human brain map where filling gaps in the resonance chain is the key. He has developed unique a quantum music measurement machine and experiments on DNA proteins, microtubules, neurons, molecular machines, cancer. He has also developed a new frequency fractal model. His group has designed and synthesized several forms of organic brain jelly that learns, programs and solves problems by itself for futuristic robots during as well as several software simulators that write complex codes by themselves.

Dr. Kavikumar Jacob has been Associate Professor in the Faculty of Applied Sciences and Technology at Universiti Tun Hussein Onn Malaysia since 2006. In 2005, he obtained his doctor's degree from Annamalai University, India. He has published more than 50 research papers including the International Journal of Fuzzy System, Physica A: Statistical Mechanics and its Applications, and the Iranian Journal of Fuzzy Systems. Most of the publications are listed on the Scopus page. His research interests are control theory, automata theory, fuzzy mathematics, and numerical analysis.

Kek Sie Long is currently working as a senior lecturer in the Department of Mathematics and Statistics, Faculty of Applied Sciences and Technology, Universiti Tun Hussein Onn Malaysia. He received his M.Sc. and Ph.D. in Mathematics from Universiti Teknologi Malaysia, Johor, Malaysia, in 2002 and 2011, respectively. He was a research associate at the Curtin University of Technology in 2009 during his Ph.D. study. His research interests include optimization and control, operation research and management science, modeling and simulation, parameter estimation, Kalman filtering, and computational mathematics. He has published more than 30 papers in refereed journals and six (6) book chapters. He is a reviewer for peer-reviewed research journals. He has led two (2) research projects supported by the Ministry of Higher Education Malaysia. He has supervised three (3) master students and three (3) Ph.D. students. During 2014–2019, he has visited Changsha University of Science and Technology, Mongolian National University, Hong Kong Polytechnic University, Chongqing Normal University, Jamal Mohamed College (Autonomous), Carleton University, Wuhan University, and Future University Hakodate.

Contributors

Ahmad Zubaidi A.L. Universiti Sultan Zainal Abidin, Kuala Terengganu, Terengganu, Malaysia

Zaid Al-Muzzammil Abdul Halim Faculty of Technology Management and Business, Universiti Tun Hussein Onn Malaysia, Parit Raja Batu Pahat, Johor, Malaysia

Adlyn Nazurah Abdul Rahman Faculty of Technology Management and Business, Universiti Tun Hussein Onn Malaysia, Parit Raja Batu Pahat, Johor, Malaysia

B. H. Ahmad FECE, University Teknikal Malaysia Melaka, Melaka, Malaysia

Md. Imtiaz Ahmed Daffodil Institute of IT, Dhaka, Bangladesh

Sayed Shifat Ahmed Department of Electrical, Electronic and Communication Engineering, Pabna University of Science and Technology, Pabna, Bangladesh

Shamim Ahmed Computer Science and Engineering, Bangladesh University of Business and Technology, Dhaka, Bangladesh

Tawsin Uddin Ahmed Department of Computer Science and Engineering, University of Chittagong, Chittagong, Bangladesh

Happy Akter University of Barishal, Barishal, Bangladesh

Shamim Al Mamun Institute of Information Technology, Jahangirnagar University, Savar, Dhaka, Bangladesh;
Applied Intelligence and Informatics Lab (AII-Lab), Jahangirnagar University, Savar, Dhaka, Bangladesh

Tanzeem Alam International Islamic University Chittagong, Chittagong, Bangladesh

Jalil Ali Asia Metropolitan University, Masai, Johor, Malaysia

Mohammad Kamarulzzaman Bin Ali Faculty of Technology Management and Business, Universiti Tun Hussein Onn Malaysia, Parit Raja, Batu Pahat, Johor, Malaysia

Radzi Ambar Department of Electronic Engineering, Faculty of Electrical and Electronic Engineering, Universiti Tun Hussein Onn Malaysia, Parit Raja, Batu Pahat, Johor, Malaysia

Karl Andersson Department of Computer Science, Electrical and Space Engineering, Luleå University of Technology (LTU), Skellefteå, Sweden

Ferdous Ara BGC Trust University Bangladesh Bidyanagar, Chandanaish, Bangladesh

Md. Sharif Uddin Azad Institute of Information Technology, Jahangirnagar University, Savar, Dhaka-1342, Bangladesh

Md Khirul Islam Badal Jagannath University, Dhaka, Bangladesh

Mahdi Bahadoran Department of Physics, Shiraz University of Technology, Shiraz, Fars, Iran

Fairuz B. Baharom Faculty of Applied Sciences and Technology, Universiti Tun Hussein Onn Malaysia, Pagoh, Johor, Malaysia

A. Bandyopadhyay National Institute for Material Science, Tsukuba, Japan

Anirban Bandyopadhyay International Center for Materials and Nanoarchitectronics (WPI-MANA) and Research Center for Advanced Measurement and Characterization (RCAMC), National Institute for Materials Science (NIMS), Tsukuba, Ibaraki, Japan

Mohammad Kamarulzzaman Bin Ali Faculty of Technology Management and Business, Universiti Tun Hussein Onn Malaysia, Parit Raja Batu PahatJohor, Malaysia

Syed Rasul G. Syed bin Hamid Department of Cardiothoracic Surgery, Hospital Sultanah Aminah, Johor Bahru, Johor, Malaysia

Ety Biswangri BGC Trust University Bangladesh, Chandanaish, Chattogram, Bangladesh

Milon Biswas Computer Science and Engineering, Bangladesh University of Business and Technology, Dhaka, Bangladesh

Sudipto Chaki Computer Science and Engineering, Bangladesh University of Business and Technology, Dhaka, Bangladesh

Farzana Haque Chowdhury Department of Information and Communication Technology, Bangladesh University of Professionals, Mirpur, Dhaka-1216, Bangladesh

Antonio Coyopol Semiconductor Devices Research Center, ICUAP, BUAP, Ciudad Universitaria, Puebla, Pue, Mexico

Promi Das International Islamic University Chittagong, Chittagong, Bangladesh

Tanusree Dutta Organizational Behavior and Human Resource Management, Indian Institute of Management, Ranchi, India

Sam Huai En Department of Electrical and Electronics, Faculty of Engineering, Universiti Malaysia Sarawak, Kota Samarahan, Malaysia

Engku Mohd Nasri Engku Mat Nasir Electronic Engineering Department, Universiti Tun Hussein Onn Malaysia, Parit Raja, Malaysia

Mohd Fahmy-Abdullah Faculty of Technology Management and Business, Universiti Tun Hussein Onn Malaysia, Parit Raja, Johor, Malaysia;
Oasis Integrated Group, Universiti Tun Hussein Onn Malaysia, Batu Pahat, Johor, Malaysia

Rahat Hossain Faisal University of Barishal, Barishal, Bangladesh

Jocelyn Faubert Faubert Lab, School of Optometry, University de Montreal, Montreal, QC, Canada

Norfaiza Fuad Electronic Engineering Department, Universiti Tun Hussein Onn Malaysia, Parit Raja, Malaysia;
Computational, Signal, Imaging and Intelligent Focus Group (CSII), Parit Raja, Malaysia

José A. García-García Directorate of Health Education and Training, Hospital General de México "Dr. Eduardo Liceaga", Mexico City, Mexico

Godofredo García Semiconductor Devices Research Center, ICUAP, BUAP, Ciudad Universitaria, Puebla, Pue, Mexico

Anita Garhwal Asia Metropolitan University, Masai, Johor, Malaysia

Subrata Ghosh Chemical Science and Technology Division, CSIR North East Institute of Science and Technology, Jorhat, Assam, India

Tapotosh Ghosh Department of ICT, Bangladesh University of Professionals, Dhaka, Bangladesh

Syed Rasul bin G. Syed Hamid Department of Cardiothoracic Surgery, Hospital Sultanah Aminah, Johor, Malaysia

Low Qiau Han Faculty of Applied Sciences and Technology, Universiti Tun Hussein Onn Malaysia, Pagoh, Johor, Malaysia

Hafiz Mahdi Hasan Daffodil International University, Dhaka, Bangladesh

Md. Nahian Imtiaz Hasan Institute of Information Technology, Jahangirnagar University, Savar, Dhaka, Bangladesh

Mohammad Shahadat Hossain Department of Computer Science and Engineering, University of Chittagong, Chittagong, Bangladesh

Mohammed Shakhawat Hossain Daffodil Institute of IT, Dhaka, Bangladesh

Sazzad Hossain Department of Computer Science and Engineering, University of Liberal Arts Bangladesh, Dhaka, Bangladesh

Hadri Hussain School of Information Technology, Monash University Malaysia, Bandar Sunway, Selangor, Malaysia;
HealUltra PLT, Taman Pulai Utama, Skudai, Johore, Malaysia

Salleh Sh Hussain HealUltra PLT, Taman Pulai Utama, Skudai, Johore, Malaysia

Md. Imran Hossain Imu Daffodil Institute of IT, Dhaka, Bangladesh

Raihan Ul Islam Department of Computer Science, Electrical and Space Engineering, LuleÃě University of Technology, SkellefteÃě, Sweden

Shariful Islam Daffodil International University, Dhaka, Bangladesh

Uwaise Ibna Islam Chittagong University of Engineering and Technology, Chittagong, Bangladesh

K. K. Jacob Faculty of Applied Sciences and Technology, Universiti Tun Hussein Onn Malaysia, UTHM Kampus Pagoh, Muar, Johor, KM, Malaysia

Kavikumar Jacob Faculty of Applied Sciences and Technology, Universiti Tun Hussein Onn Malaysia, UTHM Kampus Pagoh, Muar, Johor, Malaysia;
Faculty of Applied Sciences and Technology, Universiti Tun Hussein Onn Malaysia, Parit Raja, Malaysia

Sohely Jahan University of Barishal, Barishal, Bangladesh

M. A. Jalil Department of Physics, Faculty of Science, Unversiti Teknologi Malaysia, Skudai, Johor, Malaysia

Muhammad Arif Jalil Department of Physics, Faculty of Science, Unversiti Teknologi Malaysia, Skudai, Johor, Malaysia

Norezmi Jamal Electronic Engineering Department, Universiti Tun Hussein Onn Malaysia, Parit Raja, Malaysia

Maisarah Auni Jamaludin Faculty of Applied Sciences and Technology, Universiti Tun Hussein Onn Malaysia, Pagoh, Johor, Malaysia

Muhammad Mahadi Abdul Jamil Department of Electronic Engineering, Faculty of Electrical and Electronic Engineering, Universiti Tun Hussein Onn Malaysia, Parit Raja, Batu Pahat, Johor, Malaysia

María R. Jiménez-Vivanco Semiconductor Devices Research Center, ICUAP, BUAP, Ciudad Universitaria, Puebla, Pue, Mexico;
Institute of Physics, UNAM, Circuito de la Investigación Cientifica, Ciudad Universitaria, Mexico city, Mexico

Annie Joseph Department of Electrical and Electronics, Faculty of Engineering, Universiti Malaysia Sarawak, Kota Samarahan, Malaysia

M. Joseph Auxilus Jude Vinton Network Lab, Department of Electronics and Communication, Kongu Engineering College, Perundurai-638060, India

Tan Li Jun Faculty of Applied Sciences and Technology, Universiti Tun Hussein Onn Malaysia, Pagoh, Johor, Malaysia

Jubair Ahmed Junjun BGC Trust University Bangladesh Bidyanagar, Chandanaish, Bangladesh

M. Shamim Kaiser Institute of Information Technology, Jahangirnagar University, Savar, Dhaka, Bangladesh;
Applied Intelligence and Informatics Lab (AII-Lab), Jahangirnagar University, Savar, Dhaka, Bangladesh;

Applied Intelligence and Informatics (AII), Wazed Miah Science Research Centre (WMSRC), Jahangirngar University, Savar, Dhaka-1342, Bangladesh

M. A. I. Kamarudin School of Business Management, Universiti Utara Malaysia, Sintok, Kedah, Malaysia

Kallol Krishna Karmakar Research Fellow, School of Information and Physical Sciences, University of Newcastle, Callaghan, Australia

Md. Al-Amin Khan Daffodil International University, Dhaka, Bangladesh

Brandon Gan Yong Kien Department of Electrical and Electronics, Faculty of Engineering, Universiti Malaysia Sarawak, Kota Samarahan, Malaysia

Kuryati Kipli Faculty of Engineering, Universiti Malaysia Sarawak (UNIMAS), Sarawak, Malaysia;
Department of Electrical and Electronics, Faculty of Engineering, Universiti Malaysia Sarawak, Kota Samarahan, Malaysia

J. Eduardo Lugo Faubert Lab, École d'optométrie, Université de Montréal, Montréal, QC, Canada

Md. Kawsher Mahbub Computer Science and Engineering, Bangladesh University of Business and Technology, Dhaka, Bangladesh

Md. Julkar Nayeen Mahi Computer Science and Engineering, City University, Khagan, Bangladesh

Mufti Mahmud Department of Computer Science, Nottingham Trent University, Nottingham, NG1 4FQ, UK;
Medical Technologies Innovation Facility, Nottingham Trent University, Nottingham, UK;
Computing and Informatics Research Centre, Nottingham Trent University, Nottingham, UK

Sabrina Jahan Maisha BGC Trust University Bangladesh, Chandanaish, Chattogram, Bangladesh

Lizeth Martínez Tepeji Graduate School, Industrial Engineering, Autonomous Hidalgo State University, Tepeji del Rio, Hidalgo, Mexico

Tamara Islam Meghla Software, Web & Cloud, Faculty of Information Technology and Communications, Tampere University, Tampere, Finland

Md. Abdul Mozid Miah Computer Science and Engineering, Bangladesh University of Business and Technology, Dhaka, Bangladesh

Saiful Najmee Mohamad Faculty of Applied Sciences, Universiti Teknologi MARA, Cawangan Johor, Kampus Pasir Gudang, Masai, Johor, Malaysia

Francisco Morales-Morales Optics Research Center, A.C., León, Gto, Mexico

Choo Mun Chye Faculty of Technology Management and Business, Universiti Tun Hussein Onn Malaysia, Parit Raja Batu PahatJohor, Malaysia

Uwasila Binte Munir Department of Information and Communication Technology, Bangladesh University of Professionals, Dhaka, Bangladesh

G. Murugesan Department of Electronics and Communication, Kongu Engineering College, Perundurai-638060, India

Lutfun Nahar International Islamic University Chittagong, Chittagong, Bangladesh

Nazmun Nahar BGC Trust University Bangladesh Bidyanagar, Chandanaish, Bangladesh

Md. Arif Istiek Neloy BGC Trust University Bangladesh Bidyanagar, Chandanaish, Bangladesh

N. Nishrhutha Vinton Network Lab, Department of Electronics and Communication, Kongu Engineering College, Perundurai-638060, India

Fuad Noman School of Information Technology, Monash University Malaysia, Bandar Sunway, Selangor, Malaysia

Batrisyia Ibtisam Osman Faculty of Applied Sciences and Technology, Universiti Tun Hussein Onn Malaysia, Pagoh, Johor, Malaysia

Nurlatifah Diana Binti Pajilani Faculty of Technology Management and Business, Universiti Tun Hussein Onn Malaysia, Parit Raja, Batu Pahat, Johor, Malaysia

Anindya Pattanayak Organizational Behavior and Human Resource Management, Indian Institute of Management, Ranchi, India

Liton Chandra Paul Department of Electrical, Electronic and Communication Engineering, Pabna University of Science and Technology, Pabna, Bangladesh

P. Prabpal Department of Electrical Technology, Faculty of Industrial Technology, Institute of Vocational Education Northeastern 2, Sakon Nakhon, Thailand

Piyush Pranjal Management Department, O P Jindal Global University, Sonipat, India

Argelia Pérez-Pacheco Research and Technological Development Unit, Hospital General de México "Dr. Eduardo Liceaga", Mexico City, Mexico

Adlyn Nazurah Abdul Rahman Faculty of Technology Management and Business, Universiti Tun Hussein Onn Malaysia, Parit Raja Batu Pahat, Johor, Malaysia

Md. Mahfujur Rahman Daffodil International University, Dhaka, Bangladesh

Roksana Akter Raisa Department of Information and Communication Technology, Bangladesh University of Professionals, Mirpur, Dhaka-1216, Bangladesh

Kanad Ray Amity School of Applied Sciences, Amity University Rajasthan, Jaipur, India;
Amity School of Applied Sciences, Amity University, Jaipur, India;
Amity School of Applied Science, Amity University Rajasthan, Kant Kalwar, N11C, Jaipur, Rajasthan, India

Noortaz Rezoana Department of Computer Science and Engineering, University of Chittagong, Chittagong, Bangladesh

Jarin Tasnim Ritu Institute of Information Technology, Jahangirnagar University, Savar, Dhaka, Bangladesh

S. Z. H. Rizvi Faculty of Applied Sciences and Technology, Universiti Tun Hussein Onn Malaysia, UTHM Kampus Pagoh, Muar, Johor, KM, Malaysia

Syed Zuhaib Haider Rizvi Faculty of Applied Sciences and Technology, Universiti Tun Hussein Onn Malaysia, UTHM Kampus Pagoh, Muar, Johor, Malaysia

Sajeeb Saha Jagannath University, Dhaka, Bangladesh

Pathik Sahoo International Center for Materials and Nanoarchitectronics (WPI-MANA) and Research Center for Advanced Measurement and Characterization (RCAMC), National Institute for Materials Science (NIMS), Tsukuba, Japan

Muhamad Syazreen Md Salleh Faculty of Technology Management and Business, Universiti Tun Hussein Onn Malaysia, RajaBatuPahat, Johor, Malaysia

Sh Hussain Salleh HealUltra PLT, Taman Pulai Utama, Skudai, Johore, Malaysia

Parisa Sanati Burn and Wound Healing Research Center, Shiraz University of Medical Sciences, Shiraz, Iran

Jhimli Sarkar Department of Electronics and Electrical Communication Engineering, Indian Institute of Technology, Kharagpur, India

Soumya Sarkar Marketing Management, Indian Institute of Management, Ranchi, India

Nusrhat Jahan Sarker Daffodil Institute of IT, Dhaka, Bangladesh

Komal Saxena International Center for Materials and Nanoarchitectronics (MANA), Research Center for Advanced Measurement and Characterization (RCAMC), NIMS, Tsukuba, Ibaraki, Japan

Komal Saxena Microwave Physics Laboratory, Department of Physics and Computer Science, Dayalbagh Educational Institute, Dayalbagh, Agra, Uttar Pradesh, India

Hadrina Sh-Hussain HealUltra PLT, Taman Pulai Utama, Skudai, Johore, Malaysia

Zakarya Farea Shaaf Department of Electronic Engineering, Faculty of Electrical and Electronic Engineering, Universiti Tun Hussein Onn Malaysia, Parit Raja, Batu Pahat, Johor, Malaysia

Md. Shahadat Hossen Shakil Institute of Information Technology, Jahangirnagar University, Savar, Dhaka, Bangladesh

Mohd Nurul Al Hafiz Sha'abani Electrical Engineering Department, Center for Diploma Studies, Universiti Tun Hussein Onn Malaysia, Pagoh, Malaysia

Mohd Nurul Al Hafiz Sha'abani Microcontroller Technology for IoT Focus Group (MTIT), Pagoh, Malaysia

Fazlul Hasan Siddiqui Dhaka University of Engineering and Technology, Gazipur, Bangladesh

Pushpendra Singh International Center for Materials and Nanoarchitectronics (WPI-MANA) and Research Center for Advanced Measurement and Characterization (RCAMC), National Institute for Materials Science (NIMS), Tsukuba, Japan; Amity School of Applied Science, Amity University Rajasthan, Kant Kalwar, N11C, Jaipur, Rajasthan, India

S. F. Sufahani Oasis Integrated Group, Universiti Tun Hussein Onn Malaysia, Parit Raja, Batu Pahat, Johor, Malaysia

Suliadi F. Sufahani Faculty of Applied Sciences and Technology, Universiti Tun Hussein Onn Malaysia, Pagoh, Johor, Malaysia;
Oasis Integrated Group, Universiti Tun Hussein Onn Malaysia, Parit RajaBatu Pahat, Johor, Malaysia;
Faculty of Technology Management and Business, Universiti Tun Hussein Onn Malaysia, Parit Raja, Johor, Malaysia

Zumrotul Jannah Supaat Faculty of Technology Management and Business, Universiti Tun Hussein Onn Malaysia, Parit Raja Batu Pahat, Johor, Malaysia

Iffat Tamanna Computer Science and Engineering, Bangladesh University of Business and Technology, Dhaka, Bangladesh

Jing Ying Tee Universiti Tun Hussein Onn Malaysia, Pagoh, Johor, Malaysia

Chee-Ming Ting School of Information Technology, Monash University Malaysia, Bandar Sunway, Selangor, Malaysia

Ng Rou Ting Universiti Tun Hussein Onn Malaysia, Pagoh, Johor, Malaysia

Juwendra Tomas Universiti Tun Hussein Onn Malaysia, Parit Raja Batu Pahat, Johor, Malaysia

Ashok Vajravelu Biomedical Engineering and Measurement System Research Group, Faculty of Electrical and Electronic Engineering, Universiti Tun Hussein Onn Malaysia, Johor, Malaysia

S. K. Vijay Department of Electronics & Communication Engineering, Amity University, Jaipur, Rajasthan, India

Mohd Helmy Abd Wahab Faculty of Electrical and Electronic Engineering, Universiti Tun Hussein Onn Malaysia, Johor, Malaysia

W. N. A. Wan Ahmad Universiti Tun Hussein Onn Malaysia, Pagoh, Johor, Malaysia

Preecha Yupapin Department of Electrical Technology, Faculty of Industrial Technology, Institute of Vocational Education Northeastern 2, Sakon Nakhon, Thailand

Natasha A. M. Zailani Faculty of Applied Sciences and Technology, Universiti Tun Hussein Onn Malaysia, Pagoh, Johor, Malaysia

Wan Suhaimzan Wan Zaki Biomedical Engineering and Measurement System Research Group, Faculty of Electrical and Electronic Engineering, Universiti Tun Hussein Onn Malaysia, Johor, Malaysia

Hushairi Zen Department of Electrical and Electronics, Faculty of Engineering, Universiti Malaysia Sarawak, Kota Samarahan, Malaysia

A. L. Ahmad Zubaidi Faculty of Medicine, Medical Campus, Universiti Sultan Zainal Abidin, Kuala Terengganu, Terengganu, Malaysia

Artificial Intelligence and Soft Computing

Face Mask Detection in the Era of COVID-19: A CNN-Based Approach

Noortaz Rezoana, Mohammad Shahadat Hossain, and Karl Andersson

Abstract The global epidemic of the coronavirus COVID-19 is wreaking havoc on the world's health and according to the World Health Organization (WHO), using a face mask in crowded locations is among the most common security practices. An artificial neural network for face mask classification utilizing deep learning will be introduced in this research. As the outbreak of the COVID-19 pandemic, a remarkable development in the fields of object recognition and computer vision has been made in the identification of face masks. Many architectures and methods have been used to construct a variety of face recognition models. Face masks can be distinguished using the method proposed in this work, which makes use of deep learning, TensorFlow, Keras, and OpenCV. This approach may be evaluated for use in protection jobs due to the fact that it is quite inexpensive to execute. In fact, the GAN-generated face-masked datasets have been selected for evaluation purposes. Compared to other standard Convolutional Neural Network models, the proposed framework outscored them all, attaining a 99.73% accuracy rating.

Keywords COVID-19 · Masked face · Convolutional neural network · Data augmentation · Deep learning · Transfer learning

N. Rezoana · M. S. Hossain (✉)
Department of Computer Science and Engineering, University of Chittagong, Chittagong, Bangladesh
e-mail: hossain_ms@cu.ac.bd

K. Andersson
Department of Computer Science, Electrical and Space Engineering, Luleå University of Technology, Skellefteå, Sweden
e-mail: karl.andersson@ltu.se

M. S. Kaiser et al. (eds.), *Proceedings of the Third International Conference on Trends in Computational and Cognitive Engineering*, Lecture Notes in Networks and Systems 348, https://doi.org/10.1007/978-981-16-7597-3_1

1 Introduction

Due to the worldwide COVID-19 (coronavirus) outbreak, surgical masks in the public are becoming more common. Men and women used to carry masks to shield their well-being from environmental contamination before the coronavirus epidemic. The contagious viruses that transmit quickly between some human beings by near interaction with COVID-19 positive victims are one of the key medical concerns in COVID-19. Since its appearance in December 2019, this novel virus and illness were unknown. COVID-19 is a global epidemic that already affects every country worldwide. COVID-19 contamination may result in a massive variety of consequences ranging from moderate to serious sickness. Exhaling difficulties can result in deaths in serious situations. As per the World Health Organization 80%of illnesses are minor or undiagnosed, 15% are serious, and 5% need ventilation or oxygen. COVID-19 can disperse in three directions, e.g., interactions (direct/indirect), minuscule sprinkler splashes, and protracted vapors, commonly recognized as atmospheric delivery. Through the brief diffusion of sneezing or coughing, the disease can enter from humans to humans through granules from the nasal passage or mouth. The particles can travel up to 6 feet (180 cm) and accumulate on the surface quickly. If anyone comes into contact with certain items and then touches one's eyes, nose, or mouth, they could become contaminated. COVID-19 is transmitted primarily by intimate interaction between people. Someone who is symptom-free is much more susceptible than someone who has COVID-19 signs.

Scientific researchers have shown that using a face mask will prevent COVID-19 dissemination [20]. Coronavirus [17] is the most recent infectious virus to strike people's well-being. In recent years COVID-19 has contaminated more than 16 million people in 220 countries over the past two years. Figure 1 shows the top countries infected by coronavirus and the number of cases. The COVID-19 outbreak

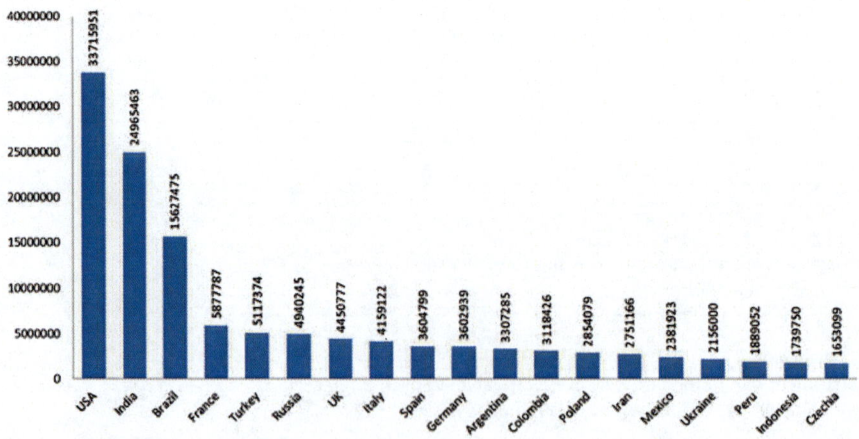

Fig. 1 Accuracy of proposed model and other traditional models

has resulted in unprecedented levels of international research collaboration. Multiple organizations already performed attempts to decrease the virus's dissemination, according to [10]. Also for time being, there is also no specific medication or vaccination available to combat this lethal virus. As a result, Bangladesh as well as other nations are relying on government-implemented measures, such as physical separation and using a face mask in public places, to prevent COVID-19 dissemination [6, 11, 15]. In several ways, Artificial Intelligence (AI) focused on machine learning techniques and deep learning approaches may assist in the battle against COVID-19. Machine learning techniques and deep learning approaches enable scientists and physicians to analyze massive amounts of data in order to predict the spread of COVID-19, act as an advanced alerting system for future epidemics, and identify endangered communities. To combat and forecast new illnesses, healthcare providers need support for advanced technologies such as deep learning. The ability of AI is being used to solve the COVID-19 global epidemic [29], including the detection of COVID-19 in surgical chest X-rays [4, 19] or CT scans, in order to effectively recognize disease patterns and to monitor and easily identify contaminants.

We present a face mask classification system dependent on deep transfer learning and Convolution neural networks in this paper. The developed model can be combined with CCTV cameras to obstruct COVID-19 transmissions by detecting individuals who are not using face masks. A new deep convolutional system architecture for recognizing and classifying human beings with and without masks has been presented, outperforming previous CNN models. For extracting features, we utilized deep transfer leering in conjunction with a fully connected layer. We compared it in order to find the most appropriate model that reached the highest accuracy while using the least amount of time during the training and detection processes.

The remainder of the paper is arranged as follows. The second section discusses prior related works. The dataset's characteristics are defined in Sect. 3. The Methodology is explained in Sect. 4. The research findings are reported and analyzed in Sect. 5, and the conclusions and possible research prospects are presented in Sect. 6.

2 Related Work

Since it is commonly implemented in many ways, object recognition from an image is perhaps the most in-depth component of computer vision [3, 7, 21]. Many deep learning techniques [12] have been proposed for different diseases detection such as skin cancer [27], COVID-19 [4]. So this segment assembles recent research articles that extend reflective research relevant to object classification using deep learning for clinical face masks. The majority of mask face detecting strategies rely on classical machine learning [1, 2, 5, 24] approaches to create and recognize faces. The purpose of this work is to identify a person wearing a face mask or not in order to potentially reduce the transmission of COVID-19. Use of Face masks has been shown to reduce the rate of infection of COVID-19 contamination, according to researchers and scientists. The researchers of [26] introduced a new facemask-wearing factor detection system. Three different types of wearing factors may be listed. Correct facemask, incorrect facemask, and no facemask wearing are the three classes. In the face detection phase. To identify the individual, Sabbir et al. [9] used Principal Component Analysis (PCA) on masked and unmasked face detection. They discovered that wearing masks has a significant impact on the performance of face desensitization using the PCA. Whenever the identified image is masked, the identification accuracy decreases to less than 70%. In addition, in [23], the authors suggested a procedure for extracting shades from a person's frontal face images using PCA. The extracted portion was rebuilt utilizing PCA restoration and reciprocal errors correction. The developers of [18] have been using the YOLOv3 technique to identify faces. The core of YOLOv3 is Darknet-53. The accuracy of the suggested approach was 93.9%. It was developed using the CelebA and WIDER FACE datasets, which contain over 600,000 photographs. The FDDB set of data was used for testing. Nizam et al. [8] suggested a modified GAN-based network that would dynamically erase masks shielding the face region and recreate the picture by filling in the gap. The suggested framework generates a full face picture that appears genuine and functional. The writers of [22] introduced a method for identifying whether or not a mandatory surgical mask is worn in the operation theater. The end goal is to reduce the number of false positive face detection and identification while avoiding missed mask sightings, so that only healthcare professionals who do not wear a surgical mask are alerted. Muhammad et al. [16] proposed MRGAN, an immersive technique. The system relies on the individual providing the speaker field, which is then rebuilt using the Generative Adversarial Network (GAN). Deep learning true face expression detection and identification was used by Shaik et al. [14]. They classified seven human gestures using VGG 16. The suggested classifier was tested on the KDEF dataset and had an accuracy of 88%.

3 Datasets Characteristics

The data collection derives from kaggle.com which includes about 20,000 photographs. All Images10,000 HD photographs in a directory without mask and another 10,000 HD photos in a filename with mask were created utilizing Style GAN-2. Each picture was allocated to a label and grouped into classes using majority class known as with_mask and without_mask. Furthermore, the set of data is widely available, and collected data binds to the level that sufficient input data for photo analyses can be provided.

Every photos are rearranged to 224×224 pixels since resizing is one of the most key aspects of data preprocessing. Some random samples from the standard dataset are seen in Figs. 2 and 3. Figures 2 and 3 show that the with mask and without mask classes vary in terms of visual characteristics. For example, in the with mask class, the mouth part is completely obscured by the mask, while in the without mask class,

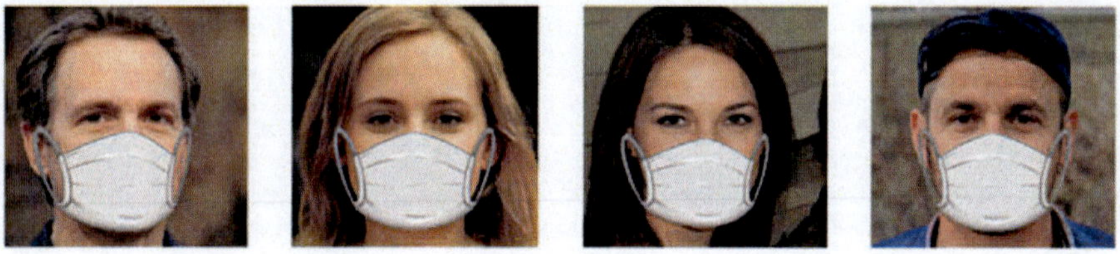

(a) (b) (c) (d)

Fig. 2 Sample images of with_mask class

(a) (b) (c) (d)

Fig. 3 Sample images of without_mask class

the visible face picture is stored. The characteristics described above will help our proposed model learn the differences between the two groups. The data collection has been proportionally divided into train, validation, and test parts.

4 Methodology

4.1 Data Augmentation

It is very well defined that a large amount of data in the sets of data is required to improve the performance of a CNN architecture. To successfully develop a Neural network, data augmentation techniques are required. This approach avoids data distortion and preserves the data's original durability. This approach is frequently implemented throughout this training period [25] to improve the modern architecture performance by tackling overfitting issues. So many characteristics can be derived from the set of data and linked to unspecified information if the datasets are huge sufficient. Data augmentation generates several photos by adding augmentation actions to train files, such as random rotation, shift, zoom, noise, flips, and so forth. Even during training process, each measurement has the potential to depict photographs in a variety of ways and emerge up with unique functionality, boosting the framework's performance. The predominant image, and also the enhanced images made from augmentation techniques, are seen in Fig. 4. In conducting to evaluate the method

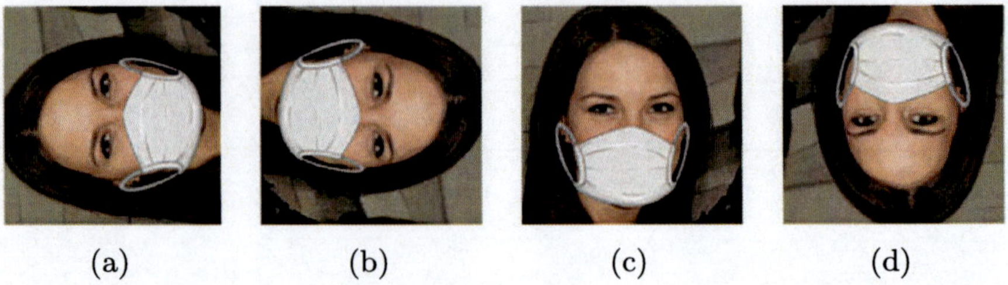

Fig. 4 Sample images after data augmentation

Table 1 Images augmentation settings

Augmentation setting	Range
Rotation	0.2
Zoom	0.1
Contast	0.1
Horizontal flip	True

during the whole experiment, we have applied 80% of the images and reused the residual 20% to validate the model. The settings for image augmentation used in our experiment can be seen in Table 1.

4.2 Model Construction

This section discusses in depth the stated convolutional network. The system introduced comprises three fundamental modules feature segmentation, detection, and categorization. In the starting, we build architectural method with simultaneous convolutional layers, activation, and max pooling to extract features from the images. The characteristics are then transferred into a layer that has been flattened. The properties or characteristics that have been flattened are then transported into dense layers. To prohibit overfitting in the dense layers, a dropout also has been used. The supreme level, that is mentioned as the softmax layer, accomplished the classification task. Figure 5 represents the system flowchart of the overall study and Fig. 6 depicts the model architecture.

Following the data enhancement techniques, the images were supplied to the framework for training purposes. The scheme includes a CBr = iconv blocks, as well as max pooling layers that is three instances in congruence in concurrent structure. Though several of the characteristics created by using a 3×3 filter dimension are overlooked, max pooling is an effective way of retrenchment the tightly-scale images, demanding maximum ratings for every category. As previously mentioned, pixellated max pooling windows may not vastly enhance overs the non-overlapping windows, so the maximum pooling layers utilized in our research study is 2×2

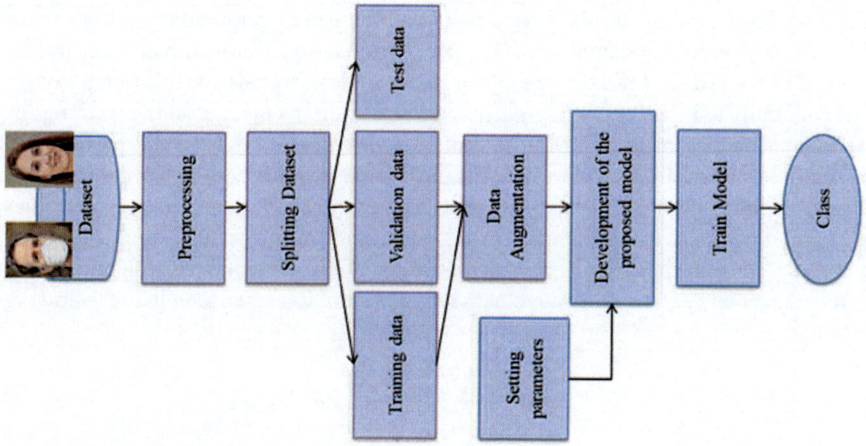

Fig. 5 System flowchart

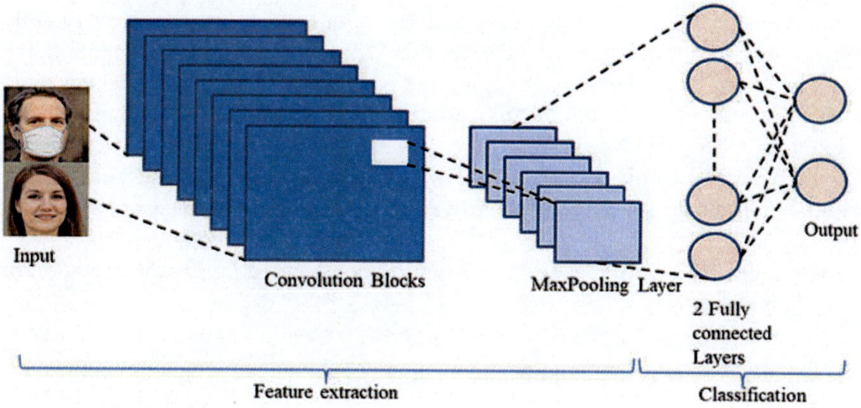

Fig. 6 Model Architecture

with stride 2. It has four convolutional layers, the initial of which has 64 filters with kernel size 3 × 3, the next of which has 128 filters, the third of which has 256 filters 3 × 3 kernel size and the last one has 512 filter with kernel size 3 × 3. The affixed activation function reuses the generated characteristics in the initial phases to create an exclusive feature map as an output right afterwards the convolution layer. Convolution over the photograph m(p, q) is described in formula utilizing a filtration n(p, q).

$$m(p,q)n(p,q) = \sum_{q=-a}^{a} \sum_{q=-b}^{b} i(x,y)j(p-x,q-y) \tag{1}$$

Every level of convolution were pursued by the ReLU activation method to produce the low-level characteristics. The ReLU activation function has been implemented in the hidden layers as well as it had been in the prior convolutional layers. The capability to rapidly spread gradients is one of ReLU's most prominent strengths. As a consequence of measuring the critical properties of CNN in the preliminary amount, the probabilities of gradient disappearance are reduced. Since the output is the same dimensional as like as the provenance, the activation function continues to conduct component-by-component interventions on this assigned input function map. As a consequence, ReLU is one of the most often implemented activation functions in several layers to illustrate that the model is not linear, as seen in the following formula:

$$ReLu(p) = \max(0, p) \tag{2}$$

To transform the feature map, the layout is revealed throughout the flattened plates. A one-dimensional feature vector has also been formed from the feature map created by the subsequent residual blocks to execute the classification process. 2 drop-out layers are attached to the fully connected layers or hidden layer, and every fully connected layer has a dimension of 2048 filter. Throughout the implementation, the dropping layers will randomly dumping the weights of the Fully connected layer to avoid excess mass and minimize overfitting. For the Neural network model, the dropout value for this layer is set to 0.5.

Finally, since the fully connected layer comprises two nodes because there are two classification categories, the softmax activation function was employed only just after Fully connected layers to divide the pictures into with mask and without mask classes, as seen in the given formulas. Figure 7 depicts the detailed layer representation of the proposed model.

$$\text{Softmax}(x) = \frac{e^i}{\sum_{i=0}^{i} e^i} \tag{3}$$

5 Result

Table 2 compares the developed framework to other typical Convolution neural networks such as Inception Net, VGG 19, Resnet50, Xception Net, DenseNet, MobileNet v2, Inception Resnet V2, and VGG 16 in terms of accuracy. In comparison to many other Neural networks, the built architecture generates a high identification accuracy considering a dataset with a modest collection of pictures. As shown in the figure Inception Net, VGG 19, Resnet50, Xception Net, DenseNet,MobileNet v2, Inception Resnet V2, and VGG 16 have 90.78, 83.95, 91.91, 90.47, 89.99, 92.38, 90.77, and 92.63% accuracy, respectively. The proposed model outperformed than these traditional CNN models with 99.73% accuracy. Based on the most recent

Layer Name	Details
1st Convolution Layer 2D	Filter Size=64,kamel size=3x3,activation=ReLU
1st MaxPooling Layer	Pooling Size= 2x2
2nd Convolution Layer 2D	Filter Size=128,kamel size=3x3,activation=ReLU
2nd MaxPooling Layer	Pooling Size= 2x2
3rd Convolution Layer 2D	Filter Size=256,kamel size=3x3,activation=ReLU
3rd MaxPooling Layer	Pooling Size= 2x2
4th Convolution Layer 2D	Filter Size=512,kamel size=3x3,activation=ReLU
4th MaxPooling Layer	Pooling Size= 2x2
Dropout Layer	Excludes 50% of the neurons at random
Flatten Layer	--
1st Fully connected Layer	Filter size=2048,activation = ReLU
Dropout Layer	Excludes 50% of the neurons at random
2nd Fully connected Layer	Filter size=2048,activation = ReLU
Dropout Layer	Excludes 50% of the neurons at random
Softmax Layer	2

Fig. 7 The proposed model's detailed layer representation

Table 2 Accuracy of other traditional CNN model with the proposed model

Model name	Accuracy
Inception net	90.78
VGG 19	83.95
Resnet50	91.91
Xception net	90.47
DenseNet	89.99
MobileNet v2	92.38
Inception-ResNet-v2	90.77
VGG 16	92.63
Proporsed model	99.73

Inception Net, VGG 19, Resnet50, Xception Net, DenseNet, MobileNet v2, Inception Resnet V2, and VGG 16 frameworks, the highest performance correlates to the one we adopted. As a consequence, it's fair to assume that the built architecture surpasses the competition. The implemented models' result, as seen in Fig. 8, shows that the approach is better compared with others.

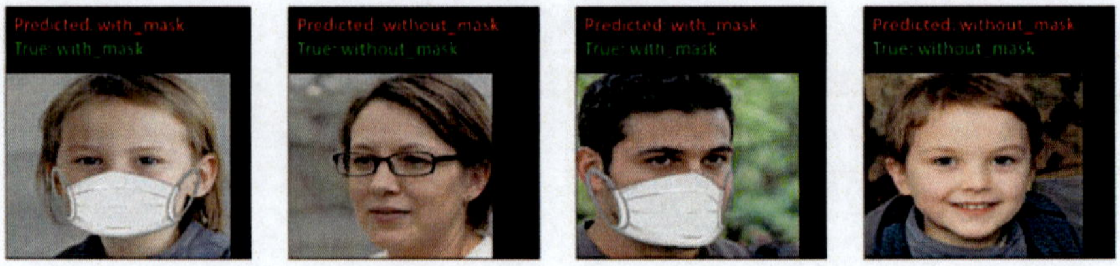

(a) (b) (c) (d)

Fig. 8 Result of detection produced by the proposed model

Fig. 9 Confusion matrix of the proposed model

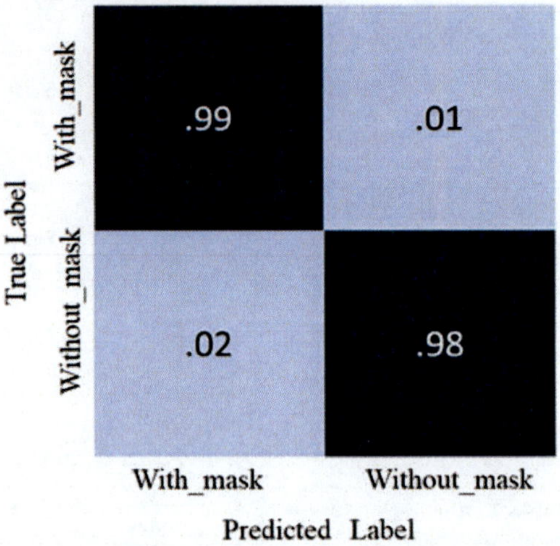

Figure 8 depicts the suggested architectures' classification success as well as the test dataset's identification results in two categories: with_mask and without_mask.40 pictures were randomly tested in contrast to the subsequently mentioned the proposed CNN versions, with 6 of them being shown in Fig. 9. The established method's detection efficiency is actually very good.

Figure 9 depicts the proposed system's confusion matrices. The diagonal quantity of the confusion matrices of both groups is larger for our modified Proposed method. In other words, the suggested framework will accurately differentiate a certain set of test cases from our test set of data as the existing Deep neural network architecture. As a consequence, our system outscored another traditional Neural network architecture in this field.

6 Epilogue and Future Work

The COVID-19 coronavirus disease outbreak is wreaking havoc on the world's well-being. Governments around the universe are fighting to keep this epidemic at bay. Based on the current news of the World Health Organization (WHO), defense against disease generated by COVID-19 is a required precaution like a face mask.

Throughout the research, a novel deep convolutional system architecture for identifying and categorizing with mask and without masked human beings was proposed. It is shown by using neural network blocks, the model can obtain essential and required characteristics. In comparison to the various possibilities, the best Convolutional neural network frameworks Inception Net, VGG 19, Resnet50, Xception Net, MobileNet v2, Inception Resnet V2, and VGG 16, the architecture has remarkable categorization accuracy.

For similar studies, a comparable finding was obtained. In terms of research accuracy, the developed approach outperformed the existing research. One of several likely upcoming activities is to utilize deeper transfer learning frameworks for extracting features and to apply the neutrosophic domain to identification and detection problems, as it has shown promising prospects. We can conduct a grid hunt on the hyper-parameters to identify the best effective numbers of parameter quantities for future research.

For the original dataset, we have been using data augmentation techniques since we required a vast volume of data for effective training and implementation of Convolutional neural network-based architecture to achieve the best results in categorizing them. It would be more successful if the applied Convolutional Neural Network (CNN) technique can be performed on a wide range of datasets.

Ultimately, several other Neural networks, such as DenseNet, EfficientNet, and others, can be considered to expand the data field, [13, 28] and the architecture's performance can be analyzed to those of other popular machine learning approaches.

References

1. Ahmed, T.U., Hossain, M.S., Alam, M.J., Andersson, K.: An integrated cnn-rnn framework to assess road crack. In: 2019 22nd International Conference on Computer and Information Technology (ICCIT), pp. 1–6. IEEE (2019)
2. Akter, M., Hossain, M.S., Andersson, K.: Hand-drawn emoji recognition using convolutional neural network. In: 2020 IEEE International Women in Engineering (WIE) Conference on Electrical and Computer Engineering (WIECON-ECE), pp. 147–152. IEEE (2020)
3. Choudhury, M.A., Hossain, M.Z., Hossain, M.S.: Estimating an ethical index of human well-being. J. Dev Areas 375–409 (2011)
4. Chowdhury, N.K., Kabir, M.A., Rahman, M., Rezoana, N., et al.: Ecovnet: An ensemble of deep convolutional neural networks based on efficientnet to detect covid-19 from chest x-rays (2020). arXiv:2009.11850
5. Chowdury, M.S.U., Emran, T.B., Ghosh, S., Pathak, A., Alam, M.M., Absar, N., Andersson, K., Hossain, M.S.: Iot based real-time river water quality monitoring system. Procedia Comput. Sci. **155**, 161–168 (2019)

6. Coccia, M.: Effects of the spread of covid-19 on public health of polluted cities: results of the first wave for explaining the dejà vu in the second wave of covid-19 pandemic and epidemics of future vital agents. Environ. Sci. Pollut. Res. **28**(15), 19147–19154 (2021)
7. Dey, P., Hossain, E., Hossain, M., Chowdhury, M.A., Alam, M., Hossain, M.S., Andersson, K., et al.: Comparative analysis of recurrent neural networks in stock price prediction for different frequency domains. Algorithms **14**(8), 251 (2021)
8. Din, N.U., Javed, K., Bae, S., Yi, J.: A novel gan-based network for unmasking of masked face. IEEE Access **8**, 44276–44287 (2020)
9. Ejaz, M.S., Islam, M.R., Sifatullah, M., Sarker, A.: Implementation of principal component analysis on masked and non-masked face recognition. In: 2019 1st International Conference on Advances in Science, Engineering and Robotics Technology (ICASERT), pp. 1–5. IEEE (2019)
10. Fadare, O.O., Okoffo, E.D.: Covid-19 face masks: A potential source of microplastic fibers in the environment. Sci. Total Environ. **737**, 140279 (2020)
11. Feng, S., Shen, C., Xia, N., Song, W., Fan, M., Cowling, B.J.: Rational use of face masks in the covid-19 pandemic. Lancet Respiratory Med. **8**(5), 434–436 (2020)
12. Hossain, M.S., Hasan, M.A., Uddin, M., Islam, M.M., Mustafa, R.: A belief rule based expert system to assess lung cancer under uncertainty. In: 2015 18th International Conference on Computer and Information Technology (ICCIT), pp. 413–418. IEEE (2015)
13. Hossain, M.S., Khalid, M.S., Akter, S., Dey, S.: A belief rule-based expert system to diagnose influenza. In: 2014 9th International Forum on Strategic Technology (IFOST), pp. 113–116. IEEE (2014)
14. Hussain, S.A., Al Balushi, A.S.A.: A real time face emotion classification and recognition using deep learning model. J. Phys. Conf. Ser. **1432**, 012087. IOP Publishing (2020)
15. Islam, M., Islam, U.S., Mosaddek, A.S.M., Potenza, M.N., Pardhan, S., et al.: Treatment, persistent symptoms, and depression in people infected with covid-19 in bangladesh. Int. J. Environ. Res. Public Health **18**(4), 1453 (2021)
16. Khan, M.K.J., Ud Din, N., Bae, S., Yi, J.: Interactive removal of microphone object in facial images. Electronics **8**(10), 1115 (2019)
17. Larsen, D.: Homemade cloth face masks to fight the covid19 pandemic; a call for mass public masking with homemade cloth masks (2020)
18. Li, C., Wang, R., Li, J., Fei, L.: Face detection based on yolov3. In: Recent Trends in Intelligent Computing, Communication and Devices, pp. 277–284. Springer, Berlin (2020)
19. Loey, M., Smarandache, F., M Khalifa, N.E.: Within the lack of chest covid-19 x-ray dataset: a novel detection model based on gan and deep transfer learning. Symmetry **12**(4), 651 (2020)
20. Ma, T., Heywood, A., MacIntyre, C.R.: Travel health seeking behaviours, masks, vaccines and outbreak awareness of australian chinese travellers visiting friends and relatives-implications for control of covid-19. Infection, Disease Health **26**(1), 38–47 (2021)
21. Nahar, N., Ara, F., Neloy, M.A.I., Barua, V., Hossain, M.S., Andersson, K.: A comparative analysis of the ensemble method for liver disease prediction. In: 2019 2nd International Conference on Innovation in Engineering and Technology (ICIET), pp. 1–6. IEEE (2019)
22. Nieto-Rodríguez, A., Mucientes, M., Brea, V.M.: System for medical mask detection in the operating room through facial attributes. In: Iberian Conference on Pattern Recognition and Image Analysis, pp. 138–145. Springer, Berlin (2015)
23. Park, J.S., Oh, Y.H., Ahn, S.C., Lee, S.W.: Glasses removal from facial image using recursive error compensation. IEEE Trans. Pattern Anal. Mach. Intell. **27**(5), 805–811 (2005)
24. Pathak, A., AmazUddin, M., Abedin, M.J., Andersson, K., Mustafa, R., Hossain, M.S.: Iot based smart system to support agricultural parameters: A case study. Procedia Comput. Sci. **155**, 648–653 (2019)
25. Progga, N.I., Rezoana, N., Hossain, M.S., Islam, R.U., Andersson, K.: A cnn based model for venomous and non-venomous snake classification. In: International Conference on Applied Intelligence and Informatics, pp. 216–231. Springer, Berlin (2021)
26. Qin, B., Li, D.: Identifying facemask-wearing condition using image super-resolution with classification network to prevent covid-19. Sensors **20**(18), 5236 (2020)

27. Rezaoana, N., Hossain, M.S., Andersson, K.: Detection and classification of skin cancer by using a parallel cnn model. In: 2020 IEEE International Women in Engineering (WIE) Conference on Electrical and Computer Engineering (WIECON-ECE), pp. 380–386. IEEE (2020)
28. Thombre, S., Islam, R.U., Andersson, K., Hossain, M.S.: Performance analysis of an ip based protocol stack for wsns. In: 2016 IEEE conference on computer communications workshops (INFOCOM WKSHPS), pp. 360–365. IEEE (2016)
29. Ting, D.S.W., Carin, L., Dzau, V., Wong, T.Y.: Digital technology and covid-19. Nat. Med. **26**(4), 459–461 (2020)

A Weighted Average Ensemble Technique to Predict Heart Disease

Md. Arif Istiek Neloy, Nazmun Nahar, Mohammad Shahadat Hossain, and Karl Andersson

Abstract In recent years, heart disease has been the cause of one out of every three human deaths. This emphasizes the critical importance of predicting heart diseases as early as possible in order to improve diagnosis chances. Researchers have been pushing the boundaries of machine learning to achieve the most accurate prediction result in order to make this endeavor possible. This study suggests a new machine learning model that combines three standard machine learning algorithms to obtain a superior outcome. Random Forest, Decision Tree, and Naive Bayes are part of this machine learning model. Combining these three algorithms, this model is named the "Weighted Average Ensemble Model." To examine the performance of the "Weighted Average Ensemble Model," the results from this model have been compared with six different classifiers (Logistic Regression, Gradient Boosting, Bagging, Decision Tree, SVM, Ada Boost). Our Proposed "Weighted Average Ensemble" had the training accuracy rate of 100% and testing accuracy rate of 93%.

Keywords Heart disease · Weighted average ensemble · Ensemble method

1 Introduction

One of the most important organs in the human body is the heart. This 300 g organ transports blood throughout the body, supplying oxygen and nutrients to the cells while also removing carbon dioxide. The heart is divided into four chambers: left and

Md. A. I. Neloy · N. Nahar
BGC Trust University Bangladesh Bidyanagar, Chandanaish, Bangladesh
e-mail: nazmun@bgctub.ac.bd

M. S. Hossain (✉)
University of Chittagong, Chittagong, Bangladesh
e-mail: hossain_ms@cu.ac.bd

K. Andersson
Lulea University of Technology, SE-931 87 Skellefteå, Sweden
e-mail: Karl.andersson@ltu.se

M. S. Kaiser et al. (eds.), *Proceedings of the Third International Conference on Trends in Computational and Cognitive Engineering*, Lecture Notes in Networks and Systems 348, https://doi.org/10.1007/978-981-16-7597-3_2

right atria on the top, and ventricles on the bottom. Valves are strategically placed between these right and left chambers to help manage blood flow in the proper direction. Because the heart is at the center of the body's whole circulation cycle, a minor problem with this organ can have a major influence on the entire body and become a life-threatening problem if not treated promptly.

According to the World Health Organization [21], heart disease is the leading cause of death worldwide, accounting for 17.9 million deaths per year. This 17.9 million represents 31% of all deaths worldwide. Some metrics are used to assess the risk factor that determines the likelihood of developing CVD. Tobacco use, high blood cholesterol, a poor diet, alcohol usage, and physiological problems such as hypertension and excessive blood sugar are just a few examples [25]. This results in frequent heart-related issues such as high blood pressure, glucose, and cholesterol levels, as well as obesity. According to studies [12], four out of every five heart attacks are silent. The damage has already been done, but the sufferer is completely ignorant of it. That is why it is critical to predict CVD or heart disease early on.

For a long time, researchers have been collecting data from heart disease patients all around the world. And that useful data is assisting in the growth of our Machine Learning field. As time goes on, stronger machine learning algorithms are being developed to forecast heart disease early on using that crucial data [1, 33]. Year after year, machine learning improves its ability to discover the correct answer with a lower possibility of error. However, more powerful machine learning will help us in recognizing patterns and learning useful information. Machine learning is often used to predict heart disease, though it has other applications in medicine.

Many researchers have been involved in using machine learning to detect diseases [22, 23] because it reduces diagnosis time and shows accuracy and usefulness. Several diseases can be detected using machine learning methods, but the primary goal of this article is to diagnose heart disease. Since heart disease is still the leading cause of death, and since early detection of heart disease is helpful to saving lives [28]. We use supervised learning methods in this paper to predict heart disease in its early stages. Weighted average ensemble supervised machine learning is employed to determine whether the people being tested have heart disease or are healthy.

The main contributions of our paper are given below

- For the prediction of heart disease, we proposed a weighted average ensemble method.
- In base classifier, we have used three classification algorithm which are Decision tree, Random Forest and Naive Bayes.
- The significance of applying our proposed ensemble method has been demonstrated.

The remainder of this paper is organized as follows: Sect. 2 summarizes the study of the literature review on current studies in this area. The proposed framework and methodology are described in Sect. 3. Section 4 presents experimental results and comparisons between classification techniques. The conclusion and future work of the paper is defined in Sect. 5.

2 Literature Review

In 2020, Escamila et al. [15] demonstrated how classification methods can be used to forecast cardiac disease via feature selection. Cleveland and Hungarian are two different datasets that were utilized. They then combined the prior two datasets to create the third. The combined dataset provided them the best accuracy (99.4%), while the other two datasets, Cleveland and Hungarian, provided them 98.7 and 99% accuracy, respectively.

Thomas et al. [38] recently published a report in which they attempted to forecast the risk rate of heart disease using several classification algorithms (KNN and ID3). The accuracy rate of their result was 40.3% at the start, but when additional attributes were included, the accuracy rate of their model significantly improved (80.6%).

Mohan et al. [24] proposed a hybrid random forest with a linear model (HRFLM), a group of researchers was able to predict cardiac illnesses with an accuracy rate of 88.7%. The HRFLM approach appears to have the highest accuracy rate among the various categorization methods.

Khan [20] used a hybrid random forest with a linear model (HRFLM), a group of researchers was able to predict cardiac illnesses with an accuracy rate of 88.7%. The HRFLM approach appears to have the highest accuracy rate among the various categorization methods.

Zafer et al. [6] proposed a method for predicting heart problems with the use of IoT is to use an IoT device to collect data on patients' heart diseases before and after treatment. After that, they use a higher order Boltzmann deep belief neural network to process the data (HOBDBNN). The deep learning model achieves 99.3% accuracy based on the data.

Repaka et al. [32] have suggested a prediction approach based on Naive Bayesian classification to reliably forecast cardiac disorders. They attained an accuracy of 89.77% using this data mining methodology, using 80 percent of the dataset for training and 20% for testing.

Amin et al. [8] have tried to properly predict Cardiovascular illness using 7 distinct classification methods (k-NN, Decision Tree, Naive Bayes, logistic regression (LR), support vector machine (SVM), Neural Network, and Vote (a hybrid technique combining Nave Bayes and Logistic Regression)) and different combinations of dataset attributes. Vote was the most successful strategy in terms of prediction, with an accuracy of 87.4%.

Saxena et al. [34] have claimed to be able to accurately forecast cardiac illnesses using data mining techniques, allowing non-specialized clinicians to make better decisions. During the testing phase, their method was able to correctly forecast heart illnesses 86.7% of the time.

Bashir et al. [9] released in which different machine learning approaches (Decision Tree, Logistic Regression, Logistic Regression SVM, Naive Bayes, and Random Forest) were used to several heart disorders datasets in order to enhance the accuracy in heart disease prediction. They also had the greatest Logistic Regression (SVM) accuracy of 84.85%.

Sultana et al. [36] employed two separate datasets (Collected and UCI standard) and utilized various data mining techniques (KStar, J48, SMO, Bayes Net, and Multilayer Perceptron). Bayes Net and SMO classifiers are the most optimal among all of these techniques.

3 Methodology

The purpose of the system methodology being proposed is to use ensemble approaches to increase heart disease prediction. The architecture of the proposed method is depicted in Fig. 1. It is divided into five stages: data collection, data preprocessing, data splitting, model training, and model evaluation. The following are the stages of the proposed methodology.

3.1 Data Collection

The dataset of heart disease [7] is required for training and model evaluation. This heart disease dataset was developed by merging five common heart disease datasets that were previously accessible separately but had never been merged. This dataset combines five heart datasets with 11 common attributes, making them the largest dataset for heart disease. The following five datasets were included in its collection:

1. Cleveland
2. Hungarian
3. Switzerland
4. Long Beach VA
5. Statlog (Heart) Data Set.

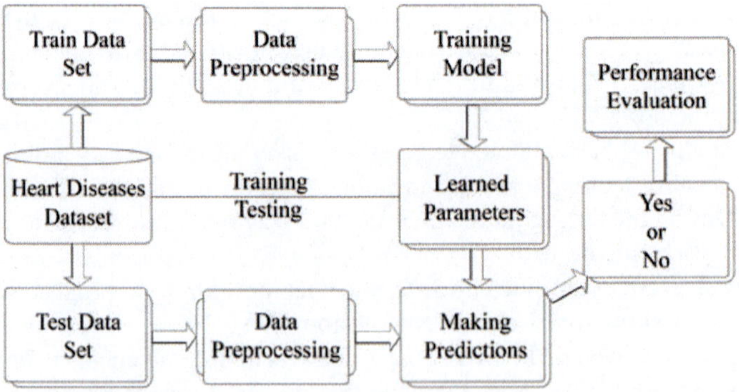

Fig. 1 Graphical illustration of proposed methodology

Table 1 Description of the dataset

SL	Attribute	Description	value
1	Age	Patients age	in years
2	Gender	Patients gender	male: 0, female:1
3	Chest Pain (CP)	Chest pain type	1:= typical angina; 2:=atypical angina ; 3:= non-anginal pain ; 4:= asymptomatic
4	Resting bp s	Resting blood pressure	in mm Hg
5	Cholesterol	Serum cholesterol	in mg/dl
6	Fasting blood sugar	Fasting blood sugar 120 mg/d	1 = true; 0 = false
7	Resting ecg	Resting electrocardiogram results	0: normal ; 1: having ST-T wave abnormality 2: showing probable or definite left ventricular hypertroph
8	Max heart rate	Maximum heart rate achieved	value from 71–202
9	Exercise angina	Exercise induced angina	1 = yes; 0 = no
10	ST slope	The slope of the peak exercise ST segment	1:= upsloping ; 2:= flat ; 3: =downsloping
11	Target	Class	1 = heart disease, 0 = Normal

It includes 1190 data, 11 attributes, and one column of the target. Two classes are in the target column: one class shows heart disease and 0 classifies non-heart disease. The descriptions of the attributes are defined in Table 1.

3.2 *Data Preprocessing*

Data preprocessing is the major phase in model building. Normalization is one of the data preprocessing techniques. In this paper, we normalize all the attributes specially high value range features such as max heart rate attribute which has the range between 71 and 202.For the normalization, the following equation is used.

$$X_{\text{norm}} = \frac{X - X_{\min}}{X_{\max} - X_{\min}} \tag{1}$$

3.3 Data Spiliting

In this phase, the heart disease dataset is separated into a training set of 80% and a testing set of 20%. The training set is used to develop the models, and the test set is used to assess the models.

3.4 Training Models

We have proposed a weighted average ensemble method to train a heart disease dataset where we used three traditional machine learning classifiers as base models. These three base classifiers are averaged, and their averaged result is used for prediction. The three traditional machine learning classifiers used as base classifiers are decision tree, random forest, and naïve bayes.

Decision Tree (DT): DT is a tree method that works on the theory of the condition. The algorithms are accurate and robust and are used for predictive models. In specific, internal nodes, branches, and a terminal node have been assigned to incorporate them. Each internal node has a function test, and the branches conclude with the test, and the class mark for a leaf node is intended. It is used for regression and for classification [14].

Random Forests (RF): RF has named random decision-making forests to play an ML role that can be used for classification and regression problems. Each function is obtained by building a varied number of DT classifications or regressors and increasing all DT performance to a single result [39].

Naïve Bayes (NB): NB is a class of basic probabilistic classification techniques that apply the Bayes theorem with the explicit assumption that attributes are independent. It is highly scalable, involving many linear parameters in a learning problem with different parameters (features or predictors) [40].

Weighted Average Ensemble Technique: Ensemble techniques are approaches that can be used to improve classification performance. It is an efficient way to classify base classifiers with a robust classifier that enhances the performance of the weak learner [13]. We have proposed a weighted average ensemble technique with three base classifiers to make a robust classifier. The average model is an ensemble learning approach, where each ensemble member contributes to the final prediction equally. The proposed weighted average ensemble technique is shown in Fig. 2 to improve heart disease prediction. Here, we combine three traditional classifiers: DT, RF and NB and from their combined result, we find the final result.

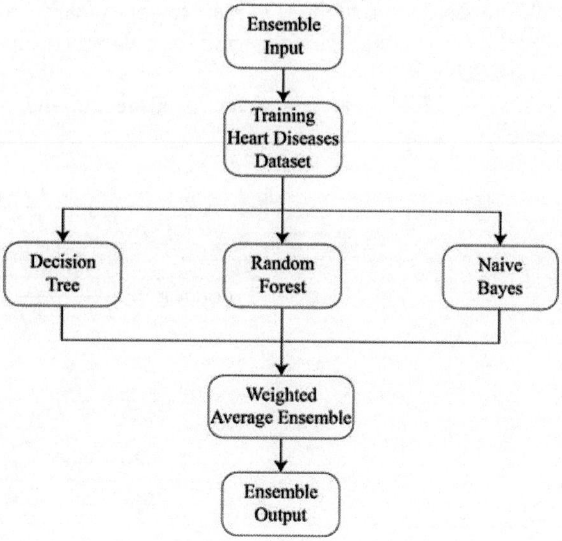

Fig. 2 Proposed weighted average ensemble method

4 Result and Discussion

In this section, we will discuss the experimental result of our proposed classification algorithm.

4.1 System Configuration

Google Colaboratory has been utilized to implement our "Average Ensemble Model." Google Colaboratory allows users to run their python-based research projects on the cloud, with free CPU and GPU assistance. Jupiter notebooks serve as the foundation for Google Colab's user interface. The PYTHON scikit-learn library [31] was used to create the "average ensemble model," which is required to run classification models such as Random Forest, Decision Tree, and Gaussian Nave Bayes.

4.2 Performence Matrix

We employed mean absolute error (MAE), mean squared error (MSE), and root mean squared error (RMSE) to get the most out of our Machine Learning model. When comparing all of these parameters to those of other well-known machine learning models. We also employed Accuracy, Precision, Recall, and F1-score [35] and compared them to those models to get a clear picture of our "Average Ensemble Model" success rate. The following formula was used to determine precision, recall, and F1-Score.

We can calculate accuracy, precision, recall, and f-score by using the following equation

$$\text{Accuracy} = \frac{TP + TN}{TP + FN + FP + TN} \tag{2}$$

$$\text{Precicion} = \frac{TP}{TP + Fp} \tag{3}$$

$$\text{Recall} = \frac{TP}{TP + FN} \tag{4}$$

$$\text{F-Score} = \frac{2 \times \text{Precision} \times \text{Recall}}{\text{Precision} + \text{Recall}} \tag{5}$$

To calculate mean absolute error (MAE), mean squared error (MSE), root mean squared error (RMSE) following formulas have been used.

$$\text{MAE} = \text{True Value} - \text{Predicted Value} \tag{6}$$

$$\text{MAE} = \frac{1}{n}\sum_{j=1}^{n}(\text{Actual Values} - \text{Predicted Values})^2 \tag{7}$$

$$\text{RMSE} = \sqrt{\text{MSE}} \tag{8}$$

4.3 Result

Table 2 shows that our proposed "Weighted Average Ensemble Model" has a training accuracy of 100 percent and a testing accuracy of 93 percent. Then, we compared our findings to six major machine learning techniques. The accuracy rate of logistic regression is 69 percent in training and 70 percent in testing. The support vector machine (SVM) has a training accuracy rate of 70% and a testing accuracy rate of 74%. The decision tree has a training accuracy rate of 100% and a testing accuracy rate of 91%. The bagging algorithm has a training accuracy of 99% and a testing accuracy of 92%. The Gradient Boosting method has a training accuracy rate of 96% and a testing accuracy rate of 90%. The Ada Boosting or Adaptive Boosting algorithm has an accuracy rate of 88% for training and 83% for testing.

The performance matrix result of the "Weighted Average Ensemble Model" can be compared to six different ML models in Table 3. In this comparison, we used the

Table 2 Training and testing accuracy of the classifier

Model	Training accuracy (%)	Testing accuracy (%)
Weighted average ensemble	100	93
Logistic regression	69	70
SVM	70	74
Decision tree	100	91
Bagging	99	92
Gradient boosting	96	90
Ada boost	88	83

Table 3 Performance comparison of different classifier

Model	Precision	Recall	F1-score	MAE error	MSE error	RMS error
Weighted average ensemble	0.93	0.93	0.93	0.07	0.07	0.27
Logistic regression	0.75	0.70	0.70	0.3	0.3	0.55
SVM	0.78	0.74	0.74	0.26	0.26	0.51
Decision tree	0.91	0.91	0.91	0.09	0.09	0.30
Bagging	0.92	0.92	0.92	0.08	0.08	0.29
Gradient boosting	0.90	0.90	0.90	0.1	0.1	0.32
Ada boost	0.83	0.83	0.83	0.17	0.17	0.41

same dataset for all the classifiers. Precision, recall, and F1-score for the average ensemble model are all 0.93. And MAE, MSE, and RMSE of 0.07, 0.07, and 0.27, respectively, which is the best performance result compared to other six algorithms.

Figure 3 shows the ROC curve of the heart disease classification. Here, we demonstrate a ROC curve with our proposed weighted average ensemble method along with other machine learning classifiers which is compared with our proposed model. The ROC curve indicates a balance between TPR and FPR. Classifiers closer to the top-left corner show a better score. A random classifier is assumed to give points that are diagonal (FPR = TPR) as a baseline. The closest the curve to the ROC space of 45 degrees the less accurate the assessment. From Fig. 1, we can see that the proposed weighted average ensemble method is closer to the top-left corner that means our method gives the better result.

In Table 4, the results of four previous research publications were compared to the findings of this publication. Using the "Weighted Average Ensemble" approach, we found 93% accuracy. A number of publications have been published in recent years that attempt to forecast heart disease in advance. We've featured four of them

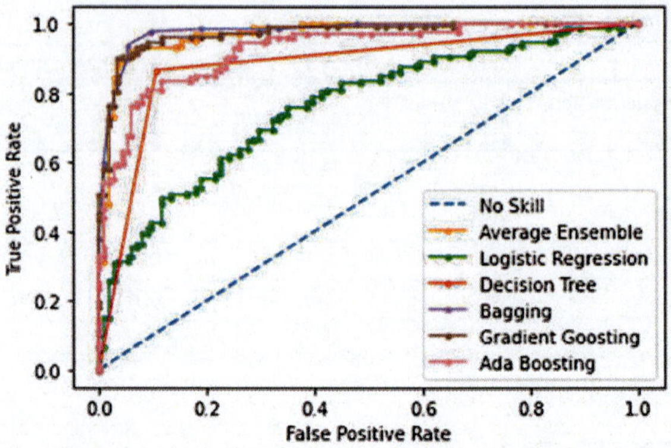

Fig. 3 ROC curve for the model evaluation

Table 4 Comparison of our proposed methodology with existing work

Author	Year	Method/Algorithm	Accuracy (%)
Thomas and Princy [38]	2016	KNN and ID3	80.6
Repaka et al. [32]	2019	Naive Bayes	89.77
Amin et al. [8]	2016	10 fold method	86.7
Bashir et al. [9]	2019	Logistic regression (SVM)	84.85
Our proposed method	–	Weighted average ensemble	93

in this table. They presented various Machine Learning and Deep Learning methods for obtaining their desired outcome. When those works are compared, it becomes clear how well "Weighted Average Ensemble" has performed.

5 Conclusion and Future Work

In the medical field, machine learning is doing an excellent job at predicting diseases. And increasing the availability of suitable datasets is a better method to make these models capable of being used in real-time prediction. Machine learning relies heavily on data preprocessing. A good dataset is crucial to having a solid prediction model. People in the medical field must be more proactive to create more accurate and dependable datasets. We were able to generate a pretty good outcome using the combined five datasets in our model. With the combination of Decision Tree, Random Forest, and Naive Bayes, "Weighted Average Ensemble" performed exceptionally well. The decision tree is quite good at predicting, and because the Random Forest has so many trees, it helps increase the average outcome. The "Weighted Average Ensemble" has exceeded our expectations when combined with Naive Bayes, which is incredibly quick and can produce good results. However, our dataset has a limited number of data. In the future, we will collect more data for predicting heart disease. We will use the deep learning and other method [2–5, 10, 11, 16–19, 26, 27, 29, 30, 37, 41] in the future.

References

1. Abdel-Aty, A.H., Kadry, H., Zidan, M., Al-Sbou, Y., Zanaty, E., Abdel-Aty, M.: A quantum classification algorithm for classification incomplete patterns based on entanglement measure. J. Intell. Fuzzy Syst. **38**(3), 2809–2816 (2020)
2. Abedin, M.Z., Akther, S., Hossain, M.S.: An artificial neural network model for epilepsy seizure detection. In: 2019 5th International Conference on Advances in Electrical Engineering (ICAEE), pp. 860–865. IEEE (2019)
3. Afroze, T., Akther, S., Chowdhury, M.A., Hossain, E., Hossain, M.S., Andersson, K.: Glaucoma detection using inception convolutional neural network v3. In: International Conference on Applied Intelligence and Informatics, pp. 17–28. Springer, Berlin (2021)
4. Ahmed, T.U., Hossain, M.S., Alam, M.J., Andersson, K.: An integrated cnn-rnn framework to assess road crack. In: 2019 22nd International Conference on Computer and Information Technology (ICCIT). pp. 1–6. IEEE (2019)
5. Ahmed, T.U., Jamil, M.N., Hossain, M.S., Andersson, K., Hossain, M.S.: An integrated real-time deep learning and belief rule base intelligent system to assess facial expression under uncertainty. In: 2020 Joint 9th International Conference on Informatics, Electronics and Vision (ICIEV) and 2020 4th International Conference on Imaging, Vision and Pattern Recognition (icIVPR). pp. 1–6. IEEE (2020)
6. Al-Makhadmeh, Z., Tolba, A.: Utilizing iot wearable medical device for heart disease prediction using higher order boltzmann model: A classification approach. Measurement **147**, 106815 (2019)
7. Alizadehsani, R., Roshanzamir, M., Abdar, M., Beykikhoshk, A., Khosravi, A., Panahiazar, M., Koohestani, A., Khozeimeh, F., Nahavandi, S., Sarrafzadegan, N.: A database for using machine learning and data mining techniques for coronary artery disease diagnosis. Sci. Data **6**(1), 1–13 (2019)
8. Amin, M.S., Chiam, Y.K., Varathan, K.D.: Identification of significant features and data mining techniques in predicting heart disease. Telemat. Inform. **36**, 82–93 (2019)
9. Bashir, S., Khan, Z.S., Khan, F.H., Anjum, A., Bashir, K.: Improving heart disease prediction using feature selection approaches. In: 2019 16th International Bhurban Conference on Applied Sciences and Technology (IBCAST), pp. 619–623. IEEE (2019)
10. Basnin, N., Nahar, L., Hossain, M.S.: An integrated cnn-lstm model for micro hand gesture recognition. In: International Conference on Intelligent Computing and Optimization, pp. 379–392. Springer, Berlin (2020)
11. Basnin, N., Nahar, L., Hossain, M.S.: An integrated cnn-lstm model for bangla lexical sign language recognition. In: Proceedings of International Conference on Trends in Computational and Cognitive Engineering, pp. 695–707. Springer, Berlin (2021)
12. CDC: Heart Disease Facts (2020). https://www.cdc.gov/heartdisease/facts.htm, [Online; Last Accessed 20 may 2021]
13. David, H.B.F.: Impact of ensemble learning algorithms towards accurate heart disease prediction
14. Fan, Q., Wang, Z., Li, D., Gao, D., Zha, H.: Entropy-based fuzzy support vector machine for imbalanced datasets. Knowl. Based Syst. **115**, 87–99 (2017)

15. Gárate-Escamila, A.K., El Hassani, A.H., Andrès, E.: Classification models for heart disease prediction using feature selection and pca. Inform. Med. Unlocked **19**, 100330 (2020)
16. Gosh, S., Nahar, N., Wahab, M.A., Biswas, M., Hossain, M.S., Andersson, K.: Recommendation system for e-commerce using alternating least squares (als) on apache spark. In: International Conference on Intelligent Computing and Optimization, pp. 880–893. Springer, Berlin (2020)
17. Islam, R.U., Hossain, M.S., Andersson, K.: A deep learning inspired belief rule-based expert system. IEEE Access **8**, 190637–190651 (2020)
18. Islam, R.U., Ruci, X., Hossain, M.S., Andersson, K., Kor, A.L.: Capacity management of hyperscale data centers using predictive modelling. Energies **12**(18), 3438 (2019)
19. Kabir, S., Islam, R.U., Hossain, M.S., Andersson, K.: An integrated approach of belief rule base and deep learning to predict air pollution. Sensors **20**(7), 1956 (2020)
20. Khan, M.A.: An iot framework for heart disease prediction based on mdcnn classifier. IEEE Access **8**, 34717–34727 (2020)
21. Khan, T.: Cardiovascular diseases (2017). https://www.who.int/health-topics/cardiovascular-diseases#tab=tab_1, [Online; Last Accessed 20 may 2021]
22. Mahmud, M., Kaiser, M.S., McGinnity, T.M., Hussain, A.: Deep learning in mining biological data. Cogn. Comput. **13**(1), 1–33 (2021)
23. Mahmud, M., Kaiser, M.S., Hussain, A., Vassanelli, S.: Applications of deep learning and reinforcement learning to biological data. IEEE Trans. Neural Netw. Learn. Syst. **29**(6), 2063–2079 (2018)
24. Mohan, S., Thirumalai, C., Srivastava, G.: Effective heart disease prediction using hybrid machine learning techniques. IEEE Access **7**, 81542–81554 (2019)
25. Nahar, J., Imam, T., Tickle, K.S., Chen, Y.P.P.: Association rule mining to detect factors which contribute to heart disease in males and females. Expert. Syst. Appl. **40**(4), 1086–1093 (2013)
26. Nahar, N., Ara, F., Neloy, M.A.I., Barua, V., Hossain, M.S., Andersson, K.: A comparative analysis of the ensemble method for liver disease prediction. In: 2019 2nd International Conference on Innovation in Engineering and Technology (ICIET), pp. 1–6. IEEE (2019)
27. Nahar, N., Hossain, M.S., Andersson, K.: A machine learning based fall detection for elderly people with neurodegenerative disorders. In: International Conference on Brain Informatics, pp. 194–203. Springer, Berlin (2020)
28. Obasi, T., Shafiq, M.O.: Towards comparing and using machine learning techniques for detecting and predicting heart attack and diseases. In: 2019 IEEE International Conference on Big Data (big data), pp. 2393–2402. IEEE (2019)
29. Pathan, R.K., Uddin, M.A., Nahar, N., Ara, F., Hossain, M.S., Andersson, K.: Gender classification from inertial sensor-based gait dataset. In: International Conference on Intelligent Computing and Optimization, pp. 583–596. Springer, Berlin (2020)
30. Pathan, R.K., Uddin, M.A., Nahar, N., Ara, F., Hossain, M.S., Andersson, K.: Human age estimation using deep learning from gait data. In: International Conference on Applied Intelligence and Informatics, pp. 281–294. Springer, Berlin (2021)
31. Pedregosa, F., Varoquaux, G., Gramfort, A., Michel, V., Thirion, B., Grisel, O., Blondel, M., Prettenhofer, P., Weiss, R., Dubourg, V., et al.: Scikit-learn: Machine learning in python. J. Mach. Learn. Res. **12**, 2825–2830 (2011)
32. Repaka, A.N., Ravikanti, S.D., Franklin, R.G.: Design and implementing heart disease prediction using naives bayesian. In: 2019 3rd International Conference on Trends in Electronics and Informatics (ICOEI), pp. 292–297. IEEE (2019)
33. Sagheer, A., Zidan, M., Abdelsamea, M.M.: A novel autonomous perceptron model for pattern classification applications. Entropy **21**(8), 763 (2019)
34. Saxena, K., Sharma, R., et al.: Efficient heart disease prediction system. Procedia Comput. Sci. **85**, 962–969 (2016)
35. Sokolova, M., Lapalme, G.: A systematic analysis of performance measures for classification tasks. Inf. Process. Manag. **45**(4), 427–437 (2009)
36. Sultana, M., Haider, A., Uddin, M.S.: Analysis of data mining techniques for heart disease prediction. In: 2016 3rd International Conference on Electrical Engineering and Information Communication Technology (ICEEICT), pp. 1–5. IEEE (2016)

37. Sultana, Z., Nahar, L., Basnin, N., Hossain, M.S.: Inference and learning methodology of belief rule based expert system to assess chikungunya. In: International Conference on Applied Intelligence and Informatics, pp. 3–16. Springer, Berlin (2021)
38. Thomas, J., Princy, R.T.: Human heart disease prediction system using data mining techniques. In: 2016 International Conference on Circuit, Power and Computing Technologies (ICCPCT), pp. 1–5. IEEE (2016)
39. Yadav, D.C., Pal, S.: Prediction of heart disease using feature selection and random forest ensemble method. Int. J. Pharm. Res. **12**(4), 56–66 (2020)
40. Yadav, S.S., Jadhav, S.M., Nagrale, S., Patil, N.: Application of machine learning for the detection of heart disease. In: 2020 2nd International Conference on Innovative Mechanisms for Industry Applications (ICIMIA), pp. 165–172. IEEE (2020)
41. Zisad, S.N., Hossain, M.S., Andersson, K.: Speech emotion recognition in neurological disorders using convolutional neural network. In: International Conference on Brain Informatics, pp. 287–296. Springer, Berlin (2020)

Nutritious Menu Planning for Sinusitis Patient in Malaysian Through Optimization Approach

Batrisyia Ibtisam Osman and Suliadi F. Sufahani

Abstract Sinuses have been blocked and filled with fluid and the germs can grow and cause sinusitis or Sinus Inflammation can be defined as the inflammation that occurs in the lining of the sinuses. Sinusitis can be from dust, seafood, and others. Most of the recommended treatment is having surgery. However, this will take a lot of cost in having the surgery and some of the individuals could not afford the cost of surgery. Furthermore, most of the people are likely to consume the antibiotic to relief the sinusitis. However, consuming the antibiotic continuously for the long term will give the side effect based on their consumption. By taking an adequate amount of nutrient requirement also helps an individual in preventing the sinusitis itself. Hence, menu planning is a design for sinusitis patient that meets the requirement of nutrient intake for sinusitis patient. The optimization approach is one of the techniques that can be used in designing the menu planning for sinusitis patient. There will be two types of approach in the optimization field that has been used which are linear programming and integer programming method. delete reshuffle algorithm has been implemented in designing the menu planning to obtain the menu for 5 days with a different menu every day and low cost. As a result, integer programming gives more practical results and the menu planning has been designed for 5 days that meet the nutrient requirement at a lower cost. In the future, the target group can be expanded to another disease that matches the characteristics of sinusitis patients with all levels of age and the list of food groups can be expanded to make more dishes that attract the patient to have a healthy lifestyle.

Keywords Sinusitis · Menu planning · Optimization approach

B. I. Osman · S. F. Sufahani (✉)
Faculty of Applied Sciences and Technology, Universiti Tun Hussein Onn Malaysia, 84600 Pagoh, Johor, Malaysia
e-mail: suliadi@uthm.edu.my

S. F. Sufahani
Oasis Integrated Group, Universiti Tun Hussein Onn Malaysia, 86400 Parit RajaBatu Pahat, Johor, Malaysia

M. S. Kaiser et al. (eds.), *Proceedings of the Third International Conference on Trends in Computational and Cognitive Engineering*, Lecture Notes in Networks and Systems 348, https://doi.org/10.1007/978-981-16-7597-3_3

1 Introduction

Hussain et al. [1] state that sinusitis is a common health problem characterized by mucosal inflammation of the paranasal sinus. Hussain et al. [1] accepted the term "Rhinosinusitis" because they believe that sinusitis coexists with rhinitis in most patients [1]. Desrosiers et al. [8] define sinusitis as the infection of the upper respiratory tract infection. However, the term "Rhinosinusitis" is more accurate to be used in define the sinus disease/infection. The sinusitis problem can be divided into two subtypes which are acute sinusitis and chronic sinusitis as stated in [1, 8].

Since the Malaysian live in an unhealthy environment such as unhealthy lifestyle which will contribute to increasing the sinusitis patient, especially in Malaysia since the air quality has significant toward health impacts. Air pollution contributes to the increase in the population of sinusitis patients. Tobacco smoke also increases the sinusitis that occurs which is more difficult to treat [7].

Sinusitis or known as sinus infection can be defined as an inflammation of the lining inside the sinuses. Sinusitis can be divided into two types of sinus infection which are acute sinusitis and chronic sinusitis which commonly happen to young children adults. In the United States, there are 31 million people affected with sinusitis which resulted from 16 million physician visits in 1985 [6]. The blocking of the sinus passage will occur due to the individual facing nose swollen due to allergies, cold condition, or something in the environment. Sinusitis can be from dust, seafood, and others. When someone is facing sinusitis, it will reduce the quality of their life itself such as did not get the proper sleeping and having difficulty doing any work. There is a natural way that the sinusitis patient must implement in their daily life which is special menu planning to relieve the symptom of sinusitis. However, there is no research regarding on daily special diet for the sinusitis problem. There is research on special diet planning for myopia and diabetes patients but it is not suitable for sinus patients due to a lot of factors that must consider.

To design the menu planning, there will be a few objectives that must be achieved in designing the menu planning. First and foremost, the value of nutrient intake for the sinusitis patient. The boundaries of each nutrient are important which is the constraint in designing the menu planning. Next, in designing the menu planning, the requirement of types of food must be identified to design the special menu planning for sinusitis patients. Last but not least, most of the articles often state that having surgery can help the sinusitis to treat sinusitis. However, it could take a huge cost to undergo the surgery. Some of the individuals are likely to consume the antibiotic in their daily life. Unfortunately, antibiotics are not recommended for the first-week infection or in the moderate level due to the risk of antibiotic resistance and other effects. Mustafa et al. [9] also mention that the antibiotic is recommended only if the symptoms did not relieve within 10 days [9].

There are three objectives in this study. First and foremost, to determine the daily nutrient intake for sinusitis patient. Secondly, to propose a special diet that fulfills all the nutrients that are necessary for sinusitis patient and the third objective is to identify the affordable cost spent by the sinusitis patient on the meal that has been proposed.

Iwuji et al. [2] used linear programming in their research to obtain daily minimum-cost diet plans that meet the nutritional requirement for controlling hypertension. Hypertension is a disease that compulsory to have a strict diet for the patient to survive. The diet planning has been designed based on the nutrient the patient could consume and the calorie intake for that individual. The calorie intake was based on their age, level of activity, and weight of individual. The serving size of food was measured in the Nutrition and Dietetics Laboratory of the Michael Okpara University of Agriculture. However, it would be better if the researcher includes the portion of the food that will help the patient to prepare the food [2, 3].

Hui and Sufahani (2019) had designed the diet menu for a high-pressure patient which believes it could help the patient in controlling and lowering the blood pressure since high blood pressure is one of the global public health issues. The diet menu planning, not just only helps the patient in lowering and controlling blood pressure. The researcher succeeds manage the lowest cost that the patient needs to spend for their diet menu [4]. The researcher used linear programming and integer programming by using LP Solve IDE Software. The researcher used integer programming since the result given in linear programming give the limitation toward the patient to know the real portion of the food itself that they should consume. Integer linear programming helped the patient in preparing the diet planning. The study is focusing on two conditions which are a female aged 54 with 80 kg and she is not a productive person but did not have any allergy or disease. Meanwhile, the other patient is a male aged 82 with 60 kg which he is a current smoker but did not have any disease. Even though both persons need different nutrient requirements, but it is not suitable for sinus disease since the nutrient intake for sinusitis is different with high blood pressure. As the result, the researcher was able to design the diet planning for both people with the lower cost. Some other researches involve optimization and analysis as well [12–14].

2 Materials and Methods

The optimization method will be applied in this research which is linear programming and integer linear programming. The collection of data related to the nutrient requirement for sinusitis patients and the current cost of the groceries. The nutrient requirement intake is referred to the dietitian or known as a consultant or nutritionist. This chapter is about the flow in developing the special diet menu planning for sinusitis patients with minimum cost and fulfill the requirement of sinusitis patients. The optimization approach will give the optimal solution that fulfills all the constraints for sinusitis patients.

2.1 Data Description

All the information regarding sinusitis patients is provided by the Taiping Medical Centre (KPJ) and Hospital Taiping, Perak. The boundary of nutrient content for the sinusitis patients was determined and consulted by the nutritionist Madam Zamila Binti Zamil. The information regarding the sinusitis patient was also referred to Dr. Lee Kean Teong which is a consultant in Ear, Nose and Throat (ENT) and Head and Neck Surgeon in Taiping Medical Centre (KPJ). However, the additional information that regarding the food content and sinusitis is referred to books, clinical journals, websites, and others. For example, Recommended Nutrient Intakes for Malaysia 2015 is used to indicate the model with suitable nutrient contents of food items for Malaysia.

2.2 Modeling

Two types of programming language will be used in this research are linear programming and integer linear programming. Linear programming can be defined as the mathematical model that is used to determine the best solution from a given set of parameters which are represented in the form of a linear relationship. In this research, linear programming is used to test whether the model can be used or not for sinusitis patient. Integer linear programming is programming that needs at least one variable that has integer values. The integer programming also acts as an optimal solution method for sinusitis patients on that menu planning. In this study, the data collection consists of 426 types of food with their nutrient composition for each food. This dataset will be used for the further analysis.

2.3 General Equation for Special Diet Model

Linear programming and integer linear programming will be used to choose the food by fulfilling the requirement of nutrient intake with the minimum cost for sinusitis patients. The objective function is shown as below:

$$\text{Minimize Cost} = \sum_{i=1}^{N}\sum_{j=1}^{P}\sum_{k=1}^{Q} c_i x_{ijk}, \tag{1}$$

where

x_{ijk} Decision variable of food items i for 10 food group, j, and 6 meals, k.
c_i Cost for each food item j.
P The number of meal per day.

Q The number of food group.
N The total number of food.

The constraint for the general nutritional requirement

$$\text{Lower bound of nutrient} \leq \sum_{i=1}^{N}\sum_{j=1}^{10}\sum_{k=1}^{5} W_i x_{ijk} \leq \text{Upper bound of nutrient}, \tag{2}$$

where

W_i The weight of nutrients for the food.

The constraint in this study are expressed as follows:

$$\text{Carbohydrate (g)} = L_C \leq \sum_{i=1}^{426}\sum_{j=1}^{10}\sum_{k=1}^{5} C_i x_{ijk} \leq U_C, \tag{3}$$

$$\text{Calcium (mg)} = L_{Ca} \leq \sum_{i=1}^{426}\sum_{j=1}^{10}\sum_{k=1}^{5} Ca_i x_{ijk} \leq U_{Ca}, \tag{4}$$

$$\text{Iron (mg)} = L_{Fe} \leq \sum_{i=1}^{426}\sum_{j=1}^{10}\sum_{k=1}^{5} Fe_i x_{ijk} \leq U_{Fe}, \tag{5}$$

$$\text{Vitamin A } (\mu g) = L_{Va} \leq \sum_{i=1}^{426}\sum_{j=1}^{10}\sum_{k=1}^{5} Va_i x_{ijk} U_{Va}, \tag{6}$$

$$\text{Vitamin C (mg)} = LVc \leq \sum_{i=1}^{426}\sum_{j=1}^{10}\sum_{k=1}^{5} Vc_i x_{ijk} \leq U_{Vc}, \tag{7}$$

$$\text{Vitamin B1 (mg)} = L_{B1} \leq \sum_{i=1}^{426}\sum_{j=1}^{10}\sum_{k=1}^{5} V_{B1_i} x_{ijk} \leq U_{V_{B1}}, \tag{8}$$

$$\text{Vitamin B2 (mg)} = L_{B2} \leq \sum_{i=1}^{426}\sum_{j=1}^{10}\sum_{k=1}^{5} V_{B2_i} x_{ijk} \leq U_{V_{B2}}, \tag{9}$$

$$\text{Niacin (mg)} = L_{\text{Niacin}} \leq \sum_{i=1}^{426}\sum_{j=1}^{10}\sum_{k=1}^{5} V_{\text{Niacin}_i} x_{ijk} \leq U_{V_{\text{Niacin}}}, \tag{10}$$

$$\text{Protein (g)} = L_{\text{protein}} \leq \sum\sum\sum \text{Protein}_i \, x_{ijk} \leq U_{\text{protein}}, \tag{11}$$

$$\text{Fat (g)} = L_{\text{Fat}} \leq \sum_{i=1}^{426} \sum_{j=1}^{10} \sum_{k=1}^{5} \text{Fat}\, x_{ijk} \leq U_{\text{Fat}}, \tag{12}$$

$$\text{Energy (Kcal)} = L_{\text{Energy}} \leq \sum_{i=1}^{426} \sum_{j=1}^{10} \sum_{k=1}^{5} \text{Energy}\, x_{ijk} \leq U_{\text{Energy}}, \tag{13}$$

where

L	Lower bound for each nutrient content.
U	Upper bound for each nutrient content.
C_i	Amount of carbohydrate in gram for the food item, i
Ca_i	Amount of Calcium in milligram for the food item, i
Fe_i	Amount of Iron in milligram for food item i.
Va_i	Amount of Vitamin A in microgram for food item i.
Vc_i	Amount of Vitamin C in microgram for food item i.
V_{B1i}	Amount of Vitamin B1 in microgram for food item i.
V_{B2i}	Amount of Vitamin B2 in microgram for food item i.
V_{Niacin}	Amount of Niacin in microgram for food item i.
$Protein_i$	Amount of Protein in gram for the food item, i.
Fat_i	Amount of Fat in gram for the food item, i.
$Energy_i$	Amount of Energy in for fkcalood item, i.

2.4 Delete Reshuffle Algorithm

The delete reshuffle algorithm [10] will be implemented in this study for advancement in designing the menu planning for 5 days. The algorithm is focusing on problem-solving in giving the optimal solution with the minimum cost. This algorithm is very important as it helps to find the optimal solution (final menu) and at the same time eliminate the food that is being chosen on Day 1 before running for Day 2. Therefore, the menu that is being produced daily is different except for plain water n plain rice (or certain preference foods) which is compulsory to have every day for Malaysian people. The program will be designed with the aid of LpSolve IDE software. The technique of looping has been implemented to produce the menu planning 5 days which each food must be assigned once (either 1 or 0) except for plain water with the minimum cost The constraint of the problem can be used to produce the optimal solution with the minimum cost. This algorithm can be adapted to solve other diet planning problems such as military, hospital nursing homes, and universities [5, 11]

3 Results and Discussion

3.1 Linear Programming Model

Simplex method is one of the branches of the linear programming approach that is used to optimize the output that satisfies all the constraints with the minimum cost. Linear programming is used to minimize the objective function that will reduce the cost for sinusitis problems. The result for 1 day of the menu structure by the linear programming is shown in Table 1. The model is designed for adults by using the optimization approach. The cost for a 1-day menu for adults is RM 6.95. The model selects the menu that is expressed in positive real numbers with decimal places. The total number of food consumed has fulfilled the food group constraints which is 18.

The result for the 1-day menu structure that used real dataset is illustrated in Table 2. The food group requirements are satisfied. The minimum cost that the output gives is RM 7.30. Based on the result, all the food items are selected in the menu that is expressed in real positive number units. The amount of food required has been fulfilled the constraint of nutrient intake which is 18.

Table 1 Meal structure per day for Adult (19 Years Old–65 Years Old) by using small data model

Meals	Food item, i	Amount
Breakfast	Milk, UHT, low-fat, recombined	0.80839
	Kuih Buah Melaka	0.07608
	Kuih lompang	0.92392
Lunch	Plain water	2
	Rice, cooked	1
	Beef, fried (Daging lembu goreng)	0.94017
	Lungs, fried (Paru lembu goreng)	0.05983
	Asam gelugor, pucuk	0.67236
Evening	Kuih ketayap	1
	Pengat keledek, gula merah	0.08064
	Bingka ubi Kayu	0.91936
Dinner	Syrup rose (Sirap ros)	2
	Rice, "oily"	1
	Shrimp, small, cooked in chili (Udang Kecil Sambal)	1
	Cucumber	1.32764
Supper	Malted milk drink, packet	0.26962
	Banana, common varieties	2
	Cookies, peanut	1
	Total number of food per day	18
	Total Cost (RM)	6.95305

Table 2 The menu planning of sinusitis patient using linear programming model with real data model

Meals	Food item, i	Amount
Breakfast	Milo	0.58612
	Kuih kasui	0.4571
	Quail egg, whole	0.25194
Lunch	Plain water	2
	Rice, cooked	1
	Fish, unspecified, dried, salt	1
	Spinach Onion (Daun Bawang)	1.30016
Evening	Sugar cane juice	1.41388
	chicken_satay	0.74806
	Yau-car-kue	1
Dinner	Coconut water	2
	Noodle, rice	1
	Celery (daun seladeri)	0.69984
	Guava	0.00377
Supper	Biscuit, coconut	0.04719
	Biscuit soda/plain	0.95281
	Rempeyek	0.5429
	Nangka	1.99623
	Candy, coconut	1
	Total number of food per day	18
	Total Cost (RM)	7.30

3.2 Integer Programming Model

In this study, Branch and Bound Method (BNB) was used to design the menu planning. BNB is an algorithm that solving the multiple linear programming and rounds them up to integer values. The result of the menu planning that has been generated by using the integer programming in Table 3. The cost for a 1-day menu is RM 7.20. Based on these results, each of the food groups has been selected from the first ten of each food group. The result has produced the amount value without the decimal point which is easy for the individual or caterer to follow the menu planning. The total number of food has fulfilled all the requirements of constraint which is 18.

The result for a 1-day menu structure is illustrated in Table 4. The food group requirements are satisfied. The minimum cost that the output gives is RM 7.30. Based on the result, all the food items are selected in the menu that is expressed in real positive number units. The amount of food required has been fulfilled the constraint of nutrient intake which is 18.

Table 3 The menu planning structure by using integer programming model (Small Dataset)

Meals	Food item, i	Amount
Breakfast	Milk, cow, fresh	1
	Kuih lompang	1
Lunch	Plain water	2
	Rice, cooked	1
	Chicken kurma (Ayam kurma)	1
	Asam gelugor, pucuk	1
Evening	Kuih ketayap	1
	Bingka ubi Kayu	1
Dinner	Syrup rose (Sirap ros)	2
	Rice, "oily"	1
	Shrimp, small, cooked in chili (Udang kecil sambal)	1
	Cucumber	1
Supper	Milk, UHT, full cream, recombined	1
	Banana, common varieties	2
	Cookies, peanut	1
	Total number of food per day	18
	Total Cost (RM)	7.2

3.3 Discussion on Linear Programming and Integer Programming Approach by Using a Small Dataset

The result generated from the two approaches yields the optimal results for a 1-day menu planning for an adult that satisfy the nutritional requirements within the range of lower bound and upper bound as shown in Table 5. Based on the menu planning, the food group requirement or the total number of food to be followed in one day as stated in Chap. 3 is also satisfied by the two programming approach. In addition, the difference in cost for developing the menu diet in one day by using linear programming and integer programming approach is small which is RM 0.24695 using a small dataset. However, by comparing the food amount served, integer programming will produce the solution which more sense to the caterer and to the sinusitis patient itself. It is due to the food portion is expressed in positive value instead of positive real number in the food preparation and easy to help the sinusitis patient to follow the menu planning.

Table 4 The menu planning of sinusitis patient using linear programming model (Real Data)

Meals	Food item, i	Amount
Breakfast	Milo	0.58612
	Kuih kasui	0.4571
	Quail egg, whole	0.25194
Lunch	Plain water	2
	Rice, cooked	1
	Fish, unspecified, dried, salt	1
	Spinach Onion (Daun Bawang)	1.30016
Evening	Sugar cane juice	1.41388
	chicken_satay	0.74806
	Yau-car-kue	1
Dinner	Coconut water	2
	Noodle, rice	1
	Celery (daun seladeri)	0.69984
	Guava	0.00377
Supper	Biscuit, coconut	0.04719
	Biscuit soda/plain	0.95281
	Rempeyek	0.5429
	Nangka	1.99623
	Candy, coconut	1
	Total number of food per day	18
	Total Cost (RM)	7.30

Table 5 Comparison of nutrient intake for linear programming and integer programming

Nutrients	Lower boundary	Linear programming	Integer programming	Upper bound
Energy (kcal)	1780	2082	2260	2460
Fat (g)	46	53.47	58.77	86
Carbohydrate (g)	180	330	330	330
Calcium (g)	800	800	800	1000
Vitamin A (μg)	500	600	600	600
Vitamin B1 (mg)	1.1	1.1	1.1	
Vitamin B2 (mg)	1	2.193	1.539	
Vitamin C (mg)	70	80	70	80
Niacin (mg)	16	16	16	30
Iron (mg)	14	19.64	18.33	29
Protein (g)	51	63.46	71.63	62

Table 6 the cost for one day menu planning (1 Day)

Menu planning cost	Cost (RM)
Linear programming	7.30
Integer programming	7.00

3.4 Comparison of Linear and Integer Programming Approach

The results generated from these two programming approaches have yielded the optimal result for a 1-day menu that satisfies the nutrient requirement within the range of lower bound and upper bound of the nutrient boundaries as shown in Table 5. From Table 5, three of the eleven fundamental nutrients generated in the integer programming model are higher than compared to the linear programming model. The total number of food to be consumed and the food group requirement is satisfied for both approaches. From Table 6, the difference in cost for developing the menu planning in one day by using the linear programming approach and integer programming approach which is RM 0.30. Based on the result, the integer programming approach has produced the result that is practical to caterer to follow the menu planning. Next, the food portion expressed in the positive integer value instead of the positive real number makes the food preparation and serving process easier and concise.

3.5 Further Analysis

To obtain the result for 5 days, the delete reshuffle algorithm [10] was used to generate the menu planning. The result have been shown in Table 1. The looping technique has been applied in delete reshuffle algorithm which is used to eliminate the variables that have been assigned the day before and reshuffle the variables to generate the menu planning for the other days. The total amount of food required every day is 18. The results show in integer unit which will ease the caterer to prepare the food as shown in Table 7.

4 Conclusion

There are three objectives in this study that has been achieved. First and foremost, the daily nutrient intake for sinusitis patients was determined with the dietitian at government hospitals. They suggested the nutrient intake follow the RNI 2017. Furthermore, seeking a specialist is also helpful based on the theoretical of the disease itself and how to overcome the disease. As for the second objective, in designing the menu planning for the sinusitis patient, there are two approaches in designing the menu

Table 7 Meal structure for 5 days using integer programming approach

Meals	Days				
	Day 1	Day 2	Day 3	Day 4	Day 5
Breakfast	Milk, UHT, chocolate flavored	Milk, cow, fresh	Malted milk drink, packet	Soya bean milk, packet	Soya bean milk, packet
	Quail egg, whole	Kuih lopes pulut	Roti canai + yellow dhal gravy	Kuih buah Melaka	Roti canai + yellow dhal gravy
				Egg Banjo	
Lunch	Plain water	Plain water	Plain water	Plain water	Plain water
	Rice, cooked	Rice, cooked	Rice, cooked	Syrup rose (Sirap ros)	Rice, cooked
	Fish, unspecified, dried, salt	Chicken, fried (Ayam goreng)	Fish, unspecified, dried, salt	Rice, cooked	Chicken curry (Kari ayam)
	Spinach Onion (Daun Bawang)	Asparagus, canned	Spinach Onion (Daun Bawang)	Spinach Onion (Daun Bawang)	Mutton curry (Kari kambing)
					Cucumber
Evening	Sugar cane juice	Sugar cane juice	Sugar cane juice	Fish, unspecified, dried, salt	Syrup rose (Sirap ros)
					Kuih ketayap
	Mee, soup	Yau-car-kue	Mee Curry	Coconut water	Bingka ubi Kayu
Dinner	Coconut water	Coconut water	Coconut water		Plain Water
	Rice porridge, fish, instant	Beef burger	Noodle, rice	Sugar cane juice	Rice, "oily"
	Celery (daun seladeri)	Cucumber	Celery (daun seladeri)	Chicken soto	Shrimp, small, cooked in chili (Udang kecil sambal)
	Guava	Guava	Guava		
Supper	Biscuit soda/plain	Biscuit soda/plain	Biscuit soda/plain	Curry puff twisted	
	Rempeyek	Egg Banjo	Rempeyek	Candy, coconut	Banana, common varieties
	Malted milk powder	Yogurt Banana	Yogurt Banana	Yogurt Oren	Milk, UHT, low-fat, recombined
	Candy, coconut	Candy, coconut	Candy, coconut	Nangka	
Total Cost (RM)	7.00	8.20	6.40	7.20	7.30

planning by using the mathematical approach which is linear programming and integer linear programming. Delete reshuffle algorithm is used to obtain menu planning for 5 Days. This approachable give the minimum cost to the menu planning and the appropriate portion to help the caterer or patient itself to follow the menu planning. The minimum cost has been determined by using these approaches and compared with these two approaches, the integer linear programming gives the most minimum value and the portion value in an integer value. By using the food dataset with the list of nutrients and the nutrient requirement intake for the sinusitis patient, the optimization approach has been developed to solve the sinusitis patient that minimizes the cost for the diet and optimize the nutrient intake. Thus, the mathematical modeling is to compute the number of nutrients in the food consumption with the nutrient boundaries that are suitable for children and adults level. In the future study, the target group can be expanded to another disease that matches the characteristics of sinusitis patients with all levels of age. Next, the list of food groups can be expanded to make more dishes that attract the patient to have a healthy lifestyle. Furthermore, the extended task is to have collaborated with the medical agency or any dietitian agency in exploring the use of menu planning for sinusitis itself and other diseases. Having a healthy life by consuming the foods also improves the quality of work by reducing the expenditures and improving the efficiency of the food planning process.

Acknowledgements Fundamental Research Grant Scheme (FRGS) No. K175 (FRGS/1/2019/STG06/UTHM/02/2) from the Ministry of Higher Education Malaysia (MOHE) and Universiti Tun Hussein Onn Malaysia (UTHM)

References

1. Hussain, S., Amilia, H.H., Rosli, M.N., Zahedi, F.D., Sachlin, I.S.: Management of rhinosinusitis in adults in primary care. Malays. Fam. Physician **13**(1), 28–33 (2018)
2. Iwuji, A.C., Nnanna, M., Ifeyinwa, N., Ndulue, C.: An Optimal DASH Diet Model for People with Hypertension Using Linear Programming Approach, pp. 14–21 (2016)
3. Iwuji, A.C., Agwu, E.U.: A Weighted Goal Programming Model for the DASH Diet Problem : Comparison with the Linear Programming DASH Diet Model, pp. 307–322 (2017)
4. Su Hui, L., Sufahani, S.F., Khalid, K., Wahab, M.H.A., Idrus, S.Z.S, Ahamd, A. & Subramaniam, T.S. (2021). Menu scheduling for high blood pressure patient with optimization method through integer programming. J. Phys.: Conf. Ser. **1874**(1), Article number 012088. https://doi.org/10.1088/1742-6596/1874/1/012088
5. Sufahani, S.F., Kamardan, M.G., Rusiman, M.S., Mohamad, M., Khalid, K., Ali, M., Khalid, K., Nawawi, M.K.M, Ahmad, A. (2018). A mathematical study on "additive Technique" versus "branch and Bound Technique" for solving binary programming problem. J. Phys.: Conf. Ser. **995**(1), Article number 012001. https://doi.org/10.1088/1742-6596/995/1/012001
6. Stammberger, H., Wolf, G.: Headaches and sinus disease: The endoscopic approach. Ann. Otol. Rhinol. Laryngol. 97(5 II SUPPL.), 3–23 (1988)
7. Mirgain, S.A.: Whole Health : Change The Conversation Advancing Skills in the Delivery of Progressive Muscle Relaxation Clinical Tool Whole Health : Change The Conversation. VHA Office of Patient Care and Cultural Transformation, pp. 1–3 (2016)

8. Desrosiers, M., Frenkiel, S., Hamid, Q.A., Low, D., Small, P., Carr, S., Davidson, R.: Acute bacterial sinusitis in adults: Management in the primary care setting. J. Otolaryngol. **31**(SUPPL. 2), 1–14 (2002)
9. Mustafa, M., Iftikhar, M., Choudhury Shimmi, S.: Acute and Chronic Rhinosinusitis, Pathophysiology and Treatment Both project View project Multi Drug View project Acute and Chronic Rhinosinusitis, Pathophysiology and Treatment, pp. 30–36 (2015)
10. Sufahani, S.F., Ismail, Z.: planning a nutritious and healthy menu for malaysian school children aged 13–18 using delete-reshuffle algorithm in binary integer programming. J. Appl. Sci. **15**(10), 1239–1244 (2015)
11. Sheng, L.Z., Sufahani, S.: Optimal diet planning for eczema patient using integer programming. J. Phys: Conf. Ser. **995**(1), 1–9 (2018)
12. Latif, M.S.A., Fahmy-Abdullah, M., Sieng, L.W.: Determinants factor of technical efficiency in machinery manufacturing industry in Malaysia. Int. J. Supply Chain Manag. **8**, 917–928 (2019)
13. Soebagyo, D., Fahmy-Abdullah, M.O.H.D., Sieng, L.W., Panjawa, J.L.: Income inequality and convergence in central java under regional autonomy. Int. J. Econ. Manag. **13**(1)
14. Idris, A.I.M., Fahmy-Abdullah, M., Sieng, L.W.: Technical efficiency of soft drink manufacturing industry in Malaysia. Int. J. Supply Chain Manag. **8**, 908–916 (2019)

Nutritious Menu and Nutrient Planning for Avoiding Stroke Disease by Using Optimization Technique

Maisarah Auni Jamaludin and Suliadi F. Sufahani

Abstract Nutrient and menu planning is important for all generations. Having proper nutrition is one of the needs to have a better lifestyle. The factor of most people that suffer from any disease is by not taking nutrients and meals that are required in their body. This research has identified a suitable range of nutrients which is lower bound and upper bound for stroke patients that can assist them to plan balance diet selection based on the guidelines of Recommended Nutrient Intake for Malaysia (RNI 2017). Optimization techniques have been widely used and applied in various areas including nutrient planning problems. Optimization techniques such as linear programming and integer programming have been discussed in this research to develop the mathematical model of meal planning to avoid and prevent stroke disease in Malaysia. This research aims to determine nutritious menu planning as the objective function is to minimize the cost. For menu planning purpose, the software tools that have been used is LPSolve IDE. The results showed that further analysis has been done by integer programming with applying the Delete-Reshuffle Algorithm in generating a 5-day structure of menu planning. The optimal cost of the meal planning structure is below RM 20.00 per day for 5 days. Hence, this research can give contributions to society especially for Malaysian people in solving the nutrient and menu problem to avoid stroke disease besides can be a guideline for them to practice a healthy lifestyle. It also can assist them to plan their cost budget in meal planning especially for Stroke Patient which is the main contribution of this paper.

Keywords Nutrient planning · Stroke · Optimization technique · LPSolve IDE · Delete-reshuffle algorithm · Minimize cost

---

M. A. Jamaludin · S. F. Sufahani (✉)
Faculty of Applied Sciences and Technology, Universiti Tun Hussein Onn Malaysia, 84600 Pagoh, Johor, Malaysia
e-mail: suliadi@uthm.edu.my

S. F. Sufahani
Oasis Integrated Group, Universiti Tun Hussein Onn Malaysia, 86400 Parit RajaBatu Pahat, Johor, Malaysia

M. S. Kaiser et al. (eds.), *Proceedings of the Third International Conference on Trends in Computational and Cognitive Engineering*, Lecture Notes in Networks and Systems 348, https://doi.org/10.1007/978-981-16-7597-3_4

1 Introduction

Stroke disease is the second cause that contributes to death and the third main reason of disability throughout the entire globe. The overall stroke burden has risen in both men and women of all ages in terms of total numbers of people affected by or who remained disabled even though the stroke occurrence, prevalence, mortality, and disability-adjusted life year rates continue to decline from 1990 to 2013 [1]. Stroke is a disease or illness that can negatively affect the arteries leading to and within the brain. The occurrence of cardiovascular disease, diabetes, and obesity is high among stroke patients. The main factors for this disease include hypertension (high blood pressure), high cholesterol, poor nutrition, and having physical inactivity. Poor nutrition can lead as a significant contributor to a variety of chronic diseases include cardiovascular disease, diabetes, and osteoporosis. Therefore, providing good nutrition can also aid in the prevention and treatment of chronic diseases [2, 14].

In Malaysia, stroke is among the causes of top diseases in death and one of the top 10 reasons that lead to hospitalization [3]. The occurrence of stroke has become an increase among the population of Malaysian. Due to the higher percentage and number of death attributed to stroke day by day, some prevention and precaution should have arisen. Mostly, stroke disease affecting people between 54 until 70 years old. Nutritional environments, awareness, and economic differences can be considered as factors of stroke. High prevalence of high-calorie foods, obesity, and lower-calorie foods are the environmental factors related to diet and nutrition in this region [4]. Due to the high fat and calories, fast food is also linked to stroke. Hankey [5] stated that poor nutrition and overnutrition can raise the risk of stroke.

Diet optimization or diet modeling has been introduced by Georges Stigler, who applied diet problems in the mathematical model. Identifying and determining the set of foods that met daily nutritional requirements at minimum cost is the aim and purpose of the diet problem [6]. Optimization techniques such as linear programming and integer programming aids in planning for dietary and nutrients needs. The study of [7] also used linear programming to solve the diet problem since there is a shortage of necessities food during the COVID-19 pandemic. The goal of the research was to add the limitation of scarcity includes the availability of food and time in the diet problem to show that even during a pandemic, practicing healthy and proper meals also can be maintained. The uses of integer linear programming (ILP) have been addressed by [8] to build the mathematical model of diet selection problem with the lowest cost for students in Universiti Utara Malaysia (UUM).

Hence, this study aimed to identify the range of nutrients which are lower bound and upper bound that should be intake by stroke patients. Besides, this study also has been conducted to apply optimization techniques which are linear programming and integer programming that can be used to solve the menu planning in this research. Thru this study, the nutritious menu planning for stroke patients with minimizing the cost can be determined with further analysis. This study focused on stroke patients in Malaysia range from 54 until 70 years to solve the nutrient and menu problem. Computer software such as LPSolve IDE is used for menu planning purposes. Besides, Recommended Nutrient Intakes for Malaysia (RNI 2017) could be the guideline and reference in this research. There is some other research that involves optimization and analysis as well [16–18].

2 Materials and Methods

Throughout this research, primary data and secondary data were used to collect the data based on the problem statement, research objectives, and literature review. The primary data including interviews with a dietitian or nutritionist. They are experts and trained professional who is qualified to conduct a nutritional assessment, nutritional diagnosis, and nutritional intervention in the prescribing and customization of a diet or nutrients to meet the needs of the individual based on medical problems. Secondary data which is Recommended Nutrient Intake (RNI 2017) is the guidelines and reference of nutrients that are required for stroke patients [9].

2.1 *Modeling*

Decision variables, objective function, and constraints are the parameters of the mathematical model in the optimization problem and will be solved by using Linear Programming (LP) and Integer Programming (IP). Generally, the optimization model aims to identify the combination of values for decision variables that produce the optimal value of maximizing or minimize the objective function while at the same time fulfilling the constraints. The objective function is to minimize the cost as well as one of the objectives in this research while meeting the nutritional requirements that should be taken by stroke patients. The model of the objective function as follows [10, 11, 15]:

$$\text{Minimize cost} = \sum_{i=1}^{N}\sum_{j=1}^{P}\sum_{k=1}^{Q} c_i x_{ijk} \quad (1)$$

where

x_{ijk} Decision variable of food items, i for 10 food group, j and 6 meals course, k
c_i cost for each of the food items, i
N number of food items, i
P number of food groups, j
Q number of food meals per day, k

The equation for the nutrient planning model included the general nutritional requirements for stroke patients. The constraints for the general nutritional requirements [11]:

$$LB_i \leq \sum_{i=1}^{N}\sum_{j=1}^{10}\sum_{k=1}^{6} w_i x_{ijk} \leq UB_i \quad (2)$$

where

LB_i Lower bound for each nutrient needed.
UB_i Upper bound for each nutrient needed.
w_i Weight of nutrient for each of the food item, i
x_{ijk} Decision variable of food items, i for 10 food group, j and 6 meals course, k

The constraints of the required nutrients are shown as follows [11]:

$$\text{Energy (kcal)} : 1550 \leq \sum_{i=1}^{426}\sum_{j=1}^{10}\sum_{k=1}^{6} E_i x_{ijk} \leq 2030 \quad (3)$$

$$\text{Protein (g)} : \sum_{i=1}^{426}\sum_{j=1}^{10}\sum_{k=1}^{6} P_i x_{ijk} \geq 50 \quad (4)$$

$$\text{Fats (g)} : 49 \leq \sum_{i=1}^{426}\sum_{j=1}^{10}\sum_{k=1}^{6} F_i x_{ijk} \leq 59 \quad (5)$$

$$\text{Calcium (mg)} : 1000 \leq \sum_{i=1}^{426}\sum_{j=1}^{10}\sum_{k=1}^{6} Ca_i x_{ijk} \leq 2000 \quad (6)$$

$$\text{Sodium (mg)} : 1500 \leq \sum_{i=1}^{426}\sum_{j=1}^{10}\sum_{k=1}^{6} Na_i x_{ijk} \leq 2300 \quad (7)$$

$$\text{Potassium (mg)} : \sum_{i=1}^{426}\sum_{j=1}^{10}\sum_{k=1}^{6} Ka_i x_{ijk} \geq 4700 \quad (8)$$

$$\text{Vitamin B3 (mg)} : 15 \leq \sum_{i=1}^{426}\sum_{j=1}^{10}\sum_{k=1}^{6} Vb3_i x_{ijk} \leq 35 \quad (9)$$

$$\text{Vitamin B9 Folate (mg)} : 400 \leq \sum_{i=1}^{426}\sum_{j=1}^{10}\sum_{k=1}^{6} Vb9_i x_{ijk} \leq 1000 \quad (10)$$

$$\text{Vitamin E (mg)} : 75 \leq \sum_{i=1}^{426}\sum_{j=1}^{10}\sum_{k=1}^{6} Ve_i x_{ijk} \leq 1000 \quad (11)$$

LB_i Lower bound for each nutrient needed.
UB_i Upper bound for each nutrient needed.

E_i	Amount of Energy in kcal for the food item, i
P_i	Amount of Protein in gram for the food item, i
F_i	Amount of Fats in gram for the food item, i
Ca_i	Amount of Calcium in milligram for the food item, i
Na_i	Amount of Sodium in milligram for the food item, i
K_i	Amount of Potassium in milligram for the food item, i
$Vb3_i$	Amount of Vitamin B3 in milligram for the food item, i
$Vb9_i$	Amount of Vitamin B9 in milligram for the food item, i
Ve_i	Amount of Vitamin E in milligram for the food item, i

In this study, the decision variables of the menu planning model consisted of different kinds of Malaysian food. The set in variations of food will be shown in the menu lists according to the six classifications of meal options. The decision variable can be written as follow:

$$X_i = \begin{cases} 1 \text{ or } 2 & \text{if menu } i \text{ appears in the menu list} \\ 0 & \text{otherwise} \end{cases} \tag{12}$$

where

i any type of menus.

2.2 *Delete-Reshuffle Algorithm*

The algorithm of this research has been expressed as a mathematical process in optimization involved Delete-Reshuffle Algorithm [12]. This algorithm is very important as it helps to find the optimal solution (final menu) and at the same time eliminate the food that is being chosen on Day 1 before running for Day 2. Therefore, the menu that is being produced daily is different except for plain water n plain rice which is compulsory to have every day for Malaysian people. Besides, the algorithm focused on problem-solving in nutrient planning for stroke patients by searching and identifying optimal solutions with minimizing the cost. The constraints of the problem can be used directly in producing optimal solutions. From the algorithm, the food groups have been assigned based on each meal. For 5 days, each food must be assigned once (either 1 or otherwise 0) except plain water and plain rice. The technique of looping has been used to run the program for 5 days by deleting the chosen variables and reshuffle again to generate the food for the next day. The program was running by using LPSolve IDE and the optimal solution cost was obtained.

3 Results and Discussion

3.1 The Lower Bound and Upper Bound of Nutrient Contents

The lower bound and upper bound of each nutrient for stroke patients is important as they can avoid taking the excess of nutrients or lack of nutrients that can affect their health. In this research, men and women who are moderately active aged between 54 until 70 years in Malaysia have been focused on. Table 1 shows the value of lower bound and upper bound of 9 nutrients for the male who is between 54 and 70 years while Table 2 shows the value of lower bound and upper bound of 9 nutrients for the female who is between 54 and 70 years. The value of upper and lower bounds was referred to RNI 2017 and also interview with the dietitian.

Table 1 Value of lower bound and upper bound of 9 nutrients for male [9]

Nutrients	Lower bound	Upper bound
Energy (kcal)	1780	2030
Protein (g)	58	–
Fats (g)	56	58
Calcium (mg)	1000	2000
Sodium (mg)	1500	2300
Potassium (mg)	4700	–
Vitamin B3 (mg)	16	35
Vitamin B9 (μg)	400	1000
Vitamin E (mg)	10	1000

Table 2 Value of lower bound and upper bound of 9 nutrients for female [9]

Nutrients	Lower bound	Upper bound
Energy (kcal)	1550	1770
Protein (g)	50	–
Fats (g)	49	59
Calcium (mg)	1200	2000
Sodium (mg)	1500	2300
Potassium (mg)	4700	–
Vitamin B3 (mg)	15	35
Vitamin B9 (μg)	400	1000
Vitamin E (mg)	7.5	1000

Table 3 Structure of meal per day using linear programming

Meals	Food items	Required amount
Breakfast	Soya bean milk, packet	1
	Biscuit, alphabet	0.5566
Morning tea	Fresh cow milk	1
	Puding jagung	1
	Curry puff	1
Lunch	Plain Water	1
	Rice porridge, fish instant	1
	Beef rendang canned	0.1860
	Carrot	0.7386
	Lemon	0.0725
Evening tea	Tea	1
	Bingka Tepung Beras	1
Dinner	Plain Water	1
	Rice, cooked	1
	Maw Satay	0.8140
	Mushroom dried	0.2614
Supper	Durian belanda	0.9275
	Biscuit, coconut	0.4434
Total number of food per day		13
Total cost per day (RM)		20.10

3.2 Linear Programming Model

$$\text{Minimize cost} = \sum_{i=1}^{426} c_i x_i \tag{13}$$

The output and result of the menu structure for one day were generated by using linear programming in Table 3. The cost for a one-day menu is RM 20.10. Based on the results, the first 10 of each food group have been selected and thru the model, it produced the amount with a decimal point that should be consumed. Then, the total number of food required has been fulfilled the constraint of nutrient requirement which is also 13.

3.3 Integer Programming Model

$$\text{Minimize cost} = \sum_{i=1}^{426} c_i x_i \tag{14}$$

The output and result of the menu structure for one day were generated by using integer programming in Table 4. The cost for a one-day menu is RM 15.10. Based on the results, the first 10 of each food group have been selected and thru the model,

Table 4 Structure of meal per day using integer programming

Meals	Food items	Required amount
Breakfast	Soya bean milk, packet	1
	Biscuit, alphabet	1
Morning tea	Fresh cow milk	1
	Puding jagung	1
	Curry puff	1
Lunch	Plain Water	1
	Rice porridge, fish instant	1
	Carrot	1
Evening tea	Tea	1
	Bingka Tepung Beras	1
Dinner	Plain Water	1
	Rice, cooked	1
	Maw Satay	1
Supper	Durian belanda	1
Total number of food per day		13
Total cost per day (RM)		15.10

it produced the amount without decimal point or integer numbers that should be consumed. Then, the total number of food required has been fulfilled the constraint of nutrient requirement which is also 13.

3.4 Comparison of Linear and Integer Programming Approach

The total cost per day for linear programming is RM 20.10 while the total cost per day for integer programming is RM 15.10 based on Table 5. The difference in cost for the 2 programming models approach was RM 5.00. Therefore, integer programming shows the best solution in minimizing the cost of meal and nutrient planning compared to linear programming for 426 food items. Besides, integer programming also has a better method in applying the amount of food that should be consumed.

Table 5 Total cost per day

Model	Total cost (RM)
Linear programming	20.10
Integer programming	15.10

Table 6 Structure of meal per day (Day 2) using integer programming

Meals	Food items	Required amount
Breakfast	Fresh Milk	1
	Biscuit Marie	1
Morning tea	Pulut Panggang	1
Lunch	Plain Water	1
	Chicken Rice	1
	Cuttlefish small fried in chili	1
	Celery	1
Evening tea	Anggur	1
	Pengat Keledek	1
Dinner	Plain Water	1
	Liver, Rendang	1
Supper	Cucur Udang	1
	Mango	1
Total number of food per day		13
Total cost per day (RM)		15.50

3.5 *Further Analysis*

Since integer programming showed better results compared to linear programming, therefore Delete-Reshuffle Algorithm [12] then was used to produce and generate the structure of meal planning for further analysis from Day 2 until Day 5. The results have been illustrated in Table 6 until Table 9 after the structure of the meal on day 1 has been generated. Delete-Reshuffle Algorithm was applied to generate food items for 5 days. The technique of looping has been used where it will eliminate the variables that have been assigned and reshuffle other variables to generate the result for the next days [12]. The food items model in Day 1 has been deleted except plain water to prevent the duplicate and same menu plan for Day 2, Day 3, Day 4, and Day 5 as shown in Tables 7, 8, and 9.

4 Conclusion

There are 9 nutrients for stroke patients according to the value of lower bound and upper bound that was recommended by RNI 2017 [9]. In this research, 426 variables of food items have been used with 10 food groups, 9 nutritional requirements, and 6 meals course. To build the model, linear programming and integer programming have been used. The amount of food intake is slightly different between the 2 methods based on results in the analysis and discussion part by running LPSolve IDE software. Integer programming is considered suitable and appropriate to use compared to linear programming in identifying the menu planning for stroke patients. The findings show that per one day, the adult people will spend below RM 20.00 based on the structure

Table 7 Structure of meal per day (Day 3) using integer programming

Meals	Food items	Required amount
Breakfast	Malted Milk Drink Biscuit, cream crackers	1 1
Morning tea	Kuih buah Melaka Doughnut	1 1
Lunch	Plain Water Fried Chicken Fish Curry	1 1 1
Evening tea	Sugar Cane Juice Rojak	1 1
Dinner	Plain Water Fried Rice Cucumber	1 1 1
Supper	Pear	1
Total number of food per day		13
Total cost per day (RM)		16.00

Table 8 Structure of meal per day (Day 4) using integer programming

Meals	Food items	Required amount
Breakfast	Soya bean milk (unsweetened) Kuih tepung pelita	1 1
Morning tea	Yoghurt Apricot Sandwich sardine	1 1
Lunch	Plain Water Rice porridge Anchovy dried, fried in chili Petai	1 1 1 1
Evening tea	Cookies, peanut	1
Dinner	Plain Water Mutton Curry Avocado	1 1 1
Supper	Mashed Potatoes	1
Total number of food per day		13
Total cost per day (RM)		14.10

of meal planning. Further analysis also has been done in structuring and developing 5 days of meal planning followed the required nutrient which can benefit people. On top of that, the expected cost of meal planning for every 5 days is also below RM 20.00 and it can help people in minimizing their budget.

This study has several limitations that have been encountered and observed. Firstly, this research only considers and focuses on adult stroke patients instead of the whole population in Malaysia. A stroke may also face by infants as the factors

Table 9 Structure of meal per day (Day 5) using integer programming

Meals	Food items	Required amount
Breakfast	Coffee mixture	1
	Bread, wholemeal	1
Morning tea	Kuih Apam	1
Lunch	Plain Water	1
	Rice coconut	1
	Sardine	1
	Tomato	1
Evening tea	Orange Flavored Drink	1
	Kuih Bom	1
Dinner	Plain Water	1
	Beef curry	1
	Coleslaw	1
Supper	Grape	1
Total number of food per day		13
Total cost per day (RM)		16.10

and causes may vary from those in young adults and persons over the age of 65 [13]. Besides, the current model only considers the structure of meal planning for 5 days since it results in some errors in analysis. Some suggestions are recommended to improve more in the related area for future work research. One of the suggestions is by taking into account different ages for stroke patients. The probability that the findings obtained from other ages or generations in Malaysia might give different results. The model also should be modified by considering for 1 week or 7 days of the structure in meal planning instead of 5 days to assist people in choosing the right meal with required nutrients and at the same time, it can minimize the cost.

This research will give some contributions to the Malaysian in preventing stroke disease as it helps the people with the motive of diet, illness, and disease control. Controlling the food intake and actively exercising is an example of having a better lifestyle. It also can be a guideline and reference for them to get proper nutrition as they are struggling with eating. Furthermore, most of them are not aware of what kind of food or required nutrition that they should intake. Having proper and good nutrition is important for stroke recovery and also can prevent themselves from having a high probability of stroke disease.

Acknowledgements Fundamental Research Grant Scheme (FRGS) No. K175 (FRGS/1/2019/STG06/UTHM/02/2) from the Ministry of Higher Education Malaysia (MOHE) and Universiti Tun Hussein Onn Malaysia (UTHM).

References

1. Feigin, V.L., Norrving, B., Mensah, G.A.: The Global Burden of Stroke, pp. 439–449 (2017)
2. Lange, K.W.: Movement and Nutrition in Health and Disease. Movement and Nutrition in Health and Disease **1**, 3–10 (2017)
3. Loo, K.W., Gan, S.H.: Burden of stroke in Malaysia. Int. J. Stroke **7**(2), 165–167 (2012)
4. Zhu, Y.L.: Community based health promotion program for stroke in Malaysia. In: 2019 International Conference on Biomedical Sciences and Information Systems, (Icbsis), pp. 91–95.
5. Hankey, G.J.: Nutrition and the risk of stroke. Lancet Neurol **11**(1), 66–81 (2012)
6. Wirths, W.: The cost of human energy in agriculture. J. Farm Econ. **39**(1), 155–161 (2010)
7. Joanis, S.T.: Family meal planning under COVID-19 scarcity constraints: a linear programming approach. J. Food Nutri. Res. **8**(9), 484–495 (2020)
8. Ahmad, N., Syazwanie, N., Sani, A.: Optimal diet selection for university students using integer linear programming. AIP Conf. Proc. 2138(1) (2019)
9. Ministry of Health Malaysia: Recommended nutrient intakes for Malaysia 2017. In: Ministry of Health Malaysia (2017)
10. Lancaster, L.M.: The history of the application of mathematical programming to menu planning. **57**(3), 339–347 (1992)
11. Ali, M., Sufahani, S., Ismail, Z.: A new diet scheduling model for malaysian school children using zero-one optimization approach. Global J. Pure Appl. Math. **12**(1), 413–419 (2016)
12. Sufahani, S.F., Kamardan, M.G., Rusiman, M.S., Mohamad, M., Khalid, K., Ali, M., Khalid, K., Nawawi, M.K.M, Ahmad, A.: A mathematical study on "additive Technique" versus "branch and Bound Technique" for solving binary programming problem. J. Phys.: Conf. Ser. **995**(1), Article number 012001 (2018). https://doi.org/10.1088/1742-6596/995/1/012001
13. Sudin, A.M., Sufahani, S.: Mathematical Approach for serving nutritious menu for secondary school student using "Delete-Reshuffle-Reoptimize Algorithm." J. Phys.: Conf. Ser. **995**(1) (2018)
14. Pathirage, M., Senadheera, C., Van Raay, L., Cellesteno, S., Wijeratne, T.: Young stroke. Cerebrovasc. Dis. **32**, 12 (2011)
15. Sheng, L.Z., Sufahani, S.: Optimal diet planning for eczema patient using integer programming. J. Phys: Conf. Ser. **995**(1), 1–9 (2018)
16. Latif, M.S.A., Fahmy-Abdullah, M., Sieng, L.W.: Determinants factor of technical efficiency in machinery manufacturing industry in Malaysia. Int. J. Supply Chain Manag. **8**, 917–928 (2019)
17. Soebagyo, D., Fahmy-Abdullah, M.O.H.D., Sieng, L.W., Panjawa, J.L.: Income inequality and convergence in central java under regional autonomy. Int. J. Econ. Manag. **13**(1) (2019)
18. Idris, A.I.M., Fahmy-Abdullah, M., Sieng, L.W.: Technical efficiency of soft drink manufacturing industry in Malaysia. Int. J. Supply Chain Manag. **8**, 908–916 (2019)

Application of Machine Learning to Traditional System for Analyzing the Conditional Results of School and College Students

Md. Imtiaz Ahmed, Mohammed Shakhawat Hossain, Nusrhat Jahan Sarker, and Md. Imran Hossain Imu

Abstract Machine learning algorithms are used for various purposes to predict, classify, or forecast by training the algorithms with the specific dataset. SVR and multiple linear regression can take numerous features to forecast or predict scores through the train-test-split. The education sector has been changed rapidly due to the pandemic of COVID-19 where online classes are being a module worldwide. However, junior schools or colleges stubbed into a position where student performance measurement is a hindrance due to the lack of taking physical examinations. During the COVID-19, student performance can be acquired using the previous achievement of individual students where multiple conditions can be applied. The aim of this paper is to train and test the conditional dataset of student's results through SVR and Multiple Linear Regression to predict and justify the results in accordance with using the proposed model in the future. As conditions have been applied to the individual subjects when calculating new results based on the previous achievement of student's performance so that each subject's score has been trained and tested individually through the machine learning algorithms.

Keywords Machine learning algorithms · SVR · Multiple linear regression · Students result calculation · Conditional data training · Score of predicted data

---

Md. I. Ahmed (✉) · M. S. Hossain · N. J. Sarker · Md. I. H. Imu
Daffodil Institute of IT, Dhaka, Bangladesh
e-mail: imtiaz_nu@diit.info

M. S. Hossain
e-mail: nup.principal@diit.info

N. J. Sarker
e-mail: nusrhat_nu@diit.info

Md. I. H. Imu
e-mail: imran_nu@diit.info

M. S. Kaiser et al. (eds.), *Proceedings of the Third International Conference on Trends in Computational and Cognitive Engineering*, Lecture Notes in Networks and Systems 348, https://doi.org/10.1007/978-981-16-7597-3_5

1 Introduction

COVID-19 pandemic has led to the rush of the education sector to a position that cannot be retained easily. The education sector has been introduced through online or blended learning during the pandemic. Traditional teaching methods subsided by blended learning but at the same time, it hampered the education model of high school or college students. Higher education or universities have advantages, but it has some disadvantages as none of the institutions provided a secure way for conducting the examinations. Development is still underway to build a stable and vast framework that can benefit every organization.

Machine learning approaches can implement to determine the students' performance. Some emerging technologies in education have already been approached through Machine Learning, and results have been checked [1]. The Caret Package with Rd has been checked for an online survey of student's performance on some specific courses [2]. Vast research has already been introduced by several researchers in this field.

Machine learning is a technique or algorithm that predicts or classifies based on a dataset. There are three kinds of techniques used in machine learning. Supervised learning refers to a model with input and output data classified as training and test data [3]. In supervised learning classification and regression, both are possible where regression is used to identify accuracy and classification is used to classify the dataset as a dependable variable [4].

Linear regression is the process where a model even predicts or forecasts based on their trained data. Furthermore, it can help whether the predicted score or the forecasting is accurate or not. When one wants to predict based on multiple features, then multiple linear regression is used. Multiple linear regression known simply as multiple regression is a statistical technique that uses several explanatory variables to predict the outcome of a response variable [5]. SVR (Support Vector Regression) has proved to be an effective tool for estimating a real value function [6]. SVR is a subset of SVM (Support Vector Machine), which is not as popular as SVM. The purpose of Multi Regression and Support Vector Regression is to predict based on multiple features and the output for dependent labels.

SVR and Multi regression both can predict the values of the dataset, but whether it performs on a conditional dataset and scores accurate or not. This paper aims to determine whether the SVR and multi regression can observe conditional datasets to provide the labeled data's accurate output.

2 Related Work

Machine learning in education has been categorized into four categories named as Grading students, improving student retention, predicting student performance, and testing students [7, 8]. In our country or on this subcontinent, we depend on the grading system to determine the students' accuracy. Due to COVID-19, it has been harsh to figure out the students' performance regarding the way of unsecured examinations or not physical examinations.

For predicting student's performance two classification algorithm: Random forest (RF) and Support Vector Machine (SVM) have been used [9]. For predicting private university student's results in machine learning, seven algorithms have been used and got the accuracy to 81.73% [10]. Student performance prediction with the supervised and unsupervised learning algorithms in machine learning helps high accuracy and k-means algorithm used for the clustering of the students [11]. Machine learning supervised algorithms are used frequently to predict student's academic performance. In this regard, Support Vector Machine and Logistic regression used to predict the student's academic outcomes [12], where Logistic regression provides the most accurate results. Secondary and Intermediate student's performance and results have been into account in Machine learning with the utilization of regression algorithm in Peshwar to justify the quality of education [13].

Publishing student's results without taking any examinations are very hard, still the pandemic, in Bangladesh, the higher secondary examination (HSC) results have been published by analyzing or calculating the students' previous two examination's achievement [14]. Due to this, the education ministry of Bangladesh must track all the student's previous achievements, and at the same time, the calculation has been done through conditions.

The machine learning algorithms are used to predict poor or good performers using two methods and six algorithms [15]. The idea was to determine whether machine learning algorithms deliver accurate results by learning the old dataset and at the same time, the performance of new students' can be measured in this process.

In this paper, two learning algorithms are trained with the dataset and predict whether the results are actual or have some difficulties as the dataset is conditional. Based on the literate analysis and the result of publishing without taking physical examinations, the paper aimed to determine the accuracy of the data with the two algorithms.

3 Data Collection

Dataset has been collected for 50 students who have passed HSC examinations without physically attending the examination. Primary School Certificate examination's data, Junior School Certificate examination's data, and the Secondary School Certificate examination's data have been collected and three subject's data have been examined in the process. Bangla, English, and Mathematics subject GPA or grade have been collected [16]. In our country, the grades of passed students are in the range of 2.5–5. Below is the table of the data explanation. It is noted that the dataset does not contain any students who failed in any of the examinations (Table 1).

Table 1 Data explanation of the collected data

Data features	Data explanation
Students ID	Students numeric ID
Students name	Students first and last name
PSC_b	Primary School Certificate Bangla Subject GPA
PSC_e	Primary School Certificate English Subject GPA
PSC_m	Primary School Certificate Math Subject GPA
JSC_b	Junior School Certificate Bangla Subject GPA
JSC_b	Junior School Certificate English Subject GPA
JSC_b	Junior School Certificate Math Subject GPA
SSC_b	Secondary School Certificate Bangla Subject GPA
SSC_e	Secondary School Certificate English Subject GPA
SSC_m	Secondary School Certificate Math Subject GPA

4 Error Analysis

The student's results have been published by the Ministry of education on 30th January 2021 and after the publication of the result, an announcement in the newspaper has been noticed by most of the people that 17043 students did not get a GPA-5 in SSC or in JSC but how they got the GPA-5 in the published result of HSC [17].

After getting the results, thoughts on finding how to accumulate the data of two specific exams can be best fitted for giving another examination's result. In this thought, three consecutive examination results of students have been collected and wanted to create a model where the student's results will be calculated through test conditions that can be accurate with the published exam's result.

The results will be calculated by dividing the subject-wise GPA where the subject-wise GPA depends on the previous examination's results for each subject and a condition applied to the predicted subject's result so that it will be well fitted to the taken specific parameter of the GPA. As conditions applied to the subjects, so the main focus is how the machine learning algorithm can predict the conditional result by training a specific dataset of subjects instead of calculating with conditions. Actually, how the conditional result (GPA) of each subject can be trained by the algorithm and how much accurate the predictions of the algorithm with the traditional conditional approach are the main focuses.

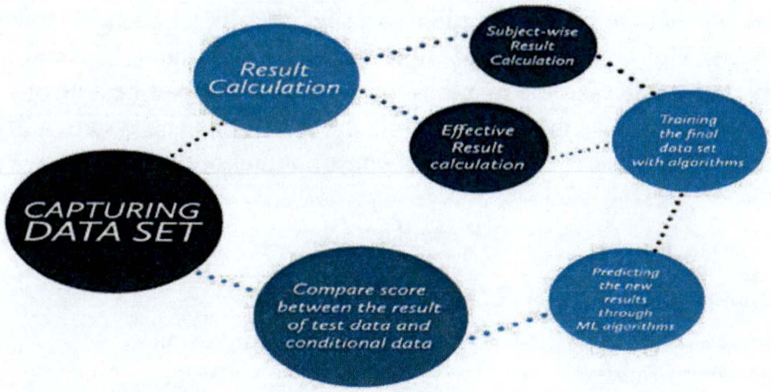

Fig. 1 Conceptual system diagram of the proposed system

5 Proposed System

5.1 Conceptual Proposed System Diagram

The conceptual system diagram is the model of the proposed system by which anyone can understand the whole process of the proposed system. The conceptual system diagram is given below (Fig. 1).

5.2 Result Publishing Method

The results of the proposed system have been checked by three consecutive subject's grades. Each subject is calculated and divided with the number of examinations that contain the subject grades so that the next examination which would not be taken physically can be calculated. The results of the 'Bangla' subject are found three times in three consecutive examinations of the taken dataset. The calculation process for the next examination of the 'Bangla' subject can be achieved by using the formula 1.

$$S_{\text{final}} = S_{\text{total}} / E_s, \tag{1}$$

S_{final} Subject-wise result for the next predicted examination.
S_{total} Sum of Achieved GPA in each subject in number of consecutive examinations.
E_s Number of examinations contains the identical subject.

After that, condition has applied to the result whether the result will be which GPA as in each subject there should be a specific parameter like 5, 4.5, 4, 3.5, 3, and 2.5. The conditions have applied in accordance with the find the nearest best fit

GPA. Let us say that the formula applied to a specific subject and a student achieved a 4.13 GPA but where the nearest possible values are 4.5 and 4. The conditions are set in a way that it will take the result as 4.5 as it is the nearest best result.

After calculating each subject's GPA with the conditions, the system will calculate the results of a student using the formula 2 where no rules or conditions have imposed.

$$R_s = S_{final}/S_n, \tag{2}$$

R_s Students final result.
S_{final} Subject-wise result for the next predicted examination.
S_n Number of Subject.

For better understanding the system, the data visualization of each subject with the respective desired results has described below (Fig. 2).

Afterward, whether the results of Rs are better than the brutal result (Rb) achieved by formula 3 has inspected. Data visualization of Rs and Rb is visualized on the dataset for a better understanding of the system. Given below the formula 3 for calculating the Rb.

$$R_b = E_r/E_n, \tag{3}$$

R_b Brutal results.
E_r Each examination's achieved result.
E_n Number of examinations taken.

The data visualization clearly indicates that after the condition applied to each subject for the calculating the overall results are different from the brutal results (Fig. 3).

5.3 Flowchart of the Result Publishing Process

Flowchart helps to find out the steps or procedures that belong to a system. The flowchart of the result publishing process stated below (Fig. 4).

5.4 Algorithm for Publishing the Result

Algorithm for the calculation of the results has given below:

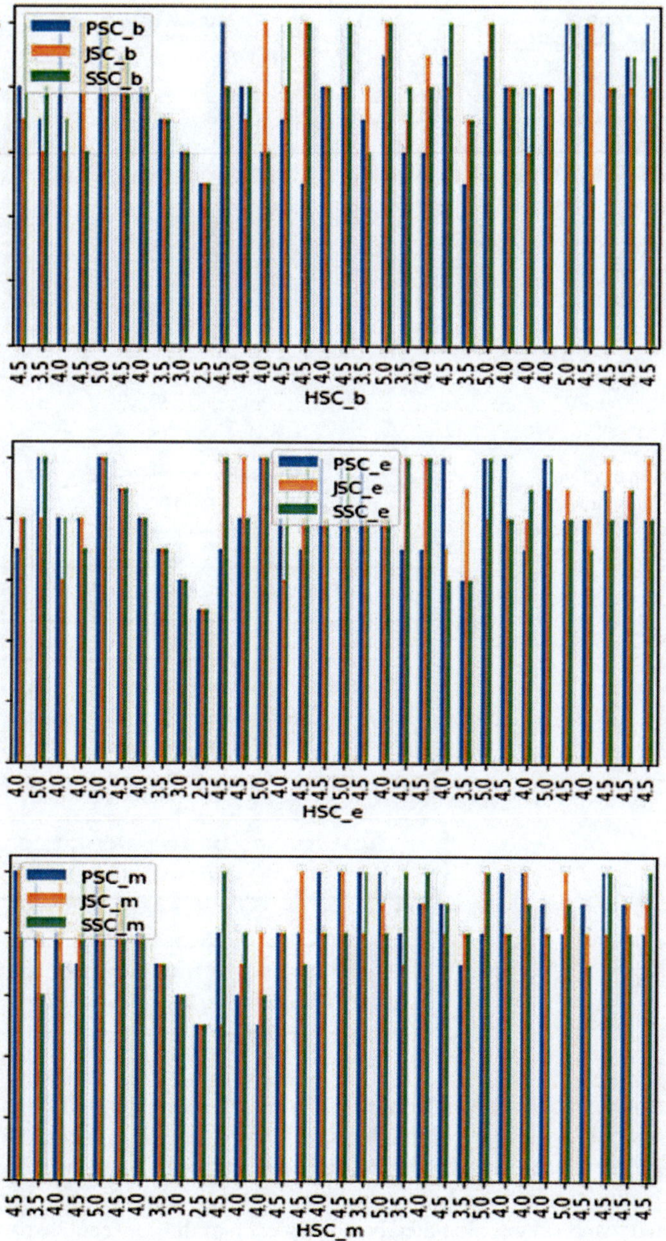

Fig. 2 Data Visualization of the subject-wise new calculated results

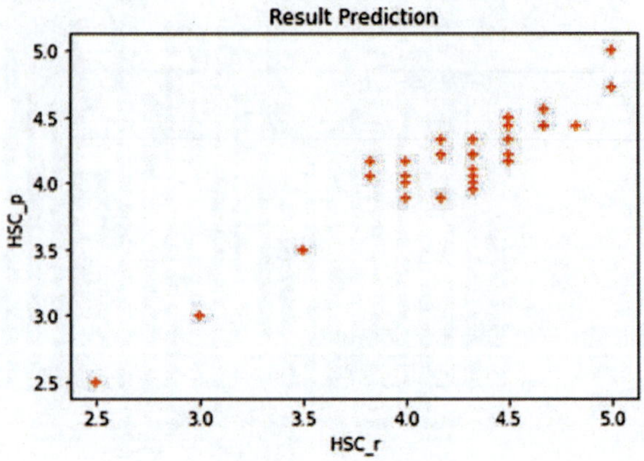

Fig. 3 Comparison of effective result versus brutal results

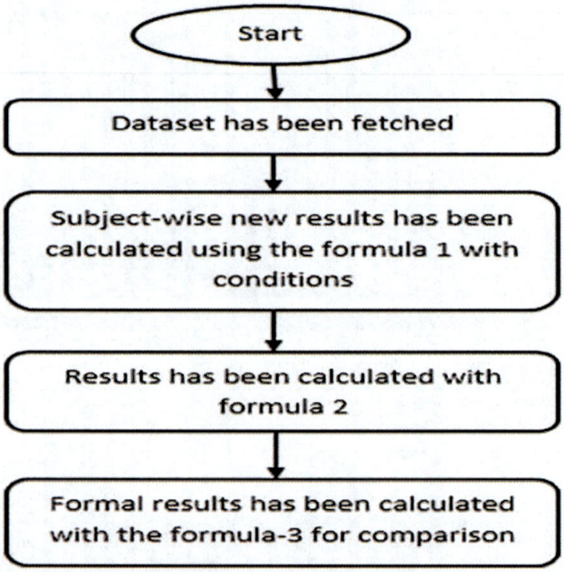

Fig. 4 Flowchart of the result publishing process

- Subject-wise data has taken from each consecutive examination.
- Using formula 1 the new column has created per subject with conditions.
- Each achieved results of the subjects have divided by the total number of subjects.
- The effective results have been achieved by the calculation of the 3rd step.
- Predicted results have calculated by adding each of the consecutive exam re- sults by the number of examinations.

5.5 *Machine Learning Approach in the Dataset*

The machine learning approach for big datasets is very appreciable nowadays as machine learning models can help to create a system easier. In this regard, the datasets that have taken in this process are being checked with the machine learning forecasting algorithms. Whether machine learning algorithms produce as accurate as of the system condition that is why the results of each subject have been trained with the algorithms.

Each examination's each subject has trained with the algorithms, and the predicted score of each subject has noted. Whether the algorithms give the exact result as the system proposes so that we can easily replace the calculations with the machine learning models.

Since the main objective of the proposed systems is forecasting or prediction depends on multiple features and one specific label. SVR **(Support Vector Regression) and Multiple Regression algorithms** have chosen in this system to justify the predictions or forecasting. All the results of each examination including the newly calculated results for each subject are merged into separate subject-specific data tables so that the prediction of the results of each subject can be found through the algorithms. 70% of the total data are used as the training data, and 30% of the total data are used as the test data. **Random state 42** has applied to each of the training data so that we can secure a specific prediction through each algorithm (Fig. 5).

5.6 *Score of the Algorithms*

The score of each algorithms prediction is given below to find out which algorithms are predicting accurate results. As a small amount of student's data has trained in this process, we cannot guarantee 100% accurate results (Table 2).

6 Error Analysis of the Proposed System

The proposed system has some errors when testing the dataset through the chosen two algorithms. Figured out initial errors are listed below.

6.1 *Prediction Accuracy*

After checking the data from the score of each algorithm, it shows that none of the predictions is 100% accurate though it shows that the predictions are above 90% accurate. The range of the score is 92–98%, and it is noted that SVR provides the best

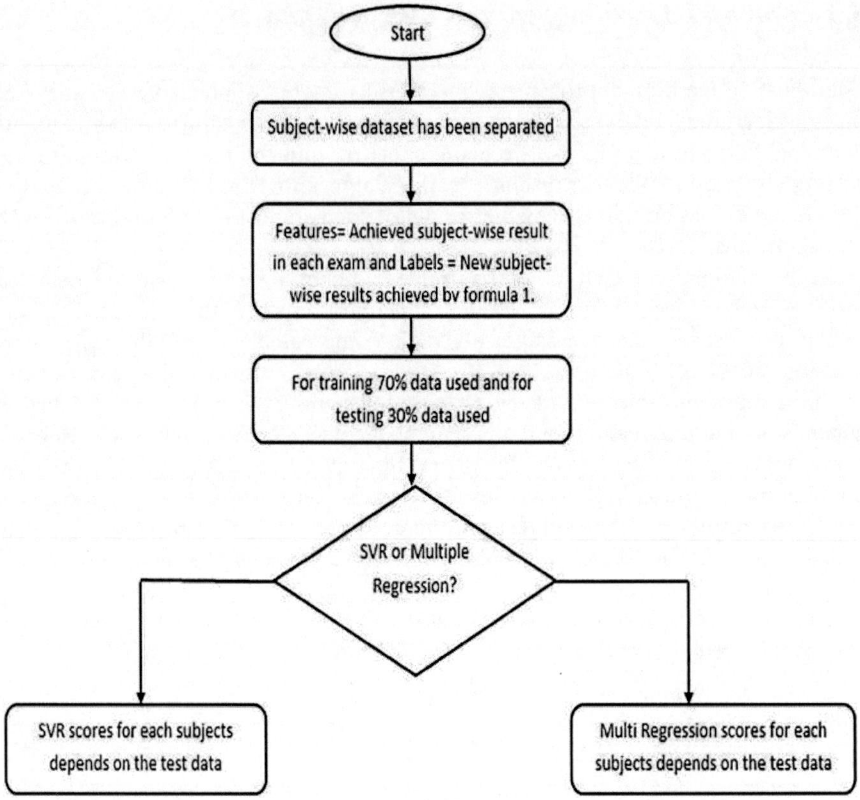

Fig. 5 Flowchart of the machine learning approach

Table 2 Score of the algorithms where condition applied

Test subject name	SVR score	Multiple regression score
Bangla	0.953395	0.934929
English	0.937353	0.987658
Math	0.986563	0.925395

score where the Multiple Linear Regression does not provide as much appropriate as the SVR score.

6.2 Result of SVR and Multiple Regression Are not Same as Conditional

After obtaining the results of the test data, it appear that some values do not match with the conditions as used in the proposed model. The output of test data through SVR and Multiple regression has a similar value of 5.13 where the result must not be above 5 GPA. The conditions that applied on the initial subject data on formula 1 must need to use for the output of the tested data therefore accuracy can be maintained.

7 Open Challenges

The main aim of this paper is to attain trust in the publication of results without taking any physical examination and publishing the result using the student's previous exam results data. As two forecasting algorithms have checked in this system, if anyone wants to implement more algorithms to the proposed system, they can try with other different machine learning algorithms for checking the other algorithms accuracy. There are some errors in the score of the two algorithms so if anyone wants to try with large datasets that can be more ethical. However, some errors such as the prediction across the maximum boundary of the GPA need to address seriously so that other conditions do not need to be applied to the predicted data of algorithms.

8 Conclusion

Machine learning techniques help the human being to work less on a vast dataset. If machine learning algorithms can be trained easily, it will be a matter of time for the machine to predict or forecast the actual values or results. In this paper, the main focus on how the conditional data set can be trained by the algorithms and how the algorithms predict based on the training data set. Whether the predicted result of the algorithms is the same as the conditional subject-wise results the main focus of this paper. It has been determined that algorithms can accurately predict more than 90% using small datasets. Whether more data can be trained to predict more accurate results as conditional data that needs to be an account for future implementation or testing. Implementation of a machine learning system will be really impressive as if we had to face issues like COVID-19 again where we had to work remotely and publish or predict students' performance based on their previous performance.

References

1. Catherine A. Bacos, "Machine Learning and Education in the Human Age: A Review of Emerging Technologies", Springer CVC 2019, December 2019.
2. Lukasz Kidzinski, Michail Giannakos, Demetrios G. Sampson, Pierre Dillenbourg, "A Tutorial on Machine Learning in Educational Science", Springer, State-of-the-Art and Future Directions of Smart Learning, pp 453–459, September 2015.
3. S. Russell, P. Norvig, "Artificial Intelligence: A Modern Approach, Third Edition.", Else- vier, Volume 175, Issues 5–6,Pages 935–937, January 2011.
4. Ethem Alpaydin, "Introduction to Machine Learning, Second Edition", MIT Press. p. 9. ISBN 978–0–262–01243–0, December 2009.
5. Will Kenton, "Multiple Linear Regression (MLR)", Investopedia, February 2021.
6. Awad M., Khanna R., "Support Vector Regression", Springer, Efficient Learning Ma- chines. Apress, Berkeley, CA., pp 67–80, April 2015.
7. Danijel Kucak, Vedran Juricic, Goran Đambic, "Machine Learning In Education - A Sur- vey Of Current Research Trends", 29th Daaam International Symposium On Intelligent Manufacturing and Automation, 2018.
8. Poornachander Gummadi, "Analysis Of Machine Learning In Education Sector", Decem- ber 2020.
9. Nespereira C.G., Elhariri E., El-Bendary N., Vilas A.F., Redondo R.P.D. (2016) Machine Learning Based Classification Approach for Predicting Students Performance in Blended Learning. In: Gaber T., Hassanien A., El-Bendary N., Dey N. (eds) The 1st International Conference on Advanced Intelligent System and Informatics (AISI2015), November 28- 30, 2015, Beni Suef, Egypt. Advances in Intelligent Systems and Computing, vol 407. Springer, Cham.
10. Md. Sabab Zulfiker, Nasrin Kabir, Al Amin Biswas, Partha Chakraborty and Md. Mah- fujur Rahman, "Predicting Students' Performance of the Private Universities of Bangla- desh using Machine Learning Approaches" International Journal of Advanced Computer Science and Applications(IJACSA), 11(3), 2020.
11. Iatrellis, O., Savvas, I.K, Fitsilis, P., et al.: A two-phase machine learning approach for predicting student outcomes. Educ Inf Technol **26**, 69–88 (2021)
12. Bhutto, E.S., Siddiqui, I.F., Arain, Q.A., Anwar, M.: Predicting Students' Academic Performance Through Supervised Machine Learning. International Conference on Information Science and Communication Technology (ICISCT) **2020**, 1–6 (2020)
13. Hussain, S., Khan, M.Q. Student-Performulator: Predicting Students' Academic Perfor- mance at Secondary and Intermediate Level Using Machine Learning. Ann. Data. Sci. (2021).
14. The Daily Stars, "HSC, equivalent exam results tomorrow", 29th January 2021, Link: https://www.thedailystar.net/bangladesh/news/hsc-equivalent-exams-results-tomorrow- 2035933.
15. S. Kotsiantis, C. Pierrakeas & P. Pintelas, "Predicting Students' Performance In Distance Learning Using Machine Learning Techniques", Applied Artificial Intelligence, pp 411- 426, August 2010.
16. Ministry of Education, "Intermediate and Secondary Education Boards Bangladesh results publishing site", Intermediate and Secondary Education Boards Bangladesh, 30th January, 2021, link: http://www.educationboardresults.gov.bd/.
17. The Daily Stars, "HSC results 2020: 17,043 of the GPA 5 achievers didn't get GPA 5 in SSC, JSC", 30th January 2021, Link: https://www.thedailystar.net/hsc-result-2020-17043- gpa-5-achievers-didnt-get-gpa-5-in-ssc-jsc-2036361.

Machine Learning Classification Algorithms for Predicting Depression Among University Students in Bangladesh

Uwasila Binte Munir, M. Shamim Kaiser, Uwaise Ibna Islam, and Fazlul Hasan Siddiqui

Abstract Depression is a dreadful mental disorder affecting negatively one's way of thinking and ability of functioning while the person may not be aware of it. The prevalence of depression is high among the young generation of developing countries because of ever-increasing academic and career-related distress, job uncertainty, and family problems. In Bangladesh, there is a dearth of information about the predictors of depression among university students, so is a model to identify them. The information is important for the prevention of depression and the promotion of mental health. The data set we used for this research was built on the data collected by a questionnaire circulated through social media. Using Pearson's chi-squared test and back elimination method, we have identified the key feature variables. We used six different ML classifiers to build the classification model that is capable of detecting the presence of depression. Among the six, the stacking classifier with 24 attributes shows the highest accuracy.

Keywords Depression · Prediction model · Machine learning · Feature selection · Classification · Ensemble

---

U. B. Munir (✉)
Department of Information and Communication Technology,
Bangladesh University of Professionals, Dhaka 1216, Bangladesh
e-mail: uwasilabinte@gmail.com

M. S. Kaiser (✉)
Institute of Information Technology, Jahangirnagar University, Savar, Dhaka 1342, Bangladesh
e-mail: mskaiser@juniv.edu

U. I. Islam
Chittagong University of Engineering and Technology, Chittagong 4349, Bangladesh

F. H. Siddiqui
Dhaka University of Engineering and Technology, Gazipur 1707, Bangladesh

M. S. Kaiser et al. (eds.), *Proceedings of the Third International Conference on Trends in Computational and Cognitive Engineering*, Lecture Notes in Networks and Systems 348, https://doi.org/10.1007/978-981-16-7597-3_6

1 Introduction

Depression is one of the most prevalent and agitating mental disorders of this era which negatively influences a person's feelings, way of thinking, and ability of functioning. More than 264 million people are affected by depression all over the world [1]. Studies show that depression is more prevalent among university students compared to the general population, and the prevalence ranges from 14% to 85% [2]. Depression is more common among the students of low- and middle-income countries as 43.7% of the Indian students, 40.9% of the Pakistani students, and 52.2% of the Bangladeshi students were reported as depressed in a recent study[3, 4]. Though depression is a serious mental health condition that needs to be handled with great concern, still, in most places depression is not taken seriously and in some places, it resembles a taboo [5]. So, people are not very comfortable sharing the reasons for their mental illness with others, not even with psychiatrists or counselors. Studies show that around two-thirds of young people do not talk about or seek help for mental health issues such as depression [6, 7]. With time, the condition often gets worse and the impact of long-term depression can be really dangerous which may also lead to increased risk of self-injury, dropping out or failing college, attempting or committing suicide, and other risky behaviors [8]. The prevalence of suicide ideation among university students is around 6.3% noted in American students and 13.8% among Bangladeshi students, and depression is present in about half of all these cases [9, 10].

According to the studies conducted, several socio-demographic and career-related issues are contributing as risk factors in depression among university students. Age, family financial crisis, gender, residence, result, university type, etc. are some major socio-demographic factors associated with depression[11, 12]. Lack of job opportunities, future unemployment worries, lack of financial support and hardships, family and peer pressure, academic dissatisfaction, career indecision, etc. are the career and job-related stress contributing toward causing depression [13–15].

Initially, we have identified the common risk factors that are responsible for depression. Then we have formed the questionnaire accordingly and circulated it for data collection. Later the collected data set was pre-processed and used in building ML models. Using Pearson's chi-squared test of independence, we have calculated the feature importance of each of the attributes and identified the most important features for building our predictive model. In this study, we have used six different classifiers; Random Forest, Multinomial Naive Bayes, Logistic Regression, Gradient Boosting, Linear SVM, and Stacking algorithm. After doing all the necessary performance comparisons between the classifiers, we have identified our proposed model [16, 17].

Depression is reported to be the single biggest contributor toward youth disability [2], and this huge upcoming workforce is going to lead the nation shortly, so identifying depression at an early stage is a must before it is too late. So, the objective of this work is to propose a model that not only detects depression but also gives an explicit idea about the key risk factors responsible for causing it. The major contributions of our research work are

- Identifying the best performing features subset for predicting Depression;
- Finding the most efficient model to offer for Depression detection that may provide valuable insights into diagnosis and prediction.

The rest of our paper is arranged as follows: Sect. 2 discusses the related works; the methodology is discussed in Sect. 3. Section 4 discusses the comparisons of the obtained results for different algorithms. Finally, Sect. 5 concludes our work.

2 Related Works

Though there has been a lot of research on analyzing depressive disorder, the prediction-related work in this field is not very popular. Choudhury et al. used a machine learning approach for depression prediction among university undergraduates using 577 samples collected from different private universities. In our study, besides considering the demographic issues we have also focused on career-related problems often faced by university students, whereas they have mostly addressed socio-demographic issues. Moreover, among 3 different classifiers, they found Random Forest to be the best performing algorithm with an accuracy, precision, and recall values of 75%, 70%, and 53%, respectively, which means in terms of performance our model outshines their model [5].

Daimi et al. predicted the patient who will possibly develop depression in near future or who are currently suffering from depression using the classification approach. Unlike us, they used synthetic data prepared through a Java program for training and testing the model. Using the Weka tool, they have implemented a decision tree algorithm for prediction and finding out the hidden patterns from the data[18]. Mahmoud et al. have analyzed and identified different important factors that are contributing toward causing anxiety, depression, and stress among undergraduates of the USA by using the multiple regression analysis. Other than demographics, they have also considered life satisfaction, coping style, etc.; on the contrary, we have addressed issues related to future worries, job uncertainty, etc. as these are the major concern among university students in Bangladesh[19]. Saunders et al. examined how depression and dysfunctional career thoughts are significantly associated with career indecision using hierarchical multiple regression analysis. Using 215 samples, they found dysfunctional career thoughts, vocational identity, locus of control, state and trait anxiety, depression, etc. which are the main contributing factors in career indecision [14]. Bhakta et al. used 5 different ML classifiers for predicting depression among older people. Later through performance comparison, they have chosen Bayes Net classifier as the best method for prediction in their case [20]. Adewuya et al. estimated the prevalence of depression and examined the factors associated with depression among university students in Western Nigeria. Also, they have found some significant factors such as very large family size, problems with accommodation, female gender, heavy cigarette smoking, and alcohol consumption that were associated with depression [21].

3 Methodology

3.1 Developing Questionnaire

Our preliminary task was to review the related literature to identify the common risk factors. As these risk factors may vary globally, consulting with a psychiatrist we have verified our findings from related literature and also added some factors that are merely found in Bangladesh. Finally, we have formed a questionnaire that consists of 30 questions in three separate sections which includes Socio-Demographic measures (Q1–Q9), Career and Job Seeking stress-related questions (Q10–Q21), and PHQ-9 scale (Q22–Q30).

Socio-Demographic Measures: Socio-Demographic questions contained some basic personal and education-related information about the participants. Personal information-related questions were **Gender**, **Age**, **Financial condition of your family**, **Permanent Residence**, etc. And education-related information includes **University type**, **Field of study**, **Year of study**, **Result**, and **Reason behind choosing current area of study**.

Career and Job-Seeking stress-related Questions: Career and Job-Seeking stress-related questions are divided into four subsections: **Job ability stress, family and surroundings environment stress, employment environment stress, and personality stress** [13]. Each of the questions can be answered using a four-point Likert scale that ranges from 0 ("Strongly Agree") to 3 ("Strongly Disagree"). **Job ability-related** questions include questions about personal ability such as skills and prior work experience (e.g., I am worried because I don't have enough skills needed for getting a job). **Family and surrounding environment**-related questions contain questions regarding the family financial crisis, family and social pressure, etc. (e.g., I need to find a job right after my graduation to solve my family's economic problems). **Employment environment-related** questions consist of questions regarding the fierce competition and scant job opportunities (e.g., I am worried because of scant job opportunities due to the potential economic recession these days). And **personality stress**-related questions contain questions about personal deficits such as lack of confidence and lack of self-knowledge (e.g., I am uncertain about my expertise and interest).

Patient Health Questionnaire (PHQ-9): The PHQ-9 scale is one of the most dependable tools to assess depressive disorder among the general population in epidemiological studies [22]. To measure depression, we have used the PHQ-9 scale which consists of 9 questions and each of them can be answered on a four-point Likert scale. The total score for this module ranges from 0 to 27 where higher scores indicate having depressive symptoms with greater severity. In our study, we have considered two levels "Depressed" and "Healthy"; those individuals scoring moderate to extremely severe (≧10) were classified as "Depressed" [23] and others are considered as "Healthy".

3.2 Data Collection and Pre-processing

After pretesting, we have circulated the questionnaire through social media to students from undergraduate and graduate programs. We could manage to collect around 520 responses from students from different institutions all over Bangladesh, who were anxious about their future careers.

The data set consists of 511 instances and 26 columns (25 feature variables and 1 target variable). Each of the socio-demographic and career-related questions (Q1–Q21) serves as a column in the data set except for Q9. Q9 takes five columns since an individual was allowed to choose more than one option for that question so each of the options takes a column in the data set. We have initiated the pre-processing by cleaning the data set and calculating the PHQ-9 score to derive our target variable. Those individuals having PHQ-9 score (≧10) were leveled as 'Depressed' and others as 'Healthy'. As answers to all of the questions were categorical, we have transformed them into numerical class values because ML algorithms cannot deal with string values. Finally, the transformed data set was split randomly into 80% and 20% for training and testing, respectively (Fig. 1).

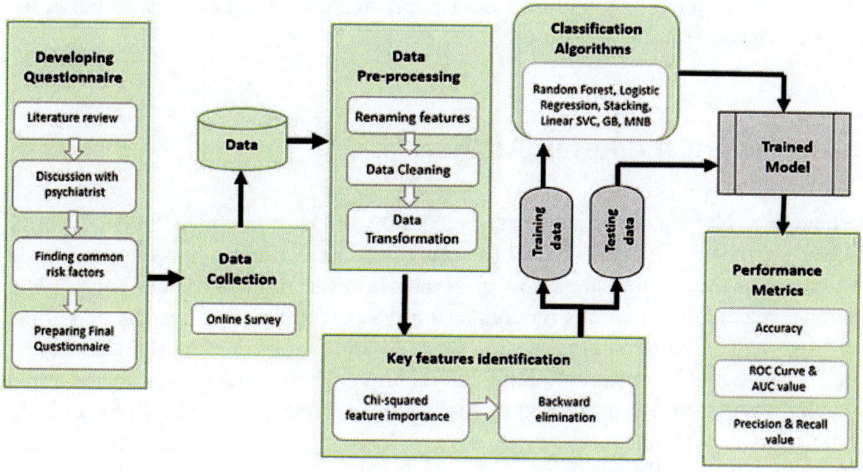

Fig. 1 Overview of research methodology

3.3 Developing Classification Predictive Model

We have used Pearson's chi-squared test of independence for measuring the degree of the coalition between two categorical variables. We have calculated the chi-square statistics between every single feature variable and the target variable to observe the existence of a significant relationship between them. Dependency between the variables is calculated by creating a contingency table. If c is considered as the number of columns in the table and r as the number of rows, then the Chi-Square Test of Independence can be calculated using the formula

$$\chi^2 = \sum_{i=1}^{r}\sum_{j=1}^{c}\frac{(O_{ij} - E_{ij})^2}{E_{ij}}$$

Here, O_{ij} is the observed value of two categorical variables and E_{ij} is the expected value of two categorical variables [24].

We have calculated the χ^2 values using the built-in function $feature\_selection$ $.chi2()$ of scikit-learn library. All these Python codes were implemented in Google Colab. Higher χ^2 value represents a higher impact on the target variable. After sorting the features in descending order based on χ^2 value, we applied the Backward elimination method for feature selection. We have started training the classifiers with all the 25 features and calculated the accuracy, then eliminated one feature in each step that has the lowest χ^2 value and trained all the classifiers without that eliminated feature and calculated the accuracy and like this, we continued the process till 1 feature was left. The performance of all of the six algorithms was recorded in each step and later was compared for identifying the best case. Initially, comparing the performance of the classifiers we have identified the best performing classifier in each iteration, then finally comparing that with the other best performing classifiers in consecutive iterations we have decided our proposed model. The best performing case is considered as our proposed model and its feature variables are selected as the best performing features subset.

4 Evaluation and Result Analysis

We have used Chi-squared statistics and the Backward elimination method to identify the best performing features subset for depression detection. In our study, we used 80 percent of the whole data set as train data in order to train the classifiers well. We have selected the features by comparing the performance of various classifiers that were trained with features starting from all 25 features coming right down to a single feature in 25 steps, eliminating the feature with the lowest chi-squared value in each progression. We have used three different performance matrices to compare the performance of the classifiers: Accuracy Score, ROC curve and AUC value, and $Precision$ and $Recall$ value.

As stated earlier, our first objective is to discern the most influential socio-demographic and career-related risk factors that play an important role in causing depression. *scarce_entrepreneurship_scope*, *lack_of_confidence*, *Lack of_Skills*, *Lack_of_work_experience*, *Fear_of_losing_prestige*, *lack_of know ledge_about_self*, and *Mismatch_between_education_&_job* are some of the best performing features as shown in Fig. 2. From the evaluation, we have identified the Stacking algorithm as the best performing algorithm when trained with 24 best chi-squared feature variables. Initially, we have started with 25 features, and we could manage to cut off only one feature through feature selection.

Since it's difficult to show the performance comparison between all the 25 steps, so, therefore, in Table 1 we have chosen 4 well-performing cases to find out the best possible case through performance comparison between these cases. Table 1 shows the accuracy scores for the models trained with 22, 23, 24, and lastly with all

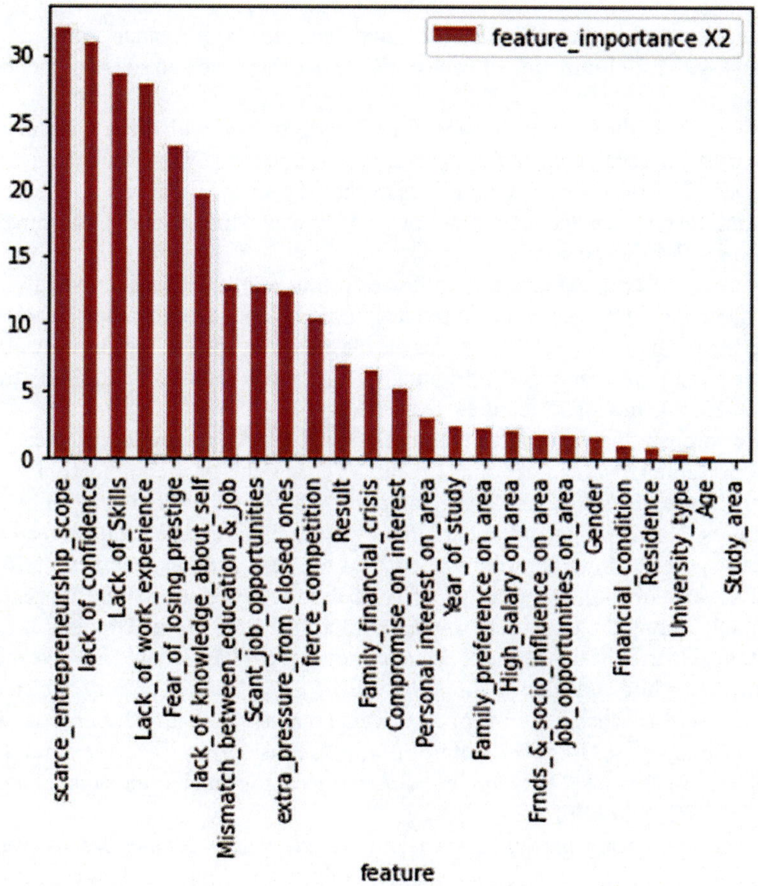

Fig. 2 Chi-squared feature importance of all the feature variables

Table 1 Accuracy of different classifiers for 22, 23, 24, and 25 features

Classifier with (%)	22 attributes (%)	23 attributes (%)	24 attributes (%)	25 attributes (%)
Logistic regression	72.54	75.49	75.49	74.50
Gradient boosting classifier	72.54	68.62	72.54	76.47
Linear SVC	73.52	72.54	74.50	75.49
Random forest	70.58	73.52	74.50	73.52
Multinomial Naive Bayes	69.60	70.58	71.56	71.56
Stacking classifier	72.54	76.47	77.45	74.50

the features. For less than 22 attributes, the Classifier Performance starts reducing gradually with the elimination of each attribute in every step and so is not mentioned here.

If we compare the accuracy score for the models trained with different numbers of features, we can see that the highest accuracy was obtained by the stacking algorithm when trained with 24 best χ^2 features, as expected because it is a powerful ensemble algorithm. It shows an accuracy score of 77.45% which means our model can detect depression 77.45% accurately.

To have a better understanding of the performance of the classifier, we also need to consider the ROC curve and AUC value along with the accuracy score while comparing the models. ROC curves for the classifiers in each of the cases mentioned in Table 1 are plotted in figures. Figures 3, 4, 5, and 6 represent classifiers trained with 22, 23, 24, and all 25 features, respectively.

If we compare the AUC of the classifiers with better accuracy scores in each case shown in Table 1, we can see that Linear SVC has the best combination with an accuracy score of 73.52% and AUC value of 0.78 when classifiers are trained with 22 features. However, when the classifiers were trained with 23 and 24 features in both cases, the stacking algorithm performed the best with an accuracy of 76.47% and AUC value of 0.80, and accuracy of 77.45% and AUC value of 0.80, respectively.

Though among the classifiers trained with all the 25 features Gradient boosting algorithm performs best in terms of accuracy score (76.47%), it could not give the best performance while considering the AUC value (0.77). On the contrary, the stacking algorithm performs best in terms of AUC value that is 0.80 but with a comparatively lower accuracy score (74.50%). But from Table 2, considering $Precision$ and $Recall$ values, we can see that $GB(Gradient\ Boosting)$ algorithm has the higher precision value than $SG(Stacking\ Gneralised)$.

Finally, comparing the $Precision$ and $Recall$ values for the best performing classifiers in each case, we can see that classifier trained with 24 features gives the best combination of $Precision$ and $Recall$ value that is 0.77 and 0.64, respectively.

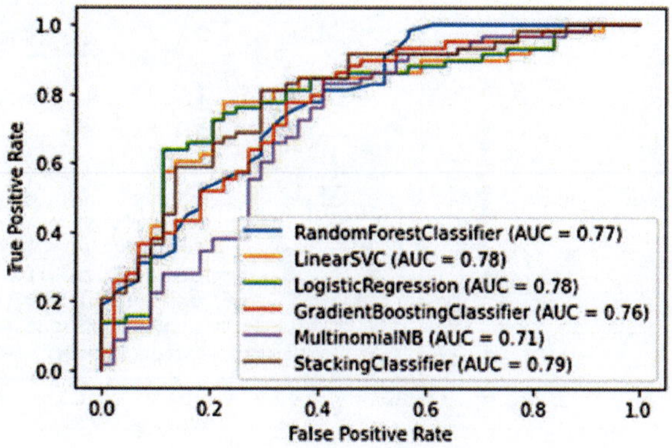

Fig. 3 *ROC* curve and *AUC* value for classifiers with 22 features

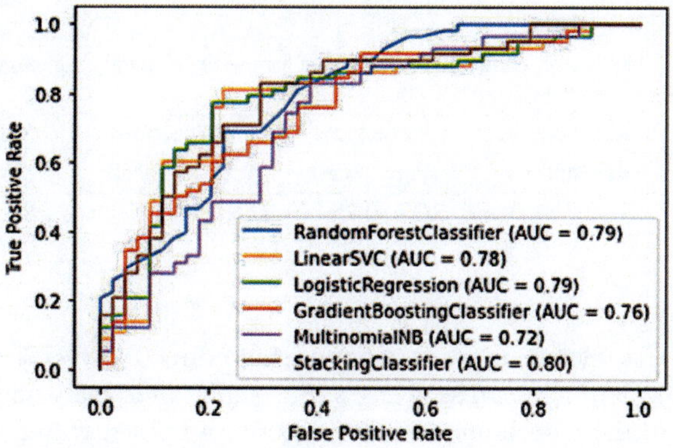

Fig. 4 *ROC* curve and *AUC* value for classifiers with 23 features

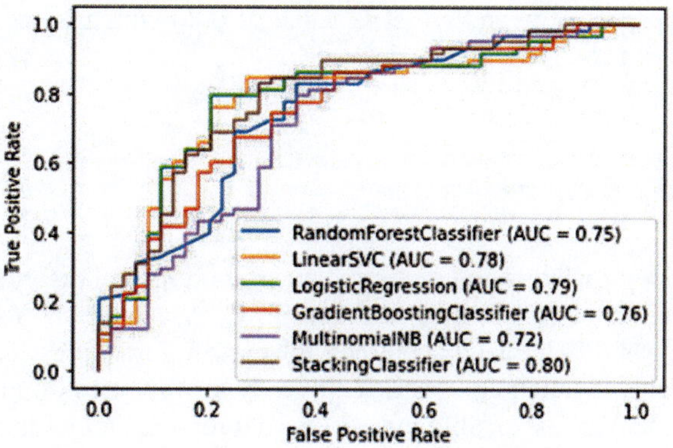

Fig. 5 *ROC* curve and *AUC* value for classifiers with 24 features

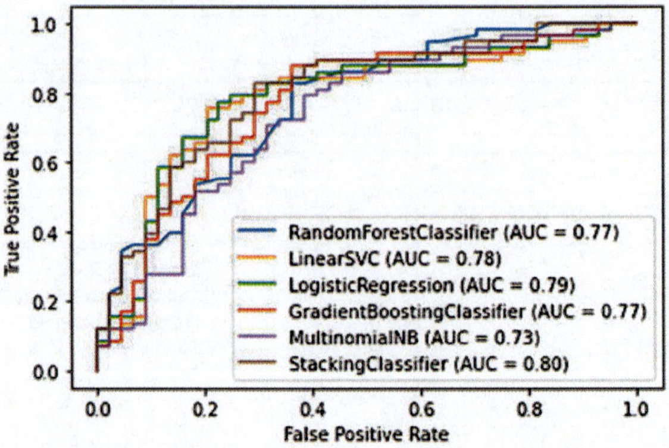

Fig. 6 *ROC* curve and *AUC* value for classifiers with 25 features

Table 2 Precision and recall values for the best performing classifier in each case considering class level "Depressed"

22 features			23 features			24 features			All 25 features					
Prec. recall			Prec. recall			Prec. recall			Prec. recall			Prec. recall		
LSVC	0.74	0.59	SG	0.75	0.68	SG	0.77	0.68	SG	0.74	0.64	GB	0.77	0.64

That means, it predicts depression correctly in 77% cases and correctly identifies 64% of all depressed cases. So finally after analyzing the result, we found among all the models the stacking algorithm trained with 24 features performed the best with an accuracy score of 77.45%, *Precision* and *Recall* value of 0.77 and 0.64, respectively, and an AUC value of 0.80 which is considered to be a good one for psychological diagnosis.

5 Conclusion

Depression is one of the most common mental disorders among young adults especially in middle-income countries like Bangladesh. So, this research was conducted to build a model that predicts the presence of depression among individuals using the classification approach. This study also focuses on identifying the key risk factors that are responsible for causing depression among the university students. Besides, it also recommends the individuals consult a psychiatrist when predicted depressed before it's too late. Our proposed model can be used as a tool by universities, psychiatrists, and counselors for identifying depression so that necessary measures can be taken to lessen the impact of depression among university students. There is a wide scope of expanding our work in the future: Using a larger data set, a greater accuracy can be obtained and a model can be built which not only predicts the presence but also the severity level of depression (mild, moderate, moderately severe, and severe).

References

1. Sadock, B.J., Sadock, V.A., Ruiz, P., Kaplan, H.I.: Kaplan and Sadocks Comprehensive Textbook of Psychiatry. Wolters Kluwer (2017)
2. Ibrahim, A.K., Kelly, S.J., Adams, C.E., Glazebrook, C.: A systematic review of studies of depression prevalence in university students. J. Psychiatr. Res. **47**(3), 391–400 (2013)
3. Islam, S., Akter, R., Sikder, T., Griffiths, M.D.: Prevalence and factors associated with depression and anxiety among first-year university students in Bangladesh: a cross-sectional study. Int. J. Ment. Health Addict. 1–14 (2020)
4. Miah, Y., Prima, C.N.E., Seema, S.J., Mahmud, M., Shamim Kaiser, M.: Performance comparison of machine learning techniques in identifying dementia from open access clinical datasets. In: Saeed, F., Al-Hadhrami, T., Mohammed, F., Mohammed, E. (eds.) Advances on Smart and Soft Computing. Advances in Intelligent Systems and Computing, pp. 79–89. Springer, Singapore (2021)
5. Choudhury, A.A., Khan, M.R.H., Nahim, N.Z., Tulon, S.R., Islam, S., Chakrabarty, A.: Predicting depression in Bangladeshi undergraduates using machine learning. In: 2019 IEEE Region 10 Symposium (TENSYMP), pp. 789–794. IEEE (2019)
6. Sarokhani, D., Delpisheh, A., Veisani, Y., Sarokhani, M.T., Manesh, R.E., Sayehmiri, K.: Prevalence of depression among university students: a systematic review and meta-analysis study. In: Depression Research and Treatment, vol. 2013 (2013)
7. Mahmud M., Kaiser M.S.: Machine learning in fighting pandemics: a COVID-19 case study. In: Santosh, K., Joshi A. (eds.) COVID-19: Prediction, Decision-Making, and its Impacts. Lecture Notes on Data Engineering and Communications Technologies, pp. 77–81. Springer, Singapore (2021)
8. Gollust, S.E., Eisenberg, D., Golberstein, E.: Prevalence and correlates of self-injury among university students. J. Am. Coll. Health **56**(5), 491–498 (2008)
9. Tasnim, R., Islam, M.S., Sujan, M.S.H., Sikder, M.T., Potenza, M.N.: Suicidal ideation among Bangladeshi university students early during the covid-19 pandemic: prevalence estimates and correlates. Child. Youth Serv. Rev. **119**, 105703 (2020)
10. Bachmann, S.: Epidemiology of suicide and the psychiatric perspective. Int. J. Environ. Res. Public Health **15**(7), 1425 (2018)
11. Bostanci, M., Ozdel, O., Oguzhanoglu, N.K., Ozdel, L., Ergin, A., Ergin, N., Atesci, F., Karadag, F.: Depressive symptomatology among university students in Denizli, Turkey: prevalence and sociodemographic correlates. Croat. Med. J. **46**(1), 96–100 (2005)
12. Bayram, N., Bilgel, N.: The prevalence and socio-demographic correlations of depression, anxiety and stress among a group of university students. Soc. Psychiatry Psychiatr. Epidemiol. **43**(8), 667–672 (2008)
13. Lim, A.Y., Lee, S.-H., Jeon, Y., Yoo, R., Jung, H.-Y.: Job-seeking stress, mental health problems, and the role of perceived social support in university graduates in Korea. J. Korean Med. Sci. **33**(19) (2018)
14. Saunders, D.E., Peterson, G.W., Sampson, J.P., Jr., Reardon, R.C.: Relation of depression and dysfunctional career thinking to career indecision. J. Vocat. Behav. **56**(2), 288–298 (2000)
15. Mamun, M.A., Hossain, M.S., Griffiths, M.D.: Mental health problems and associated predictors among Bangladeshi students. Int. J. Ment. Health Addict. 1–15 (2019)
16. Mahmud, M., Kaiser, M.S., McGinnity, T.M., Hussain, A.: Deep learning in mining biological data. Cogn. Comput. **13**(1), 1–33 (2021)
17. Kaiser, M.S., Mahmud, M., Noor, M.B.T., Zenia, N.Z., Al Mamun, S., Mahmud, K.A., Azad, S., Aradhya, V.M., Stephan, P., Stephan, T. et al.: iWorksafe: towards healthy workplaces during covid-19 with an intelligent phealth app for industrial settings. IEEE Access **9**, 13814–13828 (2021)
18. Daimi, K., Banitaan, S.: Using data mining to predict possible future depression cases. Int. J. Public Health Sci. (IJPHS) **3**(4), 231–240 (2014)

19. Mahmoud, J., et al.: The relationship among young adult college students' depression, anxiety, stress, demographics, life satisfaction, and coping styles. Issues Ment. Health Nurs. **33**(3), 149–156 (2012)
20. Bhakta, I., Sau, A.: Prediction of depression among senior citizens using machine learning classifiers. Int. J. Comput. Appl. **144**(7), 11–16 (2016)
21. Adewuya, A.O., Ola, B.A., Aloba, O.O., Mapayi, B.M., Oginni, O.O.: Depression amongst Nigerian university students. Soc. Psychiatry Psychiatr. Epidemiol. **41**(8), 674–678 (2006)
22. Martin, A., Rief, W., Klaiberg, A., Braehler, E.: Validity of the brief patient health questionnaire mood scale (PHG-9) in the general population. Gen. Hosp. Psychiatry **28**(1), 71–77 (2006)
23. Kroenke, K., Spitzer, R.L.: The PHG-9: a new depression diagnostic and severity measure. Psychiatr. Ann. **32**(9), 509–515 (2002)
24. Bolboacă, S.D., Jäntschi, L., Sestraş, A.F., Sestraş, R.E., Pamfil, D.C.: Pearson-fisher chi-square statistic revisited. Information **2**(3), 528–545 (2011)

Performance Analysis of Deep Neural Network Models for Weather Forecasting in Bangladesh

Md Khirul Islam Badal and Sajeeb Saha

Abstract Climatology and Weather forecasting play an important role to determine future climate expectations and help the farmer to make a plan for crop irrigation, fertilization, and suitable days for working in the field. Forecasting weather is a challenging task due to the uncontrolled nature of the surrounding atmosphere. Nowadays, deep learning models are widely used for weather forecasting that explores the hidden hierarchical patterns in big weather datasets to extract high-level features. In this paper, we investigate the performance of Multilayer Perceptron (MLP), ARIMA, and Bi-directional Long Short-Term Memory (BiLSTM) models for forecasting weather in agricultural applications of Bangladesh. The performance of the models is evaluated using Mean Absolute Error (MAE), Mean Square Error (MSE), and Root Mean Square Error (RMSE). The results depict that BiLSTM provides better performance compared to other state-of-the-art models to predict accurate weather in Bangladesh.

Keywords Weather forecasting · Deep neural network · Bi-directional LSTM

1 Introduction

In today's world, weather forecasting plays a significant role in our social life and in the functional operation of science and technology to predict the temperature, rainfall, pressure of the air, and other weather elements for a given location and time. A detailed weather forecast can tell a farmer the perfect time for seeding and planting. Farmers can better understand and supervise the growth status by using detailed and reliable historical, real-time, and forecast weather information [1]. Weather forecasts can be applied to ensure that fertilizer is used in the right conditions. Similarly, crop dusters, fungicidal or insecticidal chemical-spraying aircraft can be used when wind conditions are in favor [4].

M. K. I. Badal · S. Saha (✉)
Jagannath University, Dhaka, Bangladesh

M. S. Kaiser et al. (eds.), *Proceedings of the Third International Conference on Trends in Computational and Cognitive Engineering*, Lecture Notes in Networks and Systems 348, https://doi.org/10.1007/978-981-16-7597-3_7

The weather forecasting method can be divided into two categories: physical method and statistical method. A neural network with its superior nonlinear fitting and generalization ability becomes the most widely used method in these methods. The traditional neural network has some disadvantages like stringent requirements for input variables and training samples; too much or too little data inputs will affect the training effect. When the feature's dimension is too massive and challenging to extract high-quality features effectively, the neural network can hardly obtain good results. The fuzzy-rough collection method can effectively prevent the loss of essential information during dimension reduction compared to the conventional mainstream method of Principal Component Analysis (PCA). By analyzing missing information and reasoning evidence, it can uncover implied knowledge and expose the underlying rules. Deep learning could be a viable solution to the problem. It does not rely on high-quality features, can learn massive data, and has strong generalization ability. Bi-directional LSTM (BiLSTM), as an excellent variant of the RNN model, is suitable for dealing with the problems of high correlation with time series.

A critical task of the meteorological department is to forecast the daily atmosphere's status for a particular location. Numerous techniques are available to predict the daily or weekly atmospheres, including statistics, data mining, machine learning, and deep learning models. However, real-time and monthly accurate weather prediction have remained a primary challenging task for many decades because it is a nonlinear and small dataset. Crop growth, or crop yield, requires appropriate amounts of moisture, light, and temperature. Detailed and accurate historical, real-time, and forecasted weather information can help farmers better understand and track the growth status to make informed decisions. Due to this fact, the agriculture sector directly depends on the forecasted weather and monthly weather updates to estimate the production of crops.

In this paper, we focused on analyzing the performance of deep learning models to forecast weather for agricultural applications. We consider three state-of-the-art deep learning models named Multilayer Perceptron (MLP), ARIMA, and BiLSTM to predict the weather parameters. A monthly weather dataset has been used to train into the deep neural network models. The contributions of this paper are summarized as follows:

- We consider three state-of-the-art deep learning models to predict the monthly weather forecast in terms of temperature, rainfall, and humidity.
- Mean Absolute Error (MAE), Mean Square Error (MSE), and Root Mean Square Error (RMSE) metrics have been used to measure the performance of the models.
- The experimental result illustrates that BiLSTM has higher prediction accuracy than the traditional neural networks like ARIMA and MLP.

The rest of the paper is organized as follows: Sect. 2 gives a brief overview of the literature survey on weather forecasting solving techniques and their limitations. Section 3 presents our implementation details and Sect. 4 outlines the comparative analysis of different techniques in terms of different performance indicators. Finally, Sect. 5 concludes this paper with a discussion of the future scope of this work.

2 Literature Review

In the state-of-the-art works, several works already have been done on weather prediction. Holmstrom et al. utilized a linear regression model to forecast the maximum and minimum temperature of the next 7 days based on the past 2 days [5]. Four classifications such as rain, precipitation, cloudy, and very cloudy have been used to conduct the study. However, due to the small dataset, their work was unable to produce a promising prediction result. Jakaria et al. [6] implemented Extra Tree Regression, Random Forest Regression, and Ridge Regression models on Nashville city weather data. The results depict that Random Forest Regressor provides better performance as it is ensembled with multiple decision trees while making a decision. In 2017, Voyant et al. utilized a number of machine learning models to forecast the radiation [8]. In 2021, Murugan implemented a hybrid machine learning model to predict the daily weather forecasting [7].

A short-term local weather forecasting model was proposed using a deep neural network to predict rain and temperature [9]. In 2021, Bewoor used an artificial neural network model to predict the daily weather parameters [3]. In 2021, Zhang implemented a semi-supervised BiLSTM neural network model to predict air quality where the BiLSTM model performed better than the LSTM model [10]. However, the paper considered only one parameter to predict the weather. Althelaya implemented different LSTM models to predict the stock price prediction where the BiLSTM model performed well with a low error rate [2]. They concluded that the accuracy is highest in the deep neural network rather than among the several machine learning models. Based on the above discussion, it is evident that most of the existing approaches use daily weather parameters to forecast daily weather. It is challenging to predict monthly weather parameters based on monthly weather data with high accuracy and low error rate. For these reasons, we focus on investigating the performance of deep learning models in predicting the monthly weather forecast in a certain area.

3 Methodology

In this paper, weather parameters are forecasted and analyzed with three deep neural network models. The proposed study compared MLP, BiLSTM, and ARIMA models. Figure 1 shows the workflow of our proposed methodology. Initially, the dataset is collected from an authentic organization, and prepossessing of the data is performed to organize it. Then the whole dataset is divided into two parts: the training part and the testing part. The three deep neural models are implemented and the three models are trained with the training dataset. After testing the models with the test dataset, the proposed study analyzes the performance of the three models.

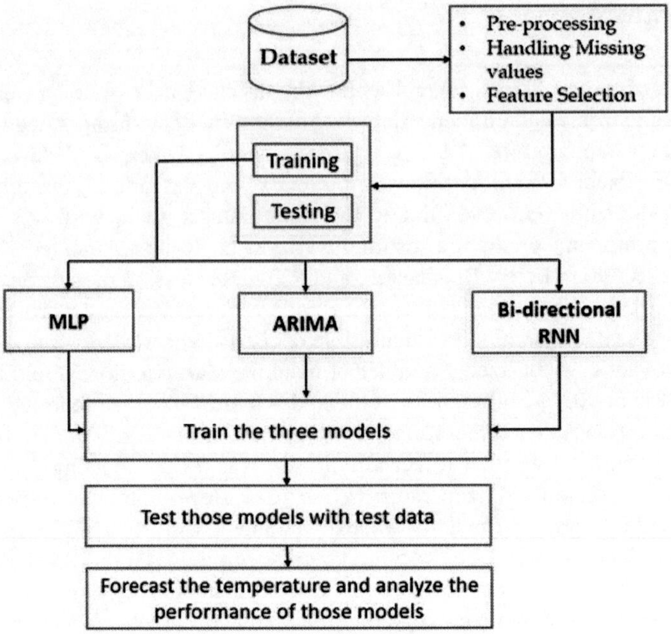

Fig. 1 Flowchart of the proposed study

3.1 Data Collection

The experiment has been conducted on a dataset collected from the Meteorological Department, Rajshahi city, Bangladesh. The dataset contains 1080 rows representing temperature, rainfall, humidity, wind speed, wind direction, and air pressure parameters in six columns from 1930 to 2019. Traditionally, hourly or daily weather parameters are used for the weather forecast, but the challenging task is to estimate the future monthly weather condition with high accuracy using monthly weather datasets. The description of the features is given below in Table 1.

Table 1 Description of the features

No.	Features name	Unit
1	Temperature	°C
2	Rainfall	mm
3	Humidity	%H
4	Air Pressure	Mb
5	Wind Speed	Km/h

3.2 Data Prepossessing

Deep learning models require a large amount of data in the training phase. Unfortunately, in most cases, the collected data contains noise, resulting in poor performance of the models. Hence, careful data collection and preparation is a prerequisite for achieving the best results. In this paper, the missing values of the parameters have been replaced according to the data type. For example, the missing numerical data has been replaced with variable mean, while categorical values are replaced with variable mode. This technique will remove the false bias created due to the missing values. We split the data into two selected groups, the training group, corresponding to 80% of the patterns, and the test group, corresponding to 20%.

3.3 Normalization

The initialization with random weights and the configuration of the models are the critical part of the deep neural network models. The distribution of the input is the main reason behind these difficulties. Batch normalization is a technique for training the models that standardize the inputs to a layer for each mini-batch. It helps to reduce the number of training epochs that are required to train the models. In this study, a min-max normalization process has been exploited to rescale the input dataset, and the equation is

$$x' = \frac{x - min(x)}{max(x) - min(x)} \tag{1}$$

In Eq. (1), x is the current value of the features, and max(x) and min(x) represent the maximum and minimum values of the features. x' is the normalized value.

3.4 Multilayer Perceptron

Multilayer Perceptron (MLP) acts as a supplement to the feedforward neural network. It consists of three sorts of layers—the input layer, output layer, and hidden layer, as shown in Fig. 2. The input layer receives the input to be processed. The output layer performs the required task such as classification and prediction [11]. A random number of hidden layers in between the input and output layers are computational truth engines of the MLP.

Similar to a feedforward network, the flows in an MLP model are in the forward direction: from the input to the output layer. The neurons in the MLP model are trained with the rear propagation learning algorithm. MLP is designed to approximate with continuous function and can solve problems that are not linearly separable. The prominent use cases of MLP are pattern classification, recognition, prediction, and

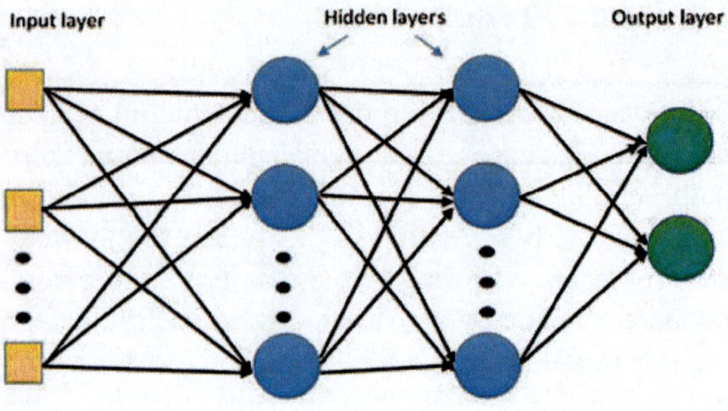

Fig. 2 Structure of Multilayer Perceptron

approximation. The computations happening at every neuron within the output and hidden layers are as follows:

$$o(x) = G(b(2) + W(2)h(x)) \tag{2}$$

$$h(x) = \phi(x) = s(b(1) + W(1)x) \tag{3}$$

Equations (2) and (3) contain bias vectors b(1), b(2); weight matrices W(1), W(2) and activation functions G and s. The set of parameters to learn is the set $\theta = W(1), b(1), W(2), b(2)$.

3.5 ARIMA

In ARIMA, 'AR' stands for autoregressive, 'I' stands for integrated, and 'MA' stands for moving average and is specified by these three order parameters: (p, d, q). The process of fitting an ARIMA model is usually known as the Box-Jenkins method.

- The 'p' parameter represents the number of autoregressive terms or the number of lag orders. It describes the result of the model by providing lagged data points.
- The 'd' parameter represents the degree of difference. The result of d indicates the number of times the lagged indicators have been subtracted.
- The 'q' parameter represents the number of forecast errors in the model and also acts as the size of the moving average window.

3.6 Bi-directional Long Short-Term Memory (BiLSTM)

BiLSTM is a variation of the recurrent neural network, and it connects two hidden layers that are coming from opposite directions to a single output. It allows the hidden layers to pursuit information from both past and future states. This structure permits

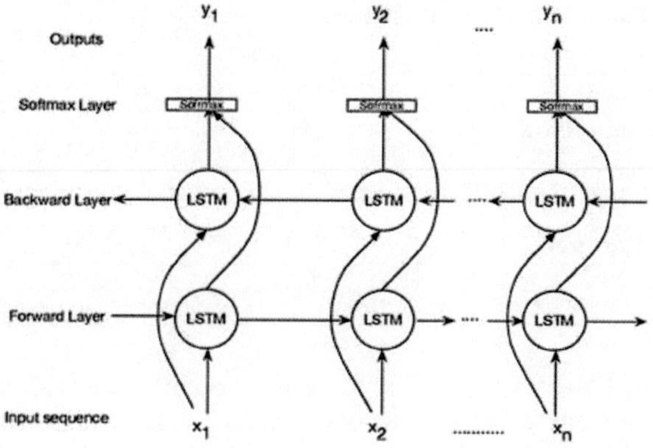

Fig. 3 Structure of bi-directional recurrent neural networks

the networks to have both backward and forward information about the sequence at every time step. BiLSTM is trained with similar algorithms as RNN since the two-directional neurons do not interact with one another. If backpropagation is mandatory, a few additional processes are needed since both input and output layers cannot be updated at once. In general training, forward and backward states are processed first in the 'forward' pass before output neurons are passed. For the backward pass, the opposite takes place; output neurons are processed first, then forward and backward states are passed next. Only after the forward and backward passes weights are updated. From Fig. 3, it shows that one layer performs the operation through the same direction of the data sequence and the other layer applies vice versa.

4 Performance Evaluation

In this section, we represent the simulation environment used to evaluate the performance of the stated deep learning models, trained with the collected weather data. To conduct the experiments, all the codes are written in Python version 3.8 and using TensorFlow and Keras library for implementing the MLP, ARIMA, and BiLSTM models. The three models are run into the Google Collab virtual environment. Mean Absolute Error (MAE), Mean Square Error (MSE), and Root Mean Square Error (RMSE) are used to validate the accuracy of these models.

4.1 Parameter Settings

In this work, several experiments were conducted to find the optimal parameters. Only the parameters that yield the best results are reported. For this purpose, the whole dataset is divided into two parts. The first one is the training set, and the size

Table 2 Simulation parameters

Parameter Name	Value	Description
learn_rate	0.01	The Learning Rate
batch_size	60	The Batch Size
max_iter	100	The Max Iteration
activation	sigmoid	The Activation Function
optimizer	adam	The Optimizer

of this set is 80% of the whole dataset, and the test set contains the remaining 20% of the dataset. The number of epochs is 70 for all the experiments. For the regularization of the neural networks and to avoid the over-fitting problem, we apply Dropout, with a dropout rate of 0.5. Table 2 shows the parameter list for the models.

4.2 *Mean Absolute Error (MAE)*

Mean Absolute Error is a model assessment metric used with regression models. The mean absolute error of a model concerning a test set is the mean of the absolute values of the individual prediction errors on overall instances in the test set.

$$mae = \frac{\sum_{i=1}^{n} abs(y_i - x_i)}{n} \tag{4}$$

Here in Eq. (4), n is the total number of observations, x_i is the real data, and y_i is the predicted data.

4.3 *Mean Squared Error (MSE)*

Mean Squared Error represents the accuracy level of a model. MSE error is the square absolute difference between real data and predicted data.

$$mse = \frac{1}{n}\sum_{i=1}^{n} abs(y_i - x_i)^2 \tag{5}$$

Here in Eq. (5), n is the total number of observations, x_i is the real data, and y_i is the predicted data.

4.4 Root Mean Squared Error (RMSE)

Root Mean Squared Error (RMSE) is the standard deviation of the errors which occur when a prediction is made on a dataset. This is the same as Mean Squared Error (MSE), but the root of the value is considered while determining the accuracy of the model.

$$rmse = \sqrt{\frac{1}{n}\sum_{i=1}^{n} abs(y_i - x_i)^2} \tag{6}$$

Here in Eq. (6), n is the total number of observations, x_i is the real data, and y_i is the predicted data.

4.5 Result Analysis

In this section, we illustrate the comparative analysis to measure the accuracy of the three models. We used three performance metrics in the experiments: Mean Absolute Error (MAE), Mean Square Error (MSE), and Root Mean Square Error (RMSE). For better prediction, MAE, MSE, and RMSE should be closer to zero. Figure 4a–c illustrated the comparison between the actual temperature and predicted temperature where the red line represents the actual value, and the blue line represents the models' output value. Here, X-axis represents the name of the month, and Y-axis represents the temperature. To validate the result, the actual temperature data of 2019 was extracted from the dataset and compared with the predicted result.

In Fig. 4a, the difference between the actual temperature and predicted temperature by the MLP model is significantly low initially as the model predicts the temperature with low uncertainty. However, from April to May, the difference between the two lines is much higher, representing more deviation from the ground truth. In Fig. 4b, the output result of the ARIMA model is not fair enough for the weather forecasting due to the high variation with the actual value for every month. Furthermore, according to Fig. 4c, the BiLSTM model can forecast closer to the ground truth compared with the ARIMA and MLP models for a given period.

The comparison result of the actual rainfall and predicted rainfall are plotted in Fig. 5. Here, the predicted rainfall values (mm) are plotted for 12 months from January to December 2019 and compared with the actual rainfall data of the same year. The graph depicts that the predicted values by the BiLSTM model almost match the actual values, whereas MLP and ARIMA models deviate slightly from the actual value. Similarly, Fig. 6 demonstrates the humidity prediction result of the three models. The graphs reveal that BiLSTM can predict the humidity close to the ground value, whereas both MLP and ARIMA models struggle to predict the humidity. It can audibly represent that the BiLSTM model's output almost meets the actual humidity values except for March.

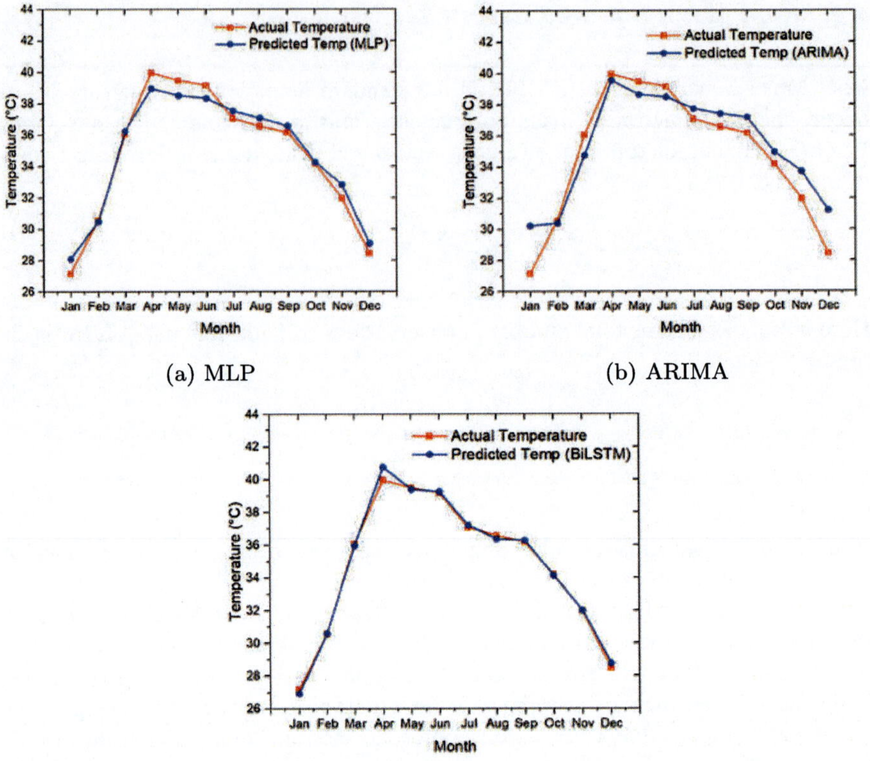

(a) MLP

(b) ARIMA

(c) BiLSTM

Fig. 4 Forecasted monthly temperature of 2019

Fig. 5 Monthly rainfall of 2019 by three models

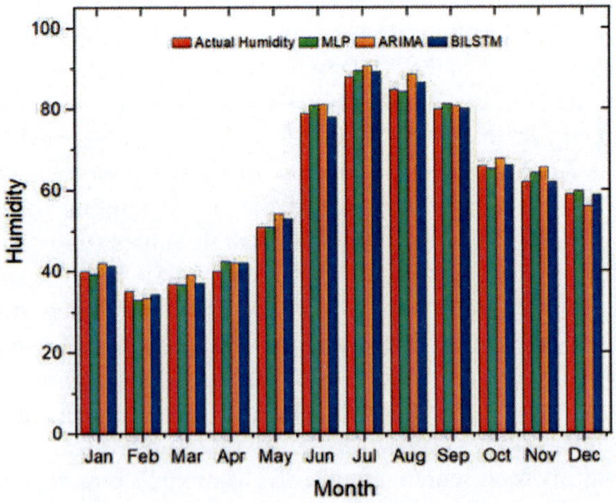

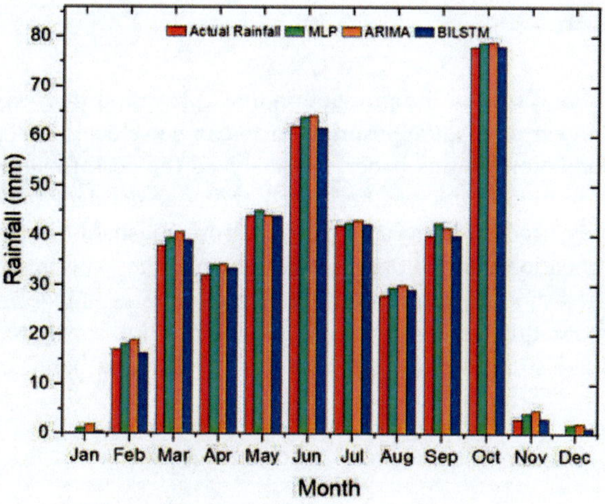

Fig. 6 Monthly humidity of 2019 by three models

Table 3 Performance table for three models

Model Name	MAE	MSE	RMSE
MLP	0.542	0.517	0.715
ARIMA	0.597	0.541	0.763
BiLSTM	0.483	0.414	0.584

Moreover, the BiLSTM model can predict the weather parameter close to the ground truth of the actual value compared to the MLP and ARIMA models in a significant number of times. This is due to the fact that the BiLSTM model can preserve the information from the past and future as it runs in two ways, whereas the ARIMA model focuses on univariate data with linear relationships.

Table 3 lists the performance of the three models based on the evaluation metrics. Experimental results point out that the ARIMA model cannot perform well compared to the other two models. The ARIMA model contains a mean absolute error of 0.597, mean square error of 0.541, and root mean square error of 0.763 where the MLP model performs better with a low error rate compared to the ARIMA model. According to the result in Table 3, it can be stated that the BiLSTM model provides superior performance in terms of accuracy than the ARIMA and MLP models as it obtains the lowest error rate.

5 Conclusion

Weather plays a vital role in the production and cultivation of crops in a country. Based on the weather data, an appropriate crop can be selected in a particular area to maximize production. In this paper, we analyzed the performance of three deep learning models to predict the weather based on the actual data and different error models. The study used a real weather dataset from Rajshahi district, Bangladesh, for model identification and validation. According to the experimental results, the BiLSTM exhibited superior results compared to other models in terms of humidity, rainfall, and temperature. In the future, we plan to develop a crop recommendation framework to suggest appropriate crops in a particular area based on the result of this work.

References

1. Abhishek, K., Singh, M., Ghosh, S., Anand, A.: Weather forecasting model using artificial neural network. Procedia Technology **4**, 311–318 (2012)
2. Althelaya, K.A., El-Alfy, E.S.M., Mohammed, S.: Evaluation of bidirectional lstm for short- and long-term stock market prediction. In: 2018 9th international conference on information and communication systems (ICICS). pp. 151–156. IEEE (2018)
3. Bewoor, L.A., Bewoor, A., Kumar, R.: Artificial intelligence for weather forecasting. In: Artificial Intelligence, pp. 231–239. CRC Press (2021)
4. Bloomfield, H., Gonzalez, P., Lundquist, J.K., Stoop, L., Browell, J., Dargaville, R., De Felice, M., Gruber, K., Hilbers, A., Kies, A., et al.: The importance of weather and climate to energy systems: A workshop on next generation challenges in energy-climate modeling. Bulletin of the American Meteorological Society **102**(1), E159–E167 (2021)
5. Holmstrom, M., Liu, D., Vo, C.: Machine learning applied to weather forecasting. Meteorol, Appl (2016)
6. Jakaria, A., Hossain, M.M., Rahman, M.A.: Smart weather forecasting using machine learning: a case study in tennessee. arXiv preprint arXiv:2008.10789 (2020)
7. Murugan Bhagavathi, S., Thavasimuthu, A., Murugesan, A., George Rajendran, C.P.L., Raja, L., Thavasimuthu, R.: Weather forecasting and prediction using hybrid c5. 0 machine learning algorithm. International Journal of Communication Systems **34**(10), e4805 (2021)
8. Voyant, C., Notton, G., Kalogirou, S., Nivet, M.L., Paoli, C., Motte, F., Fouilloy, A.: Machine learning methods for solar radiation forecasting: A review. Renewable Energy **105**, 569–582 (2017)
9. Yonekura, K., Hattori, H., Suzuki, T.: Short-term local weather forecast using dense weather station by deep neural network. In: 2018 IEEE International Conference on Big Data (Big Data). pp. 1683–1690 (2018). 10.1109/BigData.2018.8622195
10. Zhang, L., Liu, P., Zhao, L., Wang, G., Zhang, W., Liu, J.: Air quality predictions with a semi-supervised bidirectional lstm neural network. Atmospheric Pollution Research **12**(1), 328–339 (2021)
11. Zhang, P., Jia, Y., Gao, J., Song, W., Leung, H.: Short-term rainfall forecasting using multi-layer perceptron. IEEE Transactions on Big Data **6**(1), 93–106 (2018)

Instantaneous Communication Between Cerebellum, Hypothalamus, and Hippocampus (C–H–H) During Decision-Making Process in Human Brain-III

Pushpendra Singh, Komal Saxena, Pathik Sahoo, Jhimli Sarkar, Subrata Ghosh, Kanad Ray, and Anirban Bandyopadhyay

Abstract Using the available database of cerebellum, hypothalamus, and hippocampus (C–H–H) architecture of the human brain, we rebuilt the structures theoretically and experimentally using a similar dielectric material and studied their resonant communication. We also replicated humanoid brain circuits: language and conversation, thinking and intelligence, emotion, love, fear, threat, hunger and pain, and memory using special cables and Yagi antenna junctions. By comparing the experimental responses of three different brain components of the humanoid bot's functional circuits, we have identified that resonance frequencies vary widely. However, there are similarities in the ratio of resonance frequencies. Thus, a geometric language would probably be the governing machine language of future humanoid bots.

Keywords Cerebellum motor function · C–H–H connections · Clocking assembly · Structural symmetry · Clocking model of consciousness features

---

P. Singh · K. Saxena · P. Sahoo · A. Bandyopadhyay (✉)
International Center for Materials and Nanoarchitectronics (MANA), Research Center for Advanced Measurement and Characterization (RCAMC), NIMS, 1-2-1 Sengen, Tsukuba, Ibaraki 3050047, Japan

P. Singh · K. Ray
Amity School of Applied Science, Amity University Rajasthan, Kant Kalwar, N11C, Jaipur Delhi Highway, Jaipur 303007, Rajasthan, India

K. Saxena
Microwave Physics Laboratory, Department of Physics and Computer Science, Dayalbagh Educational Institute, Dayalbagh, Agra 282005, Uttar Pradesh, India

J. Sarkar
Department of Electronics and Electrical Communication Engineering, Indian Institute of Technology, Kharagpur 721302, India

S. Ghosh
Chemical Science and Technology Division, CSIR North East Institute of Science and Technology, Jorhat 785006, Assam, India

M. S. Kaiser et al. (eds.), *Proceedings of the Third International Conference on Trends in Computational and Cognitive Engineering*, Lecture Notes in Networks and Systems 348, https://doi.org/10.1007/978-981-16-7597-3_8

1 Introduction

The brain is the most complex part of our body. Signal movement in the form of the electromagnetic field through the brain components is not clear yet. The electromagnetic field distribution has been observed in the specific anthropomorphic mannequin (SAM) and spherical human brain model. Such EM field distribution depends on the shape and dielectric properties of the brain tissue [1, 2]. But previous studies suggest that such models act as cavity or dielectric resonators because there is no neural network inside. Either it is hollow or solid [3, 4]. The brain phantom model consists of two layers: brain tissue and skull. It has almost the exact dimensions of a biological brain [1], whereas the real biological brain includes all components. In that case, how would the phantom model observe true biological properties? This is still a subject of discussion. The magnitude and phase of the electric field in the brain are measured during multi-electrode transcranial electric stimulation (TES) under various stimulation conditions [5]. Several animal experimental studies have reported the effect of electromagnetic fields on blood–brain barrier (BBB) permeability [6, 7]. But a detailed discussion of generated internal field and its effect is missing, which is important to understand the complete effects of electromagnetic fields on the expansion of the blood–brain barrier. To obtain the effect of EM field on the BBB, we have to stimulate the BBC using multiple probes at its various locations.

Alternating electric fields are used for the treatment of glioblastoma. Several studies predicted the distribution of electric field in the brain during the tumor treatment field with the help of a realistic head model [2, 8, 9]. How predicted results from the modeling framework are close to the actual brain tumor is still unclear. Electromagnetic pulse transmission is characterized by a homogeneous spherical model of human and animal heads. Some parameters affect the output of the model which are unclear [10]. The influences of electric field distribution on the brain are observed by variation in its structural and biological properties but field distribution at different points on the brain has not been considered [11]. Current studies discuss only the electric field distribution while there is no discussion about the magnetic field.

Resonance is key to the functional activity of brain components and appears in the overall brain–body [12]. Resonance in a schematic way in the brain is the origin of information processing. The dielectric components of biomaterials with different symmetries are responsible for different resonance frequencies [13]. Resonance binds the complex action of neurons in a collective presentation with different regions of the brain [14, 15]. Recent studies show the resonance of brain components in low-frequency domain [16, 17], for example, cerebellum shows resonance at ~4–25 Hz in the low-frequency domain that originates in granule cell layers [18, 19]. Granule cells also exhibit resonance in the 4–10 Hz frequency range [20], although the characteristic of field potential in the cerebellar cortex is detected in 150–350 Hz [17, 21]. In the case of hippocampus, the literature is rich for hippocampus oscillations. The human hippocampus shows the theta oscillation ~ 4–10 Hz. Such oscillation is linked to the model of spatial navigation, episodic and spatial memory formation, behavioral and physiological data, etc. [22–25]. But the origin and functional activities for these three entities at high-frequency regions, kHz and MHz, are not yet clearly explored. There have been limited studies on the hypothalamus in terms of its function and anatomy [26–29]. The hypothalamus can modulate the activity of several brain components and improve memory [30]. To our knowledge, no electromagnetic study of hypothalamus has been done before based on resonance and EM field distribution.

The main objective of this paper is to explore the individual differences between electromagnetic field distribution and resonance frequency for brain components such as the cerebellum, hippocampus, and hypothalamus. Therefore, we have to build a suitable artificial model of these components. Almost all physical models of brain components are freely available on the Internet. We can study them by specifying their biological properties. But, because of the low resolution, it is difficult to separate its constituents, and in that case, numerical simulation tools are unable to find their individual constituents. The whole structure acts as a solid or hollow resonator. By simulating such structure, we can't get accurate results like biological structure [3, 4].

Here, we did not adopt any phantom model of brain components. In this study, we created brain components in CST, Computer Simulation Technology Studio Site 2012 [31], by following all of its biological properties. So we used such specific models and placed the energy sources at different locations. According to our previous study [3, 4, 32], we fixed the value of dielectric constant and other simulation parameters and changed the location of probes. The simulation parameters of the brain components are related to the dielectric properties of the material, frequency range, boundary conditions, and time or frequency domain solvers. The selection of the frequency range depends on the size of components.

The aim of this study is to explore the electromagnetic field distribution and resonance frequency on the surface and inside the C–H–H components. We have explored possible electromagnetic signal paths between them. We also analyzed here how the geometric configuration of the C–H–H components is important for establishing the relationship between them. We can build datasets for brain components. Such datasets may be useful for characterizing actual brain components and may be important for further research such as the action of electromagnetic waves on biological components. Theoretical and experimental building of C–H–H are shown in Sect. 2. Analysis of detected results is handled in Sect. 3. Finally, the paper is summarized in Sect. 4.

2 Material and Methods

Theoretically, we built the structures of C–H–H and verified them experimentally.

2.1 Theoretical Study

Cerebellum: The cerebellum model is composed of three layers, the bottom being a thick granular layer is packed by the granule cells as well as Lugaro cells, Golgi cells, and Brush cells. The central part of the cerebellum has a thin region, filled with glial cells and Purkinje cells, while its top part has a flatter pattern of dendrites and parallel filaments of Purkinje cells. The topmost layer consists of basket and stellate cells [33]. Cerebellum is comprised of a thin layer of cortex with white matter underneath. The ventricle at the base of the cerebellum is filled with fluid. There are four cerebellar nuclei in white matter. Each part of the cortex includes the same setup of these nuclei which is known as stereotyped geometry. At the intermediate level, cerebellum and its constituents can be recognized as several thousands of micro-regions. Medulla, pons, and fourth ventricle are separated from the cerebellum by a layer of dura mater. All the connections from the cerebellum to other parts of the brain pass through the pons. In the cerebellum, a fractal neural network includes 13 cavities. The cavity boundary comprises seven layers. The structure has two major folds in the horizontal cross-section and seven major folds in the vertical cross-section. The structure looks like a network of trees of neurons that read the final output from the hippocampus. The fractal wiring of neurons enables the cerebellum to learn a complex range of mechanical movement. All the sensors receive synchronized signals by the cerebellum to process the brain's decision-making process [34].

Using the literature studies of the brain anatomy and freely available 3D map data of the cerebellum, we reconstructed a cerebellum cavity theoretically and experimentally meticulously following all inner cavities. Neurons in the cerebellum are floating inside the cavities. The aim to build the hollow-shaped cerebellum cavity is to understand how the anatomy of the cerebellum without neuronal networks handles the arbitrarily propagating signal in the cavity. However, the cavities inside the cerebellum are arbitrary. Our aim is to understand how energy would be distributed without inner neurons floating around. Here, we performed the dielectric-cavity resonance studies using Yagi antennas and sensory probes by which we can verify the theoretically derived energy points or dynamic points. Studying the cerebellum without neural networks guides us to understand the role of its anatomy in energy transmission.

If we fill a detailed 3D map of neurons inside the empty cavity, a cerebellum map is obtained. To understand the projection of energy as input information through filled neuronal branches, we divided the entire cerebellum anatomy into several smaller cavities as shown in Fig. 1a. The major vertical cross-section has 3D folds A, B, and C while section B is divided into five subsections I–V. Moreover, sections A, B, and C are divided into 7 smaller vertical subsections (1–7) depending on cavity boundaries. If energy is pumped from the bottom of the cerebellum, it creates 3 local points around vertical section B. Another finding is that 7 vertical subsections have 11 (I–XI) individual folds (Fig. 1a, middle). Such sub-folds organize the energy inside the cavity in an organized manner. Sub-folds of IX (vertical cross-section of fold C) have 5 patterns. In the mid-panel of Fig. 1a, we found that there are 5 distinct layers. Consequently, neural branches in the cerebellum provide different energy

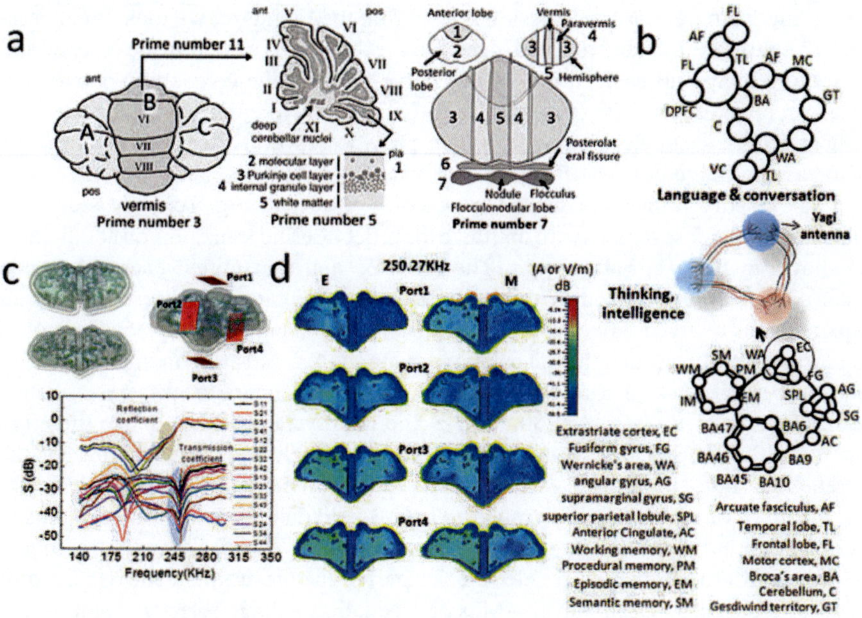

Fig.1 **a** A schematic diagram of cerebellum. In cerebellum, C3 symmetry (A, B, and C) is identified based on 3 prime regions; left lobule—A, vermis—B, and right lobule—C. Cerebellum mid-part vermis—B consists of 11 types of asymmetrical foldings I, II, III, IV, V, VI, VII, VIII, IX, X, and XI. Each folding contains 5 layers of cells; pia—1, molecular layer—2, Purkinje cell layer—3, internal granule layer—4, and white matter—5 (panel a: mid-part). Major parts of cerebellum include anterior lobe—1, posterior lobe—2, left and right hemispherical region—3, left and right paravermis—4, vermis—5, posterolateral fissure—6, and nodule and flocculonodular lobe—7 or it has C7 symmetry. **b** Two brain-conscious circuits govern language-conversion and thinking–intelligence sensing pathways. **c** A mimicked model of cerebellum is created in EM simulation software and the structure is stimulated by applying four waveguide sources; port1, 2, 3, and 4 at inferior vermis, superior vermis, left horizontal fissure, and right horizontal fissure, respectively. The resonance spectrum of reflection and transmission coefficients is observed in kHz frequency domain (panel c, bottom). d. EM field distribution of cerebellum is observed at the resonance peak 250.27 kHz for all applied waveguide ports. The field intensity scale is drawn at right side of panel. **d** Electric field and magnetic fields are noted as E and M, respectively

pathways. Primarily, there are three cavities. All 3 cavities are divided into 7 sub-cavities (Fig. 1a, right). However, 7 distinct cavities are located in the IX sub-fold and 5 molecular layers are functionally branched out around it.

We filled cerebellum cavities and sub-cavities with nerve fibers. We have equated the properties of cerebellum material to the properties of nerve fibers. The cerebellum map obtained from brain project sites show that neighboring neural branches are interconnected with unusual junctions due to a lack of proper resolutions in 3D imaging. We tried to delete all unwanted links one by one but couldn't delete all due to their extreme complexity. These links are so profound that the entire cerebellum architecture has become a single dielectric resonator. We need to identify individual nerve fiber pathways or transmission lines. From the literature, we took a real cerebellum map of the human brain and identified the actual nerve fibers in the cerebellum. Resolving lumpy unresolved bunches of nerve fibers in the cerebellum delivered a clean architecture as shown in Fig. 1c. In CST, we have carefully constructed the structure over a couple of months which provides several time improved results compared to Singh et al. 2019 [3].

Four sensory probes are placed in place as shown in Fig. 1c, right side. The simulation ran for several days to optimize the reflection and transmission coefficients as shown in Fig. 1c, bottom part. The reflected and transmitted energy between each pair of ports is noted in the kHz frequency domains. Solving the Maxwell equations for a large number of fibers requires extreme patience to place the probes at correct position around the 3D cerebellum model. Slight variation in port location and port area can lead to divergence. Four sharp resonance peaks are identified at the bottom of Fig. 1c. At 250.70 kHz, the quality factor is high. We therefore theoretically imaged the electric (E) and magnetic field (M) distributions for the cerebellum (Fig. 1d). From 3D views of the cerebellum model, one can conclude that the identities of the ports are not reflected in E-B field distribution. From an intensity scale, we found that certain regions that connect to the mid-brain or central region in the cortex are automatically selected as a part of the electromagnetic oscillations. *Simulation details:* Used solver—Maxwell equation solver; selected mode—time domain; boundary condition—open space; waveguide port dimension—40 mm × 60 mm; frequency domain—140–175 kHz range.

Hypothalamus: Since the cerebellum is triggered wirelessly by signals that pass through the spinal nerves. Its output effects are visible on mid-brain. Hypothalamus has an asymmetrical structure and consists of 13 nuclei. The names of all nuclei are depicted in Fig. 2a. All nuclei perform various functions like emotion, love, pain, fear, threat, anger, and hate (Fig. 2b). We stimulated the hypothalamus using four sensory probes at four distinct nuclei (Fig. 2d, left) and looked at the output along with the hypothalamus nuclei. Electromagnetic energy patterns (Fig. 2c) are detected at the resonance peak 497.63 kHz (Fig. 2d, left). *Simulation details:* Used solver—Maxwell equation solver; selected mode—time domain; boundary condition—open space; waveguide port dimension—300 mm × 300 mm; frequency domain—450–650 kHz range.

Hippocampus: The hippocampus has 17 cavities (Fig. 3a) wherein two spiral pathways run parallel to each other (Fig. 3a bottom). Such a pair of parallel pathways have fivefold symmetry (Fig. 3a middle). From the literature studies, we have identified distinct cavities in hippocampus structure. A schematic of hippocampus is shown in Fig. 3c where each cavity is filled with neurons. Such neurons have been purchased from various databases. We put orientations of neurons in such a way that they look exactly like the neurons in the biological hippocampus. During the creation of hippocampus, we surveyed the literature regarding the layout of branches and orientation of neurons. Here, we attempted to create a consistent output from the biological hippocampus and artificial hippocampus models. Noise-free hippocampus geometry has been directly mapped to the human brain. It was not available in the previous database. Most databases of hippocampus models have been generated from

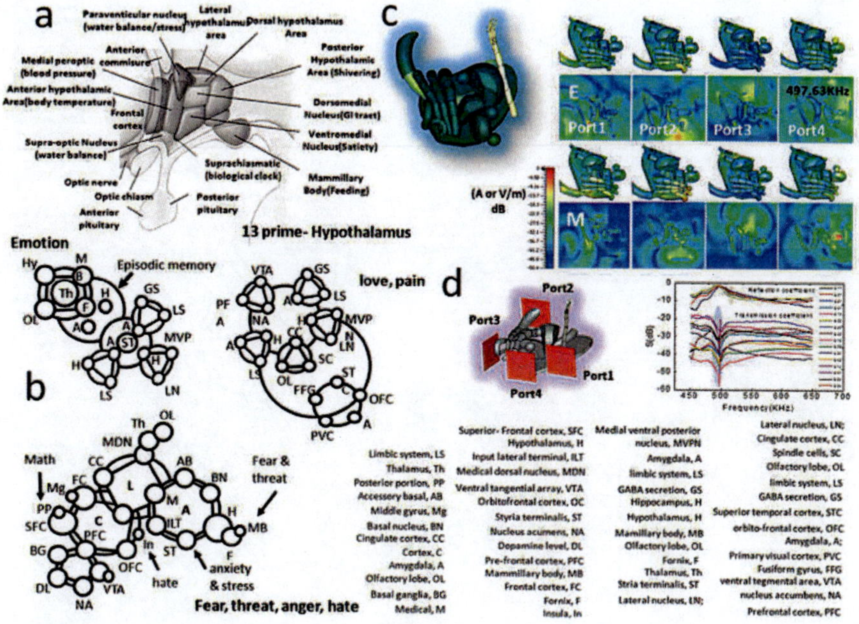

Fig. 2 **a** A schematic of hypothalamus with 13 distinct nuclei: mammillary body—1, ventromedial nucleus—2, dorsomedial nucleus—3, posterior hypothalamic nucleus—4, dorsal hypothalamus nucleus—5, lateral hypothalamus nucleus—6, paraventricular nucleus—7, medial preoptic—8, anterior hypothalamic area—9, supraoptic nucleus—10, optic chiasm—11, pituitary—12, and suprachiasmatic nucleus—13 are shown here. **b** Conscious brain circuits: emotion, love and pain, fear, threat, anger, and hate are built-in humanoid bot's brain and their rhythm maps are shown. Details about the alphabetic symbols are present at the bottom of the panel (**d**). Panel (**c**) shows the artificial hypothalamus model. d. The resonance spectrum of reflection and transmission curves is observed in the 400–700 kHz frequency range. Hypothalamus is stimulated by applying 4 waveguide ports; ports 1, 2, 3, and 4 at mamillothalamic tract, fornix, suprachiasmatic nucleus, and optic chiasm, respectively. EM field distribution over hypothalamus nuclei is detected at resonance peak 497.63 kHz for all applied energy sources: port1, 2, 3, and 4 (left of panel **c**)

similar kinds of neurons, so here we have created a distinct database. Our target here is to understand how the hippocampus projects the signal to various brain regions, and how it naturally edits decisions. However, we obtained theoretical responses in terms of resonance frequency, electric and magnetic field distribution by placing two ports at both ends of hippocampus, as shown in the right of Fig. 3c. Simulation took a long time to optimize the reflection and transmission spectrum (Fig. 3d). A sharp resonance is seen at 75.55 MHz frequency (Fig. 3d). The minor resonance peaks are generated by hippocampus. Sharp resonance peak suggests that hippocampus forms a link to different regions of the brain at that resonance frequency. *Simulation details:* Used solver—Maxwell equation solver; selected mode—time domain; boundary condition—open space; waveguide port dimension—10 mm × 10 mm; frequency domain—40–120 MHz range.

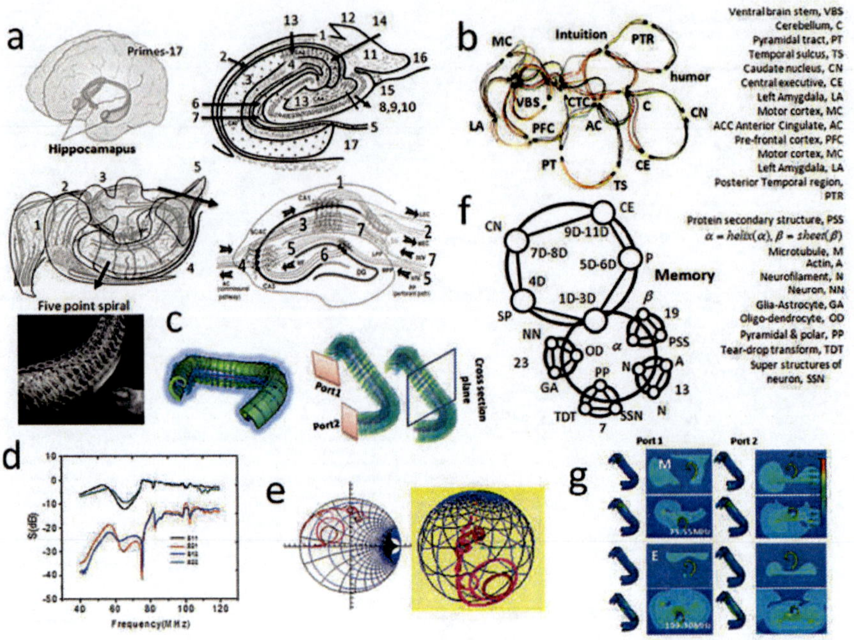

Fig. 3 A schematic diagram of hippocampus is shown in panel (**a**). Hippocampus anatomy includes five major regions 1, 2, 3, 4, and 5 (panel **a**, mid-left). Inside those regions, 7 distinct nerve bundles or layers are there; CA1—1, MEC (medial entorhinal cortex)—2, SC/EC subicular or entorhinal cortex—3, EC (entorhinal cortex) commissural pathways—4, CAII/IV(MPP)medial pathways—5, MF (Mossy fiber pathways)—6, CAIII/V (LPP) lateral pathways—7 (panel a, mid-right). Every pathway is interlocked by five spiral rings (panel (**a**), bottom left). 2D cross-section view of hippocampus is shown in the top right of panel (**a**) that has 17 nuclei (1–17). One example of the conscious circuit "creativity and humor" is shown in panel (**b**). Panel (**f**) shows a possible rhythm map that governs memory. Mimicked hippocampus model is built, as shown in panel (**c**), while panel d shows resonance spectrum of reflection and transmission curves at 75.55 MHz and 109.30 MHz resonance peaks for applied port 1, 2 at both ends of hippocampus. **e** 2D and 3D maps of reflection and transmission coefficients (S11—dark red and S21—light red) are presented in terms of Smith chart. 2D and 3D forms of the electric and magnetic fields over hippocampus model is detected at resonance peaks (panel: **g**). E and M are noted as the electric and magnetic fields

Wave port, Boundary Condition, and Simulation Time in CST. However, simulation of C–H–H models is an extremely challenging task. Proper selection of boundary conditions and probe locations provides a reduction in simulation time. Boundary conditions depend on the type of excitation sources, such as wave ports, or discrete ports. In brain components simulation, energy is pumped into the model through wave ports. Open boundary condition was kept in all directions (x-, y-, and z-axis). In the case of discrete ports, there is a need to keep 'open add space' condition in all directions. The electromagnetic simulator automatically sets the quarter wavelength in any direction with respect to available space away from the built structure. The simulation time may be changed by changing the wavelength of the incident signal into half or full wavelength. Pick face tool automatically selects the port location on the surface of created structure. Here, we kept changing the position of pick face until we actually got the truly electromagnetic resonance peaks.

2.2 *Experimental Study*

Cerebellum: To mimic cerebellum geometry, we used PVC plastic, which can be changed into the desired shape by heating. We created folds similar to the biological structure and implemented the neural network using a single cable wire. The cross-connections of wires of two channels inside the structure look like tree shape. The open ends of each wire are shaped like Yagi antennas that are suitable for absorbing and transmitting the signal from other brain components. Cerebellum is part of the hindbrain and connects to the spinal cord. 31 pairs of spinal nerves make connections to the cerebellum, and its other end is stimulated by a channel of 10 RF sources (26–35.5 MHz) programmed by an Arduino board. A 5 V power supply is used to activate the Arduino board. Photographs of schematic and actual experimental setup are shown in Fig. 4a b. An RF spectrum analyzer is used to measure the emission of signals in different directions of cerebellum (Fig. 4c).

Hippocampus and Hypothalamus: Mimicking the hippocampus geometry is a slightly challenging task due to the complexity of its internal architecture (see Fig. 4d, left). Here, a flexible Al (aluminum) wire is used to mimic two circular sets. The neural network of the hippocampus is equivalent to two bundles of semiconductor wires passing through a set of circular rings. Each ring of both channels is connected to the bundle by a single wire. Both channels are connected to each other. The entire structure is wrapped in plastic, as shown in Fig. 4d, right. Both ends of the structure are connected by helical wires. Function generator (KHz-GHz) is used to excite the structure. We measured the signal intensity at circular angles around five different positions (I, II, III, IV, and V cm) of the hippocampus by an RF spectrum analyzer. The hypothalamus is combined with the limbic system to produce an effect on the output of limbic system.

To build a programming-free artificial brain [32], we need to build the physical equivalence cerebellum–spinal cord junctions. Using a single wire cable as a spinal nerve, we have created the artificial spinal nerve system. The open ends of wires deform into 3 sharp edges that act as Yagi antennas. Yagi antenna is capable of absorbing and emitting electromagnetic signals. Since 34 pairs of spinal nerves have been prepared, all the spinal nerves make connections to the cerebellum and go to three major lobes of the brain. Principally, one lobe contains 12 spinal nerves that absorb the signals coming from other parts of the brain. There are plenty of articles on cerebellum and spinal cord creation. Literature studies show the energy distribution at the surface of the cerebellum and spinal cord, but here we don't assume that a single layer does everything or signals only flow in the outer layer. We detected the electromagnetic field distribution inside the cerebellum, hippocampus, and hypothalamus.

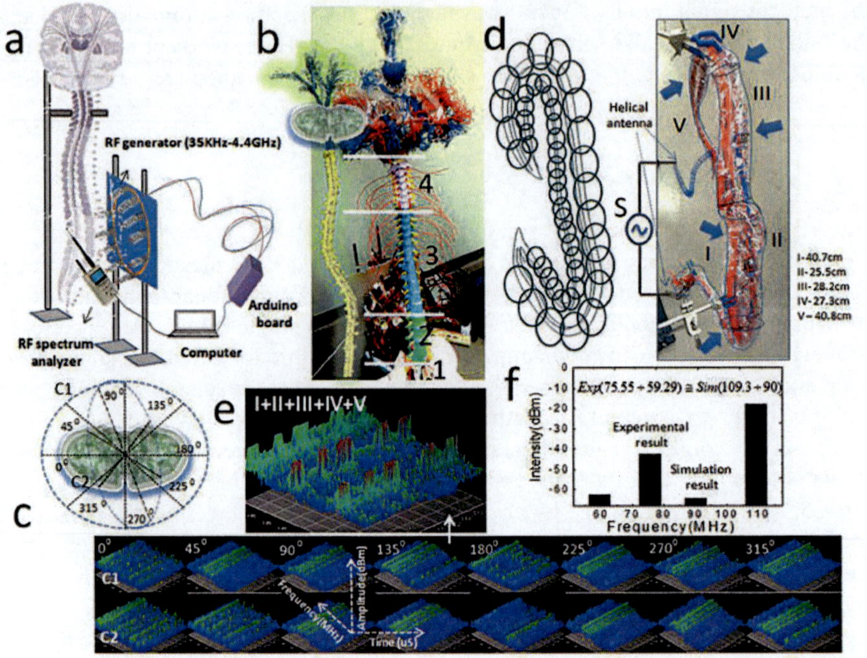

Fig. 4 A schematic diagram of the experimental setup of spinal nerve system is shown in panel (**a**). Each end of single wire that is in the spinal nerve's system has the Yagi antenna shape in cerebellum and mid-brain. The whole setup is stimulated by a circular channel that contains 10 sources of frequency (frequency range: 35 kHz- 4.4 GHz), and output is detected by an RF spectrum analyzer. The background image of experimental setup is shown in panel (**b**). 3D energy patterns of cerebellum along the transverse circles C1 and C2 (panel (a): bottom) are shown in panel (**c**). A schematic of hippocampus that has two sets of circular rings in which a bunch of wires passes every single set of rings is shown in panel (**d**). left. Every single ring of the set is connected to a single wire of bunch which goes to different regions of brain. A background image of handmade hippocampus is shown in the right of panel (**d**). The entire hippocampus length is divided into five parts: I—40.7 cm, II—25.5 cm, III—28.2 cm, IV—27.3 cm, and V—40.8 cm. 3D resonance spectrum with resonance peaks 59.29 MHz and 75.55 MHz is shown in panel (**e**) for the entire length (162.5 cm) of hippocampus. Although we measured all 3D resonance spectrums for all individual lengths (I-40.7 cm, II-25.5 cm, III-28.2 cm, IV-27.3 cm, and V-40.8 cm) of hippocampus. A comparison chart has been drawn between experimental (59.29 MHz and 75.55 MHz) and simulated (60 MHz and 109.3 MHz) data in terms of resonance peaks. Resonance peaks are widely varied in both cases, but their frequency ratio (1.2) has a similarity

3 Results and Discussion

3.1 *Cerebellum Features in Terms of Clocking Map*

A functional relationship between the constituents of the cerebellum provides the functional and anatomical connectivity of the brain. The mechanisms involved in their connections with nerve branches are unknown [35]. The size of brain wirings varies among species, but rhythm map remains the same [36]. The phenomenon of interaction can be explained in terms of symmetry breaking. A singularity is an operating clock [37], whereas rhythm map is a way of showing the undefined and collective phase relationships. The involved clocks may have a distinct dynamic nature, but they are phase-locked to each other [38]. When cerebellum learns coordination in movement, it creates a new clock [39]. Figure 1b demonstrates a 2D rhythm map of language and conversion and thinking and intelligence which are functions of the cerebellum. Symbolic details of all clocks are given in the panel. Reverse engineering of the brain circuit is possible; one practical example is shown in Fig. 3b. In our previous study, we have built the humanoid bot subject, HBS, with 20 types of conscious expressions inside its brain [34].

3.2 *Reliable Frequency Modes in Cerebellum to Communicate with Other Brain Components*

From the simulated results, we found the communication mode in cerebellum at two resonance peaks, 180 kHz and 250 kHz. At 180 kHz frequency, deep resonance peaks of the cerebellum cause a high-intensity reflection and transmission signal, meaning that the signal is transmitted to another source with high strength and back to the first source with approximately the same strength from where it is transmitted. At the second resonance peak, intensity of the transmitted signal from the energy source is higher than that of the return signal, meaning that the communication mode at the second resonance frequency is reliable. From Fig. 1d, we detected the energy distribution of the cerebellum at a frequency of 250.27 kHz and got maximum energy transmission. However, there are some losses due to energy sources 1 and 2. In other words, the cerebellum acts as a relay station allowing energy to easily travel to other regions of the brain.

3.3 *Hypothalamus: Emotion Rhythm Under Regulated Breathing*

Figure 2a shows the schematic of the hypothalamus, wherein each constituent is linked by its own name. Hypothalamus is a small part of the human brain that is used to generate rhythm and control vital motor functions. Hypothalamus comprises 13 cavities (Fig. 2a). Hypothalamus regulates electromagnetic signal bi-directionally which is evidenced by the regulation of clock of metabolism, body temperature, emotion, and physiological cycles [27]. Hypothalamus plays a crucial role in the functions of emotion, love and pain, threat, anger, fear, hate, etc. Their rhythm maps are shown in Fig. 2b. Two cross circles in the rhythm map show the junction of two constituents of a brain component. From resonance spectrum, hypothalamus has only a single communication mode at 497.63 kHz frequency; the signal transmits very well at this frequency. Although due to the asymmetrical structure of hippocampus, different energy patterns have been observed for each port. The nature of the resonance curve is almost the same as the cerebellum.

3.4 *Circular Pathways of Signal in Hippocampus*

The hippocampus is a helix shape with a large number of fibers extending radially along different lengths. It assigns a phase by evaluating the signals [40]. Time delay signal means signal travels along the long length of the wire [41]. Following this concept, we can fuse different frequency signals inside the hippocampus. Neurons present inside the cerebellum periodically form antenna-like shapes that alter signals within the rings of the hippocampus. Such rings project signals to different regions of the brain. Hippocampus simultaneously isolates and identifies signals and sends them back to different regions of the brain. These two features are automatically adjusted by the hippocampus. However, phase adjustment is accomplished by pyramidal neurons [41] that interact with inhibitory neurons to modulate the phase [42]. Helical paths or number of loops increase the delay between signals that go out. A set of signals pass through helical loops of the hippocampus. After passing through a few helical loops, the signal can be shown as a clock with a large delay or large diameter. When a signal passes through the next loop, a larger delay signal comes out and this process is continuous. The length of both channels is suitable for generating a particular delay signal. 'Hippocampal place cells' control neurons firing to modulate the timing [43] and phase of signals [44]. The electromagnetic signal is pumped through the helix antennas to the hippocampus and the output is taken after every single loop, so various phase difference signals emerge.

3.5 *Cross-Talk Between Hippocampus and Human Brain: Rule of Signal Propagation Along the Loop Pathways in Hippocampus*

Signals from all over the body have two tasks in the hippocampus: First, temporal code generation on the brain [44], and second, long-term memory storage tasks [45]. The hippocampus performs the different temporal codes for each type of memory [46]. Although the rhythm is originated from long-term memory [47], the basic operation of the brain is performed in the hippocampus because of its helix shape. A copy of the information stores in the hippocampus [48], and a copy of the information is projected into the upper brain to match the information already stored in the hippocampus [49]. It maintains the reliability of information stored and a selected frequency band can be activated spontaneously [50]. The hippocampus continuously captures the modified signals from the brain, identifies some additional signals, and filters the entire set of input signals. Filtered signals that are not found in the hippocampus are sent by the hippocampus back to certain regions of the brain for permanent storage [51, 52]. The hippocampal region CA1 edits the slowest and fastest signals for adjusting the rhythm [53]. Rhythms of the hippocampus suggest the learning rate of the brain [54].

3.6 Phase Modulation of Signals in Loop Paths

From Fig. 3d, some minor resonance peaks are generated through the hippocampus. One sharp peak shows the link of a single circular ring of the hippocampus to a functional region of the brain. The circular rings of the hippocampus project the signal to the brain regions at that frequency. The hippocampus requires additional dimensions to project the signal. Its reflection and transmission coefficients go beyond 360° 2D circular area. The presence of a 3D Smith chart is evidence of an additional phase that would have arisen in various signals passing through rings of spiral channels (Fig. 3e). So there is a variation of phase and time between signals, because of different diameters of rings, and different lengths of nerve fibers that pass through those rings. Signal from the thalamus is preserved in one channel of the hippocampus as memory (Fig. 3f), while another channel projects a copy of that signal to the brain.

3.7 Signal Synchronization in Hippocampus

The energy distribution is almost silent in the hippocampus (see Fig. 3g) because a single ring of both spiral channels makes the resonance. Here, we have only one resonance peak. Every ring of both channels resonate at an individual resonance frequency. This is the reason how the hippocampus synchronized different signals coming from hypothalamus. Here, on a limited scale, we are trying to understand if the hippocampus changes signal projection on brain regions, then how does it synchronize input signal from the whole body. The resonant rings of the hippocampus and layers of the cerebellum have different materials, and their structural configurations are very different from each other so for coherent energy transfer, matching of the ratio of resonance frequencies for both entities is necessary.

Hypothalamus and cerebellum have a common resonance mode. Cerebellum conveys the signal to the hypothalamus. The hypothalamus adjusts the resonance of its constituents in a way that synchronizes the signals coming from the cerebellum and transmits those signals to the hippocampus. The hippocampus synchronizes individual frequency signals through circular rings of spiral channels and stores them as memory.

3.8 *Clocking Assembly of C–H–H*

Brain components: C–H–H equivalent hardware is possible using the clocking assembly (Fig. 5a–c). If we build the circuits in the same way, it would follow the features of brain components. The clocking assemblies of major components of the brain are presented by tracing the clocks from brain standard books [55, 56]. The temporal map of the human brain consists of 537 clocks [32]. When the human brain makes the decision, all clocks run parallel.

Here, we are focusing on the mutual correlation between C–H–H by their resonance behavior. We have constructed a large replica of C–H–H components and found resonance frequencies in the MHz or kHz domains. In the theoretical and experimental cases, brain components always maintain similarity in the ratio of resonance frequencies. Signal transfers from the cerebellum to hippocampus, hypothalamus, and other brain components; after that, it propagates to brain regions. From our results, resonance frequency of the cerebellum differs by 2 (multiplication or division) with respect to the hippocampus and hypothalamus, while resonance frequency of the hypothalamus and hippocampus is differing by 3. So we can conclude that for signal transfer between brain components: (1) the ratio of resonance frequencies is

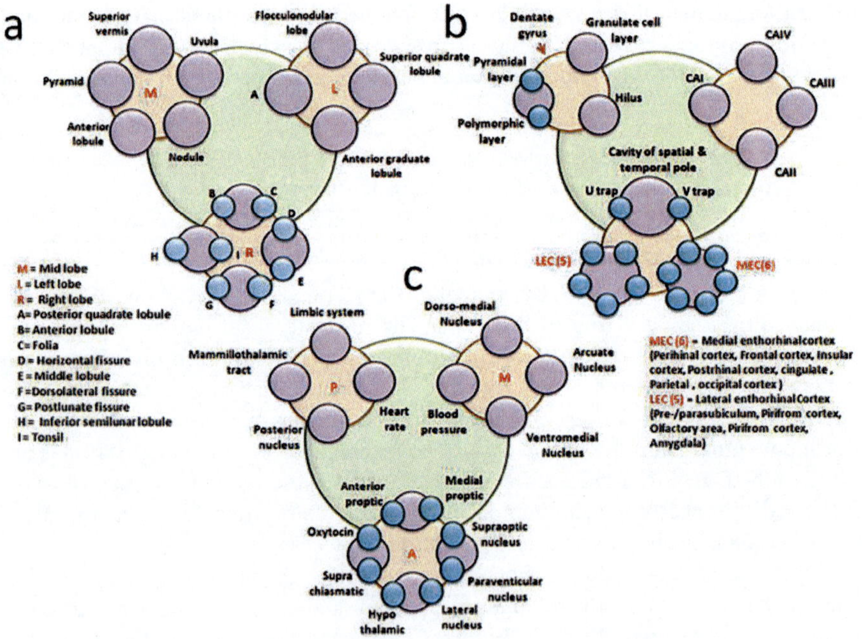

Fig. 5 Compiling all databases from literature works, we constructed nested clock maps of C–H–H as shown in panels (**a**), (**b**), and (**c**), respectively

the same. (2) The resonance frequencies of one component of the brain are arranged in an integral multiple of integer numbers with respect to another brain component.

4 Conclusion

Here, we have detected the cross-talking between C–H–H through electromagnetic resonance in terms of frequency and field distribution and suggested that their geometries are sufficient to establish their connections. Neural networks of C–H–H components resonate in almost same frequency domain during the decision-making process, and their internal geometries are arranged in the prime order. The C–H–H communicate with each other instantaneously on a common resonance mode. All three entities resonate at different frequencies but for energy transmission among themselves, they always maintain same resonance frequency ratio. All theoretical studies are experimentally verified. How do C-H-H have a combined effect on the brain? To explore this, we integrated the entire human brain-body nervous system in HBS and observed the projection of its characteristics such as emotion, love, anger, memory, etc., as an energy density distribution on the HBS brain. If we mimic the real brain component geometry, it is possible that they will follow the actual properties. We do not need to go for technical details of the structure. Such kinds of studies could have applications in robotics and biomedical areas.

Competing Interests

The authors declare that there is no competing interest.

Acknowledgements The authors acknowledge the Asian Office of Aerospace R&D (AOARD) a part of United States Air Force (USAF) for the grant no. FA2386-16-1-0003 (2016-2019) on electromagnetic resonance-based communication and intelligence of biomaterials.

References

1. Wang, S., et al.: Numerical simulation and analysis of effects of individual differences on the field distribution in the human brain with electromagnetic pulses. Sci. Rep. **11**, 16504 (2021)
2. Wanger, C., et al.: The electric field distribution in the brain during TT Fields therapy and its dependence on tissue dielectric properties and anatomy: a computational study. Phys Med Biol. **60**(18), 7339–7357 (2015)
3. Singh, P., Ray, K., Fujita, D., Bandyopadhyay, A.: Complete dielectric resonator model of human brain from MRI data: a journey from connectome neural branching to single protein. In: Ray K., Sharan S., Rawat S., Jain S., Srivastava S., Bandyopadhyay A. (eds.) Engineering Vibration, Communication and Information Processing. Lecture Notes in Electrical Engineering, vol. 478, pp.717–733. Springer, Singapore (2019)
4. Singh, P., Sahoo, P., Ray, K., Ghosh, S., Bandyopadhyay, A.: Building a non-ionic, non-electronic, non-algorithmic artificial brain: cortex and connectome interaction in a Humanoid Bot Subject (HBS). In: Kaiser M.S., Bandyopadhyay A., Mahmud M., Ray K. (eds.), Proceedings of International Conference on Trends in Computational and Cognitive Engineering. Advances in Intelligent Systems and Computing, vol. 1309, pp. 245–278. Springer, Singapore (2021)

5. Alekseichuk, I. et al.: Electric field dynamics in the brain during multi-electrode transcranial electric stimulation. Nat. Commun. **10**(2573) (2019)
6. Qiu, L.B., et al.: Effects of electromagnetic pulse on blood-brain barrier permeability and tight junction proteins in rats. Zhonghua Lao Dong Wei Sheng Zhi Ye Bing Za Zhi **27**(9), 539–543 (2009)
7. Wang, Q., et al.: To study of dose-response relationship of pulsed electromagnetic radiation on rat blood-brain-barrier. Chin. J. Dis. Control Prevent. **7**(13), 401–404 (2003)
8. Miranda, P.C., et al.: Predicting the electric field distribution in the brain for the treatment of glioblastoma. Phys. Med. Biol. **59**(15), 4137–4147 (2014)
9. Wenger, C., Salvador, R., Basser, P. J., Miranda, P.C.: Modeling tumor treating fields (TTFields) application within a realistic human head model. Annu. Int. Conf. IEEE Eng Med Biol Soc. PMID- 26736813, 2555–2558 (2015)
10. Lin, J.C., Wu, C.L., Lam, C.K.: Transmission of electromagnetic pulse into the head. Proc. IEEE **63**, 1726–1727 (1975)
11. Wang, S., Song, Z.G., Wu, D.C., Pu, Y.R.: Numerical simulation and analysis of the effect of individual differences on the field distribution in human brain with electromagnetic pulse. Sci. Rep. **11**, 16504 (2021)
12. Basar, E.: Chaotic dynamics and resonance phenomena in brain function: progress, perspectives and thoughts. In: Basar, E. (ed.) Chaos in Brain Function, pp. 1–30. Springer, Heidelberg (1990)
13. Saxena, K., Singh, P., Sahoo, P., Sahu, S., Ghosh, S., Ray, K., Fujita, D., Bandyopadhyay, A.: Fractal, scale free electromagnetic resonance of a single brain extracted microtubule nanowire, a single tubulin protein and a single neuron. Fractal Fract. **4**(11), 1–16 (2020)
14. Buzsáki, G., Draguhn, A.: Neuronal oscillations in cortical networks. Science **304**(5679), 1926–1929 (2004)
15. Buzsaki, G.: Rhythms of the Brain. Oxford University Press, New York (2006)
16. Goyal, A., et al.: Functionally distinct high and low theta oscillations in the human hippocampus. Nat. Commun. **11**(1), 2469 (2020)
17. Lu, H., Hartmann, M.J., Bower, J.M.: Correlations between purkinje cell single-unit activity and simultaneously recorded field potentials in the immediately underlying granule cell layer. J. Neurophysiol. **94**(3), 1849–1860 (2005)
18. Courtemanche, R., Pellerin, J.P., Lamarre, Y.: Local field potential oscillations in primate cerebellar cortex: modulation during active and passive expectancy. J. Neurophysiol. **88**(2), 771–782 (2002)
19. Hartmann, M.J., Bower, J.M.: Oscillatory activity in the cerebellar hemispheres of unrestrained rats. J. Neurophysiol. **80**(3), 1598–1604 (1998)
20. D'Angelo, E., Nieus, T., Maffei, A., Armano, S., Rossi, P., Taglietti, V., Fontana, A., Naldi, G.: Theta-frequency bursting and resonance in cerebellar granule cells: experimental evidence and modeling of a slow k+-dependent mechanism. J. Neurosci. **21**(3), 759–770 (2001)
21. Middleton, S.J., et al.: High-frequency network oscillation in cerebellar cortex. Neuron **58**(5), 763–774 (2008)
22. Jacobs, J.: Hippocampal theta oscillations are slower in humans than in rodents: implications for models of spatial navigation and memory. Philos. Trans. R. Soc. Lond. B Biol. Sci. **369**(1635), 20130304 (2014)
23. Maidenbaum, S., et al.: Grid-like hexadirectional modulation of human entorhinal theta oscillations. Proc. Natl. Acad. Sci. **115**(42), 10798–10803 (2018)
24. Hasselmo, M.E. et al.: Theta rhythm and the encoding and retrieval of space and time. Neuroimage **85**, 656–66 (2014).
25. Lega, C.B., et al.: Human hippocampal theta oscillations and the formation of episodic memories. Hippocampus **22**(4), 748–761 (2012)
26. Saleem, S.N., et al.: Lesions of the hypothalamus: MR imaging diagnostic features. Radio Graph. **27**(4), 1087–1108 (2007)
27. Stanley, S., et al.: Bidirectional electromagnetic control of the hypothalamus regulates feeding and metabolism. Nature **531**, 647–650 (2016)

28. Fujisawa, I.: Magnetic resonance imaging of the hypothalamic-neurohypophyseal system. J. Neuroendocrinol. **16**(4), 297–330 (2004)
29. Ono, D., Yamanaka, A.: Hypothalamic regulation of the sleep/wake cycle. Neurosci. Res. **118**, 74–81 (2017)
30. Hamani, C. et al.: Memory enhancement induced by hypothalamic/fornix deep brain stimulation. Ann Neurol. **63**, 119–123 (2008)
31. CST Studio Suite. CST Studio Suite 3D EM Simulation and Analysis Sofware Network. https://www.3ds.com/products-services/simulia/products/cst-studio-suite/
32. Singh, P., Saxena, K., Singhania, A., Sahoo, P., Ghosh, S., Chhajed, R., Ray, K., Fujita, D., Bandyopadhyay, A.: A self-operating time crystal model of the human brain: can we replace entire brain hardware with a 3D fractal architecture of clocks alone? Information **11**(5), 238 (2020)
33. Llinas, R.R., Walton, K.D., Lang, E.J. (2004) Cerebellum. In: Shepherd, G.M. (eds). The Synaptic Organization of the Brain. Oxford University Press, New York (2004)
34. Eccles, J.C.: Review lecture: the cerebellum as a computer: patterns in space and time. J. Phygiol. **229**, 1–32 (1973)
35. Reddy, S., et al.: A brain-like computer made of time crystal: could a metric of prime alone replace a user and alleviate programming forever? In: Ray, K., Pant, M., Bandyopadhyay, A. (eds.) Soft Computing Applications. Studies in Computational Intelligence, vol. 761, pp. 1–43. Springer, Singapore (2018)
36. Ringo, J.L., Doty, R.W., Demeter, S., Simard, P.Y.: Time is of the essence: a conjecture that hemispheric specialization arises from interhemispheric conduction delay. Cereb Cortex **4**, 331–343 (1967)
37. Matthews, P.C., Strogatz, S.H.: Phase diagram for the collective behavior of limit-cycle oscillators. Phys. Rev. Lett. **64**, 1701–1704 (1990)
38. Wang, X.J.: Multiple dynamical modes of thalamic relay neurons: rhythmic bursting and intermittent phase-locking. Neuroscience **59**, 21–31 (1994)
39. Graybiel, A.M.: The basal ganglia: learning new tricks and loving it. Curr. Opin. Neurobiol. **15**, 638–644 (2005)
40. Kamondi, A., Acsády, L., Wang, X.J., Buzsáki, G.: Theta oscillations in somata and dendrites of hippocampal pyramidal cells in vivo: activity-dependent phase-precession of action potentials. Hippocampus **8**, 244–261 (1998)
41. Harris, K.D., Henze, D.A., Hirase, H., Leinekugel, X., Dragoi, G., Czurkó, A., Buzsáki, G.: Spike train dynamics predicts theta-related phase precession in hippocampal pyramidal cells. Nature **417**, 738–741 (2002)
42. Lytton, W.W., Sejnowski, T.J.: Simulations of cortical pyramidal neurons synchronized by inhibitory interneurons. J. Neurophysiol. **66**, 1059–1079 (1991)
43. Royer, S., et al.: Control of timing, rate and bursts of hippocampal place cells by dendritic and somatic inhibition. Nat. Neurosci. **15**, 769–775 (2012)
44. Bliss, T.V., Collingridge, G.L.: A synaptic model of memory: long-term potentiation in the hippocampus. Nature **361**, 31–39 (1993)
45. Buzsáki, G. et al.: Oscillatory and intermittent synchrony in the hippocampus: rele-vance to memory trace formation. In: Buzsáki G., Llinás R., Singer W., Berthoz A., Christen Y. (eds.), Temporal Coding in the Brain. Research and Perspectives in Neuro-sciences. pp. 145–175. Springer, Berlin (1994)
46. Leutgeb, S., Leutgeb, J.K., Barnes, C.A., Moser, E.I., McNaughton, B.L., Moser, M.B.: Independent codes for spatial and episodic memory in hippocampal neuronal ensembles. Science **309**, 619–623 (2005)
47. Ferbinteanu, J., Shapiro, M.L.: Prospective and retrospective memory coding in the hippocampus. Neuron **40**, 1227–1239 (2003)
48. Wallenstein, G.V., Eichenbaum, H., Hasselmo, M.E.: The hippocampus as an associator of discontinuous events. Trends Neurosci. **21**, 317–323 (1998)
49. Jonas, P., Bischofberger, J., Fricker, D., Miles, R.: Interneuron diversity series: Fast in, fast out—temporal and spatial signal processing in hippocampal interneurons. Trends Neurosci. **27**, 30–40 (2004)

50. Buño, W., Jr., Velluti, J.C.: Relationships of hippocampal theta cycles with bar pressing during self-stimulation. Physiol. Behav. **19**, 615–621 (1977)
51. Cenquizca, L.A., Swanson, L.W.: Spatial organization of direct hippocampal field CA1 axonal projections to the rest of the cerebral cortex. Brain Res. Rev. **56**, 1–26 (2007)
52. Ciocchi, S., et al.: Brain computation. Selective information routing by ventral hippocampal CA1 projection neurons. Science **348**, 560–563 (2015)
53. Rotstein, H.G., Pervouchine, D.D., Acker, C.D., Gillies, M.J., White, J.A., Buhl, E.H., Whittington, M.A., Kopell, N.: Slow and fast inhibition and an H-current interact to create a theta rhythm in a model of CA1 interneuron network. J. Neurophysiol. **94**, 1509–1518 (2005)
54. Berry, S.D., Thompson, R.F.: Prediction of learning rate from the hippocampal electroencephalogram. Science **200**, 1298–1300 (1978)
55. Carter, R.: The Human Brain Book: An Illustrated Guide to its Structure, Function, and Disorders. DK: London, UK (2014)
56. Bandyopadhyay, A.: In Nanobrain. The Making of an Artificial Brain from a Time Crystal, 1st edn. Taylor & Francis Inc., Bosa Roca (2020)

Bangla Depressive Social Media Text Detection Using Hybrid Deep Learning Approach

Tapotosh Ghosh and M. Shamim Kaiser

Abstract Depression is considered as one of the major reasons of suicide which also affects personal and professional life of an individual. As people nowadays prefer to use social media, analysis of the contents generated by a person can lead us to get an insight of his mental health. For this, automated systems are essential due to huge explosion of social-media-generated data and privacy issue. In this paper, a deep learning approach based on Long Short-Term Memory (LSTM), Gated Recurrent Unit (GRU), Convolutional Neural Network (CNN), and Word2Vec embedding has been proposed to recognize depressive Bangla social media texts, which obtained 92.25% accuracy, 94.46% sensitivity, and 91.15% specificity. A dataset of 4000 Bangla social media posts was also created. The proposed approach was found better performing than LSTM, GRU, and classical machine learning models that proved its effectiveness in Bangla depressive text detection.

Keywords Depression · Deep learning · Machine learning · Sentiment analysis · Suicide prevention

1 Introduction

Depression is an emotional and mental state characterized by feelings of anxiety, sadness, or guilt, as well as a diminished capacity to enjoy life. Depression lessens the work rate, hampers personal and professional life, affects the relationships, and can even lead toward suicidal thoughts. 264 million people suffer from depression, which equates to 3.4% of the world's population [1]. Around 800,000 people die by

---

T. Ghosh (✉)
Department of ICT, Bangladesh University of Professionals, Dhaka, Bangladesh

M. S. Kaiser
Institute of Information Technology, Jahangirnagar University, Savar, Dhaka, Bangladesh
e-mail: mskaiser@juniv.edu

M. S. Kaiser et al. (eds.), *Proceedings of the Third International Conference on Trends in Computational and Cognitive Engineering*, Lecture Notes in Networks and Systems 348, https://doi.org/10.1007/978-981-16-7597-3_9

suicide every year [2]. Suicide is one of the major reasons of death among the youths [3]. These statistics clearly show the spread of depression-related disorders and its impact on our society.

Social media has already become part and parcel of our life. There are around 3.96 billion social media users which is around 50.6% of the population of the world [4, 5]. We share almost everything in social media. Even nowadays, news on posting suicidal notes on social media websites before committing suicide are found very often. There is a noticeable difference in writing pattern of posts and comments between depressive and non-depressive users. Therefore, analyzing social media contents can lead us to detection of a potential depressive person. But, social media generates a huge amount of data in every second which is not possible to analyze manually. Machine learning (ML) and deep learning (DL)-based frameworks are widely used in fall detection [6–9], earthquake prediction [10–12], autism detection [13, 14], biomedical applications [15–17], and so on. DL-based systems are capable of analyzing social media posts very efficiently. This system can also be used for opinion and sentiment mining, even in finding insights of mental state of a society. For this, several researches on depressive text detection have been conducted which are mostly focused on English language.

A combination of Long Short-Term Memory (LSTM) and Convolutional Neural Network (CNN) was adopted by Mumu et al. [18] in Bangla depressive text detection. LSTM layer was used to extract feature, where CNN layer was incorporated for feature mapping purpose. Billah et al. [19] took a classical ML approach, where they extracted Term frequency–Inverse Document Frequency (TF-IDF) feature from Bangla text, and SGD classifier was proposed to detect depressive texts. Uddin et al. explored LSTM [20] and GRU [21] models for Bangla depressive text detection. They found that five-layered LSTM and another five-layered GRU performed the best in their depressive text detection experiments. A bidirectional LSTM model was proposed by Ahmad et al. [22] in English depressive text detection which performed better than several LSTM, Recurrent Neural Network (RNN), and Gated Recurrent Unit (GRU) models. A Naive Bayes (NB)-based approach was taken by Deshpande et al. [23] in depressive English text detection. Zogan et al. [24] experimented a BERT-BART hybrid model for feature extraction, and a combination of GRU-CNN in classification of depressive Reddit Users. English Reddit posts were analyzed for identifying depressive attitude by Tadesse et al. [25]. They combined the extracted features by Latent Dirichlet Allocation (LDA) and Linguistic Inquiry and Word Count (LIWC) tool, which was further given input to classical ML models for predicting depressive person. Islam et al. [26] extracted features by using LIWC tool from English social media texts and detected depressive texts by using classical ML models. Several embeddings and 17 metadata extracted from writing of depressive people were explored by Trotzek et al. [27]. The extracted features were further given input to a CNN model to classify depressive text. An LSTM-CNN approach was proposed by Ghosh et al. [28]. The proposed model was further used to explore the impact of COVID-19 in mental health. A mobile application was developed by Suman et al. [29], where a transformer-based model was in the backend of the app that was trained on English tweets of positive and negative sentiments.

Bangla is a very popular language, with seventh largest community in the world [30]. Bangladesh has around 45 million social media users, from which a large portion of the users prefer to use Bangla in social media [31]. A lot of research endeavors are taken in depressive text detection from English or Chinese language, but it is very few in the case of Bangla. Therefore, in this paper, an LSTM-GRU-CNN-based Bangla depressive text detection method has been proposed. The main contribution of this paper is given below:

- Developing a Bangla depressive social media text detection dataset.
- Proposed an LSTM-GRU-CNN hybrid approach for detecting depressive Bangla text.
- Comparing performance of the proposed model with two other LSTM- and GRU-based models and classical ML models.

This paper is organized as follows: Sect. 2 describes the methodology of this research work. Result obtained by the proposed model is discussed in Sect. 3. Section 4 concludes the paper by discussing some research scopes in this field.

2 Methodology

The whole framework can be divided into several sub-phases. At first, a dataset was created for this task. Then, the texts were pre-processed and input tensor was generated from it. Then, a pre-trained Word2Vec was used to replace the elements of input tensors with pre-trained embedding vector which was the input to the proposed LSTM-GRU-CNN hybrid model. The proposed model predicted the class label of the input text.

2.1 Dataset Creation

At first, 4000 Bangla Facebook posts, Tweets, and YouTube comments were collected using APIs. Then, three native speaker labeled these samples into depressive or non-depressive. After labeling, 2752 statuses were found as non-depressive, where 1248 statuses were marked as depressive. This dataset was splitted at an 80:20 ratio. 3200 statuses were used for training purpose, and 800 texts were utilized as the testing set for evaluating the proposed model. There were 529 non-depressive and 271 depressive samples in the testing set.

2.2 Dataset Preprocessing

In the preprocessing phase, at first, the texts were converted to a list of words by performing tokenization. Then, punctuation, emoticons, unknown characters, and words of other languages except Bangla were removed. After that, a stemming was performed using a rule-based stemmer to convert the words into its base form. After that, stopwords that do not have any significant importance or meaning are removed to reduce the dimension of the text. Then, a dictionary is created by taking unique words from the dataset, and vectors were created by replacing the words with dictionary key. The input was created by adding padding 0s to the vectors to make it equi-length.

2.3 Word Embedding

The process of word embedding involves the development of representations that are similar for words with the same meaning. A real-valued numerical vector is used to represent words in a predefined vector space [32–34]. In Word2Vec, two-layered shallow neural networks are used to reassemble linguistic contexts. It generates a vector space by taking a large corpus of words, and assigning each word of the corpus an appropriate vector in the space. It can be achieved in two ways: Skip Gram and Continuous Bag of Words (CBOW).

The neural network predicts the corresponding word based on the context of the input hot vector of multiple words in CBOW. Training is based on the difference between the predicted and actual one-hot encoding vectors. This procedure results in the creation of a vector representation for the target word. In Skip Gram, the model takes one word as input and generates multiple words out of a probability distribution. CBOW provides better representation in common words, while Skip Gram is good for rare words and small data. In this work, a pre-trained 100-dimensional Word2Vec trained on Wiki dump dataset has been used [35, 36].

2.4 Proposed Model

A hybrid model (Fig. 1) consisting of LSTM, GRU, and CNN has been used in this work. LSTM and GRU layers were invented to handling larger sequences which create gradient vanishing problem in RNNs. Convolutional layer is capable of extracting more deep features from the feature set. At first, the proposed model replaced the elements of the input tensor with embedding matrix by using a embedding layer. Then, the output of it went through an LSTM layer of 100 neurons. It also goes through a GRU layer of 100 neurons. Output of the LSTM layer and GRU layer was concatenated, and the concatenated feature vector was given input to a convolutional layer of 512 filters. The convolutional layer extracted deep features from it. A max-

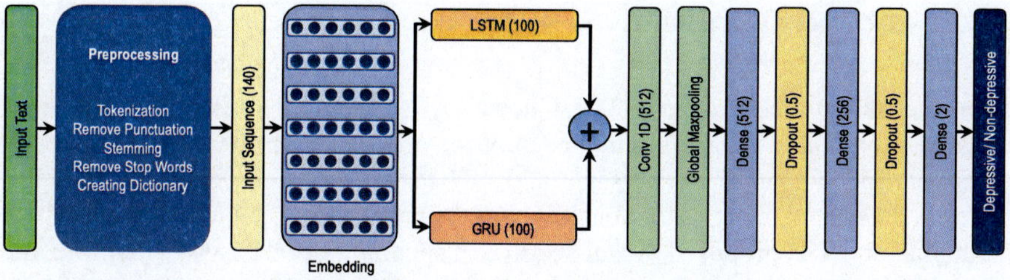

Fig. 1 The proposed model. LSTM and GRU layers took the input and output of the layers were concatenated which was the input to the convolutional layer. Dense layers were used to predict the text as non-depressive or depressive from the extracted features of the LSTM, GRU, and convolutional layers

pooling layer was stacked on it to reduce dimensionality of the feature matrix which was given input to a dense layer with 512 neurons and Rectified Linear Unit (ReLU) activation function. There were 2 more dense layers with 256 and 2 neurons. The last layer contained Softmax activation function provided the prediction of the proposed model. 50% dropouts were introduced between dense layers which helped the model to deal with overfitting problem.

The reason of using LSTM and GRU layers is handling the larger sequences. The output of these two layers were merged, which was the input to the convolutional layer that was introduced to extract more meaningful features that would help the dense layers to predict the class label. This architecture was found out after trying out several combinations of these layers using Keras tuner. Neuron number and activation function were selected based on the obtained performance in several tests. Performance of the proposed models was also compared with three LSTM-GRU-based DL models and classical ML models, such as Support Vector Machine (SVM), Decision Tree (DT), K-nearest Neighbors (KNN), Logistic Regression (LR), and Random Forest (RF). The proposed model performed better than all the explored models.

2.5 *Training and Testing*

The proposed model was trained for 30 epochs with a learning rate of 0.001, batch size of 64, and Adam optimizer. This model took around 22 s for an epoch which was quite fast. The proposed approach was tested with the separated test set of 800 samples after the completion of 30 epochs.

3 Result and Discussion

The proposed model acquired 92.25% accuracy in Bangla depressive social media text detection. Table 1 describes the result of the proposed hybrid model. It correctly classified 256 out of 271 depressive and 487 out of 529 non-depressive samples which is illustrated in Fig. 2(a). Sensitivity of the proposed model was 94.46% which is satisfactory. The proposed model acquired a specificity of 91.15%. Therefore, the proposed model was better in recognizing depressive posts, than recognizing non-depressing post. It found an AUC score of 0.93 which is shown in Fig. 2(b).

To compare the performance of the proposed model with similar and simpler DL models, three other models were created. In the first model, there was an LSTM layer with 100 neurons which took input from the embedding layer. The output of the LSTM layer was gone through a flatten layer and a three-layered dense layer which was similar as the proposed model. A GRU-based model was also created which had a GRU layer with 100 neurons, a flatten layer, 3 dense layers, and 2 dropout layers. Another LSTM-GRU hybrid model was created where LSTM-GRU were sequential layers. The LSTM-based model achieved 90.62% accuracy, 90.03% sensitivity, and 90.93% specificity. On the other hand, the GRU-based model acquired 89.75% accuracy, 89.66% sensitivity, and 89.79% specificity. LSTM-GRU hybrid model achieved 90.88% accuracy, where the sensitivity was 82.11%. This model achieved the highest specificity (94.16%). The proposed LSTM-GRU-CNN hybrid model provided more generalized result and outperformed these explored models in terms of accuracy and sensitivity. Figure 3 shows the performance comparison between the proposed model and the explored models.

DL models are computationally expensive compared to ML models. Performance of the ML models in Bangla depressive text detection has been explored. The input of all the ML models was the vector obtained after Word2Vec embedding layer. Among these models, SVM acquired the highest accuracy (81.39%). LR performed

Table 1 Performance of the proposed model

Metrics	LSTM-GRU-CNN hybrid model
Accuracy	0.9225
True negative (TN)	482
False positive (FP)	47
False negative (FN)	15
True positive (TP)	256
Sensitivity	0.9446
Specificity	0.9115
Negative predictive value (P_0)	0.9698
Positive predictive value (P_1)	0.8448

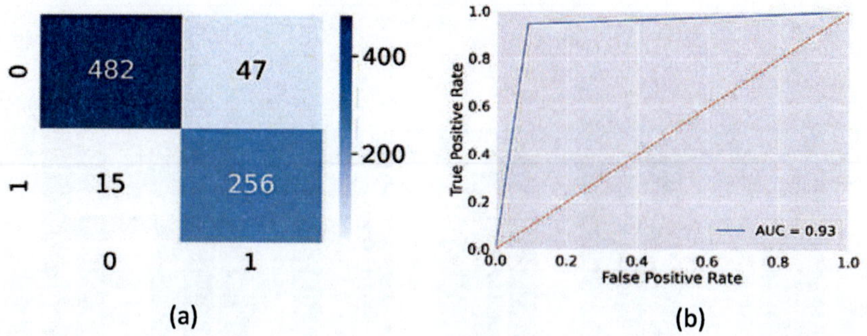

Fig. 2 **(a)** Confusion matrix **(b)** ROC-AUC curve of the proposed model. The proposed hybrid model correctly classified 256 out of 271 depressive samples. ROC score of the model was 0.93

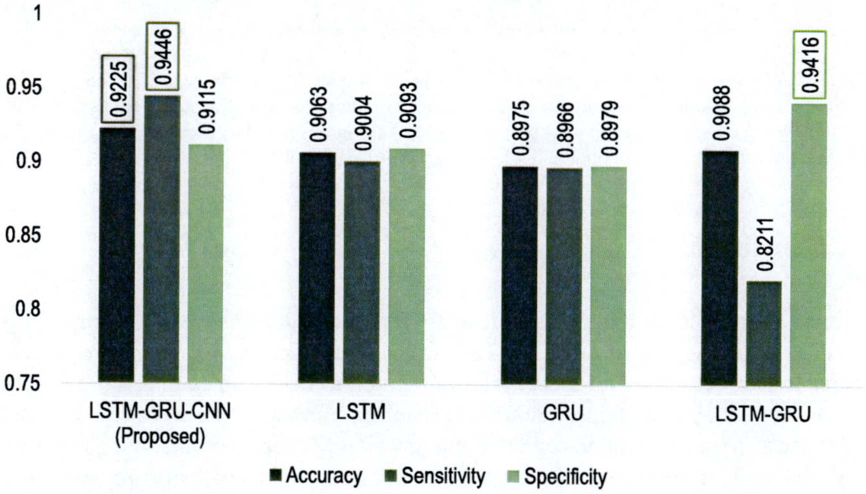

Fig. 3 Performance comparison between the proposed model and the explored models. The proposed hybrid model outperformed all the explored models in terms of accuracy and sensitivity. Word2Vec embedding was used in all the models

quite similar to SVM. KNN, DT, and RF could achieve more than 80% accuracy. The proposed model outperformed these classical ML models in terms of accuracy and sensitivity. RF achieved better specificity than the proposed model, but it was not enough generalized to consider as the model achieved a very poor sensitivity (27.811%). Figure 4 depicts the performance comparison between the proposed model and the classical ML models.

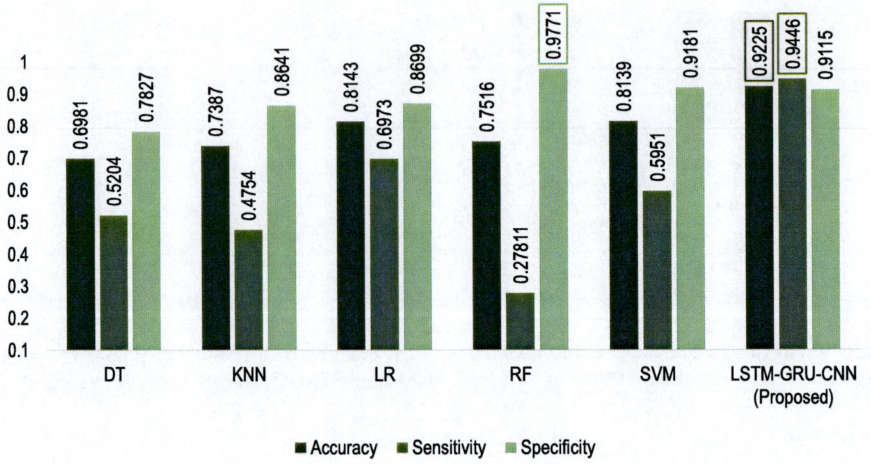

Fig. 4 Performance comparison between the proposed model and the classical ML models. The proposed hybrid model outperformed all the explored models in terms of accuracy and sensitivity. RF achieved the highest specificity, but it was not enough generalized to consider. All the results mentioned in this figure were obtained while using Word2Vec embedding

4 Conclusion

Depression-related problems are increasing day by day due to excessive competitive mindset, negativity, change of lifestyle, less social interaction, etc. People nowadays prefer to interact via social media sites compared to face-to-face interaction. They love to share their life events, emotions, griefs, etc. through these websites. Suicidal notes are also being found in social media nowadays. Therefore, analyzing contents of social media through an automated system has become a necessity. To contribute in this field, an LSTM-GRU-CNN-based Bangla depressive text detection model has been proposed which achieved over 92% accuracy with more than 94% sensitivity. This model can be incorporated with social media to prevent depression-related accidents. Government can use this framework in opinion and sentiment mining from social media text. Though the performance is satisfactory, there is enough scope of research in this field. Dataset that has been created in this paper was not large enough. Robustness is also needed to be ensured which was out of the scope of this research work. Attention- and transformed-based frameworks can be explored in future to achieve even better performance in depressive text detection.

Acknowledgements This research received funding from the ICT division of the Government of the People's Republic of Bangladesh for 2020-21 financial year (tracking no: 20FS13595).

References

1. Global, regional, and national incidence, prevalence, and years lived with disability for 354 diseases and injuries for 195 countries and territories, 1990–2017: a systematic analysis for the global burden of disease study 2017. Lancet **392**(10159), 1789–1858 (2018)
2. Organization, W.H., et al.: Preventing suicide: a global imperative. World Health Organization (2014)
3. Organization, W.H., et al.: Suicide worldwide in 2019: global health estimates (2021)
4. Global social media stats - datareportal – global digital insights, https://datareportal.com/social-media-users (2021)
5. Dean, B.: How many people use social media in 2021? (65+ statistics). https://backlinko.com/social-media-users (2021). Accessed 29 June 2021
6. Al Nahian, M.J., Ghosh, T., Uddin, M.N., Islam, M.M., Mahmud, M., Kaiser, M.S.: Towards artificial intelligence driven emotion aware fall monitoring framework suitable for elderly people with neurological disorder. In: International Conference on Brain Informatics, pp. 275–286. Springer (2020)
7. Al Nahian, M.J., Ghosh, T., Al Banna, M.H., Aseeri, M.A., Uddin, M.N., Ahmed, M.R., Mahmud, M., Kaiser, M.S.: Towards an accelerometer-based elderly fall detection system using cross-disciplinary time series features. IEEE Access **9**, 39413–39431 (2021)
8. Al Nahian, M.J., Ghosh, T., Al Banna, M.H., Uddin, M.N., Islam, M.M., Taher, K.A., Kaiser, M.S.: Social group optimized machine-learning based elderly fall detection approach using interdisciplinary time-series features. In: 2021 International Conference on Information and Communication Technology for Sustainable Development (ICICT4SD), pp. 321–325. IEEE (2021)
9. Nahian, M., Raju, M.H., Tasnim, Z., Mahmud, M., Ahad, M.A.R., Kaiser, M.S.: Contactless fall detection for the elderly. In: Contactless Human Activity Analysis, pp. 203–235. Springer (2021)
10. Al Banna, H., Ghosh, T., Taher, K.A., Kaiser, M.S., Mahmud, M., et al.: An earthquake prediction system for bangladesh using deep long short-term memory architecture. In: Intelligent Systems, pp. 465–476. Springer (2021)
11. Al Banna, M.H., Ghosh, T., Al Nahian, M.J., Taher, K.A., Kaiser, M.S., Mahmud, M., Hossain, M.S., Andersson, K.: Attention-based bi-directional long-short term memory network for earthquake prediction. IEEE Access **9**, 56589–56603 (2021)
12. Al Banna, M.H., Taher, K.A., Kaiser, M.S., Mahmud, M., Rahman, M.S., Hosen, A.S., Cho, G.H.: Application of artificial intelligence in predicting earthquakes: state-of-the-art and future challenges. IEEE Access **8**, 192880–192923 (2020)
13. Al Banna, M.H., Ghosh, T., Taher, K.A., Kaiser, M.S., Mahmud, M.: A monitoring system for patients of autism spectrum disorder using artificial intelligence. In: International Conference on Brain Informatics, pp. 251–262. Springer (2020)
14. Ghosh, T., Al Banna, M.H., Rahman, M.S., Kaiser, M.S., Mahmud, M., Hosen, A.S., Cho, G.H.: Artificial intelligence and internet of things in screening and management of autism spectrum disorder. In: Sustainable Cities and Society, p. 103189 (2021)
15. Mahmud, M., Kaiser, M.S., McGinnity, T.M., Hussain, A.: Deep learning in mining biological data. Cogn. Comput. **13**(1), 1–33 (2021)
16. Mahmud, M., Kaiser, M.S., Hussain, A., Vassanelli, S.: Applications of deep learning and reinforcement learning to biological data. IEEE Trans. Neural Netw. Learn. Syst. **29**(6), 2063–2079 (2018)
17. Shamma, Z.S., Ghosh, T., Taher, K.A., Uddin, M.N., Kaiser, M.S.: Towards social group optimization and machine learning based diabetes prediction. In: 2021 International Conference on Information and Communication Technology for Sustainable Development (ICICT4SD), pp. 422–427. IEEE (2021)
18. Mumu, T.F., Munni, I.J., Das, A.K.: Depressed people detection from bangla social media status using LSTM and CNN approach. J. Eng. Adv. **2**(01), 41–47 (2021)

19. Billah, M., Hassan, E.: Depression detection from bangla facebook status using machine learning approach. Int. J. Comput. Appl. **975**, 8887
20. Uddin, A.H., Bapery, D., Arif, A.S.M.: Depression analysis from social media data in bangla language using long short term memory (LSTM) recurrent neural network technique. In: 2019 International Conference on Computer, Communication, Chemical, Materials and Electronic Engineering (IC4ME2), pp. 1–4. IEEE (2019)
21. Uddin, A.H., Bapery, D., Arif, A.S.M.: Depression analysis of bangla social media data using gated recurrent neural network. In: 2019 1st International Conference on Advances in Science, Engineering and Robotics Technology (ICASERT), pp. 1–6. IEEE (2019)
22. Ahmad, H., Asghar, D.M., Alotaibi, F., Hameed, I.: Applying deep learning technique for depression classification in social media text. J. Med. Imaging Health Inform. **10**(6), 2446–2451 (08 2020). https://doi.org/10.1166/jmihi.2020.3169
23. Deshpande, M., Rao, V.: Depression detection using emotion artificial intelligence. In: 2017 International Conference on Intelligent Sustainable Systems (ICISS), pp. 858–862. IEEE (2017)
24. Zogan, H., Razzak, I., Jameel, S., Xu, G.: Depressionnet: a novel summarization boosted deep framework for depression detection on social media (2021). arXiv:2105.10878
25. Tadesse, M.M., Lin, H., Xu, B., Yang, L.: Detection of depression-related posts in reddit social media forum. IEEE Access **7**, 44883–44893 (2019)
26. Islam, M.R., Kamal, A.R.M., Sultana, N., Islam, R., Moni, M.A., et al.: Detecting depression using k-nearest neighbors (KNN) classification technique. In: 2018 International Conference on Computer, Communication, Chemical, Material and Electronic Engineering (IC4ME2), pp. 1–4. IEEE (2018)
27. Trotzek, M., Koitka, S., Friedrich, C.M.: Utilizing neural networks and linguistic metadata for early detection of depression indications in text sequences. IEEE Trans. Knowl. Data Eng. **32**(3), 588–601 (2018)
28. Ghosh, T., Al Banna, M.H., Al Nahian, M.J., Taher, K.A., Kaiser, M.S., Mahmud, M.: A hybrid deep learning model to predict the impact of covid-19 on mental health form social media big data (2021)
29. Suman, S.K., Shalu, H., Agrawal, L.A., Agrawal, A., Kadiwala, J.: A novel sentiment analysis engine for preliminary depression status estimation on social media. arXiv:2011.14280 (2020)
30. Hasan, M.: Bangla ranked at 7th among 100 most spoken languages worldwide | dhaka tribune (2020). https://www.dhakatribune.com/world/2020/02/17/bengali-ranked-at-7th-among-100-most-spoken-languages-worldwide. Accessed 21 Oct 2021
31. UNB: Bangladesh charts 9m new social media users | dhaka tribune (2021). https://www.dhakatribune.com/bangladesh/2021/04/26/bangladesh-charts-9m-new-social-media-users. Accessed 21 Aug 2021
32. Ghannay, S., Favre, B., Esteve, Y., Camelin, N.: Word embedding evaluation and combination. In: Proceedings of the Tenth International Conference on Language Resources and Evaluation (LREC'16), pp. 300–305 (2016)
33. Lai, S., Liu, K., He, S., Zhao, J.: How to generate a good word embedding. IEEE Intell. Syst. **31**(6), 5–14 (2016)
34. Levy, O., Goldberg, Y.: Neural word embedding as implicit matrix factorization. Adv. Neural Inf. Process. Syst. **27**, 2177–2185 (2014)
35. Bengali natural language processing(BNLP)—BNLP latest documentation. https://bnlp.readthedocs.io/en/latest/#word-embedding. Accessed 07 Feb 2021
36. Index of /bnwiki/latest/. https://dumps.wikimedia.org/bnwiki/latest/. Accessed 07 Feb 2021

Implementation of Real-Time Automated Attendance System Using Deep Learning

Hafiz Mahdi Hasan, Md. Mahfujur Rahman, Md. Al-Amin Khan, Tamara Islam Meghla, Shamim Al Mamun, and M. Shamim Kaiser

Abstract In comparison to general manual operations, contemporary technology always saves time and is often more hassle-free when it comes to verifying human authenticity using their biometrical components. However, despite the fact that face recognition technology has been used in a variety of sectors such as human identification systems, this work is the first to describe how the Face Recognition Technique can be integrated with a deep learning approach. Advanced deep learning techniques can make the attendance system completely automated, highly secure, easier to use, and faster to implement than older systems. Nowadays, the Attendance System is becoming increasingly automated, resulting in time-saving, effective, and beneficial solutions that reduce the burden on administration and organizations. In this paper, we suggest an automatic attendance mechanism that is based on Deep Convolutional Neural Networks (DCNN). SeetaFace, a deep convolutional neural network-based face detection system, is employed in this research effort to detect faces in real-time video capture. This implementation is a VIPLFaceNet implementation, to be more specific. AlexNet, which is also a DCNN, is used for image categorization. The experimental results bring four short similarity situations of the classroom such as absence, delayed appearances, early leave, and unauthorized entry during class or session along with the name, student id, and section and passes this information to the attendance sheet which will evaluate the students/persons in the classroom. This methodology saves time when compared to the traditional method of attendance marking, as well as allows organizations to conduct stress-free observations of students and staff.

H. M. Hasan · Md. M. Rahman (✉) · Md. A.-A. Khan
Daffodil International University, Dhaka 1207, Bangladesh

T. I. Meghla
Software, Web & Cloud, Faculty of Information Technology and Communications, Tampere University, Tampere, Finland

S. Al Mamun · M. S. Kaiser
Jahangirnagar University, Savar, Dhaka 1342, Bangladesh

M. S. Kaiser et al. (eds.), *Proceedings of the Third International Conference on Trends in Computational and Cognitive Engineering*, Lecture Notes in Networks and Systems 348, https://doi.org/10.1007/978-981-16-7597-3_10

Keywords Bio-metric identification · DCNN · SeetaFace · VIPLFaceNet · AlexNet · DLA · Open CV

1 Introduction

Face recognition is a unique bio-metric authentication technology which is widely used in security, financial, and military fields. In recent years, face recognition has attracted great interest and has motivated various academics to not only develop new algorithms but also improve existing systems. Due to the deployment of CCTV surveillance systems almost everywhere, real-time video recognition can be achieved by considering a succession of frames, where each frame is regarded as a single picture. We must first confirm the presence of a face in the frame before facial recognition can be performed. This can be done by the detection of the face and separating it from the identification image, eliminating redundant data not needed to recognize your face. This decreases the number of pixels the model needs to work on and so increases the overall efficiency.

The attendance system is a part of human resource management and authentication. It's vastly needed in the classroom or workspace. The attendance system is important to classroom evaluation, population, and workforce management. Caused by the event of technology, computer scientists share a keen interest in this area. Currently, there are several technology-based attendance systems, but we always discuss more sufficient systems saving time and toil. In the traditional attendance system, there are used paper sheets or registry books. Due to time-consuming, it demands modern technology to invent a new model of a smart attendance system. Modern technologists always try to research and analyze new problems and fix them with a smarter solution by modern technology. A lot of automated attendance systems are proposed and implemented, but they have some limitations like multiple and real-time image capturing, obstacles in front of the face, low light, etc. So, we need to develop an automated attendance system and also overcome these limitations without increasing the complexity and decreasing the accuracy.

This paper proposed a tentative solution with the help of a Deep Convolutional Neural Network (DCCN) besides this research studies about limitations and challenges in this field. The proposed method will be integrated into a camera setup. Real-time video capture as an output will be matched with a dataset, and it saves the information in the ERP database. This paper presented how to build an automatic attendance system with the assistance of deep learning, a branch of ML within a sensible site of technology. It's a computer-aided system for automatically identifying an individual from a real-time image or video frame with high accuracy.

2 Literature Review

Deep Learning forms a subset of the Machine Learning world, wherein the data is passed through multiple layers of non-linear mappings to allow the algorithm to learn a representation. A particularly hard and hotly researched topic in Deep Learning is the topic of Face Detection and Face Recognition in automatic attendance systems. There are several systems that have been proposed regarding this. Akay et al. [1] proposed a deep learning-based convolutional neural networks (CNNs) to identify the students in the classroom with the mask checking feature and compared performances with other proposed models, but there was a limitation that the systems are not efficient for the correct identification of the distance of seating ones. The 3D facial reconstruction is used to learn facial features from a single image [8]. When people entered the room, MTCNN detected the person automatically and then they used the C2D-CNN model for face recognition with the combined features learned originally inputted image features from CCTV footage. They claimed that the students' images having partial faces should also be detected and recognized. Darapaneni [3] says they proposed a Face Recognition system that is capable of identifying a human face automatically. They used architectural solutions to their system using YOLO, MTCNN, and FaceNet with the application of multiple augmentations. The proposal contained methods to get a better system with less maintenance like de-noising. There is a study of FaceNet system developed by Google, and it's an open-source implementation and pre-trained models can be used to extract high-quality features from faces, called face embedding, that can be used to train a face identification system, but during the implementation some corrupted data might be sent along with the original information about attendance due to mismatch of the network layers. Pre-trained CNN was evaluated for a single face recognition problem in Filippidou and Papakostas's proposal [5]. They claim to prove that the single face recognition accuracy can be reached up to 99% with the model of DenseNet121. The author said it is the best model for single face detection. They used the C2D-CNN model for face recognition with the combined features learned originally inputted image features from CCTV footage. So, they had faced some limitations to make an advanced attendance system that can't recognize multiple faces at a time. Another proposition was also revealed by Viola-Jones et al. [13]: HOG features together with the SVM classifier. It is highly susceptible to feature extraction, face classification, face recognition, image acquisition, and face detection to make a smart and modern attendance system. The quantitative evaluation was also added on the basis of PSNR value in this proposal. The Eigenface and the Fisherface techniques are linearly projection based on the PCA algorithm [2]. PCA can remove correlated features from the images and when the dataset contains too many variables, it helps to reduce overfitting issues by reducing the number of features. It is also quite good at the visualization of 2D. Dimension plots are reduced to 2D visualization. There are some abridgements that original features are highly interpretable instead of independent (Principal Components) variables and before applying PCA data should be standardized. Though the PCA ensures max variance among the features in a dataset, in case of lack of selection of the number of the PCA components carefully it may miss some information as compared to the original features. Linear Discriminant Analysis (LDA) matches biometric information and makes predictions using logistic regression and various powerful classification algorithms [9]. LDA is a diversified variation to eigenfaces being less sensitive to variation while matching techniques related to the computation of geometrical features extraction from the images. The author applied LDA and achieved impressive accuracy in their methodology. DCNN is highly accurate to detect objects and human faces. In their comparison, they used a simple DCNN with very few layers compared with two pre-trained networks, [7] AlexNet which has eight layers and GoogLeNet which has 22 layers. They observed that GoogLeNet learns the features and gets trained very soon with less iteration. The ImageNet dataset helps to evaluate residual learning framework instead of training by deep neural networks. It has a depth of max 152 layers so that it is much deeper than [6] VGG Net with lower complexity. But the human face cannot be properly identified while the eyes are closed due to the inconsistent linearity of HOG features. It performed well on the ILSVRC 2015 classification task also. The analysis showed CIFAR-10 with 100 and 1000 layers. The depth of representations is of central importance for many visual recognition tasks. According to the author, images are affected by illumination and attitude angle, and the traditional expression extraction method canâŁ™t effectively extract the features, [11] they proposed a deep learning expression recognition algorithm model based on multi-model. The author said "Experiments show that this method has achieved good recognition results in the KDEF database". However, the majority of the models [4, 10, 12, 14] struggle to reach a near-optimal solution. To address the issues, this research presented DCNN, which provides rapid and accurate results when several objects must be identified in a short period of time.

3 Methodology

The main goal of this paper is very simple such that students in the class or people stationed at any Human Assembly will continue their normal activities and at the same time, their attendance will be automatically saved in the database or ERP. Research like this type of human authentication might be developed more efficiently and more accurately. The proposed method overcomes the limitations of other existing research works by implementing new ideas in image datasets that the increasing number of a person's images and setting a big count for train data and test data can ensure maximum accuracy. We figured out the schedule of workflow and summarized our proposed methodology in Fig. 1. Figure 1 depicts the whole structure of the proposed model where we collect the dataset, analyze the features, and train the data to determine who are present or who are absent.

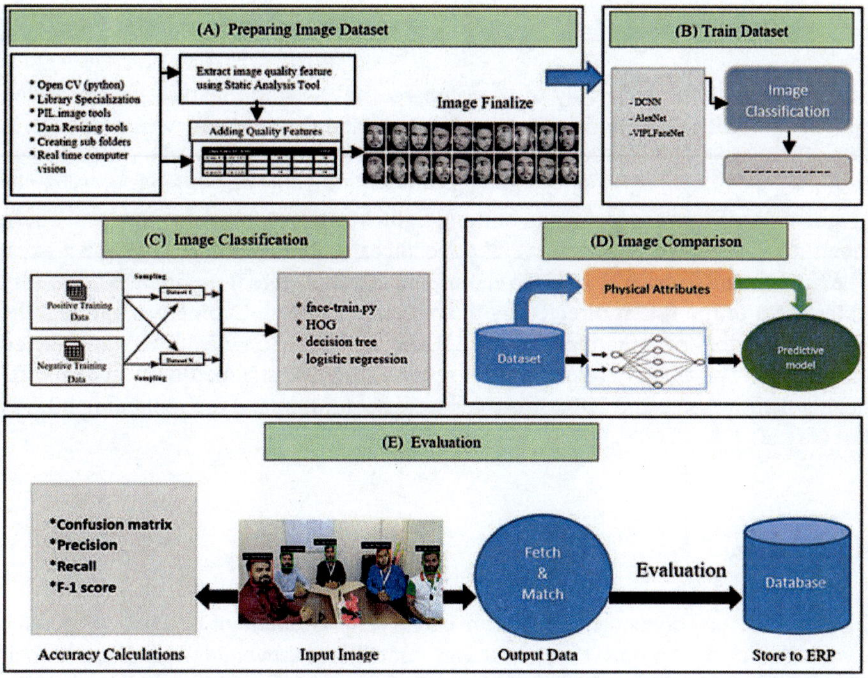

Fig. 1 The workflow of proposed methodology

3.1 Preparing Face Image Dataset

Real-time computer vision needs a predefined dataset to identify actual faces. OpenCV library has programming functions (including deep learning algorithms) that help to create and process a dataset with its optimization of multi-core processing. In this work, OpenCV is used for making a dataset. To prepare the dataset, the system needs a camera set up to capture images to involve the person in the system. For every person, we have taken 18–20 images that are going to be captured at different angles with a feature to require random still images for a person's faces by executing the program. To reduce time complexity and convert from RGB to grayscale, the images have been cropped and resized automatically on the model to 110*90 pixels. When we prepared to train and test the dataset with the OpenCV library, we defined dataset paths.

3.2 Train Dataset

Stored raw data are not ready to be compared unless some features are extracted and added during the training dataset. We imported the OpenCV version which is OpenCV-Python=3.4 and also imported the cv2 module, then defined the dataset directory, and added some labels of folders for the sake of differentiation. We have given our label name as top, front, bottom, right, and left of individual images. These labels are distributed in three several appendices according to the x, y, and z-axes. The final and most crucial part is to capture images and store those images according to the label of the folder directed by the path we declared. During the training, the library extracted some features that are not good enough. Some features were labeled to raw data. The Main dataset contains several subfolders named with students' ID having their images, and 70% of images were set to train and the remaining 30% to test (Fig. 2).

3.3 Image Classification

Images into small connected regions are called cells. Gradient orientation discredited each cell into angular bins. Pixel values are contributed as weighted gradients to their corresponding bin. The spatial region is made with a gaggle of adjacent cells then this group of adjacent cells is normalized as a histogram. As we can mention the process that we used Keras to demonstrate image classification using DCNN in this project. The deep learning API Keras is an excellent framework to find out more detail from an object image. Flatten the dataset images dimension and normalize these images with pixel value, encoding the categorical column; after that, a sequential model with a dense layer is created for each sub-folder. That process helps in retaining the "spatial" properties of images while classifying them. In the faces-train.py file, the system trains the dataset and transforms images into grayscale matrices as shown in Fig. 3.

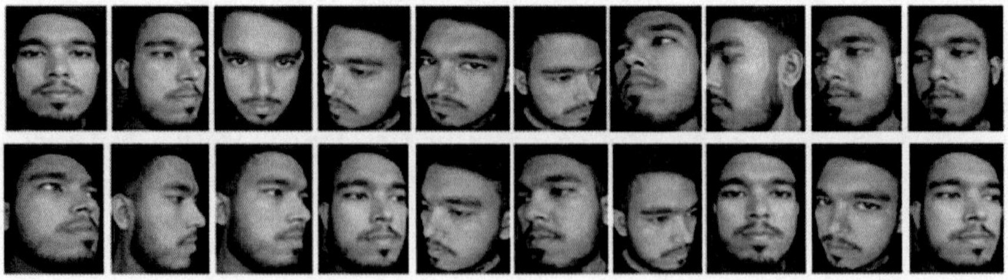

Fig. 2 Images in dataset

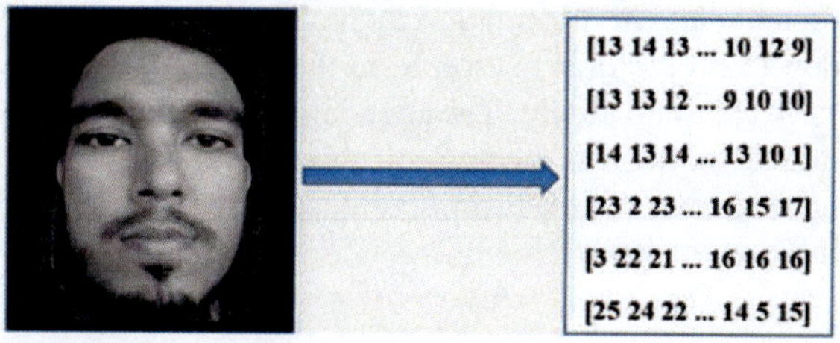

Fig. 3 Grayscaling before classification

3.4 Comparison and Recognition

VIPLFaceNet has some procedural connecting layers to identify objects. Among them, the first and last layers are most crucial for image recognition. The first layer is mostly responsible for the study of image filtering. VIPLFaceNet is deeper by replacing the 5×5 convolutional kernel layer with two 3×3 kernel layers, and it also reduces the number of kernels in each layer. It introduces the fast normalization layer for faster convergence of the network optimization. The evaluation shows that, compared with the original AlexNet, the error rate of VIPLFaceNet is reduced by 40% just in case of an equivalent training set, while its training time is reduced by 80% and its computational cost for feature extraction is reduced by 40%. While the image classification is done then the duty begins for VIPLFaceNet to find matching edges from real-time capturing.

One of the most useful implementations of VIPLFaceNet is SeetaFace. We set the seeta:FaceIdentification to the directory of the model file. After a face image is read, one must pack the image data with seeta:ImageData. It is mentioned that the pixel values should be stored during a continuous 1D array in a row-major style. The class for face identification is included in the seeta namespace. To use the function of SeetaFace Identification, one should first instantiate an object. While deep neural network layers match with the decision layer for a particular face, the system recognizes the face. To calculate the similarity between two faces, call similarity function with the corresponding feature vectors.

3.5 Collection of Attendance

The DCNN and SeetaFace will be starting their operation with the starting of video capturing. The camera captures the faces of the person appearing in the classroom or assembly. The VIPLFaceNet matches network layers for input images and finds trained regions of an individual with the dataset by neural networking mechanism and temporarily stores the attributes for the person, as we know that SeetaFace is an implementation of VIPLFaceNet; after matching the faces, SeetaFace labels some additional information about person's activity in front of the cameras as absence, delayed appearances, early leave, unauthorized entry before sending the attendance related data, or information to the REP models of administration server. If attributes don't find the information from the trained dataset, it can't recognize an individual from captured videos. The attendance collection system monitors some parameters of evaluation such as the present, absent, delayed appearance, early leave, unauthorized entry during the class time, or declared session using a person's activity in front of the camera setup.

3.6 *Evaluation of Attendance*

Collected attendance data will be stored in the ERP database to evaluate. Although the main goal and objective of our project are whether the person is present or not, we implemented some additional features provided by SeetaFace which can evaluate the person's activity during the class or meeting. DCNN recognizes the faces with a trained dataset and SeetaFace traces the student activities with their movement. When finding information about students' appearance in the classroom, the system sends an information message to the ERP database "162-15-1078 is present" and similarly for the students who do not appear in the classroom system through a message that "162-15-1080 is absent" was tested in the editor terminal window. The DCNN cannot recognize the student as the savior fairer. For the early departure, late arrival, or unauthorized entry of students, a bad impact (or ratings) will be hit on the ERP system detected by the define in SeetaFace by VIPLFaceNet module. Besides, we designed a data table (Fig. 4) to store and evaluate attendance data in the ERP database.

s_id	s_name	date	time_slot	attendance_status	late_arrival	early_departure
1002310	al-amin	2/23/2021	9:00-9:05am	present	-	√
1002311	mahdi	2/23/2021	9:00-9:05am	present	√	-
1002312	saklain	2/23/2021	9:00-9:05am	present	-	-
1002313	nahian	2/23/2021	9:00-9:05am	absent	-	-
1002314	russel	2/23/2021	9:00-9:05am	present	√	-
1002315	fayez	2/23/2021	9:00-9:05am	pesent	-	√
1002316	anik	2/23/2021	9:00-9:05am	absent	-	-
1002317	nayem	2/23/2021	9:00-9:05am	absent	-	-
1002318	imran	2/23/2021	9:00-9:05am	present	-	√
1002319	khalid	2/23/2021	9:00-9:05am	present	√	-
1002320	nabil	2/24/2021	9:00-9:05am	present	-	-
1002321	mukit	2/25/2021	9:00-9:05am	present	√	√
1002322	emon	2/26/2021	9:00-9:05am	present	√	-
1002323	rayhan	2/27/2021	9:00-9:05am	present	√	-
unauthorized	---	2/23/2021	9:00-9:05am	---	---	---
unauthorized	---	2/23/2021	9:00-9:05am	---	---	---
unauthorized	---	2/23/2021	9:00-9:05am	---	---	---

Fig. 4 Attendance information stored in database

4 Evaluation and Result Analysis

4.1 Experimental Environment and Output

The experiment is conducted with the technologies of Python, Anaconda, MySQL, Deep Neural Network, Keras, SeetaFace facial recognition algorithm, VIPLFaceNet, OpenCV library, Google Colab and pyCharm IDE, webcam or external camera setup (min 720px), etc. We took 5 people named Jewel, Alamin, Anik, Russel, and Nahian. Along with their predefined attributes and preset label, the system treats their images as five different classes and can recognize the face based on the DCNN model. These class attributes were frontal, top, right, bottom, and left facial regions. After successfully running the program connected with a camera setup, a window will appear in the UI bringing information we want to get for collecting attendance. If a person sits in front of the camera whose photos are already saved in the image dataset without any flaws, is properly classified, and well-trained, the information of identification/recognition will be presented through a text with a bordered box. We do not need any external UI device during implementation except database administration end though we use our computer window as UI. We tested and verified the output with 5 people which will be explained in the part of the confusion matrix. We test the system output in several situations, conditions, and environments such as with a single person at daylight for the initial, with multiple people at indoor and outdoor daylight, with single and multiple people at artificial lighting, as well as with single and multiple people at low lighting. All of them had shown the output with a very satisfying detection except the testing at low light (Figs. 5) and 6).

Due to low light, it may take some interruption to show more accuracy for multiple faces but in a sorta good camera setup can be resolved for a single face.

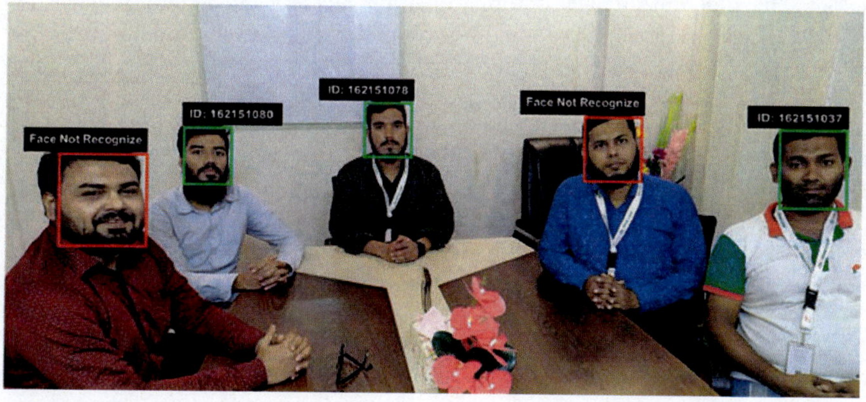

Fig. 5 Person Identification with their ID

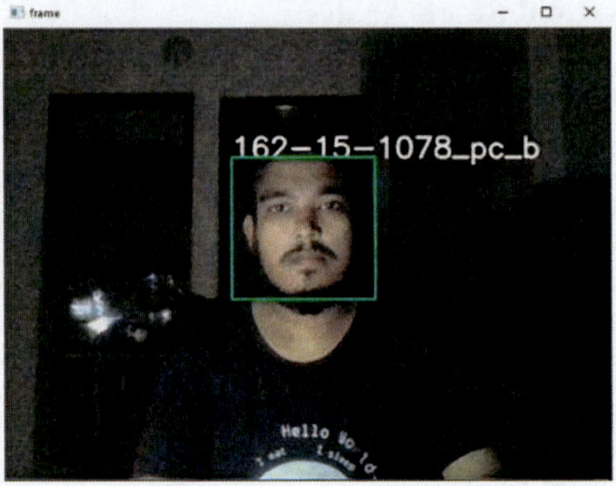

Fig. 6 Person Identification with low light

4.2 Result Analysis and Comparative Study

To analyze the performances of the executed models, the Confusion Matrix is measured. Figure 7a shows the regions with a graphical evaluation between predicted class and actual class. The cells having the lightest color represent the higher value of matching input data between actual and predicted classes; on the other hand, the cells having the darkest region represent the lower matching. From the confusion matrix, Precision, Recall, and F1-score have also been measured to evaluate the model. These metrics depend on four basic qualitative model quality indicators, namely true positive, true negative, false positive, and false negative. This study also measured the individual images of top, front, right, and left, and bottom labels match between actual and predicted data; it displays the efficient results and improves the overall accuracy. Confusion Matrix of individual images is shown in Fig. 7b. When we created a dataset with 10 images for each person, we didn't get the accuracy as we expected, then we increased the number of photos from 10 to 20 and got 15 percent more accuracy. Significant improvements have also been made in the results of precision, recall, and F-1 score only by increasing the number of images for individual persons' image dataset. Accuracy along with precision, recall, and F-1 score is compared to other methods which are mentioned in Fig. 7c. And here, the proposed study has achieved an accuracy of 95% and outperforms the other models.

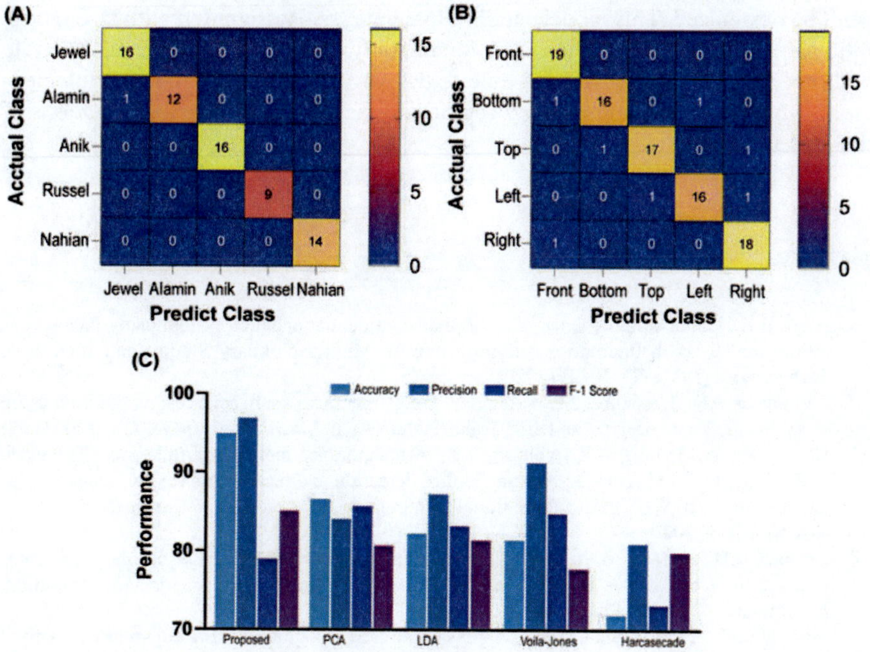

Fig. 7 **a** Confusion Matrix for face identification; **b** Confusion Matrix for individual feature identification; **c** Accuracy, Precision, Recall, and F1-score comparison among the researches mentioned in the state-of-the-art and presented systems which showcase DCNN to be most efficient and accurate in attendance system

5 Conclusion

The proposed model has a strong ability in the shaking of facial recognition with detection in this research. The proposed model has to acquire the attendance information and make the process easier for the organizer with an optimal result and evaluation. Based on this, the proposed automatic attendance system will be more modern and smarter. At the same time, some problems have been overcome such as giving one's attendance to another, identifying someone who is not a student of the institution, saving time, etc. In this paper, we designed and developed a real-time face tracking and recognition system, but this system still has some shortcomings. To affect the matter of low-light video capturing, there could be some difficulties to get proper accuracy, though this model tried to capture high-resolution real-time video for getting better performance. The problem can be raised due to wearing the hijab that may cover some of the parts of the face of a person. Due to wearing the mask as the same as the present epidemic situation in the classroom or production line, face can't be recognized. This model couldn't integrate Iris-Recognition with face recognition and keep continuing concentration of study on how this model will integrate both iris and face recognition at a time or the way to develop the model implementation which may detect human iris region and face together to settle limitations of the proposed model.

References

1. Akay, E.O., Canbek, K.O., Oniz, Y.: Automated student attendance system using face recognition. In: 2020 4th International Symposium on Multidisciplinary Studies and Innovative Technologies (ISMSIT). IEEE (2020)
2. Belhumeur, P.N., Hespanha, J.P., Kriegman, D.J.: Eigenfaces vs. fisherfaces: recognition using class specific linear projection. IEEE Trans. Pattern Anal. Mach. Intell. **19**(7), 711–720 (1997)
3. Darapaneni, N., Evoori, A.K., Vemuri, V.B., Arichandrapandian, T., Karthikeyan, G., Paduri, A.R., Babu, D., Madhavan, J.: Automatic face detection and recognition for attendance maintenance. In: 2020 IEEE 15th International Conference on Industrial and Information Systems (ICIIS). IEEE (2020)
4. Ferdous, R.H., Arifeen, M.M., Eiko, T.S., Mamun, S.A.: Performance analysis of different loss function in face detection architectures. In: Advances in Intelligent Systems and Computing, pp. 659–669. Springer, Singapore (2021)
5. Filippidou, F.P., Papakostas, G.A.: Single sample face recognition using convolutional neural networks for automated attendance systems. In: 2020 Fourth International Conference On Intelligent Computing in Data Sciences (ICDS). IEEE (2020)
6. He, K., Zhang, X., Ren, S., Sun, J.: Deep residual learning for image recognition. In: 2016 IEEE Conference on Computer Vision and Pattern Recognition (CVPR). IEEE (2016)
7. Kalaiarasi, P., Esther Rani, P.: A comparative analysis of AlexNet and GoogLeNet with a simple DCNN for face recognition. In: Advances in Intelligent Systems and Computing, pp. 655–668. Springer, Singapore (2021)
8. Ki Chan, C.C., Chen, C.C.: Continuous real-time automated attendance system using robust C2D-CNN. In: 202020 3rd IEEE International Conference on Knowledge Innovation and Invention (ICKII). IEEE (2020)
9. Kumar, N., Madhavan, S.: Incremental weighted linear discriminant analysis for face recognition. In: Lecture Notes in Electrical Engineering, pp. 677–687. Springer, Singapore (2021)
10. Moshin Reza, S., Mahfujur Rahman, M., Parvez, H., Badreddin, O., Al Mamun, S.: Performance analysis of machine learning approaches in software complexity prediction. In: Advances in Intelligent Systems and Computing, pp. 27–39. Springer, Singapore (2021)
11. Qi, A., Wei, J., Bai, B.: Research on deep learning expression recognition algorithm based on multi-model fusion. In: 2019 International Conference on Machine Learning, Big Data and Business Intelligence (MLBDBI). IEEE (2019)
12. Rahman, M.M., Mamun, S.A., Kaiser, M.S., Islam, M.S., Rahman, M.A.: Cascade classification of face liveliness detection using heart beat measurement. In: Advances in Intelligent Systems and Computing, pp. 581–590. Springer, Singapore (2021)
13. Rathod, H., Ware, Y., Sane, S., Raulo, S., Pakhare, V., Rizvi, I.A.: Automated attendance system using machine learning approach. In: 2017 International Conference on Nascent Technologies in Engineering (ICNTE). IEEE (2017)
14. Tabassum, T., Tasnim, N., Nizam, N., Al Mamun, S.: Anonymous person tracking across multiple camera using color histogram and body pose estimation. In: Advances in Intelligent Systems and Computing, pp. 639–648. Springer, Singapore (2021)

Cognitive Science and Computational Biology

Cognitive Engineering for AI: An Octave Drawing Test for Building a Mathematical Structure of a Subconscious Mind

Anindya Pattanayak, Tanusree Dutta, Piyush Pranjal, Pushpendra Singh, Pathik Sahoo, Soumya Sarkar, and Anirban Bandyopadhyay

Abstract Since the 1920s, drawing tests have been used to measure intelligence, maturity, and personality: in the later parts of the twentieth century, drawing tests were extended to estimate depression, schizophrenia, and the degree of Alzheimer's. However, despite extensive literature on cognitive engineering, there is no grammar as generic rules for explaining drawing tests. Most inferences are drawn based on experiences, intuitions, and medical test confirmations. Here, we propose a set of eight tests to correlate fundamental rules of subconscious minds of a human subject follows based on which it constructs logical constructs for solving a problem. Based on the studies carried out on human subjects, we have proposed invariants or conservations across human subjects to interpret the drawing results and build technologies. Thus far, cognitive technologies of subconscious minds were limited to explore criminal minds. Our test protocol reveals generic rules of subconscious minds for engineering cognitive AI machines.

Keywords Cognitive technology · Subconscious mind · Artificial intelligence · Time crystal · Sociology

---

The original version of this chapter was revised: The author's "Piyush Pranjal" affiliation has been updated. The correction to this chapter is available at https://doi.org/10.1007/978-981-16-7597-3_46

A. Pattanayak · T. Dutta
Organizational Behavior and Human Resource Management, Indian Institute of Management, Ranchi 834008, India

P. Singh · P. Sahoo · A. Bandyopadhyay (✉)
International Center for Materials and Nanoarchitectronics (WPI-MANA) and Research Center for Advanced Measurement and Characterization (RCAMC), National Institute for Materials Science (NIMS), 1-2-1 Sengen, Tsukuba, Japan

P. Pranjal
Management Department, O. P. Jindal Global University, Sonipat 131001, India

S. Sarkar
Marketing Management, Indian Institute of Management, Ranchi 834008, India

M. S. Kaiser et al. (eds.), *Proceedings of the Third International Conference on Trends in Computational and Cognitive Engineering*, Lecture Notes in Networks and Systems 348, https://doi.org/10.1007/978-981-16-7597-3_11

1 Introduction

Identifying the psychological attributes for decision-making is an open challenge in cognitive engineering. There is no known principle for brain functioning nor does we know the information-processing structure of the brain. We need to build a structure of the inner mind for cognitive engineering that eventually takes all decisions. Drawing tests are one of the trickiest routes to unveil mysteries of cognitive processes in the human brain. Tree drawing test reveals subconscious minds of generic depression [1, 2]. While human figure drawings have been used as a projective measure of intelligence [3] since 1926 [4]. Cognitive ability is proportional to the complexity and details of the drawings [5]. Finish the drawing from hints reveals personality. There are enormous versions of personality tests starting from interpretation of complex patterns, choosing favored ones from multiple choices to transform random lines into a picture. However, for cognitive engineering (CE), one of the primary requirements is to understand the mysteries of brain functions [6, 7] only then the techniques from cognitive psychology could be used to the design of human–machine systems [8].

Moment-to-moment decision-making to respond efficiently to complex systems is governed by situation awareness, which depends on the mental representation of the environment around the subject at any given time [9]. The subconscious mind conceptualized by Freud (1893) [10] that works beneath focal awareness is mostly involved in constructing the mental presentation of the environment. However, studies on the subconscious mind have not progressed due to the lack of falsifiability and testability of the claims. However, the quest for representative symbols went on [11]. Symbolism culminated into the autonomous microdynamics of the trance state of mind that builds the structure of the subconscious mind as suggested by Erickson [12]. Structuring the subconscious mind by metaphors and symbols requires key premises of its operational dynamics. Based on the Jung–Freud debate that Erikson and others advanced arguments, the subconscious mind has eight fundamental features. First, information structure is an image; second, it's always active and autonomous; third, emotion delivers its dynamics. Fourth, desire for attraction creates its events; five, past present and future superpose; sixth, confusion is a key constituent driving force for microdynamics of the subconscious mind, seventh, fear, reproduction and food creates feedforward cycle; eighth, suppression of desire, emotion fear reproduction desire, and food inflates the occupied domain of subconscious in the mental structure.

Structuring the subconscious mind was also attempted by Hindu philosophers, eliminating the mathematical part, we are interested in here. Hindu philosophers consider that every event in the universe is cyclic (Chhanda = rhythm [13]), namely Rita; and the smallest pixel that constitutes a cycle is also a cycle or Rita [14]; Fig. 1a. Thus, there is a within and above the network of cycles (layer = Loka). The smallest cycle of the universe is a clock, represented by a mechanical vibration called Brahman and the composition of cycles builds light and all other forms of waves (Asat) and matter (Sat). The entire universe is a large clock or cycle on whose perimeter other clocks or cycles reside, namely Purush, it is a life form called Virata Purusha [15], the guest clock structures that interact among themselves are called

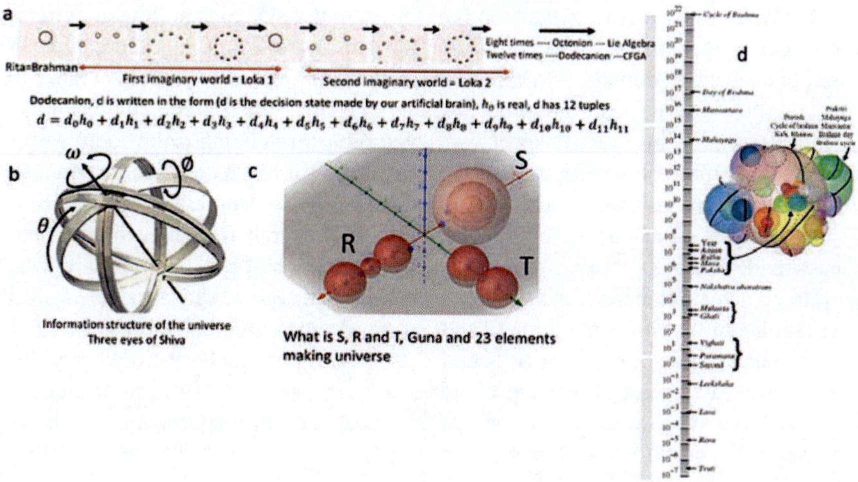

Fig. 1 *Introduction to the foundation of subconscious minds using time crystal*: **a** A pixel in a circle is a circle (Circle = Clock = Rita), this concept is shown by zooming. Zoomed 8 times we get octonions, if zoomed 12 times we get dodecanions. Twelve tuples of a dodecanion is shown in a 12D vector expression. **b** Three eyes of Shiva, θ, $\varnothing$, ω, together builds units of information for consciousness. We connect it with geometric phase triplet of vortices responsible for decision-making in a neural network. **c** Concentric spheres making concentric clocks represent S or Satva, Two large clocks connected by a very small clock is Transformation or Rajah or R; two spheres connected by tama, T. **d** Crystalization of clocks or synthesis of Kala and Kali conjecture from the Hindu units of time. Smallest unit of time is 0.1 microseconds or 10 ns or truti and longest unit is 10^22 s or one brahma cycle. Four primary domains are shown (equivalent to four faces of Brahma), all units of times are put together in an integrated geometry. Vertical projection of this crystal is used to build Mandala. This is an empty information structure or pot (aadhar) that holds information

Prakriti. Mind is a 3D clock assembly (Fig. 1b, c), a class of Prakriti, conscious mind exchanges clocks with the other Prakriti, or guests on the Purusha or clocks. The subconscious mind holds fundamental symmetries of Purusha–Prakriti conjecture [16]. These developments completed the pre-Vedic era, before 2000 BC and debates continued finding the maximum number of symmetries that constitute the clocks, identifying basic spatio-temporal patterns and rhythms that regulate the integration of clocks [17].

At this stage, post 2000 BC, when basic constructs were built, the advanced mathematics and concept of elements poured in, becoming the primary form of millions of literature and debates.

1.1 Structure and Model for a Subconscious Mind

We have used thousands of years old mathematical structure for the units of times to find the integration of clocks from 0.1 microseconds (truti) to 10^{22} years [17–20];

Fig. 1d. We have noted the critical temporal folds in the 3D clock architecture to find conservation laws used by scholars to find the universe's temporal architecture. We have put each clock domain on a sphere (to make it complete = purna) to create a 3D clock architecture (Raga or dynamic Clock or Prakriti + Ragini or static or Purusha [21]). The second important aspect for developing a mathematical structure for a subconscious mind is developing clock architecture for the 23 elementary qualities (unit of quality = tanmatra, Guna = quality [22]; Fig. 1c. There are three qualities, T = bonding of clocks or periodic events (Tama), R = transformation of geometric shapes made of clocks or periodic events (Raja), and S = reflection from infinity when infinite geometric transformations of clocks (Kala and Kali or Bhairavi; Satva). Kali = crystallization of universal time [23]; ten fundamentals of clock architecture [24]. Continuously, three parameters S, R and T are also represented by clocks, and they change their diameters. The state of mind is a composition of different elements made by the composition of S, R, and T. Origin of Kali or crystallization of time or clocks or bits and pieces of Rita.

1.2 Our Background Studies in Neuroscience and Mathematics

We have developed integrated clock architecture by measuring resonant oscillation frequencies of neuron, microtubule, etc. [25], and discovering a fractal network of clocks. There are many hidden circuits in the brain, each in a particular time domain [26, 27]. Based on these findings, we have already created a self-operating mathematical universe, SOMU model for understanding the human brain [28, 29]; eventually developing a complete time crystal model of the human brain [30]. Thus, Kala and Kali's concept is the crystallization of clocks into one unified architecture of the whole universe, mostly applied in astrophysics and consciousness of the human mind. At the same time, our research work has strictly been neuroscience. However, our output of human brain hardware is similar to the form argued for thousands of years. Now we have to functionalize this particular time crystal.

Here using SRT features, we have looked into the patterns created by human subjects and found geometric similarity, symmetry or asymmetry, integration, fractality, convergence or divergence, engagement or disengagement, complementarity, boldness or cleanliness, hint for a dynamic, sensitivity to intricate details as revealed from the shapes (Fig. 2a). Then, we created 3D clock assembly, which delivers the functional map of a human subject's subconscious mind (Fig. 2b).

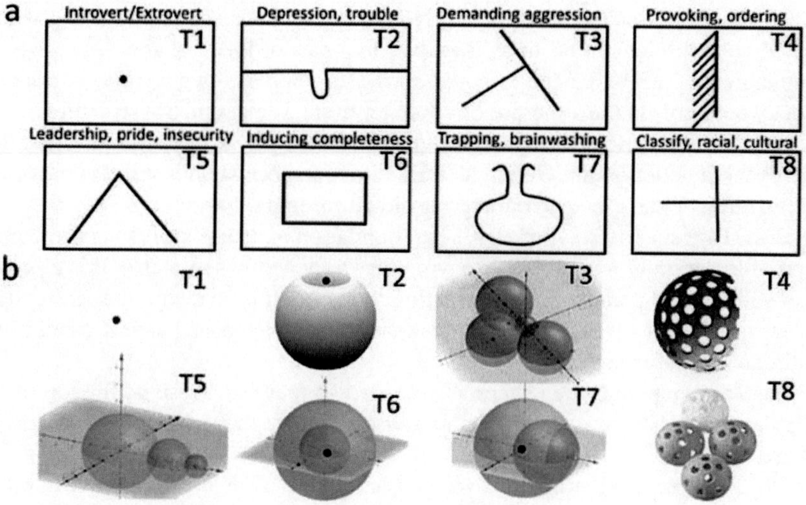

Fig. 2 **a** Octave test given to human subjects and **b** equivalent clocks spheres that possibly generate in the mind of a human subject when he/she looks at the image

2 Material and Methods

2.1 Experiment on Human Subjects

Consequently, we derive eight seed geometries to probe a subconscious mind as shown in Fig. 2. Here, the human subjects are asked to draw whatever they like including the seed pattern, considering it as an integral part of whatever they draw. Eight seed geometries are integrated as part of unique, distinct patterns by the subject. We analyze the patterns for subtle responses as we do not allow subjects to think a lot and change their drawing pattern. We do not care about drawing, we suggest subjects not to draw nice picture, only the minimum number of lines rapidly that they think would give them comfort. The test cannot be done twice. Only the first attempt, because it has been argued that every test would synchronize the Purusha with the Prakriti.

First, T1, a point is to be given to a human subject and he would be asked to draw whatever. It is defined that the universe is embedded inside a point following Vedic philosophy, so a subconscious mind always seeks a point in the emptiness. The human subject's subconscious mind is trapped in a single point, seeing an empty world. Thus, we check whether the subject's mind requires a line to start from the mind or a single point to connect to the external world or surrounds the point with a loop. The communication modes with the inner mind with the external world are revealed.

Second, T2, a U-shaped pattern, is given to the subconscious mind along with a plane. A subconscious mind finds it at the inner part of the U-shape. It has a hope to come out or create a bridge to cover up. Based on these two basic constructs, a subject would unfold the structure of the inner mind as a geometric structure.

Third, T3, a 45° tilted T, is sent as a query to the subconscious mind. The mind acquires the junction point. Given the trail, either a subconscious mind is provoked to break through and create symmetry or hide it through.

Fourth, T4, generating symmetry is the foundation of nature's working principles. We provide the mind with a simple asymmetry to generate symmetry. It is a provocation with several parallel lines suggesting the same. It is an arrow, has a direction. It filters minds that love to follow, adjust, and always oppose. Three distinct kinds of attitude responses are provoked.

Fifth, T5, a mountain-like shape is provided so that mind rests at the top of the triangular shape. The mind is fed with a sense of pride, a fear, testing the quality of a leader.

Sixth, T6, a rectangular box is given. It is complete satisfaction, happiness, either to be inside or use it to explore the unknown. Therefore, the mind has two choices, either explore inside or adventure outside.

Seventh, T7, is a pot with deep trouble, difficult to come out. A mind entered inside finds a circular barrier, it is frustrating to destroy confidence, and then, we monitor the path how a mind gets over it.

Eight, T8, it is a confusing structure. It is complete, asks nothing, mind harvests noise, so we test the response of completion. A free flow of the subconscious mind.

Declaration on Human Subjects: The local ethics committee approved the study (Indian Institute of Management, IIM Ranchi). It was performed following the ethical standards in the Declaration of Helsinki (last modified, 2004). All 89 subjects were above 18 years and written informed consent was taken from all of them.

Human Subjects: For the experiment, 20 people aged between a maximum of 37 years and a minimum of 21 years were selected, all of whom were healthy, their body temperatures around 98 deg F, initial pulse rate between 78 and 132, brought down to 80–85; blood pressure 85–130; and the oxygen level in their blood between 95 and 97%. For matching the subjects' circadian rhythm, all experiments were performed in the evening, after the students were done with their classes for the day. Subjects with high fever, neurological disorders, major diseases in the past, or any forms of abnormal responses were rejected.

2.2 *Temporal Restrictions and Dim Light in the Room*

All subjects were asked to draw anything. They were specifically mentioned that we are not interested in nice drawing, they should put some lines only to their satisfaction at the earliest. Most subjects completed the tasks within 5 min. A few subjects took up to 10 min. Therefore, we made an effort to acquire the first response of the mind without using significant memory and cognitive analysis. The dim light ensures less optical interference.

3 Results and Discussion

We have analyzed here ten subjects in detail, as shown in Fig. 3.

Subject 1, S1 had spiraled out on the first test, a conservative mind that slowly socializes, uncovering the self slowly in an organized manner. In the second test, the S1 does not want to transform a problem, resolve the crisis. Rather, digs deeper into it to go with the wind. By creating multi-storied houses on the flatlands, the subject has proved that inflating a problem from a smaller U to a larger U could be an easy choice. In the third test, the subject confirms the lack of killer instinct. It wants to live within the boundaries given, so it created a flag and remained within one of the three sides that could be chosen. The subject is not interested in generating symmetry. It sees asymmetry as a mirror and creates a reflection on the other side, adopting an alternative route to generate symmetry. The very obvious solutions are avoided in all tests thus far. In the fifth test, the subject has decided to remain within

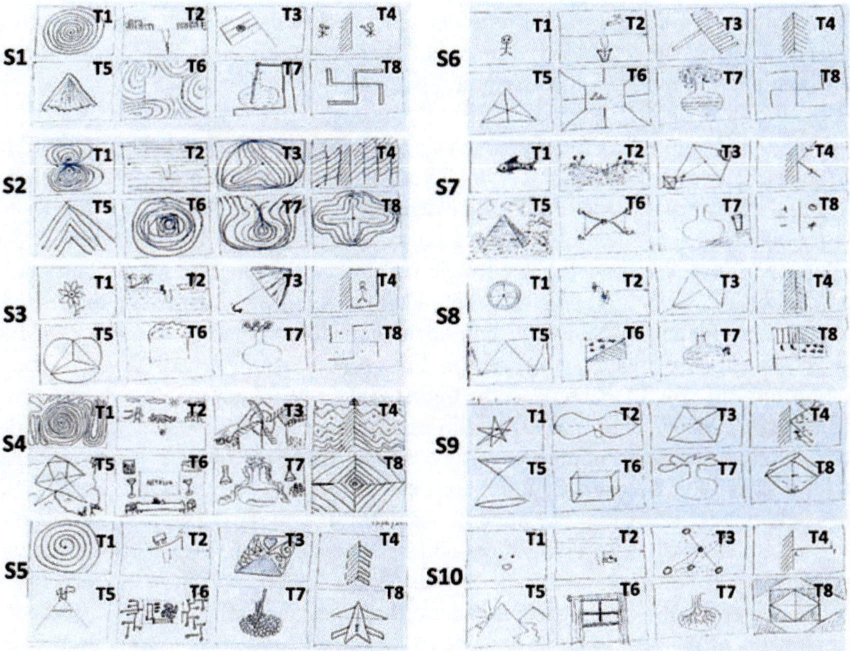

Fig. 3 *Ten selected results for human subjects S1–S10*. Ten selected subjects drew response to the query made in Fig. 2. Eight tests are numbered T1–T8

the mountain, not adopting leadership challenges. Rather, confine himself within the territory. In the sixth test, when we check expression at happiness, the subject finds external subconscious minds or points and expects them to be like him and slowly build a relation meeting his expectations. In the seventh test, claustrophobia triggers coming out of the bottleneck in the simplest possible way. The subject may lose balance as it perceives running out of all but one option. In the eighth test, the mind should be at the center of + sign, but, the insecurity of residing on the edges of four points drives encircling the whole area. The swastika is the outcome.

Subject 2 or S2 is a classic case of an insecure mind. In the first test T1, S2 divides the world into two parts only to create a cover around itself. However, it does not dare to cover the single point fully, as it wants to remain in contact with the root and then unfold to the world. In the second test, S2 ignores the existence of a deep trap and wants to create several parallel lines to create an artificial sense that the deep trap does not exist. S2 continues its introvert and compromising attitude when in T3 it was provoked for showing aggression. Test three has one junction and three endpoints, S2 makes an effort to nullify the risky triplets and enclose everything inside a nearly circular structure. In the fourth test, the subconscious mind is provoked for generating a complementary symmetry. Instead, S2 has replicated the query, repeating the query multiple times, covering the entire space. The subject does not respond, interact, engage, or analyze. It blindly replicates. A similar trend is followed in T5, which replicates a mountain. The sixth test reveals a very interesting aspect. While S2 is a blind replicator of the questioner, it understands a closed-loop and responds with a circle. For T7, S2 did not try to come out of the pot, merely built nested enclosures, then continued to replicate the problem.

The third subject S3 responds to first test T1 in two folds. First, it creates a cover around its true nature that it defines as a single point given as a test. Then S3 builds symmetric networking with the external world, a feedforward network between the inner and the external world is reflected in a petal. In the second test T2, S3 crawls out of the hollow dip and, after that, creates a world of his own on the flat surface. In T3, S3 understands the necessity to break through. His mind at the junction senses it properly and creates a cover that shows that the edge points are included in the aggressive move against the barrier. S3 adopts unique methods for generating symmetry for T4. The asymmetry we created using parallel lines is resolved by creating a box, where S3 puts himself. For T5, S3 repeats the trend to integrate all edges into one structure. Moreover, S3 has created a central point as a hierarchical observer of the network. For T6, S3 senses the completeness; a rectangle inside a rectangle is the ultimate match and delivers a flourishing, unbalanced expression of celebration. For T7, S3 does not prefer to be inside. The approach is to resolve directly. The path coming out of the pot is decorated, reaching there without tracing a path. For T8, S3 has made an effort to build an enclosure made of four edges.

The subject 4, S4, is too careful in unfolding the inner self in T1 as differences in diameters of concentric circles are very low. Surprisingly, frustration inflates and saturates to such an extent that S4 creates two new points. The creation of multiple identities reflects insecurities. For the T2, S4 confirms its existence deep inside the shallow tube by drawing a point and enlarging it. Then creates a bridge to connect both sides. Always, excessive carefulness prompted S4 to construct a bridge made of circular arcs. For T3, S4 understood the imbalance and his need to be aggressive and complete the cross (+) by breaking through. By constructing a dynamic rotor, S4 has made an effort to encompass the edge or endpoint of lines. Note that edge or endpoints are split parts of the "self", creating multiple existences and conflicts. S4 has perfectly responded to T4, where it has generated the symmetry, removed imbalance, and replicated the evolved mountain everywhere, covering the entire space. Comparatively, T5 shows a unique effort to generate symmetry and balance out the leadership query we made. S4 drew the mirror inverse of the pattern, resulting in a unified structure, to which S4 made an effort to deliver dynamics. Generating symmetry and encoding dynamic movements are two key features of the S4. However, T6 shows that S4 keeps completeness or satisfaction as an isolated symbolic structure that is multidimensional in expression, and all are distinct. Creating conceptual replicas as a mode to respond to queries is fundamental to S4. For T7, we find just that. While acknowledging the deep trouble and finding an exclusive artistic way-out is deep-rooted in S4, he replicates the query as a mark of celebration. Expression and campaigning for achievements are inevitable. For T8, connecting the edges holding his split "self" states is fundamental to him. Possibly integrating all identities into one is deep-rooted.

Fifth subject, S5 is an introvert that spirals out to the external world, as seen in the test 1, T1. For T2, we find that S5 has discovered the most difficult solution, S5 detects the problem, analyzes it properly, but over effort or amplifying the problem, complicating the environment for solving it seems trademark of S5. However, in T3, S5 accepts the challenge, shows aggression, resolves the junction barrier, and creates a closed conservative loop with mirror symmetry. The affection for developing a hierarchical symmetry is profound in T4. We see that S5 has created a balance by regenerating the parallel lines and coupled the parallel lines in pairs. In the leadership test of T5, S5 secures the top position providing a base and creates environmental protection to the top point. T6 repeats T2. In T6, S5 has decorated happiness with complex pathways, trying out options randomly while finding introspection patterns inside. T7 response is similar to T2, complicating the problem and finding the most difficult solution possible in a given scenario. T5 knows what to do, but it inflates the problem quantitatively (T7), trying the most difficult solution (T2) and manipulating the structure to inflate only one aspect or direction of thoughts (T8).

The sixth subject S6 has a dual perspective. Seeing one point, S1 creates another one in T1. Paired point creation requires resolving loneliness, the insecurity of starting an adventure alone. However, as soon as S6 creates two points side by side, it covers it with a loop, which suggests working together with a virtual singular identity. For T2, S6 accepts it as absolute truth. Using it as an object metaphor becomes S6 priority. Not by bridging the obstacle but by using it as the positive instrument that could deliver life-like functions becomes a sole target. The use of obstacles keeps the problem as is and represents the problem as the outcome or solution of another problem. Compromising weakness? The answer lies in the T3 for S6 where the junction is extended with many parallel lines. Therefore, it is not about weakness. The aggressive generation is evident in T3 and T4, possibly hidden in T2. For T5,

S6 left the top of the mountain and created another junction point inside. Once again, confining himself within a protected enclosure reflects signs of a conservative structure while amplifying appreciation of completion it connects with environment and self as shown in T6. Then finds himself inside. S6 blooms perfectly, reaching out to global players enriching from within in T7. S6 makes an effort to dynamically arrange four edge points or four self-identities in order to integrate them.

The subject S7 is extremely particular about the absoluteness of its identity in T1. It is never about interacting with the external world but generating an absolute identity with bold lines to follow. The absoluteness of self reflects more profoundly when S7 disgraces the hole by chopping off both sides to create a smooth downfall towards the problematic hole in T2. The decorating tendency without addressing the real problem remains there even in T3 when the pattern takes the mind to the junction, trapping there while making an effort to split it into three parts. The subject is expected to generate the symmetry by converting it into a + structure and avoiding that S7 has covered it up. The missing imbalance has been regenerated with a tiny replica at the end. S7 responds to another provocative asymmetry in T4 with uncompromising creative symmetry of reflection. A robust attitude of ignoring the given asymmetry and regenerating symmetry with complementary patterns has been a foundation of S7, and we see it in T5. S7 creates a point, an identity deep inside when completeness and a trigger of satisfaction is provided in T6. S7 does not sense the rectangle inside a rectangle due to the large hysteresis of previous tests. On the contrary, S7 finds the setup as a shrinking rectangle. In T7, S7 does not make any contact with the pot, rather finds a buffer medium to interact. Reluctance to engagement is a great obstacle for S7 as it does not engage with T8 and consider pocketing by + as resting sites.

The eighth subject S8 creates an inner world and encodes various dynamics of its core values within as shown in the T1. The effort to come out of trouble is hope or optimism that the problem will be solved via some mysterious power, as shown in T2. S8 understands the totality of the situation. However, it does not try to solve the problem. Rather, it creates a totality review even when an aggressive symmetry generation is demanded is shown in T3. Isolating self and create alternative, independent existence is one of the foundations of S8, and it is reflected in T4 when an independent isolated line is generated. For T5, S8 has provided amplification of opportunities, minimizing the typical prize given to S8 by placing him at the top of a mountain. The rectangle of T6 is not sensed globally or locally due to a lack of seeing the totality under complete satisfaction. Therefore, S8 makes an effort to see the local totality. However, if there is a global symmetry, it prefers to isolate itself under two triangles or the smallest arm-covered areas. In T7, we find optimism without liberation. Hope is there, with imagination but not aggressive. In T8, S8 has once again decided to accept the rooms given, neither to generate dynamics nor to play with it, rather accept the fate and live with it as gifted.

The ninth subject, S9, is ambitious in expanding pride and fame boldest possible ways, as seen in T1. In T2, S9 has received the hint of trouble. Rather than directly engaging with the problem, it has made an effort to depict it as a universal generic problem and then tried to create a situation is if the problem does not exist. In T3, S9 has approached aggressively and resolved the symmetry, finally covered the entire area building an integrated architecture. In T4, the essence of parallel asymmetry is sensed by S9. He creates parallel lines without directly engaging with the problem geometry. S9 has created a mirror symmetry of the pyramid in T5. It is securing the top position as an effort to justify reaching the top position. A similar sense of consolidation is reflected in T6 where S9 has solidified the happiness or sense of complete satisfaction. In T7, S9 has flourished from within, merely a wish or dream as he does not show any path. In T8, we see another consolidation without adding dynamics, which suggests a fear of losing that repeated over multiple tests.

The tenth subject S10 loves duality similar to S6, creates an alternative self by adding a new point and pairing them as an integrated unit for interacting. However, S10 is different because, without detailed engagement with self, the S6 has responded interacting with the environment as an integrated one. In T2, S6 has engaged with the problem; however, it bridged it carefully with a distinction from the plane of problem. In T3, direct engagement with the problem aggressively broke through the barrier to creating symmetry, encompassed the entire system to integrate split selves, and identified edge points where insecurities were encoded, added circles to identify them. S10 avoids forced asymmetry of T4 only to keep a distance. For S10, insecurity of leadership is visible as S10 acquires alternative parallels to work hand in hand. S10 understands nested rectangles of spiritual completeness as he responded by creating more similar geometric shapes. S10 has created branches inside the pot of T7 only to engage within the problem deep, addressing every single concern more intricately. In T8, S10 responded by creating a hierarchical network of complementary symmetry; when a symmetric environment is provided, S10 enjoyed deeply flourishing whatever was received as a gift without creating new dynamics.

As observed above, we do not engage with what subjects want to paint or their objective. We only notice the geometric similarity, symmetry or asymmetry, integration, fractality, convergence or divergence, engagement or disengagement, complementarity, boldness or cleanliness, hint for dynamics, sensitivity to intricate details as revealed from the shape. Thus, we concentrate on the areas and how sections of areas were created. From these sections or loops, we build the spatiotemporal architecture. 3D clock assembly is generated for each response, and we have a generic physical significance associated with it. The differential deviation is the mathematical architecture of the subject's subconscious mind.

4 Conclusion and Mathematical Model for Cognitive Engineering

Analyzing the ten subjects above could vary only when we assign significance to patterns. However, we do not do that. Several psychological tests assign basic meaning to a geometric shape. Here, we have only one significance. A point represents the self, the test subject's existence. During analysis, we restrict ourselves to a single assumption that the mind is trapped at a point and test patterns are designed

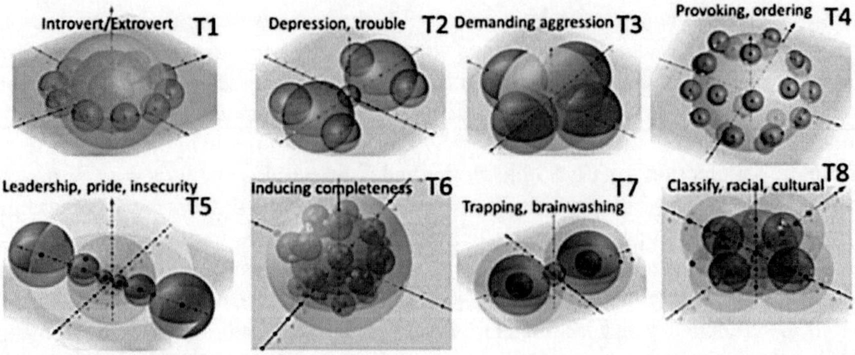

Fig. 4 *Time crystal structure or invariants of a generic expression of a subconscious mind*: The crystallization fundamentals of a subconscious mind is shown here in terms of SRT parameters. T = bonding of clocks or spheres. R = Transformations of clocks or spheres and S = Projection to Infinity and feedback from infinity. T1 shows many concentric clocks growing one above another. And side by side there are qualities or gunas all around the external sphere. T2 shows a generic tensor structure for two different worlds, one for the deep trouble and the other for the solution. Two different modes of transformations are adopted for both the worlds. T3 shows integration of four worlds using a hypothetical imaginary central sphere is the fundamental structure how mind responds to a challenge. T4 shows the drive to create 2D surface of crystals on the surface of a sphere. T5 shows mirroring of a nest of a set of clocks, and then a set of concentric spheres or clocks covering them. T6 shows growth of 2D surface made of clocks between two concentric spheres or clocks. T7 shows growth of concentric clocks in two distinct centers. T8 shows concentric growth of clocks in four points

to trap the subconscious mind at junctions, peaks, edges, and allow the mind to explore the empty space with respect to the trapped mind. Thus, analysis, as it may appear subjective, turns quantitative, and here we have outlined the grammar and mathematical estimation of different mental states and emotions as shown in Fig. 4. Decision-making is often governed by emotion, desire, and ignoring logic. Thus, current AI technologies artificially induce noise to replicate the brain. There is no mathematical protocol to estimate and induce effects for emotional states, attitudes, degree of fear, love, hate, and essential parameters that are part of the subconscious mind that regulates decision-making. Here in Fig. 4, we have created a model for the emotional states using mathematical formulae. One could configure functional states and build complex theories by editing them. We begin the first journey to read the subconscious mind, model it using quantitative parameters. We do not claim that our model is an absolute truth, it is primitive yet the first effort to develop an organized architecture of a subjective mind. Instead of using noise to create artifacts in decisions so that AI appears like a human, one could use our model that enables tuning parameters in an organized manner, verify protocols, and improving formulae by systematic experiments.

Acknowledgements We thank Dave Sonntag and Martin Timms for the independent test and verification of our device as part of patent US9019685B2. Authors acknowledge the Asian office of

Aerospace R&D (AOARD), a part of the United States Air Force (USAF), for the Grant no. FA2386-16-1-0003 (2016–2019) on the electromagnetic resonance-based communication and intelligence of biomaterials. Authors also acknowledge the financial assistance of Scheme for Promotion of Academic and Research Collaboration (SPARC) an MHRD, Govt. of India initiative for the project titled "Management of Fractal Time in High-level Decision Making" (Govt of India, MHRD; project number P 524; Start date: 15.03. 2019–14.03.2021; Duration: 2 years).

Contributions A.B. and T.D. conceptualized the idea, A. P.; P. P.; P. Sa.; and P. Si. did the experiment, A. B. analyzed the data and wrote the paper with A.P. and P. Si. S.S. verified experimental protocol and helped in human subject study, M.K.M. reviewed the validity of the data.

Competing Interests The authors declare that there is no competing interest.

References

1. Gu, S., Liu, Y., Liang, F., Feng, R., Li, Y., Liu, G., Gao, M., Liu, W., Wang, F., Huang, J.H.: Screening depressive disorders with tree-drawing test. Front Psychol. **11**, 1446 (2020)
2. Kaneda, A., Yasui-Furukori, N., Saito, M., Sugawara, N., Nakagami, T., Furukori, H., Kaneko, S.: Characteristics of the tree-drawing test in chronic schizophrenia. Psychiatry Clin. Neurosci. **64**(2), 141–148 (2010)
3. Imuta, K., Scarf, D., Pharo, H., Hayne, H.: Drawing a close to the use of human figure drawings as a projective measure of intelligence. PLoS ONE **8**(3), e58991 (2013)
4. Goodenough, F.: Measurement of intelligence by drawings. Chicago (1926)
5. Harris, D.B.: Children's drawings as measures of intellectual maturity. New York (1963)
6. Norman, D. A.: Steps towards a cognitive engineering (Tech. Report). San Diego: University of California, Program in Cognitive Science (1981)
7. Hollnagel, E., Woods, D.D.: Cognitive systems engineering: New wine in new bottles. Int. J. Man Mach. Stud. **18**, 583–591 (1983)
8. Woods, D.D., Roth, E.M.: Cognitive engineering: human problem solving with tools. Hum. Factors **30**, 415–430 (1988)
9. Endsley, M. R.: A construct and its measurement: the functioning and evaluation of pilot situation awareness (No. NOR DOC 88-30). Hawthorne, CA: Northrop Corporation (1988)
10. Freud, S., Miss Lucy, R.: Case histories from studies on Hysteria. The standard edition of the complete psychological works of Sigmund Freud **2**, 106–124 (1893)
11. Jung, C.: Man and his symbols. Approaching the unconscious. Doubleday. p. 37 (1964)
12. Rossi, E.: The collected works of Milton H. Erickson. The Milton H. Erickson Foundation Press, pp. 229–260 (2008)
13. https://religion.wikia.org/wiki/Chandogya_Upanishad
14. James, E.O.: Creation and cosmology: a historical and comparative inquiry. In: Leiden, E.J. Brill (1969)
15. Visvanathan, M.: Cosmology and critique: charting a history of the Purusha Sukta. In: Roy, K. (ed.) Insights and Interventions: Essays in Honour of Uma Chakravarti, pp. 143–168 (2011).
16. Larson, G.J.: Classical sāṃkhya: an interpretation of its history and meaning, Motilal Banarsidass (2011)
17. Gupta, S.V.: Ch. 1.2.4 Time Measurements. In: Hull, R., Osgood, Jr., Richard M., Parisi, J., Warlimont, H. (eds.). Units of Measurement: Past, Present and Future. International System of Units. Springer Series in Materials Science. Springer (2010)
18. Jha, G.: Discourse I—origin of the work—creation of the world—summary of contents of the book. Manusmriti with the Commentary of Medhatithi. Ist edn, 1920. Wisdom Library. Adhyāya 1. Motilal Banarsidass Publishers Pvt. Ltd (1999).
19. https://en.wikipedia.org/wiki/Hindu_units_of_time

20. Klostermaier, K.: Time in Patañjali's Yogasūtra. Philos. East West **34**(2), 205–210 (1984)
21. Raffé, W.: Rāgas and Rāginīs: a key to Hindu aesthetics. J. Aesthet. Art Critic. **11**(2), 105–117 (1952)
22. Larson, G.J., Bhattacharya, R.S.: The encyclopedia of indian philosophies—samkhya, a dualist tradition in Indian philosophy. In: Potter, K.H. (ed.), vol. 4, pp. 65–66. Princeton University Press (2014)
23. Coburn, T.B. Devī-Māhātmya—crystallization of the goddess tradition. Motilal Banarsidass, Delhi (1984)
24. Kinsley, D.: Tantric visions of the divine feminine: the Ten Mahavidyas. University of California Press, Berkeley (1997)
25. Agrawal, L., et al.: Inventing atomic resolution scanning dielectric microscopy to see a single protein complex operation live at resonance in a neuron without touching or adulterating the cell. J. Integr. Neurosci. **15**(04), 435–462 (2016)
26. Singh, P., et al.: Electrophysiology using coaxial atom probe array: live imaging reveals hidden circuits of a hippocampal neural network. J. Neurophysiol. **125**(6), 2107–2116 (2021)
27. Singh, P., Sahoo, P., Saxena, K., Manna, J.S., Ray, K., Ghosh, S., Bandyopadhyay, A.: Cytoskeletal filaments deep inside a neuron are not silent: they regulate the precise timing of nerve spikes using a pair of vortices. Symmetry **13**, 821 (2021)
28. Singh, P. et al.: Quaternion octonion to dodecanion manifold: stereographic projections from infinity lead to a self-operating mathematical universe. In: Singh, P., Gupta, R.K., Ray, K., Bandyopadhyay, A. (eds.) Proceedings of International Conference on Trends in Computational and Cognitive Engi-neering 2020, Advances in Intelligent Systems and Computing, vol. 1169, pp. 55–77. Springer, Singapore (2020)
29. Singh, P. et al.: A Space-Time-Topology-Prime, stTS Metric for a Self-operating Mathematical Universe Uses Dodecanion Geometric Algebra of 2–20 D Complex Vectors. In: Ray, K., Roy, K.C., Toshniwal, S.K., Sharma, H., Bandyopadhyay, A. (eds.) Proceedings of International Conference on Data Science and Applications 2020, Lecture Notes in Networks and Systems, vol. 148, pp.1–31. Springer, Singapore (2020).
30. Singh, P., Saxena, K., Singhania, A., Sahoo, P., Ghosh, S., Chhajed, R., Ray, K., Fujita, D., Bandyopadhyay, A.: A self-operating time crystal model of the human brain: can we replace entire brain hardware with a 3D fractal architecture of clocks alone? Information **11**, 238 (2020)

A Hybrid CNN-LSTM-Based Emotional Status Determination using Physiological Signals

Nazmun Nahar, Ferdous Ara, Jubair Ahmed Junjun, Mohammad Shahadat Hossain, and Karl Andersson

Abstract Automatic real-time emotion recognition based on GSR and ECG signals becomes an effective computer-aided tool for emotional recognition as a challenge to pattern recognition. Traditional machine learning methods require the development and extraction of various features dependent on extensive domain knowledge. As a result, non-domain experts can find these methods challenging. On the other hand, deep learning methods have been widely used in several current studies to learn features and identify various types of data. In this paper, to characterize human emotion states, we proposed a hybrid neural network that combines 'Convolutional Neural Network (CNN)' and 'Long-Term Short-Term Memory (LSTM)'. Our dataset consists of four types of emotions which are happy, sad, fear, angry. We have trained our model with CNN-LSTM. Our proposed CNN-LSTM model gives 100% training accuracy and 99.05% validation accuracy with RMSProp optimizer. We also compare our result with machine learning algorithms: Random forest, Logistic Regression, Support Vector Machine, and Naïve Bayes. The comparison result clearly shows that our proposed CNN-LSTM gives the best result among the other classifiers.

Keywords Emotion Recognition · ECG · GSR · CNN-LSTM

N. Nahar · F. Ara · J. A. Junjun
BGC Trust University Bangladesh Bidyanagar, Chandanaish, Bangladesh
e-mail: nazmun@bgctub.ac.bd

M. S. Hossain (✉)
University of Chittagong, Chittagong, Bangladesh
e-mail: hossain_ms@cu.ac.bd

K. Andersson
Lulea University of Technology, SE-931 87 Skellefte â, Sweden
e-mail: Karl.andersson@ltu.se

M. S. Kaiser et al. (eds.), *Proceedings of the Third International Conference on Trends in Computational and Cognitive Engineering*, Lecture Notes in Networks and Systems 348, https://doi.org/10.1007/978-981-16-7597-3_12

1 Introduction

Emotion is an integral part of human everyday life. It represents the human emotions of things. The state of mental health also affects human interaction and decision-making. Patients' emotional states can be used as an indication of such functional emotional conditions, such as psychological disorders after trauma and severe depression. Recently, researchers used EEG signals to examine the emotional features of a mobile overuse community and a healthy group [19].

The facial expressions [6], speech [27], eye blinking [34], and physiological signals [39] are able to detect human emotions. However, the first three techniques are vulnerable to subjective effects of the participants, i.e., participants may hide their emotions on purpose. As physiological signals such as electroencephalogram (EEG), electrocardiogram (ECG), the pressure of the blood volume (BVP), Galvanic Skin Response (GSR) is automatically released by the human body.

Therefore, in capturing genuine emotional states of person, the physiological signal is more accurate and objective. Among these physiological signals, The ECG and GSR are specifically important in assessing various pathological and psycho-physiological disorders. The ECG is one of the most useful signals for assessing the electrical signals of the heart, and the GSR may give useful information for assessing the function of the sweet glands [31]. Easy, accurate, low-cost, non-invasive, and continuous recordings are provided by these methods. Automatic perception, on the other hand, is critical for identifying preferred patterns associated with distinct mental and physiological states.

For better recognition output, a multi-modal approach blends EEG, ECG, and GSR techniques. Deep-Neural Networks (DNN) for physiological cues are being used in new ways to compute emotion detections, with better detection speeds. Cheng et al. [10] used Wavelet Transform and Backpropagation Neural Networks (BPNN)-based emotion recognition from Electromyogram (EMG) signals to identify four emotions: excitement, sorrow, anger, and enjoyment. For an emotion elicitation mission, an early study suggested [43] Radial Basis Function (RBF) Neural Network. However, all methods used proprietary databases to compute emotions; the DEAP dataset, which is freely accessible, is also incorporated using DNN architectures. Long Short-Term Memory (LSTM) was used for EEG-based emotion elicitation [21] with recall scores of 72.06 percent and 74.12 percent for binary classification of valence and arousal, respectively, using the DEAP dataset. The arousal and valence spaces seem to be the most often used dimensions for expression identification since they characterize emotions in terms of their stimulation or non-activation and their positivity or negativity [13]. The feature vector dimensions are typically serving to describe different emotion information when fed to the machine learning algorithm that deals with and vector dimension. These methods are well researched in the art of classifying and predicting which features work best in this domain. This expands on the notion of simple emotions and characterizes a variety of subtle emotions that can be distinguished by two dimensions [40] .On the basis of the previous research, the following few points must be dealt with current work contributions. In literature,

ECG and GSR are less studied than EEG in spite of its higher emotional significance. A high-performance non-invasive emotional detection system needs to be developed using cost-effective wearable sensors for larger emotion classes. To understand and construct more stable neural network architectures, connectivity between spectral and deep-neural networks features have to be examined. ECG and GSR are high-memory continuous signals that could be used by LSTM for improved performance. The LSTM algorithm has the ability to keep patterns for a long time. The proposed architecture of 1D-CNN + LSTM is then employed with data from the ECG and GSR time series.

The remainder of the paper is divided into several parts. The literature review is presented in Sect. 2 and the methodology is presented in Sect. 3. Section 4 examines the outcomes of the experiments, Sect. 5 concludes and describes the future work.

2 Literature Review

Physiological data is gathered through the use of ECG and sensor oximetry components that have a significant influence on how an individual responds to stimuli from a human body. The data must be pre-processed before feature is extracted and classifiers must be used to gather specific features to facilitate the classification process. The usage of physiological cues has been shown to be more efficient for emotional recognition than either voice or body language. Emotional state; for example, ECG, GSR, and respiration rate (expiratory lab) was used to infer the various emotional states from the valence-anger, endocrine, and respiratory patterns. Time and frequency domain properties were differently derived, and a certain time or frequency is no longer enough to define [7, 41].

There are a wide variety of physiological cues that help us interpret the human emotional state. All of them aid in the study of emotion recognition. Signals typically used to calculate cardiac operation, temperature, vibration, and conductivity (i.e., electrical stimulation of the skin activity) are used. Images often contract the size of the emotional reaction of the human pupils, and therefore this can indicate emotions (especially arousal), however in order to be automatic, it needs to be examined and identified first [17, 18, 28].

Zhang and his collaborators suggest a nonlinear paradigm to capture the features of emotions utilizing multi-channel physiological signals. In the high-dimensional space, the data is isolated, and then plotted out in the space where the dimensions overlap. The methods that come from Kernel Principal Component Analysis (KPCA) were applied. In addition they used 'Facial EMG' and 'PPG' as a metabolic and electrical-coupling variable [44]. The investigators chose 29 subjects, who have different emotional problems, and the accuracy of four separate emotional classifications for those individuals came out to be 93.42 percent. There are some parameters that cannot be well modeled by the classifier: most notably efficiency of the classifiers for reducing physiological signals.

Siddharth et al. [32] experiments that used the DREAMER and AMIGOS datasets found a precision of 79.95% when judging binary arousal and valence with regard to DREAMER results. They registered overall accuracies of 83.94% and 82.76% in multi-modal integration in the AMIGOS dataset of physiological signals.

Several feature extractors such as logistic regression (LR), discriminative Graph regularized extreme learning machine (GELM), k nearest neighbors (KNN) were proposed for emotion classification on DEAP and SEED dataset [42]. KNN, LR, and GELM were proposed using video as stimuli for recognition of positive, negative, and neutral with the accuracy of 69.67% and 91.07% [45].

The DEAP 3D-CNN model was used for multi-channel emotion detection to boost the model's performance with an accuracy of around 86.44% and 88.49% with valence and arousal [29]. In order to categorize human emotions on the DEAP dataset using the hybridization of CNN and RNN, the system obtained an accuracy of 90.83 for the mean level and 91.03 for the maximum-entropy method [38].

The deep learning technique, which was discovered by the Santaria-Gado et al. [30] (AMIGOS), was successfully used to recognize AMIGOS dataset emotion. For the first datasets, 40 subjects (about 250 seconds), the experimenter showed subjects 16 video recordings; for the second, it required 17 people to explain the subject images, but only five groups of people each containing 4. These people were shown 4 videos (approx. 14 min) and the signals from the test subjects and data subjects were identified in the physiological experiments. Biology may reveal these signs as either the subject's overall level of arousal, the valence of that exists with it, or other sources of the above. It uses a CNN to pull out physiological features to forecast emotional condition through the input of additional data.

3 Methodology

In this part, the methodology used for this analysis is illustrated. Each component demonstrated how the study was conducted to determine the various emotions. The methodology used for this study is depicted in Fig. 1.

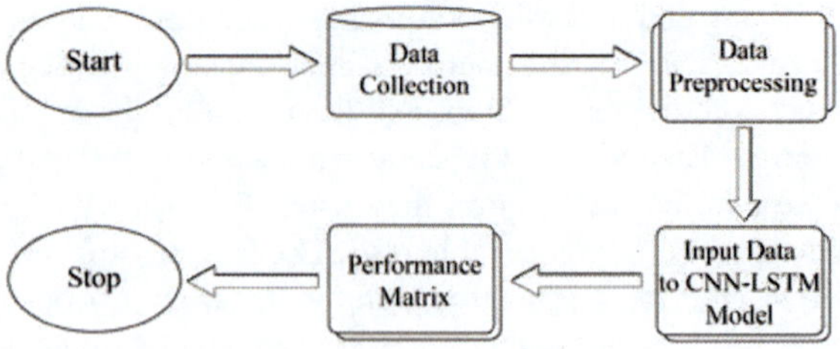

Fig. 1 Methodology steps

3.1 Dataset Preparation and Preprocessing

Data is obtained by using MySignals Healthcare Toolkit Galvanic Skin Response Sensor and Electrocardiogram Sensor. The Arduino Uno board and various sensor ports make up the MySignals toolkit. The sensors were attached to the hardware kit's various ports, which were operated by the Arduino SDK. An arrangement was made in a closed and silent space for the experiment, including the selection of subjects, the configurations of experiment equipment, the selection of emotion triggered recordings, and the design of the experiment scene, in order to obtain a quality database for physiological signals. In order to elicit different emotions, 17 videos were chosen: anger, happiness, sadness, and fear.

The following are the stages in the dataset preparation process:

- • At the beginning, the participant was told about the entire trial set-up and was briefed about the task it has to accomplish.
- • After confirming their understanding of the experimental procedure, we assisted them in properly wearing the sensors and confirming if the readings from the sensors were accurate or not.
- • The subjects must fill out a questionnaire with their personal information at the outset of the experiment.
- • The raw data was gathered and entered into a computer.
- • The raw data from the sensors contained some noise data that could not be processed, so it was cleaned before feature extraction.
- • With the support of regular expressions, noise in the data was observed and eliminated.
- • GSR raw data included time stamps for each reading, which were omitted because they were worthless.
- • The mean of the data for that particular video was also used to correct some missing readings.
- • The files were cleaned before being prepared for feature extraction.
- • A final integrated data set was prepared after a comprehensive data analysis.

Figure 2 shows the happy state of GSR data and Fig. 2 shows the happy and sad state of ECG data.

3.2 Convolutional Neural Network (CNN)

CNN is a deep learning algorithm, also known as ConvNet, that is mainly used for classifying images. CNN architecture is similar to the pattern of connectivity in the human brain and was influenced by the Visual Cortex. Convolution, max-pooling, full connection, and fully connected-Relu are some of the phases for classification of the dataset. Convolution is essential for feature extraction and data resizing after several stages.

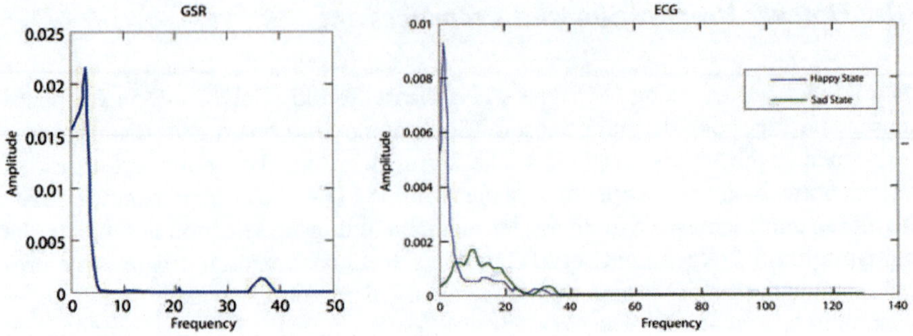

Fig. 2 GSR data happy state and ECG data happy and sad state

3.3 Long-Term Short-Term Memory

Long-Term Short-Term Memory (LSTM) is a variant of the deep learning artificial recurrent neural network (RNN). Unlike regular neural networks, LSTM has input connections. It can process whole data streams as well as individual data points (such as images). For example, non-segmented handwriting recognition, speech recognition, and the detection of anomalies in network traffic or emotion recognition are examples of LSTM. A standard LSTM unit includes a cell, an entry, an exit gate, and a door to forgetting. The cell has unpredictable time values, and the three gates control the information flows within and out of the cell. Based on time series data, the LSTM networks are suitable for classification, analysis, and forecasting because delay of unknown lengths. The breakdown and vanishing gradient problems that can be observed in traditional RNN have been addressed with LSTMs.

3.4 Proposed Hybrid CNN-LSTM Model

In contrast with the standard CNN, RNN helps create an association between the input series and thus gives a modern approach to hybrid features [12, 37]. Researchers have developed methods to integrate features by using LSTM (a version of the RNN), which can derive the long-term dependence of the data in order to increase the accuracy of the identification. In the function of this paper, a recent but similar approach has been established by using several convolutionary kernels to features extracted. Our technique comprises two phases, the first section of which is CNN-based feature extraction, and the second is an LSTM-based feature fusion part.

Our proposed CNN-LSTM model has been shown in Fig. 3. The proposed model consists of two 1D convolutional layers followed by a max-pooling layer then another 1D convolutional layer. After the third convolutional layer there is a LSTM layer followed by a dropout layer and at the end there is a dense layer or output layer. The

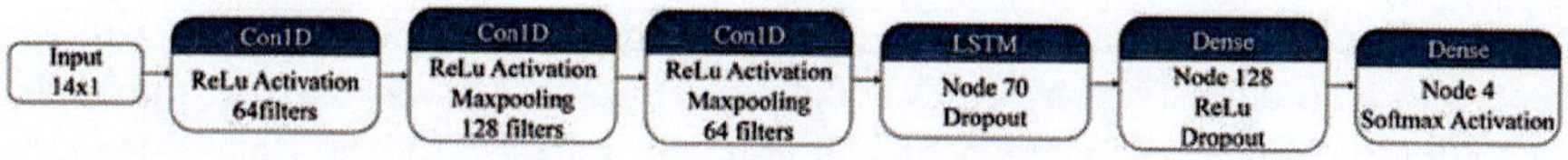

Fig. 3 Proposed CNN-LSTM model

output layer is accompanied by parameters for loss and accuracy, such as optimizer functions, learning rates, and matrices.

The first convolution layer comprises 64 filters with a kernel size of 3 by 3. Data are received from the input layer. In addition, the convolution layers have a Rectified Linear Unit activation function (ReLU). ReLU supports the model in decreasing the vanishing gradient problem [20]. A max-pooling and dropout layer is included after convolution layer. The feature map extracts the maximum features with a max-pooling layer. To save essential features in the feature map, this max-pooling layer fits the features into a two-by-two slot. Some convolution layer is dropped out by the dropout layer, which removes 25% of the nodes. Dropout layers reduce model loss as well as preventing from overfitting [35]. Then, the CNN model is combined with an LSTM layer. There are 70 nodes in the LSTM layers. Its objective is to interpret the features extracted by the CNN layer. Finally, the model is integrated in a layer of output. There are four nodes in the output layers. The four nodes relate to four different emotional categories. The output layer also includes a softmax activation function.

4 Result and Discussion

In this part, we'll discuss how the proposed model determines various emotions. We have classified emotion into four types which are happy, sad, fear, and angry. In this study the data was collected from the GSR and ECG. The dataset has been divided into two parts, each with an 80–20 percent train-test ratio. We used 80% of our data to train our CNN-LSTM model, and the remaining 20% was used to validate the model after it was trained.

4.1 System Configuration

For the implementation of CNN-LSTM, we have used the tensor flow enabled GPU [1]. A good GPU is needed for training on the CNN-LSTM model because the neural network is full of matrix multiplication operations. The proposed CNN-LSTM architecture is intended for training on the Google Colaboratory cloud server because of the computing power limitations of a CPU. The Google Colab's GPU and Jupyter Notebook environment is specifically developed to assist computer learners in overcoming the processing unit challenge. Google Colab was used to train and validate our proposed model. The PYTHON scikit-learn library [26] was used for the implementation of standard classifiers such as Random Forest, Logistic Regression, and Support Vector Machine.

4.2 Hyperparameters Tuning

Hyperparameters are needed because they have a direct impact on the model's actions. We have trained our proposed model with various optimizers like Adam, SGD, RMSProp, and Adamax because we want to see how our model behave with different optimizer. We have used 8 batch sizes and 200 epochs for training the model, and we also set the learning rate of 0.001. The categorical cross entropy loss function also measures the probability loss of the class expected by the softmax function.

4.3 Performance Matrix

In this study, we used Accuracy, Loss, and MAE (mean absolute error) [33] to evaluate emotion recognition. We also compared the proposed model performance with other machine learning classifiers using three metrics: precision, recall, and F1-score.

4.4 Result

Table 1 shows our proposed model result with different optimizers. From the table we can see that Adam optimizer training accuracy is 99.72% and validation accuracy 98.14%. The training loss of adam algorithm is 0.037 and validation loss is 0.0563. RMSProp optimizer has the highest training accuracy and validation accuracy among all the optimizer. The training and validation accuracy of RMSProp algorithm is 100% and 99.05% respectively. Adamax algorithm has the lowest training and validation accuracy among all other optimizer. The difference among these optimizer is less than 3%.

We have compared our model with the four other machine learning algorithms which are random forest, naïve Bayes, logistic regression, and Support Vector Machine (SVM). We have compared these with six performance matrix. These six performance matrices are accuracy, precision, recall, F-score, RMSE, and MSE. Table 2 shows this comparison. From Table 2 we can see that the accuracy of our proposed CNN-LSTM is 99.05% and RMSE and MSE of the proposed algorithm are 0.099 and 0.009, respectively. The accuracy of our proposed algorithm is the highest among the other four algorithms.

Table 1 CNN-LSTM model performance

Optimizer	Training accuracy	Training loss	Validation accuracy	Validation loss
Adam	99.72%	0.0357	98.14%	0.0563
RMSProp	100%	0.047	99.05%	0.0164
SGD	97.82%	0.0572	96.08%	0.1398
Adamax	97.01%	0.1004	94.12%	0.0981

Table 2 Comparisons among proposed CNN-LSTM and machine learning classifier

Algorithm	Accuracy	Precision	Recall	F-Score	RMSE	MSE
CNN-LSTM	99.05%	0.99	0.98	0.98	0.099	0.009
Random forest	96.32%	0.97	0.96	0.96	0.198	0.039
Naïve Bayes	82.35%	0.83	0.82	0.82	0.524	0.352
Logistic regression	86.84%	0.87	0.87	0.87	0.452	0.211
SVM	78.43%	0.86	0.78	0.76	0.792	0.627

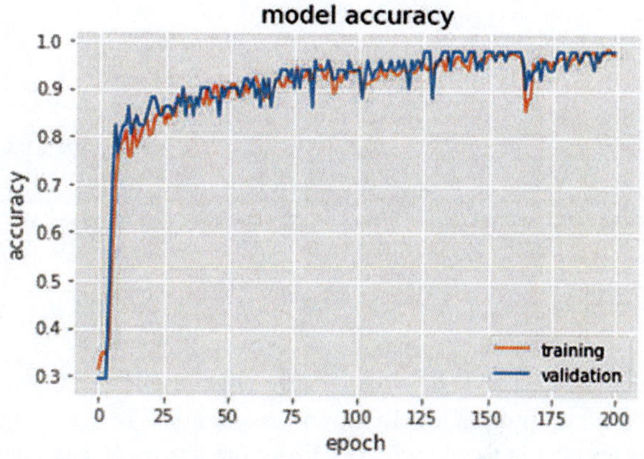

Fig. 4 Training and validation accuracy of the proposed CNN-LSTM

Figure 4 shows the training and validation accuracy of our proposed CNN-LSTM algorithm. Red line shows the training accuracy and the blue line shows the validation accuracy. From the figure we can see that the training accuracy and validation accuracy are almost same in every epoch. From this figure we can see that training and validation accuracy is nearly 30% at the first epoch. Accuracy is increasing while the number of epoch is increased. At the epoch 50 the accuracy of the training and validation accuracy is about 90%. The training accuracy is above 99% at the epoch 200.

Figure 5 shows the training and validation loss of our proposed CNN-LSTM algorithm. Red line shows the training loss and the blue line shows the validation loss. The training and validation loss of the proposed CNN-LSTM algorithm is less

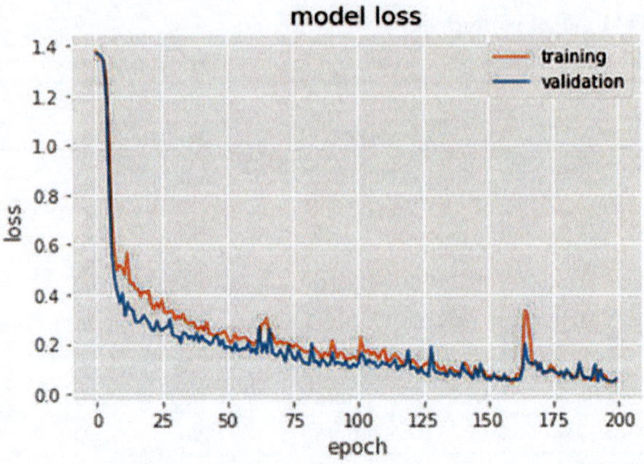

Fig. 5 Training and validation loss of the proposed CNN-LSTM

Table 3 Comparisons among CNN-LSTM and existing work

Author	Emotions	Method	Accuracy
E. S.Salama, et al. [29]	Valence, Arousal	3D-CNN	87.44% for valence, 88.49 % for the arousal
W.L. Zeng [45]	Positive, Negative, Neutral	KNN, LR, and GELM	91.07%
Y. Yang et al. [38]	Valence, Arousal	CNN-RNN	90.80% for valence, 91.03% for the arousal
Our proposed method	Fear, Angry, Happy, Sad	CNN-LSTM	99.05%

than 1.40. The loss is decreasing when the epoch number is increased. At the epoch 50 the training and validation loss is less than 0.25. The training and validation loss is less 0.1 at the epoch 200. From the figure we can also see that in the last 25 epoch the training and validation loss is almost same.

We have also compared our proposed CNN-LSTM with other existing works. Table 3 shows the comparison with other existing work. From the table we can see that the proposed method has the highest accuracy among the other work.

5 Conclusion and Future Work

In this article, we have suggested a new approach in which the emotion can be recognized by GSR and the ECG. A hybrid CNN-LSTM method is employed to recognize four types of emotion which are happy, sad, angry, and fear. Experimental results show that the proposed method have the high accuracy. We have also compared our proposed method with other machine learning classifier. Our results were also compared with other existing work. However, data is very limited. In Future ,we will train our model with more data. We will be using a some other method [2–5, 8, 9, 11, 14–16, 22–25, 36, 46] to recognize the emotion in an automated system in real time.

References

1. Abadi, M., Barham, P., Chen, J., Chen, Z., Davis, A., Dean, J., Devin, M., Ghemawat, S., Irving, G., Isard, M., et al.: Tensorflow: a system for large-scale machine learning. In: 12th USENIX Symposium on Operating Systems Design and Implementation (OSDI 16). pp. 265–283 (2016)
2. Abedin, M.Z., Akther, S., Hossain, M.S.: An artificial neural network model for epilepsy seizure detection. In: 2019 5th International Conference on Advances in Electrical Engineering (ICAEE). pp. 860–865. IEEE (2019)
3. Afroze, T., Akther, S., Chowdhury, M.A., Hossain, E., Hossain, M.S., Andersson, K.: Glaucoma detection using inception convolutional neural network v3. In: International Conference on Applied Intelligence and Informatics. pp. 17–28. Springer (2021)
4. Ahmed, T.U., Hossain, M.S., Alam, M.J., Andersson, K.: An integrated cnn-rnn framework to assess road crack. In: 2019 22nd International Conference on Computer and Information Technology (ICCIT). pp. 1–6. IEEE (2019)
5. Ahmed, T.U., Jamil, M.N., Hossain, M.S., Andersson, K., Hossain, M.S.: An integrated real-time deep learning and belief rule base intelligent system to assess facial expression under uncertainty. In: 2020 Joint 9th International Conference on Informatics, Electronics & Vision (ICIEV) and 2020 4th International Conference on Imaging, Vision & Pattern Recognition (icIVPR). pp. 1–6. IEEE (2020)
6. Anderson, K., McOwan, P.W.: A real-time automated system for the recognition of human facial expressions. IEEE Trans. Syst. Man Cybern. Part B (Cybernetics) **36**(1), 96–105 (2006)
7. Ayata, D., Yaslan, Y., Kamasak, M.E.: Emotion based music recommendation system using wearable physiological sensors. IEEE Transactions on Consumer Electronics **64**(2), 196–203 (2018)
8. Basnin, N., Nahar, L., Hossain, M.S.: An integrated cnn-lstm model for micro hand gesture recognition. In: International Conference on Intelligent Computing & Optimization. pp. 379–392. Springer (2020)
9. Basnin, N., Nahar, L., Hossain, M.S.: An integrated cnn-lstm model for bangla lexical sign language recognition. In: Proceedings of International Conference on Trends in Computational and Cognitive Engineering. pp. 695–707. Springer (2021)
10. Cheng, B., Liu, G.: Emotion recognition from surface emg signal using wavelet transform and neural network. In: Proceedings of the 2nd International Conference on Bioinformatics and Biomedical Engineering (ICBBE). pp. 1363–1366 (2008)
11. Gosh, S., Nahar, N., Wahab, M.A., Biswas, M., Hossain, M.S., Andersson, K.: Recommendation system for e-commerce using alternating least squares (als) on apache spark. In: International Conference on Intelligent Computing & Optimization. pp. 880–893. Springer (2020)
12. Greff, K., Srivastava, R.K., Koutník, J., Steunebrink, B.R., Schmidhuber, J.: Lstm: A search space odyssey. IEEE Trans. Neural Networks Learn. Syst. **28**(10), 2222–2232 (2016)
13. Hanjalic, A., Xu, L.Q.: Affective video content representation and modeling. IEEE Trans. Multimedia **7**(1), 143–154 (2005)
14. Islam, R.U., Hossain, M.S., Andersson, K.: A deep learning inspired belief rule-based expert system. IEEE Access **8**, 190637–190651 (2020)
15. Islam, R.U., Ruci, X., Hossain, M.S., Andersson, K., Kor, A.L.: Capacity management of hyperscale data centers using predictive modelling. Energies **12**(18), 3438 (2019)

16. Kabir, S., Islam, R.U., Hossain, M.S., Andersson, K.: An integrated approach of belief rule base and deep learning to predict air pollution. Sensors **20**(7), 1956 (2020)
17. Katsis, C.D., Katertsidis, N., Ganiatsas, G., Fotiadis, D.I.: Toward emotion recognition in car-racing drivers: a biosignal processing approach. IEEE Trans. Syst. Man Cybern. Part A Syst. Hum. **38**(3), 502–512 (2008)
18. Kim, J., André, E.: Emotion recognition based on physiological changes in music listening. IEEE Trans. Pattern Anal. Mach. Intell. **30**(12), 2067–2083 (2008)
19. Kim, S.K., Kang, H.B.: An analysis of smartphone overuse recognition in terms of emotions using brainwaves and deep learning. Neurocomputing **275**, 1393–1406 (2018)
20. Li, S., Li, W., Cook, C., Zhu, C., Gao, Y.: Independently recurrent neural network (indrnn): Building a longer and deeper rnn. In: Proceedings of the IEEE Conference on Computer Vision and Pattern Recognition. pp. 5457–5466 (2018)
21. Li, X., Song, D., Zhang, P., Yu, G., Hou, Y., Hu, B.: Emotion recognition from multi-channel eeg data through convolutional recurrent neural network. In: 2016 IEEE International Conference on Bioinformatics and Biomedicine (BIBM). pp. 352–359. IEEE (2016)
22. Nahar, N., Ara, F., Neloy, M.A.I., Barua, V., Hossain, M.S., Andersson, K.: A comparative analysis of the ensemble method for liver disease prediction. In: 2019 2nd International Conference on Innovation in Engineering and Technology (ICIET). pp. 1–6. IEEE (2019)
23. Nahar, N., Hossain, M.S., Andersson, K.: A machine learning based fall detection for elderly people with neurodegenerative disorders. In: International Conference on Brain Informatics. pp. 194–203. Springer (2020)
24. Pathan, R.K., Uddin, M.A., Nahar, N., Ara, F., Hossain, M.S., Andersson, K.: Gender classification from inertial sensor-based gait dataset. In: International Conference on Intelligent Computing & Optimization. pp. 583–596. Springer (2020)
25. Pathan, R.K., Uddin, M.A., Nahar, N., Ara, F., Hossain, M.S., Andersson, K.: Human age estimation using deep learning from gait data. In: International Conference on Applied Intelligence and Informatics. pp. 281–294. Springer (2021)
26. Pedregosa, F., Varoquaux, G., Gramfort, A., Michel, V., Thirion, B., Grisel, O., Blondel, M., Prettenhofer, P., Weiss, R., Dubourg, V., et al.: Scikit-learn: Machine learning in python. J. Mach. Learn. Res. **12**, 2825–2830 (2011)
27. Petrushin, V.: Emotion in speech: recognition and application to call centers. In: Proceedings of Artificial Neural Networks in Engineering. vol. 710, p. 22 (1999)
28. Rattanyu, K., Ohkura, M., Mizukawa, M.: Emotion monitoring from physiological signals for service robots in the living space. In: ICCAS 2010. pp. 580–583. IEEE (2010)
29. Salama, E.S., El-Khoribi, R.A., Shoman, M.E., Shalaby, M.A.W.: Eeg-based emotion recognition using 3d convolutional neural networks. Int. J. Adv. Comput. Sci. Appl **9**(8), 329–337 (2018)
30. Santamaria-Granados, L., Munoz-Organero, M., Ramirez-Gonzalez, G., Abdulhay, E., Arunkumar, N.: Using deep convolutional neural network for emotion detection on a physiological signals dataset (amigos). IEEE Access **7**, 57–67 (2018)
31. Shahani, B.T., Halperin, J., Boulu, P., Cohen, J.: Sympathetic skin response-a method of assessing unmyelinated axon dysfunction in peripheral neuropathies. J. Neurol. Neurosurg. Psychiatry **47**(5), 536–542 (1984)
32. Siddharth, S., Jung, T.P., Sejnowski, T.J.: Utilizing deep learning towards multi-modal biosensing and vision-based affective computing. IEEE Trans. Affect. Comput. (2019)
33. Sokolova, M., Lapalme, G.: A systematic analysis of performance measures for classification tasks. Inf. Process. Manage. **45**(4), 427–437 (2009)
34. Soleymani, M., Pantic, M., Pun, T.: Multimodal emotion recognition in response to videos. IEEE Trans. Affect. Comput. **3**(2), 211–223 (2011)
35. Srivastava, N., Hinton, G., Krizhevsky, A., Sutskever, I., Salakhutdinov, R.: Dropout: a simple way to prevent neural networks from overfitting. J. Mach. Learn. Res. **15**(1), 1929–1958 (2014)
36. Sultana, Z., Nahar, L., Basnin, N., Hossain, M.S.: Inference and learning methodology of belief rule based expert system to assess chikungunya. In: International Conference on Applied Intelligence and Informatics. pp. 3–16. Springer (2021)

37. Wang, J., Zhang, J., Wang, X.: Bilateral lstm: A two-dimensional long short-term memory model with multiply memory units for short-term cycle time forecasting in re-entrant manufacturing systems. IEEE Trans. Industr. Inf. **14**(2), 748–758 (2017)
38. Yang, Y., Wu, Q., Qiu, M., Wang, Y., Chen, X.: Emotion recognition from multi-channel eeg through parallel convolutional recurrent neural network. In: 2018 International Joint Conference on Neural Networks (IJCNN). pp. 1–7. IEEE (2018)
39. Yin, Z., Zhao, M., Wang, Y., Yang, J., Zhang, J.: Recognition of emotions using multimodal physiological signals and an ensemble deep learning model. Comput. Methods Programs Biomed. **140**, 93–110 (2017)
40. Zeng, Z., Pantic, M., Roisman, G.I., Huang, T.S.: A survey of affect recognition methods: Audio, visual, and spontaneous expressions. IEEE Trans. Pattern Anal. Mach. Intell. **31**(1), 39–58 (2008)
41. Zhang, J., Yin, Z., Chen, P., Nichele, S.: Emotion recognition using multi-modal data and machine learning techniques: A tutorial and review. Information Fusion **59**, 103–126 (2020)
42. Zhang, J., Zhou, Y., Liu, Y.: Eeg-based emotion recognition using an improved radial basis function neural network. J. Ambient Intell. Humanized Comput. pp. 1–12 (2020)
43. Zhang, S., Zhao, X., Lei, B.: Spoken emotion recognition using radial basis function neural network. In: International Conference on Computer Science, Environment, Ecoinformatics, and Education. pp. 437–442. Springer (2011)
44. Zhang, X., Xu, C., Xue, W., Hu, J., He, Y., Gao, M.: Emotion recognition based on multichannel physiological signals with comprehensive nonlinear processing. Sensors **18**(11), 3886 (2018)
45. Zheng, W.L., Zhu, J.Y., Lu, B.L.: Identifying stable patterns over time for emotion recognition from eeg. IEEE Trans Affect. Comput. **10**(3), 417–429 (2017)
46. Zisad, S.N., Hossain, M.S., Andersson, K.: Speech emotion recognition in neurological disorders using convolutional neural network. In: International Conference on Brain Informatics. pp. 287–296. Springer (2020)

Effect of Sample Volume in Escherichia Coli Detection in Water Using Double-Decker Resonator

Parisa Sanati, Mahdi Bahadoran, and Saiful Najmee Mohamad

Abstract Rapid pathogenic detection of Escherichia coli (E. coli) bacteria is vitally important in medical and pharmaceutical companies, environmental monitoring, and biomedical research. In this work, a double-decker ring resonator (DDRR) with resonant mode numbers of 9 and 11 is fabricated from a nano-core slab waveguide from Si_3N_4 placed on a coring 7980 silica substrate is used for detection of E. coli bacterium in water. The sensor's performance is studied for four waveguide layouts, and the effect of evanescent field penetration depth on volume sample is studied. Results simulated using the signal flow graph method and MATLAB software. The maximum sensitivity of the DDRR sensor is measured to be 605 nm/RIU, which corresponds to high-resolution sensing of 1.82×10^{-6} RIU. Results show that increasing the height of the superstrate sensing window waveguide to four times the height of the core layer contributes to improve the sensitivity.

Keywords Biosensor · Double-decker ring resonator · Escherichia coli · Pathogenic sensing method

1 Introduction

Quick and early detection of the foodborne pathogens (FBPs) is subject of continuing interest in food industries, medicinal companies, pollutant detection, medical diagnostic laboratories and patients' medical examinations; particularly those suffering

P. Sanati
Burn and Wound Healing Research Center, Shiraz University of Medical Sciences, 71345-1978 Shiraz, Iran

M. Bahadoran
Department of Physics, Shiraz University of Technology, 31371555 Shiraz, Fars, Iran
e-mail: bahadoran@sutech.ac.ir

S. N. Mohamad (✉)
Faculty of Applied Sciences, Universiti Teknologi MARA, Cawangan Johor, Kampus Pasir Gudang, 81750 Masai, Johor, Malaysia
e-mail: snajmee@uitm.edu.my

M. S. Kaiser et al. (eds.), *Proceedings of the Third International Conference on Trends in Computational and Cognitive Engineering*, Lecture Notes in Networks and Systems 348, https://doi.org/10.1007/978-981-16-7597-3_13

from sepsis-1 or infections [1] that bacteria strain required to be identified immediately to find the best and suitable treatment. The FBPs can cause toxic and infectious diseases upon eating polluted food or water. Escherichia coli is known as one of the FBPs that contributes to severe complications and may cause fatal health problems, mainly among newborns, kids, and the elderly. The E. coli is naturally found in the intestinal tract of warm-blooded animals and includes various groups of bacteria. Most E. coli strains are harmless and are a useful part of a healthy intestinal tract. However, some strains are pathogenic and may cause diseases, such as food poisoning and severe diarrhoea [2].

Bacterial infections cause a number of diseases per year. Thus, designing biosensors for direct detection of whole bacteria cells, rather than isolated biological components (enzymes, outer membrane proteins, or lipopolysaccharides) has several advantages since the whole cells are more tolerant to environmental changes like temperature, pH, rather than isolated structures [3]. Moreover, isolating the whole microorganisms from natural sources can be done by a simple approach, and there is no need for extensive preparation of the sample before measurements.

Today, several classical techniques have been used to detect the presence of bacteria in a solution. Traditional methods like culture collection are effective. However, the testing process may take several days to end up with results and have limitations in identifying various bacterial strains. Several methods have been developed in the detection of bacteria that take advantage of different techniques, including gen probes and polymerase chain reaction (PCR) [4], mass spectrometry [5], RNA biosensor [6], the fluorogenic probe technique [7], nucleic acid sequence-based amplification, electrochemiluminescence analysis [8], genomic sequencing [9], label-free surface-enhanced Raman scattering and mapping [10], microarrays, and enzyme-linked immunosorbent assay (ELISA) [11, 12]. The weak points of these methods are in the pretreatment of samples prior to analysis that is sophisticated, expensive, and requires expensive equipment and equipped laboratory as well as expert personnel [13].

The bacteria sensors in medical laboratories still follow classical microbiology methods that work based on bacteria isolation and growth in selective media. Although these methods are quite effective, however, the process of microorganism culturing is slow; thus, the method is time-consuming, besides it requires expert technicians for lab work and interpreting the results. To overcome the conventional limitations of these sensors, designing a new biosensor with low cost, small size, high sensitivity, and user-friendly is desirable. Nowadays, numerous bacteria sensors have been developed based on various mechanisms that can be exemplified by Amperometric sensors [14], magnetic biosensors [15], optical sensors [16–18], and electrochemical biosensors [19]. These sensors work based on measuring mass, a shift in resonance peaks, pH, conductivity, and mobility of ions. Among these sensors, the optical-based ones have superiorities of quick response, high precision, cost-effective, multiple detections by one device, high sensitivity, and immunity to electromagnetic interference, which make them a quite desirable sensor [13]. The label-free sensors have been developed in interferometric sensors [20, 21], electrochemical biosensors [22], photonic crystal sensors, cell biology detection, and refractive index-based sensors [16, 18]. Among different types of label-free sensors, the ring resonator-based devices [23–27] have attracted a lot of interest in sensing applications because of the tiny dimension, cost-effective, high-quality factor, and sensitivity and capability with conventional CMOS technology [28–30].

In this study, a double-decker ring resonator (DDRR) was used to detect E. coli bacteria in polluted water. The merits of Vernier-based systems [31–33] are in the achievement of high sensitivity and realizing extended free spectral range and sharp resonance peaks without requiring a narrow line-width tunable laser or employing a highly resolute OSA [34]. In practice, the superstrate layer of the DDRR sensing window was loaded with the antibodies of the E. coli, and the presence of E. coli bacteria in test samples brings about antibody–antigen interactions. It can vary the effectiveness and the group indices of the sensor's waveguide that consequently cause a spectral shift in output light.

In this paper, a nano-core slab from Si_3N_4 on a coring silica substrate is used in the DDRR system and applied for the detection of E. coli bacterium in water. The sensor's performance is studied for two categories of DDRR sensors with different heights of top cladding layers. The presence of E. coli bacterium in the test sample causes a change in the group and effective indices of the sensing ring and emerges a spectral shift of output light through the system.

2 Double-Decker Ring Resonator Layout and Mathematical Background

Applying a narrow linewidth tunable laser besides hiring a high-resolution OSA [34] is a conventional approach in achieving high sensitivity in optical sensors. Indeed, providing this experimental setup is costly and may not be available. One alternative is using the Vernier-based resonator that includes two ring resonators with different ring radii. In the Vernier-based ring resonator systems, each resonator generates its own FSR and the output resonance peaks form by the superposition of two FSRs from each ring. It delivered a high-quality factor resonance with broadening FSR, which may end up with a sensor with high sensitivity. For the Vernier-based systems, the total FSR of the device (FSR_{tot}) is an integer coefficient of the FSR of each ring; this number is known as the resonant mode numbers (RMNs) [35, 36]. The FSR of each ring is given by [37]

$$FSR = \lambda_0^2 / n_g L \tag{1}$$

that λ_O is the wavelength between two consecutive sharp peaks, L denotes for the circumference of each ring (optical track), and n_g shows the group index of the optical track (here ring's waveguide). The group index, n_g, is related to the effective index, n_{eff}, by $n_g = n_{eff} - \lambda_0 (dn_{eff}/d\lambda)_{\lambda_0}$ [38]. For a double-decker ring resonator (DDRR), the total FSR is given by [32, 39]

$$FSR_{tot} = N_1 . FSR_1 = N_2 . FSR_2 \tag{2}$$

that $N_m(m = 1,2)$ represents the integer RMNs and FSR_m $(m = 1,2)$ is the free spectral range of each ring of the DDRR system. The double-decker ring resonator (DDRR) sensor waveguide layout is illustrated in Fig. 1a. The ratio of output to input light fields is defined as an optical transfer function (OTF) [40–42]. For DDRR with the input and output ports of $E_{in} = E_1$ and E_{drp}, respectively, the OTF at the drop and the through ports of DDRR are given by [32]

$$OTF^{drop} = \frac{-i\sqrt{\Gamma_1\Gamma_2\Gamma_3 k_1 k_2 k_3}\, e^{-\frac{\alpha(L1+L2)}{4}} e^{-\frac{iKn_g(N1.L1+N2.L2)}{2}}}{\{1-\sqrt{\Gamma_1\Gamma_2(1-k_1)(1-k_2)}\, e^{-\frac{\alpha L1}{2}} e^{-iKn_g N_1 L_1} - \sqrt{\Gamma_2\Gamma_3(1-k_2)(1-k_3)}\, e^{-\frac{\alpha L2}{2}} e^{-iKn_g N_2 L_2} + \Gamma_2\sqrt{\Gamma_1\Gamma_3(1-k_1)(1-k_3)}\, e^{-\frac{\alpha(L1+L2)}{2}} e^{-iKn_g(N1.L1+N2.L2)}\}} \tag{3}$$

and

$$OTF^{thr} = \frac{\sqrt{\Gamma_1(1-k_1)} - \Gamma_1\sqrt{\Gamma_2(1-k_2)}\, e^{-\frac{\alpha L1}{2}} e^{-iKn_g N1.L1} - \sqrt{\Gamma_1\Gamma_2\Gamma_3(1-k_1)(1-k_2)(1-k_3)}\, e^{-\frac{\alpha L2}{2}} e^{-iKn_g N2.L2} + \Gamma_1\Gamma_2\sqrt{\Gamma_3(1-k_3)}\, e^{-\frac{\alpha(L1+L2)}{2}} e^{-iKn_g(N1.L1+N2.L2)}}{\{1-\sqrt{\Gamma_1\Gamma_2(1-k_1)(1-k_2)}\, e^{-\frac{\alpha L1}{2}} e^{-iKn_g N_1 L_1} - \sqrt{\Gamma_2\Gamma_3(1-k_2)(1-k_3)}\, e^{-\frac{\alpha L2}{2}} e^{-iKn_g N_2 L_2} + \Gamma_2\sqrt{\Gamma_1\Gamma_3(1-k_1)(1-k_3)}\, e^{-\frac{\alpha(L1+L2)}{2}} e^{-iKn_g(N1.L1+N2.L2)}\}} \tag{4}$$

where $k_m(m = 1,2,3)$ shows the coupling coefficient of each coupler[43], $\Gamma_m = 1-\gamma_m$ that γ_m is the intensity loss in each coupler [44], L_m denotes the perimeter of each ring, K represents the wave number in a vacuum, α is the average ring loss per unit length, and N_n is the RMN for each ring of DDRR system.

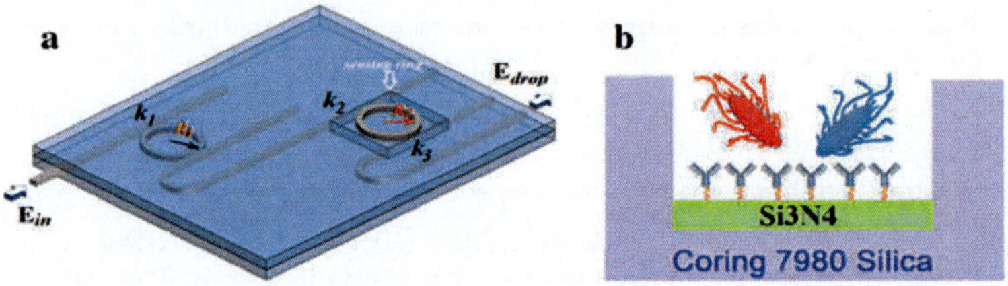

Fig. 1 **a** Waveguide configuration of double-decker ring resonator (DDRR) **b** waveguide cross-section

3 Simulated Results

Compared to silicon waveguides, the waveguides from Si_3N_4 have expressed lower crosstalk and less insertion loss [45]. In this work, we benefit from nano-size core layer slab waveguide from Si_3N_4 martial for E. coli bacterium detection. The DDRR sensor formed by 2 μm height of a substrate from coring 7980 silica, and a 100-nm-thick Si_3N_4 layer as the core[45]. The L-pyrenebutanoic acid succinimidyl ester is used as a linker of E. coli antibody to the core surface. One side of the pyrene group with strong π–π interaction bound by Si_3N_4 surface and the other side the succinimidyl ester group covalently reacts the amino group of the antibodies of E. coli [46]. Based on the antibody–antigen interaction (Fig. 1b), the E. coli bacterium in the contaminated test sample selectively get caught by the *loaded* antibodies film on the waveguide core. This interaction can alter the effective group indices of the ring of the sensing window and consequently contributes to a spectral shift of the output resonance peak. Assuming cylindrical bacterial cells for E. coli with the approximate size about 1 μm [47], a layer of E. coli's antibody–antigens with the approximate height of 1 μm can be added on the top cladding on the waveguide surface in the contaminated samples. Based on the top cladding of the sensors waveguides, four different bulk top claddings are selected to check the performance of the DDRR sensor. The width of the applied waveguide is 800 nm. The bacterial cell index of refraction can be determined from the spectral attenuation data [48]. A top cladding layer as high as 200 nm of water with an index of refraction 1.333 (at 589 nm) used as the baseline in the first category. For this waveguide layout (Fig. 2a), the effective refractive index is calculated to be 1.509799 using Finite Difference Eigen mode solver (FDE). In sensing applications, the index of aqueous sample (water) is 1.333 (at 590 nm), and indices of the Si_3N_4 core layer and coring silica substrate are 2.0275 and 1.4584, respectively.

Based on the spectral extinction measurement method, the E. coli cells index of refraction at 589 nm was measured to be 1.397 [48]. In the second approach, the contaminated aqueous sample is loaded in the sensing window. The E. coli compound is attached to the top cladding of the sensing waveguide because of the antibody–antigen reaction. The effective refractive index of the sensing waveguide (Fig. 2b) is measured to be 1.520623. In order to study the effect of evanescent filed on the top cladding layer of the sensor, the top cladding layers are used for the third and fourth layouts are two times larger than those applied in the first two cases. Illustrated in Fig. 2c is a layer with a height of 400 nm of drinking water used for the top cladding of the waveguide as the baseline of the third category. For this layout, the effective refractive index is 1.512197. The last configuration (Fig. 2d) is considered for the contaminated water with a higher concentration than the second layout. The effective refractive index of this waveguide's configuration is calculated to be 1.529604 at 589 nm. To obtain the resonant peaks in the visible region around 589 nm, an asymmetric configuration of Vernier DDRR from Si_3N_4–SiO_2 material waveguide with a top cladding of water is used, and the FSR extended into 84.3 nm using a Vernier double stage of microrings resonator with lengths of $L_1 = 26.74$ μm and

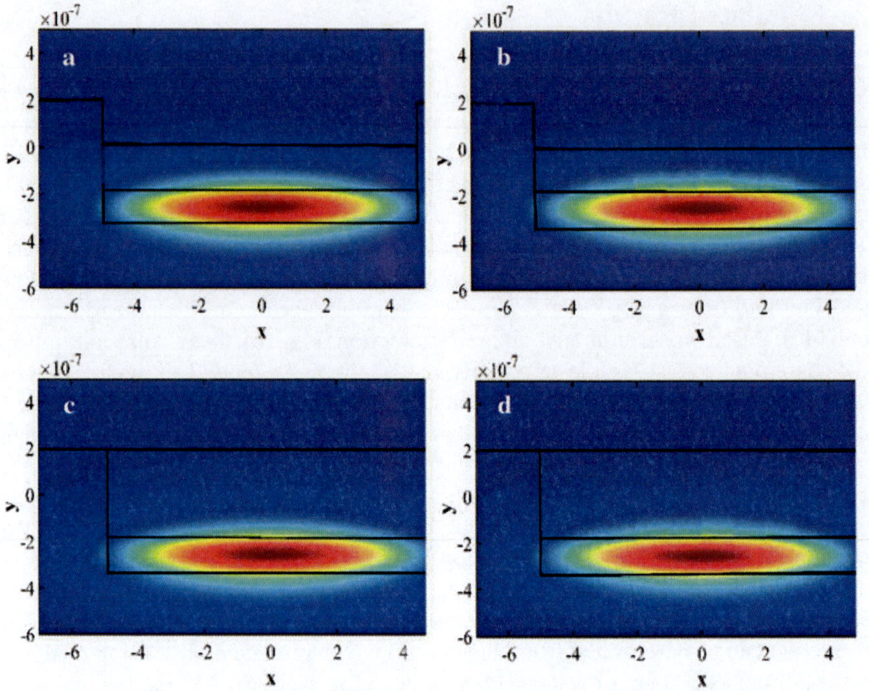

Fig. 2 The TE mode for different categories of top cladding **a** 200 nm drinking water. **b** 200 nm of contaminated water by E. coli. Considering the effect of evanescent filed penetration depth on **c** 400 nm water **d** 400 nm contaminated water by E. coli

$L_2 = 21.88$ μm. Based on Eq. (2), the resonant mode numbers (RMNs) for these microrings are calculated to be $N_1 = 11$ and $N_2 = 9$, respectively. Applying critical coupling conditions [49], $OTF^{thr} = 0$, Eq. (4), [50, 51], the following optical parameters are achieved: the symmetric lateral coupling coefficient and coupling loss $k_1 = k_3 = 0.1$, $\gamma_1 = \gamma_3 = 0.001$, respectively. The critical coupling coefficient, waveguide loss and the intensity insertion loss coefficients between the rings in DDRR system are $k_2 = 0.0028$, $\alpha = 20$ dBcm^{-1} and $\gamma_2 = 0.005$, respectively.

Here, we consider the practical evanescent field penetration depth as height as the length of the E. coli bacterium. In the presence of the E. coli bacterium in the test sample, the top cladding layer will divide into two parts, including the E. coli bacterium layer and water layer, which are demonstrated by E and W labels.

The W + W, E + W legends correspond to the top cladding layers of the third and fourth layouts. Based on the top cladding, the simulated results for all categories are shown in Fig. 3. Depends on the type of laser used for light sources in the input and drop ports, different wavelength shifts can be measured. A specified change in the effective refractive index of the waveguide of the sensor's mirroring brings about different wavelength shifts in DDRR resonant peaks since the FSR depends on

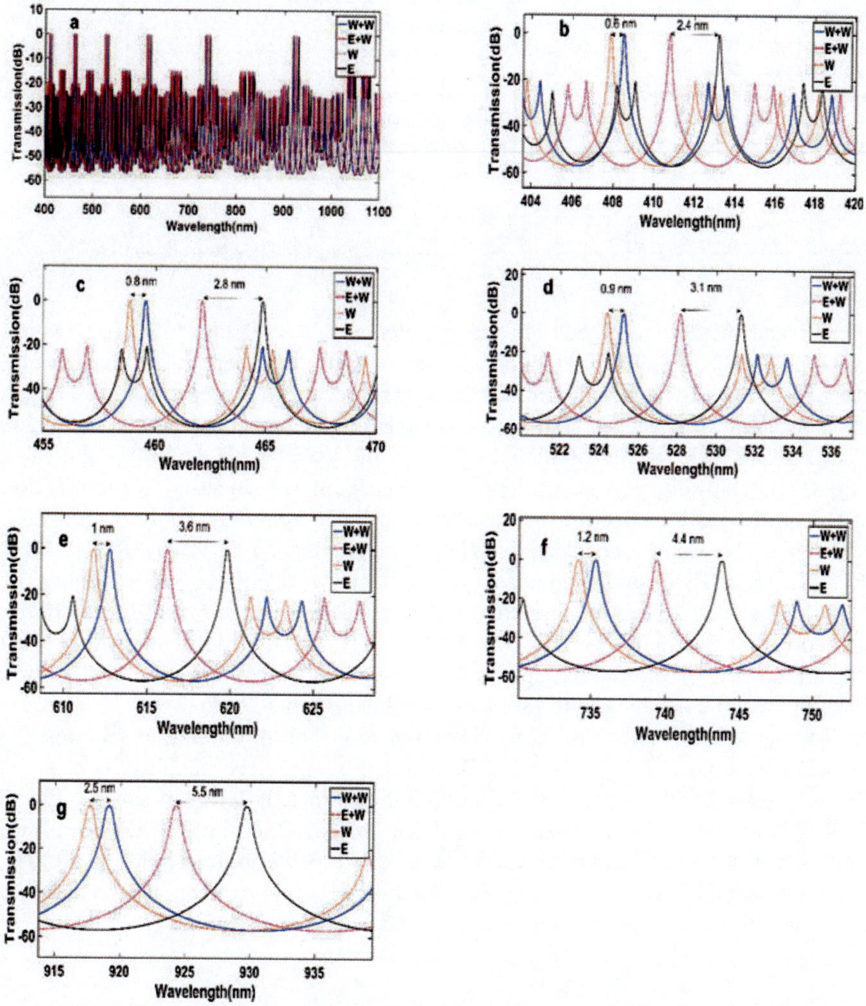

Fig. 3 OTF of DDRR sensor against wavelength for different volumes of top cladding layers. The optical parameters used in the simulation of the DDRR system are as follows: $k_1 = k_3 = 0.1$; $k_2 = 0.0028$; $\gamma_1 = \gamma_3 = 0.001$; $\gamma_2 = 0.005$; $\alpha = 20$ dB/cm **a** sensor response versus the wavelength spectrum of 400–1100 nm. The resonant peaks between **b** 408–413 nm, **c** 459–465 nm, **d** 524–531 nm, **e** 612–620 nm, **f** 734–744 nm, **g** 918–930 nm

wavelength, $\Delta\lambda_{FSR} = \lambda^2/(n_g L)$. The sensor response versus wavelength spectrum shown in Fig. 3a. As shown in Fig. 3b–g, for the change of, $\Delta n_g = 1.082 \times 10^{-3}$ in the waveguide effective index, the resonant peaks in the wavelengths of 408 nm, 459 nm, 524 nm, 612 nm, 734 nm, and 918 nm undergo the wavelength shifts of 0.6 nm, 0.8 nm, 0.9 nm 1.0 nm, 1.2 nm, 2.5 nm, respectively. Considering the evanescent field penetration depth leads to a change in the refractive as small as $\Delta n_g = 1.741$

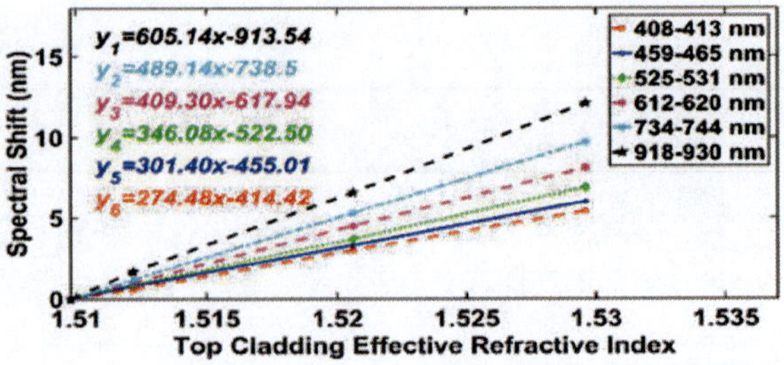

Fig. 4 The DDRR sensitivity line for E. coli bacterium detection in different wavelength range

$\times 10^{-3}$ and the resonant peaks in the wavelengths of 411 nm, 465 nm, 531 nm, 620 nm, 744 nm, and 930 nm, which contributed to the wavelength shifts of 5.4 nm, 6.0 nm, 6.9 nm 8.1 nm, 9.7 nm, 12.15 nm, respectively. Based on these results, the sensor lines for each region of the wavelength spectrum are plotted in Fig. 4. The DDRR sensitivity is calculated based on the wavelength region, as shown in Fig. 4. By taking the wavelength region into account, the sensitivity of the DDRR sensor in the visible spectrum is calculated to be 275 nm/RIU to 409.3 nm/RIU, which give rise to a limit of detection (LOD) in a range of 1.82×10^{-5} RIU–1.22×10^{-5}. The DDRR sensitivity increases in the larger wavelengths, and it reaches the maximum sensitivity of 605 nm/RIU for the wavelength range from 918 to 930 nm that corresponds to a minimum LOD of 1.82×10^{-5} RIU. Simulation results reveal that an increase in volume of superstrate sensing window enhance the light–matter interaction and contribute to the sensitivity enhancement.

Compared to reported results for detection of E. coli in the fluidic samples, the DDRR sensor has a superiority of fast detection process and is capable of working in different wavelength ranges with different light sources. In terms of sensitivity, the DDRR sensitivity is higher than the sensitivity of 221–354 nm/RIU achieved from gold–silver nano-disk arrays placed on a substrate of periodic sapphire [13]. The DDRR reached a higher sensitivity than 400 nm/RIU reported for surface plasmon resonance(SPR) biosensors made of gold and silvers dielectric layouts [52].

4 Conclusion

A double-decker Vernier ring resonator is applied to detect the E. coli bacterium in water-based solutions, considering the volume of the top cladding sensing layer. The sensor performance over the range of wavelength is studied based on the effect of evanescent field penetration depth to the aqueous medium for four different categories. The maximum sensitivity of 605 nm/RIU and small LOD of 1.82×10^{-6} RIU were realized for the DDRR sensor. It is found that increasing the height of the sensing window to four times the height of the core waveguide can improve the sensitivity of the sensor.

Acknowledgements The authors would like to acknowledge the research facilities of Shiraz University of Medical Sciences, Iran, and Universiti Teknologi MARA, Cawangan Johor, Malaysia

References

1. Choi, V.M., et al.: Activation of Bacteroides fragilis toxin by a novel bacterial protease contributes to anaerobic sepsis in mice. Nat. Med. **22**(5), 563–567 (2016)
2. Kalate Seyfari, A., Zareian, A.H., Bahadoran, M.: Design and simulation of four stage antiresonant reflecting plasmonic microring biosensor for detection of Ecoli Bacterium-O157 in water. J. Res. Many-body Syst. **10**(4), 47–62 (2021)
3. Janik, M., et al.: Live E. coli bacteria label-free sensing using a microcavity in-line Mach-Zehnder interferometer. Sci. Rep. **8**(1), 1–8 (2018)
4. Bej, A.K., et al.: Detection of Escherichia coli and Shigella spp. in water by using the polymerase chain reaction and gene probes for uid. Appl. Environ. Microbiol. **57**(4), 1013–1017 (1991)
5. Zhu, J., Hill, J.E.: Detection of Escherichia coli via VOC profiling using secondary electrospray ionization-mass spectrometry (SESI-MS). Food Microbiol. **34**(2), 412–417 (2013)
6. Baeumner, A.J., et al.: RNA biosensor for the rapid detection of viable Escherichia coli in drinking water. Biosens. Bioelectron. **18**(4), 405–413 (2003)
7. Frahm, E., Obst, U.: Application of the fluorogenic probe technique (TaqMan PCR) to the detection of Enterococcus spp. and Escherichia coli in water samples. J. Microbiol. Methods **52**(1), 123–131 (2003)
8. Min, J., Baeumner, A.J.: Highly sensitive and specific detection of viable Escherichia coli in drinking water. Anal. Biochem. **303**(2), 186–193 (2002)
9. Lyons, E., et al.: Using genomic sequencing for classical genetics in E. coli K12. PloS one **6**(2), e16717 (2011)
10. Yang, D., et al.: Reproducible E. coli detection based on label-free SERS and mapping. Talanta **146**, 457–463 (2016)
11. Daly, P., Collier, T., Doyle, S.: PCR-ELISA detection of Escherichia coli in milk. Lett. Appl. Microbiol. **34**(3), 222–226 (2002)
12. Johnson, R.P., et al.: Detection of Escherichia coli O157: H7 in meat by an enzyme-linked immunosorbent assay. EHEC-Tek. Appl. Environ. Microbiol. **61**(1), 386–388 (1995)
13. Cinel, N.A., Bütün, S., Özbay, E.: Electron beam lithography designed silver nano-disks used as label free nano-biosensors based on localized surface plasmon resonance. Opt. Express **20**(3), 2587–2597 (2012)
14. Liu, X., et al.: Amperometric glucose biosensor with remarkable acid stability based on glucose oxidase entrapped in colloidal gold-modified carbon ionic liquid electrode. Biosens. Bioelectron. **25**(12), 2675–2679 (2010)
15. Llandro, J., et al.: Magnetic biosensor technologies for medical applications: a review. Med. Biol. Eng. Compu. **48**(10), 977–998 (2010)
16. Sanati, P., et al.: Detection of Escherichia coli K12 in water using slot waveguide in cascaded ring resonator. Silicon 1–7 (2021)
17. Bahadoran, M., et al.: Detection of Salmonella bacterium in drinking water using microring resonator. Artif. Cells Nanomed Biotech. **44**(1), 315–321 (2016)
18. Sanati, P., et al.: Label-free biosensor array comprised of Vernier microring resonator and 3 × 3 optical coupler. Eur. Phys. J. Plus **135**(11), 869 (2020)
19. Zhao, W., Xu, J.J., Chen, H.Y.: Electrochemical biosensors based on layer-by-layer assemblies. Electroanalysis: Int. J. Devoted Fundam. Pract. Aspects Electroanalysis **18**(18), 1737–1748 (2006)
20. Gao, J., et al.: Vapor sensors based on optical interferometry from oxidized microporous silicon films. Langmuir **18**(6), 2229–2233 (2002)

21. Schmitt, K., et al.: Interferometric biosensor based on planar optical waveguide sensor chips for label-free detection of surface bound bioreactions. Biosens. Bioelectron. **22**(11), 2591–2597 (2007)
22. Liu, X., et al.: A label-free electrochemical biosensor based on ligand-receptor interaction. In: 2018 IEEE International Conference on Manipulation, Manufacturing and Measurement on the Nanoscale (3M-NANO), pp. 341–344. IEEE (2018)
23. Seyfari, A.K., Bahadoran, M., Aghili, A.: Ultra-sensitive pressure sensor using double stage racetrack silicon micro resonator. Opt. Quant. Electron. **52**(9), 408 (2020)
24. Aziz, M., et al.: Theoretical study on slow-light generated by integrated microring resonator with wide bandwidth and high gain. Jurnal Teknologi **78**(3), 293–300 (2016)
25. Amiri, I., Afroozeh, A., Bahadoran, M.: Simulation and analysis of multisoliton generation using a PANDA ring resonator system. Chin. Phys. Lett. **28**(10), 104205 (2011)
26. Bahadoran, M., et al.: Realizing unique bifurcation model in a cascaded microring feedback circuit. Opt. Quant. Electron. **52**(216), 1–14 (2020)
27. Amiri, I., et al.: Dual wavelength optical duobinary modulation using GaAs–AlGaAs microring resonator. Results Phys. **11**, 1087–1093 (2018)
28. Claes, T., Bogaerts, W., Bienstman, P.: Experimental characterization of a silicon photonic biosensor consisting of two cascaded ring resonators based on the Vernier-effect and introduction of a curve fitting method for an improved detection limit. Opt. Express **18**(22), 22747–22761 (2010)
29. Bahadoran, M., et al.: Modeling and analysis of a microresonating biosensor for detection of Salmonella Bacteria in human blood. Sensors **14**(7), 12885–12899 (2014)
30. Sirawattananon, C., et al.: Analytical vernier effects of a PANDA ring resonator for microforce sensing application. IEEE Trans. Nanotechnol. **11**(4), 707–712 (2012)
31. Seyfari, A.K., Bahadoran, M., Yupapin, P.: Design and modeling of double Panda-microring resonator as multi-band optical filter. Nano Commun. Netw. **29**, 100352 (2021)
32. Bahadoran, M., et al.: Nano force sensing using symmetric double stage micro resonator. Measurement **58**, 215–220 (2014)
33. Bahadoran, M., et al.: Bifurcation behaviors generated by Panda-ring control circuit. Microw. Opt. Technol. Lett. **61**(7), 1783–1787 (2019)
34. Jin, L., Li, M., He, J.-J.: Highly-sensitive silicon-on-insulator sensor based on two cascaded micro-ring resonators with vernier effect. Opt. Commun. **284**(1), 156–159 (2011)
35. Bahadoran, M., Ali, J., Yupapin, P.: Ultrafast all-optical switching using signal flow graph for PANDA resonator. Appl. Opt. **52**(12), 2866–2873 (2013)
36. Bahadoran, M., Ali, J., Yupapin, P.: Graphical approach for nonlinear optical switching by PANDA vernier filter. IEEE Photonics Technol. Lett. **25**(15), 1470–1473 (2013)
37. Bahadoran, M., et al.: Novel Approach to determine the young's modulus in silicon-on-insulator waveguide using microring resonator. Dig. J. Nanomater. Biostruct. **9**(3), 1095–1104 (2014)
38. Aziz, M., et al.: Light pulse in a modified add-drop optical filter for optical tweezers generation. J. Nonlinear Opt. Phys. Mater. **21**(04), (2012)
39. Boeck, R., et al.: Series-coupled silicon racetrack resonators and the Vernier effect: theory and measurement. Opt. Express **18**(24), 25151–25157 (2010)
40. Bahadoran, M., Yupapin, P., Amiri, I.S.: Optimum light transmission via microring resonator under a lossy-coupler critical coupling condition. Microw. Opt. Technol. Lett. **63**(2), 653–661 (2021)
41. Shafiee, A., Bahadoran, M., Yupapin, P.: Analytical microring stereo system using coupled mode theory and application. Appl. Opt. **58**(30), 8167–8173 (2019)
42. Bahadoran, M., Yupapin, P.: Butterfly-like phase shift: a novel gauge for critical coupling of add–drop resonator. J. Theoret. Appl. Phys. **12**(2), 127–134 (2018)
43. Nawi, I., et al.: A theoretical model of all-optical switching induced by a soliton pulse in nano-waveguide ring resonator. Journal of Physics: Conference Series, vol. 431, pp. 012029. IOP Publishing (2013)
44. Bahadoran, M., et al.: Slow light generation using microring resonators for optical buffer application. Opt. Eng. **51**(4), 044601–044608 (2012)

45. Dai, D., et al.: Low-loss Si 3 N 4 arrayed-waveguide grating (de) multiplexer using nano-core optical waveguides. Opt. Express **19**(15), 14130–14136 (2011)
46. Akbari, E., et al.: Escherichia coli bacteria detection by using graphene-based biosensor. IET Nanobiotechnol. **9**(5), 273–279 (2015)
47. Zibaii, M.I., et al.: Measuring bacterial growth by refractive index tapered fiber optic biosensor. J. Photochem. Photobiol. B **101**(3), 313–320 (2010)
48. Balaev, A.E., Dvoretski, K.N., Doubrovski, V.A.: Determination of refractive index of rod-shaped bacteria from spectral extinction measurements. In: Saratov Fall Meeting 2002: Optical Technologies in Biophysics and Medicine IV, vol. 5068, pp. 375–380. International Society for Optics and Photonics (2003)
49. Bahadoran, M., Amiri, I.S.: Double critical coupled ring resonator-based add–drop filters. J Theor. Appl. Phys. **13**, 213–220 (2019)
50. Yariv, A.: Critical coupling and its control in optical waveguide-ring resonator systems. IEEE Photonics Technol. Lett. **14**(4), 483–485 (2002)
51. Yariv, A.: Universal relations for coupling of optical power between microresonators and dielectric waveguides. Electron. Lett. **36**(4), 321–322 (2000)
52. Roy, S.K., Sharan, P.: Design of ultra-high sensitive biosensor to detect E. Coli in water. Int. J. Inf. Technol. **12**(3), 1–6 (2019)

A Probit Regression in Identifying the Risk Factors of Cervical Cancer in Malaysian Private Hospital

Tan Li Jun, Suliadi F. Sufahani, and Mohd Fahmy-Abdullah

Abstract Cervical cancer is the second largest cancer listed as the major cause of death for women worldwide. Generalized linear models (GLMs) were conducted to identify the risk factor of cervical cancer and to build a suitable model. This research is to identify the risk factors and investigate the relationship between cervical cancer and some risk factors based on the 854 samples from the private hospital. Probit regression was used to identify the risk factors of cervical cancer. Out of seven independent variables, two variables were given a significant relationship with cervical cancer. There was a significant relationship between cervical cancer with age and STDs which the p-value is less than 0.05. Thus, age and STDs influence the presence of cervical cancer in a private hospital. Women who have older age are more likely to have cervical cancer. For the woman who had STDs, the probability will be higher compared with no had STDs. Preliminary analysis with some cross-tabulation analysis is presented in this research. The probit regression model was built and the prediction was predicted with different cases of observation and this is the main contribution of the paper. The risk of getting cervical cancer can be reduced effectively if women have more knowledge of cervical cancer and the risk factors of cervical cancer can be identified.

Keywords Cervical cancer · Risk factor · Probit regression

T. L. Jun · S. F. Sufahani (✉)
Faculty of Applied Sciences and Technology, Universiti Tun Hussein Onn Malaysia, 84600 Pagoh, Johor, Malaysia
e-mail: suliadi@uthm.edu.my

M. Fahmy-Abdullah
Faculty of Technology Management and Business, Universiti Tun Hussein Onn Malaysia, 86400 Parit Raja, Johor, Malaysia

S. F. Sufahani · M. Fahmy-Abdullah
Oasis Integrated Group, Universiti Tun Hussein Onn Malaysia, Parit Raja, 86400 Batu Pahat, Johor, Malaysia

M. S. Kaiser et al. (eds.), *Proceedings of the Third International Conference on Trends in Computational and Cognitive Engineering*, Lecture Notes in Networks and Systems 348, https://doi.org/10.1007/978-981-16-7597-3_14

1 Introduction

God created females with a unique and complex reproductive system in their bodies. The internal parts of female sexual anatomy like the vulva, vagina, cervix, uterus, ovaries, endometrium, and fallopian tubes work together to form a complete function of the reproductive system. In the female reproductive system, the ovum is produced through the ovaries. The development of a baby from the fertilization stage to a mature fetus occurs in this entire system. Cervix is one of the parts of the female reproductive system too. It is placed at the lower part of the reproductive organ and is between the uterus and vagina. There is a small hole in the middle of the cervix. The hole is expandable until 10 cm to allow the deliverance of the baby. One function of the cervix is to allow the sperm to enter into the fallopian tubes, while another function is to allow the blood flow from the uterus to the vagina [1, 2, 5, 9, 11, 23]. Figure 1 shows the female reproductive system.

A female can get cervical cancer easily if her cervix is unhealthy [7]. Cervical cancer is listed as the major cause of death for women worldwide [8]. It is second largest cancer for women in the world. Human papillomavirus, which is known as HPV is a type of virus that usually grows in the cervix. The infection caused by HPV is the main cause of cervical cancer. Cervical cancer will start to occur in the cell of the cervix when the cell of the cervix change. When there is a virus in the cell of the cervix, the cell will be change and the condition will get worse from time to time. When the infection is small, it will be easier to treat. Due to the immune system of the body, the infection will not stay long in the cervix as the immune system will fight for it [14, 15]. However, long-term infection may lead to cervical cancer as the immune system of the body may not be able to fight it off. Cervical cancer in women can be divided into two main types, which are squamous cell carcinoma and adenocarcinoma. The ability to differentiate the type of cervical cancer can help doctors to provide effective treatment to the patients [16, 18]. Figure 2 shows the normal cervix and cervix with cervical cancer.

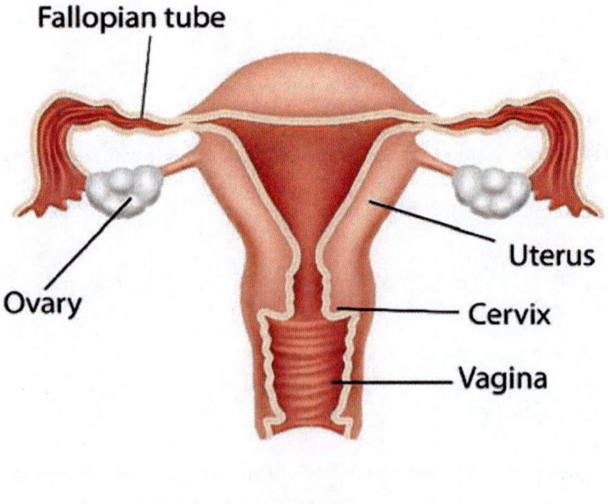

Fig. 1 Female reproductive system

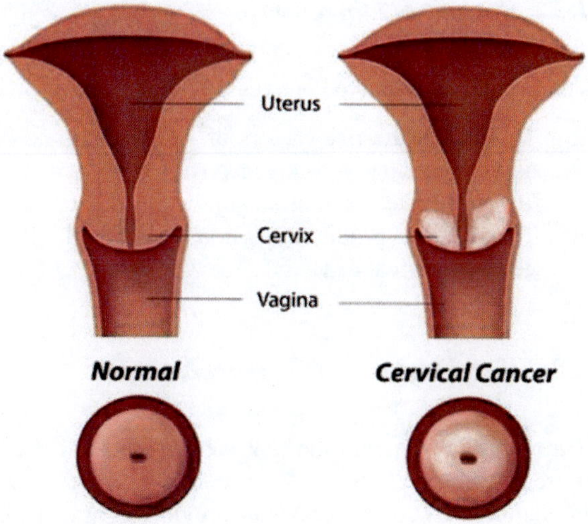

Fig. 2 Normal cervix and cervix with cervical cancer

1.1 Problem Statement

World Health Organization (WHO) predicts that people who are having cancer will increase up to 16 million yearly in 2020 [3]. Moreover, the number of new cervical cancer estimated by the Global Cancer Observatory (GLOBOCAN) is around 569,847 cases, while the number of death due to cervical cancer is 311, 365 cases globally in the year 2018 [4]. From Fig. 2, cervical cancer is a very serious sickness and it is life-threatening that can be easily noticed. Thus, it is very important to concern and obtains more information about cervical cancer. The risk of getting cervical cancer can be reduced effectively if women have more knowledge of cervical cancer and the risk factors of cervical cancer can be identified. By knowing the risk factor of cervical cancer, it can help to reduce the risk of getting cervical cancer effectively [10, 13].

However, different countries and places have different main risk factors of getting cervical cancer as different countries have different living cultures, knowledge, economy, and lifestyles. For example, some third-world countries maybe lack the knowledge and ability to carry out the testing for cervical cancer. Besides, the equipment and facilities for testing cervical cancer in third-world countries may be inadequate or their citizens are not able to afford the cost of cervical cancer testing [22]. Unfortunately, the research related to cervical cancer or its risk factors is so limited, not to mention in every country or town. Research regarding risk factors of cervical cancer in the country of each state is very less and thus research is conducted to identify the risk factor of cervical cancer in a private hospital. Throughout this research, the main factors of cervical cancer will be able to identify and reduce the chances of getting cervical cancer [17, 19, 21].

1.2 Research Objectives

The following are the objectives of this research:

1. To identify the risk factors of getting cervical cancer in a private hospital by using probit regression analysis.
2. To investigate the relationship between cervical cancer and its risk factors.
3. To build the probit regression model and predict the probability of women getting cervical cancer.

1.3 Significance of Research

Through this research, the risk factors of cervical cancer will be identified. To be more specific, the risk factors of cervical cancer in women in a private hospital will be identified. This research can provide significant information regarding cervical cancer. This enables the private hospital to identify the potential cervical cancer patients earlier and give extra attention to them or be able to provide treatment to the patients at an early stage. This research can contribute to minimizing the risk of female patients in the private hospital getting cervical cancer when they know the risk factors of cervical cancer.

As mentioned before, a regression model of risk factors in cervical cancer will be formed and the model can be used to predict the probability of women getting cervical cancer. Once the model is formed, the only information needs to be collected are the variables required in the regression model or the variables that contribute significant result to the model. The regression model might be very helpful for every woman in the future to help them in identifying their probability of getting cervical cancer.

2 Methodology

The methodology describes how the research is carried out. In this part, the methods used to carry out the analysis in this research will be discussed in detail.

2.1 Flow Chart of Research

The materials and methods section, otherwise known as methodology, describes all the necessary information that is required to obtain the results of the study as shown in Fig. 3.

Fig. 3 Flow chart of the research

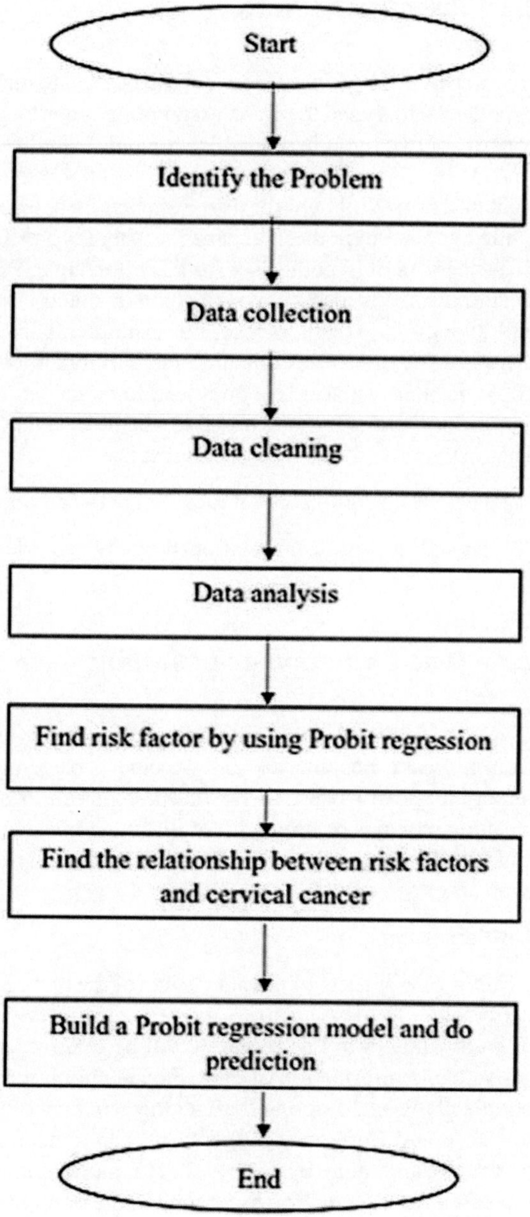

2.2 Test of Hypothesis

Hypothesis testing is a method that can be used to make a statistical decision about the hypothesis. It is used to test an assumption regarding a parameter. Hypothesis testing consists of two hypotheses, which are null hypothesis, H0, and alternate hypothesis, Ha/H1. Based on these two hypotheses, only one of them can choose to make the right decision. Null hypothesis means that there is no difference between parameters while alternate hypothesis means that there is a difference between the parameters.

A decision of hypothesis testing is referring to the level of significance. The level of significance is also known as alpha or α and it is the probability of rejecting the null hypothesis when it is true. For example, 0.05 of significant level indicates that there is a 5% of error of rejecting the null hypothesis. The reason is that there is no 100% accuracy in deciding on rejecting or accepting the null hypothesis.

The hypothesis is made based on the objects that have been discussed in Chap. 1. The assumption of the hypothesis test is:

H_0: There is no significant relationship between selected factors and cervical cancer.

H_1: There is a significant relationship between selected factors and cervical cancer.

2.3 Data Collection and Cleaning

Under the scope of identifying risk factors of cervical cancer, the data was obtained from a private hospital and only focusing on female patients. Data cleaning is an important part of research for the preparation of an analysis. The process of data cleaning includes dealing with outliers and missing data, removing duplicated data, and identify incomplete data. It can provide a higher quality of information for the analysis.

Outliers

Dealing with outliers of the data is one of the parts of data cleaning. Outliers mean an extremely high or low value in a set of data and the value is outside an overall pattern of data. Outliers in this research will be removed from the data. Although outliers are valuable information of data, the data point may be at an extremely high or low point and this condition will affect the accuracy of the result and the result might be low accuracy. Removing outliers can help to decrease the error of the result. Based on the research done by Li et al., 2015, showed that the accuracy of the data can be increased from 63 to 76% by removing the outliers of the data. Outliers are removed in this research [12].

Missing Data

Missing data is the missing value that occurs in a set of data. It may occur when the respondent chose not to answer certain questions due to privacy. Missing data can occur for many other reasons. Missing data can be categorized into three categories, which are Missing Completely and Random (MCAR), Missing at Random (MAR), and Missing Not at Random (MNAR). The listwise deletion method is used in this research [20].

2.4 Probit Regression Model

The probit regression model is one of the binary link functions in GLMs [6]. A binary probit model is a regression model in which its dependent variables are dichotomous. Binary dependent variables consist of two outcomes variables, is either yes or no or with disease or no disease. A binary probit model is used in this research and the dependent variables in this research are

$$y = \begin{cases} 1, y > 0 \\ 0, otherwise \end{cases} \tag{1}$$

where 1 indicated having cervical cancer, while 0 indicates no cervical cancer.

The dependent variable, Y, of dichotomous needed to transform into a continuous variables Y′ $(-\infty, \infty)$, the transformed Y' is

$$F(Y) = Y^{'} = X\beta \tag{2}$$

Since the link function follows the standard normal distribution, it can say that

$$\mathrm{Y} = \Phi(X\beta) \tag{3}$$

The probit model is thus

$$\Phi^{-1}(Y) = \beta_0 + \beta_1 X_1 + \beta_2 X_2 + \cdots + \beta_p X_n \tag{4}$$

where

β_0 Intercept of the model,

β_p Sope of independent variables, and

X_n Variables of independent.

3 Results and Discussion

This research of analysis is done and the results are discussed in this chapter. The research objectives have been done and write accordingly. The significance of parameters and the goodness-of-fit test is determined. The risk factors of cervical cancer and the relationship between cervical cancer and its risk factors have been analyzed and the result has also been interpreted in each section. The probit regression model has been built and predicted the probability of getting cervical cancer by using the model.

3.1 Probit Regression Analysis

In this research, three objectives are set out to achieve in the analysis of probit regression. Three objectives have been carried out and the results are interpreted. The binary dependent variable is the presence of cervical cancer, which is a dichotomous variable, where 0 is denoted as no cervical cancer, while 1 is denoted as cervical cancer. There are seven independent variables which include continuous and binary variables. Age is a continuous independent variable while sexual partner, the first age of sexual intercourse, pregnancy, smoke, hormonal contraceptives, and STDs are binary independent variables. From the Eq. (4).

$$\Phi^{-1}(Y) = \beta_0 + \beta_1 X_1 + \beta_2 X_2 + \cdots + \beta_p X_n \tag{5}$$

where Φ^(−1) (Y) is probit function. Y is cervical cancer or no cervical cancer, where

Y = {1 = cervical cancer, 0 = no cervical cancer}, β_0 is constant, X_1 is age, X_2 is sexual partner, X_3 is the age of first sexual intercourse, X_4 is pregnant, X_5 is smoking, X_6 is hormonal contraceptives, and X_7 is STDs.

Risk Factors of Cervical Cancer

The analysis aims to identify the risk factors of cervical cancer. In the seven independent variables, only the variables which bring significance to cervical cancer will be chosen.

From Table 1, only the parameter of age and STDs have a significant effect on cervical cancer. Since the p-value of age is $0.043 < 0.05$, which indicates that it has a significant relationship between age and cervical cancer. The p-value of STDs is $0.006 < 0.05$, which also indicates that it has a significant relationship between STDs and cervical cancer. For the parameters of sexual partner, age of first sexual intercourse, pregnancy, smokes, and hormonal contraceptives are not significant to cervical cancer as the p-value is greater than 0.05. It concludes that there is no significance to cervical cancer.

Table 1 The slope of parameter and significance

Parameter	Slope	P-value	Significance
Age	0.017	$0.043 < 0.05$	Significant
Sexual partners	−0.058	$0.742 > 0.05$	Insignificant
Age of first sexual intercourse	−0.350	$0.094 > 0.05$	Insignificant
Pregnant	0.038	$0.940 > 0.05$	Insignificant
Smokes	0.318	$0.067 > 0.05$	Insignificant
Hormonal contraceptives	−0.154	$0.289 > 0.05$	Insignificant
STDs	0.534	$0.006 < 0.05$	Significant

Relationship of Cervical Cancer and the Risk Factors

This objective aims to investigate the relationship between cervical cancer and its risk factors. Out of seven independent variables, two variables bring the significant result to cervical cancer. The relationship of these two variables with cervical cancer will be discussed.

Table 2 shows the parameter estimates. The intercept and slope of the parameter, standard error, confidence interval, and hypothesis test are shown. Two independent variables give a significant result to cervical cancer which is STDs and age. The independent variable of STDs contributes the most to the model which is 7.705 and the second contributes the most is the age which is 4.093. The slope of the STDs is 0.534 which indicates that a person who has STDs, will increase to 0.534 in cervical cancer. For the odd ratio (1.706) of STDs, it can be said that a person who had STDs is more likely to have cervical cancer than a person who had no STDs. The slope of the age is 0.017 which indicates when every one unit of the age change, will increase to 0.017 in cervical cancer. For the odd ratio (1.017) of age, it indicates that older women are more likely to have cervical cancer compared to younger women.

Regression Model and Prediction

The objective aims to build the regression model, thus the regression model can be used to predict the probability of women getting cervical cancer. The prediction of cervical cancer will also be done in this objective.

The probit regression model in Table 4.12 can be modeled as

$$\Phi^{-1}(Y) = -1.951 + 0.017Age + 0.534STDs \tag{6}$$

From Eq. (1), the negative coefficient (−1.951) means that it will lead to a decrease in the predicted probability of cervical cancer. It is the constant term in the model which means the variables of age and STDs are zero, the predicted probability of cervical cancer is $F(-1.951) = 0.025528$.

The prediction for the probability of getting cervical cancer is predicted by using the probit regression model in Eq. (1).

Table 2 Parameter estimates

Parameter	B	Std. error	95% wald C. I		Hypothesis test			Exp (B)	95% wald C. I. for exp(B)	
			Lower	Upper	Wald chi-square	df	Sig		Lower	Upper
(Intercept)	−1.951	0.5463	−3.022	−0.880	12.752	1	0.000	0.142	0.049	0.415
Age	0.017	0.0084	0.001	0.033	4.093	1	0.043	1.017	1.001	1.034
Sexual partners	−0.058	0.1765	−0.404	0.288	0.108	1	0.742	0.944	0.668	1.334
Age of first sexual intercourse	−0.350	0.2088	−0.759	0.060	2.804	1	0.094	0.705	0.468	1.061
Pregnant	0.038	0.4988	−0.940	1.015	0.006	1	0.940	1.038	0.391	2.760
Smokes	0.318	0.1732	−0.022	0.657	3.364	1	0.067	1.374	0.978	1.929
Hormonal contraceptives	−0.154	0.1450	−0.438	0.130	1.124	1	0.289	0.858	0.645	1.139
STDs	0.534	0.1924	0.157	0.911	7.705	1	0.006	1.706	1.170	2.487
(Scale)	1[a]									

Table 3 Example of 5 observations dataset and the predicted probability

Patient	Age	STDs	$\Phi^{-1}(Y)$	Predicted probability
1	49	1	−0.584	0.2796
2	35	0	−1.356	0.0876
3	33	1	−0.856	0.1960
4	28	0	−1.475	0.0701
5	50	0	−1.101	0.1355

Table 3 shows there are five different observations and the probability of getting cervical cancer is identified. The age of patient 1 is 49 with had sexually transmitted disease, the probability of having cervical cancer is 0.2796. It indicates that out of 100%, there is 27.96% of the probability that having cervical cancer. If the woman is younger, the probability of getting cervical cancer will be lower. For example, the age of patient 4 is 33 and she had STDs, the probability that she gets cervical cancer is 19.60% which is lower than the higher age.

Moreover, women had STDs will increase the probability of getting cervical cancer. Inpatients 1 and 5 with the age of 49 and 50, respectively, the probability of getting cervical cancer in the first patient with STDs is higher than patient 5 with no STDs.

4 Conclusion

The first objective of this research is to identify the risk factors of getting cervical cancer in a private hospital by using probit regression analysis. Seven independent variables were used in the hypothesis testing to determine the significant variable to cervical cancer. The result showed only two independent variables have a significant effect on cervical cancer which were the variables of age and STDs.

The second objective is to investigate the relationship between cervical cancer and its risk factors. Since the risk factors of cervical cancer had been found, the relationship between cervical cancer and the risk factors was investigated in this objective. According to the result of Wald chi-square, the independent variable of STDs contribute the most and the second is age. There is a positive relationship with cervical cancer. It indicated when the women are more older, the chance of getting cervical cancer will be higher than younger women. For the women who had STDs, the chance of getting cervical cancer will be higher than the women who had no having STDs.

The third objective is to build the regression model and predict the probability of women getting cervical cancer. The probit regression model has been built in Eq. 4.1. It has a negative intercept toward the equation. When the variables are in constant terms, the predicted probability of cervical cancer is $F(-1.951) = 0.025528$. The predicted probability of women getting cervical cancer was stated in Table 4.13. The predicted probability is based on Eq. 4.1 with the different observations of women. The result of the probability is matched in objective 2 which the relationship of age and STDs affect the probability of getting cervical cancer in different observations.

This research data is given by a third party; however, there is room for further research regarding some of the issues. Future research is recommended to cover all cases in the country, not just in a specific place. More research should be done on the variables age and STDs associated with cervical cancer with probit regression analysis so that it can use to be more confirm their actual relationship. Future research should discover more variables regarding the risk factors of cervical cancer, such as family historical disease, blood type, and living lifestyle in different other techniques.

Acknowledgements Fundamental Research Grant Scheme (FRGS) No. K175 (FRGS/1/2019/STG06/UTHM/02/2) from the Ministry of Higher Education Malaysia (MOHE) and Universiti Tun Hussein Onn Malaysia (UTHM).

References

1. Abduljabbar, D., Al-Rawahi, F., Faqihi, F., Al-Khayat, M., Al-Mahmeed, M., Al-Khazali, M., Al-Sayed, N., AlGhaffar, S., AlNasir, F.: Types and risk factors of cervical cancer. Bahrain Med. Bull. **36**(2), 94–96 (2014)
2. Agarwal, S.S., Sehgal, A., Sardana, S., Kumar, A., Luthra, U.K.: Role of male behavior in cervical carcinogenesis among women with one lifetime sexual partner. Cancer **72**(5), 1666–1669 (1993)
3. Bessler, P., Aung, M., Jolly, P.: Factors affecting uptake of cervical cancer screening among clinic attendees in Trelawny. Jam. Cancer Control **14**(4), 396–404 (2007)
4. Bruni, L.: Human papillomavirus and related diseases report. HPV Information Centre (2016). https://www.hpvcentre.net/statistics/reports/XWX.pdf
5. Danaei, G., Vander Hoorn, S., Lopez, A.D., Murray, C.J.L., Ezzati, M.: Causes of cancer in the world: comparative risk assessment of nine behavioural and environmental risk factors. Lancet **366**(9499), 1784–1793 (2005)
6. Dascălu, C.G., Cozma, C.D.: The principal components analysis—method to reduce the collinearity in multiple linear regression model; application in medical studies. In: International Conference on Multivariate Analysis and Its Application in Science and Engineering, pp. 140–145 (2009)
7. Fillon, M.: Engineered T cells improve pancreatic cancer outcomes in Mice. J. Natl. Cancer Inst. **108**(8), 8–9 (2016)
8. Ghim, S.J., Basu, P.S., Jenson, A.B.: Cervical cancer: Etiology, pathogenesis, treatment, and future vaccines. Asian Pac. J. Cancer Prev. **3**(3), 207–214 (2002)
9. Harmon, M.P., Castro, F.G., Coe, K.: Acculturation and cervical cancer: knowledge, beliefs, and behaviors of Hispanic women. Women Health **24**(3), 37–57 (1996)
10. Hu, K., Wang, W., Liu, X., Meng, Q., Zhang, F.: Comparison of treatment outcomes between squamous cell carcinoma and adenocarcinoma of cervix after definitive radiotherapy or concurrent chemoradiotherapy. Radiat. Oncol. **13**(1), 1–7 (2018)
11. Huang, S.H., Loh, J.K., Tsai, J.T., Houg, M.F., Shi, H.Y.: Predictive model for 5-year mortality after breast cancer surgery in Taiwan residents. Chin. J. Cancer **36**(1), 1–9 (2017)
12. Kang, H.: The prevention and handling of the missing data. Korean J. Anesthesiol. **64**(5), 402–406 (2013)

13. Lai, B.P.Y., Tang, C.S.K., Chung, T.K.H.: Age-specific correlates of quality of life in Chinese women with cervical cancer. Support. Care Cancer **17**(3), 271–278 (2009)
14. Li, W., Mo, W., Zhang, X., Squiers, J.J., Lu, Y., Sellke, E.W., Fan, W., DiMaio, J.M., Thatcher, J.E.: Outlier detection and removal improves accuracy of machine learning approach to multispectral burn diagnostic imaging. J. Biomed. Opt. **20**(12), 121305 (2015)
15. Liu, S., Semenciw, R., Mao, Y.: Cervical cancer: the increasing incidence of adenocarcinoma and adenosquamous carcinoma in younger women. CMAJ **164**(8), 1151–1152 (2001)
16. Low, E.L., Simon, A.E., Lyons, J., Romney-Alexander, D., Waller, J.: What do British women know about cervical cancer symptoms and risk factors? Eur. J. Cancer **48**(16), 3001–3008 (2012)
17. Pourhoseingholi, A., Pourhoseingholi, M.A., Rostami-Nejad, M., Rostami, K., Mirsatari, D., Zojaji, H., Solhpour, A., Zali, M.R.: Implementation of statistical analysis in the clinical research of coeliac disease: use of probit and logit analysis. East Afr. J. Public Health **7**(2), 168–170 (2010)
18. Pourhoseingholi, A., Pourhoseingholi, M.A., Vahedi, M., Safaee, A., Moghimi-Dehkordi, B., Ghafarnejad, F., Zali, M.R.: Relation between demographic factors and type of gastrointestinal cancer using probit and logit regression. Asian Pac. J. Cancer Prev. **9**(4), 753–755 (2008)
19. Sabates, R., Feinstein, L.: The role of education in the uptake of preventative health care: the case of cervical screening in Britain. Soc. Sci. Med. **62**(12), 2998–3010 (2006)
20. Sedgwick, P., Marston, L.: Statistical question: odds ratios. BMJ (Online) **341**(7769), 407 (2010)
21. Sharma, P., Pattanshetty, S.M.: A study on risk factors of cervical cancer among patients attending a tertiary care hospital: a case-control study. Clin. Epidemiol. Glob. Health **6**(2), 83–87 (2018)
22. Shields, T.S., Brinton, L.A., Burk, R.D., Wang, S.S., Weinstein, S.J., Ziegler, R.G., Studentsov, Y.Y., McAdams, M., Schiffman, M.: A case-control study of risk factors for invasive cervical cancer among U.S. women exposed to oncogenic types of human papillomavirus. Cancer Epidemiol. Biomark. Prev. **13**(10), 1574–1582 (2004)
23. Suarez, S.S., Pacey, A.A.: Sperm transport in the female reproductive tract. Hum. Reprod. Update **12**(1), 23–37 (2006)
24. Wang, J., Wang, F., Liu, Y., Xu, J., Lin, H., Jia, B., Zuo, W., Jiang, Y., Hu, L., Lin, F.: Multiple linear regression and artificial neural network to predict blood glucose in overweight patients. Exp. Clin. Endocrinol. Diabetes **124**(1), 34–38 (2016)

Binary Programming for Primary School Diet Among Autism Children in Malaysia

Fairuz B. Baharom, Natasha A. M. Zailani, and Suliadi F. Sufahani

Abstract Parents of children diagnosed with autism or autism spectrum disorder (ASD) often deal with many kinds of food-related challenges. These can include many things, like allergies or may be the child has a hard time swallowing. The children may be picky eaters or hate certain foods and refuse to eat any of them. Autism is a complex brain disorder. While it may seem that cutting out certain foods could relieve autism child's symptoms, it might cause more harm. In Malaysia, the use of mathematical modeling in creating a healthy menu for autism children is limited. Moreover, manually planning a menu is complicated, inefficient, and inaccurate, lacks variety, no consumer preferences, and flexibility, no local recipes incomplete number of nutrient and food groups, does not meet the nutrient's boundaries and is time-consuming. This study proposes a new mathematical model for solving menu planning issues using the optimization method that increases the necessary nutrient intake. Balanced nutrients are required by the autism children. Minimizing the budget will also help to overcome all the problems mentioned. Three optimization methods will be used, namely, Linear Programming, Integer Programming, and Binary Programming as a proposal to solve the diet problem for Autism children aged 7–12 years old.

Keywords Autism · Diet problem · Integer programming · Operational research

1 Introduction

According to the National Autism Society of Malaysia (NASOM) to recent statistics, there are 300,000 people on the autism spectrum in Malaysia [26, 27]. Autism patients faced their problems in choosing the diet that meets the nutrition requirements.

F. B. Baharom · N. A. M. Zailani · S. F. Sufahani (✉)
Faculty of Applied Sciences and Technology, Universiti Tun Hussein Onn Malaysia, 84600 Pagoh, Johor, Malaysia
e-mail: suliadi@uthm.edu.my

S. F. Sufahani
Oasis Integrated Group, Universiti Tun Hussein Onn Malaysia, 86400 Parit RajaJohor, Malaysia

M. S. Kaiser et al. (eds.), *Proceedings of the Third International Conference on Trends in Computational and Cognitive Engineering*, Lecture Notes in Networks and Systems 348, https://doi.org/10.1007/978-981-16-7597-3_15

There is no research about conducting an optimization approach in solving diet for autism patients. Planning adequate menus involves consideration of several types of constraints: the desired nutritional content, the amount of food to be consumed, and their food should be avoided [29].

A balanced diet is vital for every humankind to maintain good health and body condition. This diet includes a variety of foods to provide essential nutrients such as carbohydrates, proteins, vitamins, minerals, fat, and fibers to our body. Autism or Autism Spectrum Disorders (ASD) is a complex biological disorder characterized by difficulties with speech, abnormalities of posture or gesture, problems with understanding the feelings of others, sensory and visual misperceptions, fears and anxieties, and behavioral abnormalities such as compulsive/obsessive behavior and ritualistic movements [5, 8, 10]. The National Autistic Society estimates that more than half a million people have autism in the United Kingdom with four times more boys than girls affected.

The word "autism" is originated from the Greek language "autos", which means "self". It describes the conditions in which a person is removed from social interaction. In other words, he becomes an "isolated self'. Eugen Bleuler, a Swiss psychiatrist, was the first person to use the term. He started using it around 1911 to refer to one group of symptoms related to schizophrenia. In the 1940s, researchers in the United States began to use "autism" to describe children with emotional or social problems. Children with ASD often repeat behaviors and have narrow, obsessive interests. These types of attitudes can affect eating habits and food choices, which can lead to the following health things [2]:

- Limited food selection or strong food dislikes
- Not eating enough food
- Constipation
- Medication interactions.

Dairy, gluten, corn, sugar, and artificial ingredients are five types of food that autism children should avoid [1]. The reason is that the different consumption of nutrition will make different ASD conditions that are wildly overactive or often due to an inflammatory. Besides that, according to the website "Food for the brain", with the title about autism, no single cause has been confirmed, although genetic and environmental factors are involved. There is developing proof that nourishing treatment can truly have a major effect on children with autism. Many have extremely disturbed digestion, so restoring balance in the gut is a key focus for nutritional treatment. Also important is balancing blood sugar, checking for brain-polluting heavy metals, excluding food additives, identifying food allergies and possible nutrient deficiencies, and ensuring an optimal intake of essential fats [9]. Autism is already well known in Malaysia, but most parents do not realize that their children are facing autism that causes them to face difficulty in daily activities and children's growth. Recommended nutrient intake has been listed as the information on good nutrition for the process of growing for Malaysian society in RNI 2017 [15]. The body weights for different age/gender groups were important to standardize in developing the recommendations. There are four objectives of this proposed study:

- To determine the nutritional requirement for autism children at the optimum cost.
- To establish mathematical modeling as the optimal solution to diet problems for autism children.
- To develop a new mathematical algorithm for the optimal diet solution problem for autism children.
- To develop a decision support system based on the proposed algorithm in objective 4.

Through this study, linear programming, integer programming, and binary programming will be used to solve the diet problem for autism disorder patients. Computer software such as Matlab with LPSolve will be used for computation purposes. This study only involves the Malaysian recipes, which are included in the book Nutrient Composition of Malaysian Foods [25]. This diet menu is for autism disorder patients and specifically for primary school children-aged 7 until 12. All of them have different kinds of nutritional requirements regarding their age, activity, allergens, and others. Few studies have been done, which relates to solving diet problem through mathematical approach [21–24] and other research on optimization and analysis [11, 17].

2 Literature Review

Numerous approaches have been implemented in developing mathematical diet models. In this approach, there are two techniques (Linear programming and Integer programming). The diet problem involves the establishment of the mathematical model, application of optimization method to deal with the problem, and formulation of strategies and solutions. This is an interactive case study, which is originated from Optimization Technology Center (OTC) [3].

One of the example on diet problem through optimization approach is for the athlete. Two main objectives are the prepared menu that meets the nutritional requirements of athletes and keeps those expenditures allocated for athletes staying with the predetermined budget. The mathematical models were built in this research. The model is aimed to prepare the weekly menu for enhancing the athletes' performance during the tournament period. The establishment of strong athletes' physical heavily relies on the well-planned diet menu. The athletes carried out intense training and practices around the corner of the competition period. These diets are especially important to boost the athletes' potential at the best [12]. An optimization study is done by Darmon, Ferguson, and Briend in 2006, to investigate the food choice of French women to prepare a balanced meal with the cost constraint [5]. The method used is linear programming modeling to find that the food combination is optimal in nutrition with the lowest cost. The sample for this study consists of 476 French women from different socioeconomic levels. The experiment was conducted in a district in Paris. Based on the analyses, this model included 64 food items, which collated into 21 food groups and excluded some drinks such as coffee, tea, and alcohol. The diets were developed by linear programming technique, which is the Simplex procedure of the Premium Solver Platform 5.0 for Excel. The researcher proposed that the diet includes some affordable food choices (fish, fruits, and vegetables) instead of animal fats or cheese to minimize the costs and enhance the balance of nutrients in the diet [5]. Fourer, Gay, and Kernighan in 2003 generated a linear programming model to fulfill a week of predetermined nutritional requirements and cheapest packages combination of food such as chicken, macaroni, spaghetti, and so on [7]. The result obtained is a monotonous diet that included only 46.667 packages of macaroni and cheese with the lowest expenditure at $88.20. This was not his desired solution. He then modified and developed an AMPL model for this diet problem. The adjusted diet are differ from previous results but still acceptable. The diet now comprises 19.5 packages of chicken, 16.3 packages of macaroni and cheese, and 4.3 of meatloaf. However, the cost is increased from $88.20 to $89.99. Such modification to the linear program will generate a more acceptable solution in solving the diet problem. The constraints for this model:

- Protein (not less than 38%)
- Lipid (not less than 6%)
- Carbohydrate (not less than 26%)
- Calcium (not less than 1%)
- Phosphorous (not less than 0.6%)
- Amount needed (The sum for the food item as fish feed is 100 kg).

In this case, the linear programming method to generate the feed formulation is more efficient and effective than the trial-and-error method that results in lower productivity of fishes [7]. The different types of programming approaches have been discussed and summarized in Table 1.

3 Methodology

The method used or the research framework (Fig. 1) is discussed thoroughly in this section. This research part begins by observation on the arising problem (autism primary school). The relevant data regarding the problem are collected from the dietician and nutritionist. The diet problem is concerned in the area of operational research. This study would help the autism primary school to develop a balanced diet menu that meets the nutritional requirements for daily life and minimize the cost. The operational research approach is used to find the optimal solution for solving the diet problem.

Table 1 Summary of optimization study

Study	Technique	Research problems
Sklan and Dariel [19]	Mixed integer-linear programming	To determine nutrition at minimum costs for people's diet. However, this study does not determine an optimum in choosing their diet to make the meals more palatable.
Fourer et al. [7]	Linear programming	To develop a linear programming model to achieve a week of fixed nutritional requirements and cheapest packages combination of food. The solution for the linear programming model could not fulfill all the constraints that lead to an optimal solution.
Darmon et al. [4]	Linear programming	To construct a linear programming model to achieve a week for of fixed nutritional requirements and cheapest packages combination of food. The solution for the French women from low economic level may not practical; the meal planning is not familiar and not designed according to their preference.
Sufahani and Ismail [20]	Integer programming	To develop a mathematical model for diet planning that meets the necessary nutrient intake for the secondary school student as well as minimizing a budget. The planning menus for a secondary school based on the cost of food items and RDA for children aged between 13 and 18 years old are not considered in this study.
Nath and Talukdar [14]	Linear programming	To come out fish feed compounds and in return to raise the productivity of fishes and the profit earning by the fishes by using linear programming to develop the feed formulation. The upper bounds for the nutrient contents do not consider in the research.
Patil and Kasturi [15]	Integer programming	Implemented the optimization research about the nutritional ingredient needed by humans aged 40–45 and minimize the total diet cost at the same time. The food items involved in the research are limited.

(continued)

Table 1 (continued)

Study	Technique	Research problems
Sheng and Sufahani [18]	Integer programming	To undergo various diet problems, and mathematical models were built to decide a diet plan as optimal solution, which satisfy all the requirements and restrictions. The feasible solution generated for eczema patients with the diet problem.

3.1 Data Collection

Primary data and secondary data are used in this study. The primary data are taken from the interviews with the dietician, autism center, and the Malaysian Ministries. The data obtained are nutrient contents, food group requirements, and also the lower bound or upper bound of each nutrient. Next, the secondary data are extracted from books, journals, reports, and others. In this study, the food items that taken into the model only involve the Malaysian recipes in the book Nutrient Composition of Malaysian Foods [13, 25]. The food items are in standardized portion size and with their respective nutrient contents have stated. In developing a diet planning model, some data should take into consideration, for instance, the cost for the standardized portion size of the food items, the nutrient contents of each food item, and Reference Nutrient Intake (RNI) for Malaysians at any age.

Reference Nutrient Intake (RNI) indicates that the recommended amount of nutrients such as minerals or vitamins is needed by different ages of individuals. RNI ensures that the sufficient nutrients are provided to the people. RNI satisfies the nutrient requirements of up to 97.5% for healthy individuals. Upper tolerable nutrients (ULs) are the maximum nutrient intake from the food. A normal person should meet his nutritional requirements in the range of RNI and UL to prevent deficiency and toxicity. The distribution of requirements to prevent deficiency and toxicity is shown in Fig. 2.

3.2 Data Collection

In this study, three types of programming approaches will be used; linear programming, integer programming, and binary programming. Linear programming is the technique taking account of some linear inequalities in solving a certain problem. Then, the optimal solution will be obtained, which satisfies these conditions or restrictions. While integer programming is a mathematical optimization that turns the variables to be integers, as well as linear programming, integer programming searches for the optimal solution under the constraints. Furthermore, in this study, 11 types of nutrients and 10 types of food groups will be considered.

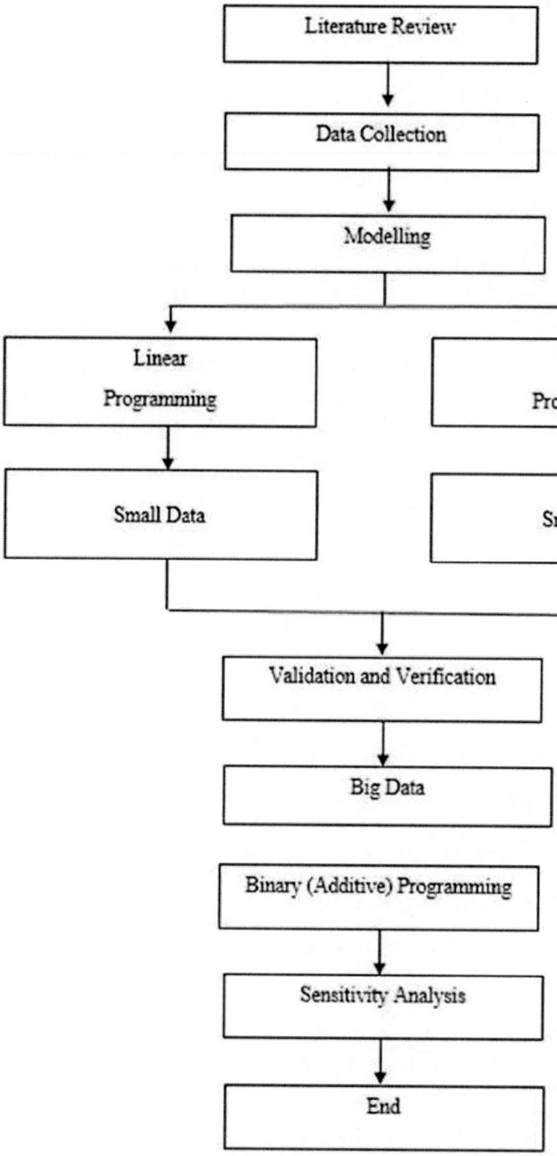

Fig. 1 Research flow: a proposed theoretical model

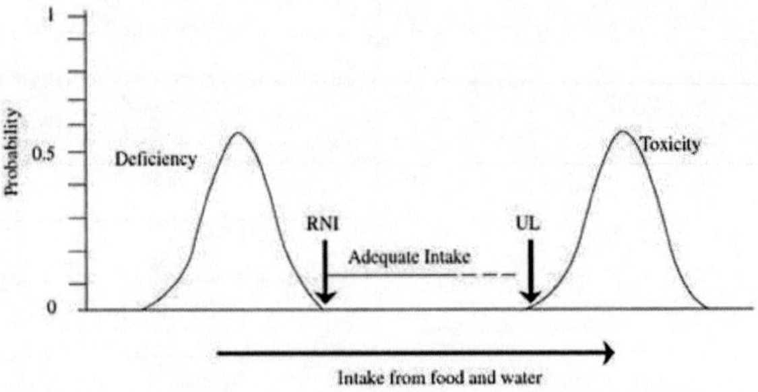

Fig. 2 Conceptual framework for FAO/WHO recommended nutrient intake (2002)

There are two types of the model generated in this study, which is Small Data model, which consists of 100 variables (100 types of food) from 10 food groups and the Big Data model, which consists of 426 variables (426 types of food) from 10 food groups.

General Equation for Menu Planning Model

$$\text{Minimize cost} = \sum_{i=1}^{N}\sum_{j=1}^{P}\sum_{k=1}^{Q} c_i x_{ijk} \tag{1}$$

where;

x_{ijk} decision variables of food items *i* for 10 food group, *j* and 6 meals, *k*

c_i cost for each food item *i*.

P the number of meals per day.

Q the number of food groups.

The Constraint for the General Nutritional Requirements

$$\text{Lower bound of nutrients} \le \sum_{i=1}^{N}\sum_{j=1}^{10}\sum_{k=1}^{6} w_i x_{ijk} \le \text{Upper bound of nutrients} \tag{2}$$

where w_i is the weight of nutrients for the food.

The constraints in this study are expressed as follows:

$$\text{Energy (kcal)} : L \le \sum_{i=1}^{426}\sum_{j=1}^{10}\sum_{k=1}^{6} E_i x_{ijk} \le U \tag{3}$$

$$\text{Protein (g)} : L \le \sum_{i=1}^{426}\sum_{j=1}^{10}\sum_{k=1}^{6} P_i x_{ijk} \le U \tag{4}$$

$$\text{Fats (g)}: L \le \sum_{i=1}^{426}\sum_{j=1}^{10}\sum_{k=1}^{6} F_i x_{ijk} \le U \tag{5}$$

$$\text{Carbohydrates (g)}: L \le \sum_{i=1}^{426}\sum_{j=1}^{10}\sum_{k=1}^{6} C_i x_{ijk} \le U \tag{6}$$

$$\text{Calcium (mg)}: L \le \sum_{i=1}^{426}\sum_{j=1}^{10}\sum_{k=1}^{6} Ca_i x_{ijk} \le U \tag{7}$$

$$\text{Iron (mg)}: L \le \sum_{i=1}^{426}\sum_{j=1}^{10}\sum_{k=1}^{6} Fe_i x_{ijk} \le U \tag{8}$$

$$\text{Niacin (mg)}: L \le \sum_{i=1}^{426}\sum_{j=1}^{10}\sum_{k=1}^{6} Nia_i x_{ijk} \le U \tag{9}$$

$$\text{Vitamin A } (\mu g): L \le \sum_{i=1}^{426}\sum_{j=1}^{10}\sum_{k=1}^{6} Va_i x_{ijk} \le U \tag{10}$$

$$\text{Vitamin B}_1\text{(mg)}: L \le \sum_{i=1}^{426}\sum_{j=1}^{10}\sum_{k=1}^{6} Vb1_i x_{ijk} \le U \tag{11}$$

$$\text{Vitamin B}_2\text{(mg)}: L \le \sum_{i=1}^{426}\sum_{j=1}^{10}\sum_{k=1}^{6} Vb2_i x_{ijk} \le U \tag{12}$$

$$\text{Vitamin C (mg)}: L \le \sum_{i=1}^{426}\sum_{j=1}^{10}\sum_{k=1}^{6} Vc_i x_{ijk} \le U \tag{13}$$

L	Lower bound for each nutrient content.
U	Upper bound for each nutrient content.
E_i	Energy in Kcal for the food item, *i*.
P_i	Amount of protein in gram for the food item, *i*.
F_i	Amount of fat in gram for the food item, *i*.
C_i	Amount of carbohydrate in gram for the food item, *i*.
Ca_i	Amount of calcium in milligram for the food item, *i*.
Fe_i	Amount of iron in milligram for the food item, *i*.
Nia_i	Amount of niacin in milligram for the food item, *i*.
Va_i	Amount of Vitamin A in microgram for the food item, *i*.
$Vb1_i$	Amount of Vitamin B1 in microgram for the food item, *i*.
$Vb2_i$	Amount of Vitamin B2 in microgram for the food item, *i*.
Vc_i	Amount of Vitamin C in microgram for the food item, *i*.

The Constraints of Food Group Requirements

$$\text{Beverages}: \sum_{i=1}^{37} x_{i,1} = r,\ \text{where plain water} = 2 \tag{14}$$

$$\text{Cereal Flour Based}: \sum_{i=38}^{85} x_{i,2} = r \tag{15}$$

$$\text{Rice Flour Based}: \sum_{i=86}^{113} x_{i,3} = r \tag{16}$$

$$\text{Cereal Meal Based}: \sum_{i=114}^{126} x_{i,4} = r \tag{17}$$

$$\text{Meat}: \sum_{i=127}^{158} x_{i,5} = r \tag{18}$$

$$\text{Vegetables}: \sum_{i=159}^{212} x_{i,6} = r \tag{19}$$

$$\text{Fruits}: \sum_{i=213}^{261} x_{i,7} = r \tag{20}$$

$$\text{Wheat Flour Based}: \sum_{i=262}^{286} x_{i,8} = r \tag{21}$$

$$\text{Seafood}: \sum_{i=287}^{324} x_{i,9} = r \tag{22}$$

$$\text{Miscellaneous} \sum_{i=325}^{426} x_{i,10} = r \tag{23}$$

where r is the daily requirement for each food group.

Definition of Decision Variable

The decision variables involved in the menu planning model are various Malaysian styles of recipes. These combinations of food will be shown up in menu lists according to six categories of dishes. It can be written as below:

$$X_i = \begin{cases} 1 \text{ or } 2 \text{ if menu } i \text{ appear in the menu list} \\ 0 \quad \text{otherwise} \end{cases}$$

i = any type of menus

The parameters include the nutritional content (such as Energy, Calories, and so on.) and the price for each menu. The lower and upper bound for every nutrient is based on the Malaysian RNI 2017 [16] along with the consultation of the dietician. Therefore, along with the decision-making in the menu planning model, there are assumptions to be made, for example:

The food items in each menu are in standardized portions. The nutrients contained in each menu are assumed to be consumed as a whole unit without change during the process of preparing, cooking, and serving the menus.

The effectiveness of the methods in solving this menu planning problem has been proven in past studies [6, 28].

3.3 Solution Approach

The model will search for an optimal solution integer programming. The coding will be programmed using Mathlab with LPSolve. Finallly, an automated system will be established whereby the user key in the data to generate a list of menus for one day as an output.

3.4 Develop the Model

Linear Programming approach, Integer Programming, and Binary Programming approach are applied in this study to construct the model that can yield the best solution. A computer program will be developed using Mathlab with LPSolve where users produce a menu list for 1-day menus that consist of Malaysian recipes for autism primary school.

3.5 Stopping Criteria

Linear Programming, Integer Programming, and Binary Programming will find the most optimal solution as the best solutions.

4 Expected Result

The five objectives in this proposal are expected to be achieved throughout the study:

- To prepare a menu to meet certain nutritional requirements and cost constraints at the lowest level for autism children.
- To determine the balanced nutrient required by the autism children.
- To establish mathematical modeling as the optimal solution to diet problems for autism children.
- To develop a mathematical modeling and a new algorithm for autism children.
- To develop a decision support system based on the proposed mathematical model and new algorithm.

By seeking the dietician and referring to the nutritional guidelines, the balanced nutrient that required by the autism primary school will be identified. About 426 food items are selected into the daily menu planning for this study. In addition, there are some results to be expected in this research such as the menus can be prepared to meet certain nutritional requirements also can determine the balanced nutrient that is required by the autism primary school. Besides, this research also will help to provide a complete menu planning for 1 day including all the essential vitamins and minerals in the meal, and all the range of nutritional requirements is defined according to their food groups. Subsequently, mathematical modeling will be established, which serves as the optimal solution to the diet problem for autism primary school. Not just that, the total cost of daily food intake also can be minimized throughout this study.

5 Conclusion

Planning a balanced and nutritious diet for autism children through decision-making and optimization model study will contribute a better approach to preparing a balanced diet menu that meets nutritional requirements and is suitable for autism children of all ages or other diseases that required diet control. Furthermore, it is a good to explore the knowledge of food intolerance, food, and their respective nutritional value and in order to promote a healthy lifestyle and at the same time try to help autism children for a better life.

Acknowledgements Fundamental Research Grant Scheme (FRGS) Number K175 (FRGS/1/2019/STG06/UTHM/02/2) from the Ministry of Higher Education Malaysia (MOHE) and GPPS H419 Universiti Tun Hussein Onn Malaysia (UTHM).

References

1. Foods That Can Make Autism Worse: (2019). Retrieved from https://www.amenclinics.com/blog/5-foods-can-make-autism-worse/
2. About Autism (2020). Retrieved from https://www.foodforthebrain.org/nutrition-solutions/autism/ about-autism.aspx
3. Czyzyk, J., Wisniewski, T., Wright, S.: Optimization case studies in the NEOS Guide. Soc. Ind. Appl. Math. **41**(1), 148–163 (1999)
4. Darmon, N., Ferguson, E.L., Briend, A.: Impact of a cost constraint on nutritionally adequate food choices for French women: an analysis by linear programming. J. Nutr. Educ. Behav. **38**(2), 82–90 (2006)
5. Dover, C.J., Couteur, A.L.: How to diagnose autism. Arch. Dis. Child. **92**(6), 540–545 (2007)
6. Fletcher, L.R., Soden, P.M., Zinober, A.S.: Linear programming techniques for the construction of palatable human diets. J. Oper. Res. Soc. **45**(5), 489–496 (1994)
7. Fourer, R., Gay, D.M., Kernighan, B.W.: AML: a modelling language for mathematical programming, 2nd edn. Thomson/Brooks/Cole & Duxbury Press, Pacific Groove, California (2003)
8. Grabrucker, A.M.: Environmental factors in Autism. Front. Psychiatry **3** (2013)
9. Gluten Free/Dairy Free Diet for Autism: My Experience. (2017). Retrieved from https://autismawarenesscentre.com/my-experience-with-the-gfcf-diet/
10. Hertz-Picciotto, I., Croen, L.A., Hansen, R., Jones, C.R., Water, J.V., Pessah, I.N.: The CHARGE study: an epidemiologic investigation of genetic and environmental factors contributing to Autism. Environ. Health Perspect. **114**(7), 1119–1125 (2006)
11. Idris, A.I.M., Fahmy-Abdullah, M., Sieng, L.W.: Technical efficiency of soft drink manufacturing industry in Malaysia. Int. J. Supply Chain Manage. **8**, 908–916 (2019)
12. Magdić, D., Kljusurić, J.G., Matijević, L., Frketić, D.: Analysis of diet optimization models for enabling conditions for hypertrophic muscle enlargement in athletes. Croat. J. Food Sci. Technol. **5**(1), 18–28 (2013)
13. Malaysian Dietary Guidelines for Children and Adolescents: Technical working group on Nutritional Guidelines (for National Coordinating on Food and Nutrition) (2013)
14. Nath, T., Talukdar, A.: Linear programming technique in fish feed formulation. International Journal of Engineering Trends and Technology **17**(3), 132–135 (2014)
15. Patil, A.N., Kasturi, S.: Optimal diet decision using linear programming. International Research Journal of engineering and technology (IRJET) **3**(8), 2197–2199 (2016)
16. Recommended Nutrient Intak, RNI (2017). http://nutrition.moh.gov.my/wp-content/uploads/2017/05/FA-Buku-RNI.pdf
17. Soebagyo, D., Fahmy-Abdullah, M.O.H.D., Sieng, L.W., Panjawa, J.L.: Income inequality and convergence in Central Java under regional autonomy. Int. J. Econ. Manage. **13**(1), (2019)
18. Sheng, L.Z., Sufahani, S.: Optimal Diet Planning for Eczema Patient Using Integer Programming. J. Phys. Conf. Ser. **995**, 012049 (2018)
19. Sklan, D., Dariel, I.: Diet planning for humans using mixed-integer linear programming. Br. J. Nutr. **70**(1), 27–35 (1993)
20. Sufahani, S., Ismail, Z.: A new menu planning model for Malaysian secondary schools using optimization approach. Appl. Math. Sci. **8**, 7511–7518 (2014)
21. Sufahani, S., Kamardan, M.G., Rusiman, M.S., Mohamad, M., Khalid, K., Ali, M., Khalid, K., Nawawi, M.K.M., Ahmad, A.: A Mathematical Study on "Additive Technique" Versus "Branch and Bound Technique" For Solving Binary Programming Problem. J. Phys. Conf. Ser. **995**(1), (2018)
22. Sudin, A.M., Sufahani, S.: Optimization Technique with Sensitivity Analysis on Menu Scheduling for Boarding School Student Aged 13–18 Using "Sufahani-Ismail Algorithm". J. Phys. Conf. Ser. **995**(1), (2018)
23. Sudin, A.M., Sufahani, S.: Mathematical Approach for Serving Nutritious Menu for Secondary School Student Using "Delete-Reshuffle-Reoptimize Algorithm". J. Phys. Conf. Ser. **995**(1), (2018)

24. Sufahani, S., Mohamad, M., Roslan, R., Kamardan, M. G., Che-Him, N., Ali. M., Khalid, K., Nazri, E.M., Ahmad, A.: Applied mathematical optimization technique on menu scheduling for boarding school student using delete-reshuffle-reoptimize algorithm. J. Phys. Conf. Ser. **995**(1), (2018)
25. Tee, E.S., MohdISmai, N., MohdNasir, A. Khatijah, I.: Nutrient composition of Malaysian foods, 4th edn, Kuala Lumpur, Malaysia, Institute for Medical Research (1997)
26. The Star Online Newspaper: Giving support to children with autism (2017). https://www.thestar.com.my/metro/community/2017/04/29/giving-support-to-children-with-autism-it-is-an-uphill-struggle-for-those-diagnosed-with-the-disorde
27. The Star Online Newspaper: More kids diagnosed with autism (2019). https://www.thestar.com.my/news/nation/2019/09/15/more-kids-diagnosed-with-autism
28. Valdez-Pena, H., Martinez-Alfaro, H.: Menu planning using the exchange diet system. SMC03 Conference Proceedings. 2003 IEEE International Conference on Systems, Man and Cybernetics. Conference Theme - System Security and Assurance (Cat. No.03CH37483) (2003)
29. Zailani, N.A.M., Sufahani, S.F., Mamat, M.: Mathematical research on optimization technique for diet planning problem: case research autism paralympic athlete. Int. J. Recent Technol. Eng. **8**(2), 387–391

Menu Planning and Scheduling for Vegetarian Breast Cancer Patients in Malaysia Using Optimization Approaches

Jing Ying Tee and Suliadi F. Sufahani

Abstract Breast cancer (BC) is a disease of tumor-forming on the breast, it is life-threatening among women, especially middle-aged females. This study aimed to develop 7 days vegetarian menu for breast cancer patients. The methods involved for food menu development in the study were Linear Programming (LP) and Integer Programming (IP) with the aid of AMPL. The results of the study showed that LP and IP could be adopted for menu planning for 7 days. In the future study, the researchers could consider designing other types of vegetarian menu using a larger food dataset.

Keywords Breast cancer · Vegetarian menu · Linear programming · Integer programming

1 Introduction

Breast cancer is the most common cancer among women in Malaysia where 1 in 30 females will be diagnosed to have breast cancer in their lifetime [1]. Diet plays a role in surviving a breast cancer diagnosis; however, there are only relatively few studies on diet and survival after breast cancer diagnosis [2]. Moreover, women with higher consumption of meat in their meals are at slightly increased risk to have breast cancer. Dairy food is perceived to be a cause of breast cancer since dairy food contains potential cancer-promoting hormones of Estrogen and insulin-like growth factor [2].

Hence, a plant-based diet could prevent people from getting breast cancer. Furthermore, it could be a health guarantee for those breast cancer patients during their treatments indirectly. The female breast cancer patients with the age from 50 to 55 years old in Malaysia were targeted in this study. This was a study about proposing the

J. Y. Tee · S. F. Sufahani (✉)
Universiti Tun Hussein Onn Malaysia, 84600 Pagoh, Johor, Malaysia
e-mail: suliadi@uthm.edu.my

S. F. Sufahani
Oasis Integrated Group, Universiti Tun Hussein Onn Malaysia, Parit Raja, 86400 Batu Pahat, Johor, Malaysia

M. S. Kaiser et al. (eds.), *Proceedings of the Third International Conference on Trends in Computational and Cognitive Engineering*, Lecture Notes in Networks and Systems 348, https://doi.org/10.1007/978-981-16-7597-3_16

vegetarian menu that contains the optimal nutrients required by breast cancer patients with the minimum cost of food using optimization approaches such as LP and IP. In this study, the economic vegetarian menu planning for 7 days that consists of optimal nutrients for breast cancer patients was proposed. There are some other research that involve optimization and economical analysis as well [3–7].

1.1 Breast Cancer

Generally, breast cancer is a tumor-forming on the breast, and it is a common life-threatening cancer among women [8]. Breast cancer rates were found to be higher in countries with high-fat diets compared with Japan and less developed countries with lower fat intake [9]. Hence, this situation led to the hypothesis that high-fat diets increase breast cancer risk. Health experts defined the term "antioxidant" as substances that protect our body cells such as beta-carotene, vitamin C, vitamin E, and selenium [10].

1.2 Linear Programming

In a study conducted by Darmon and Ferguson, they adopted the LP method to select food that satisfied nutritional and food-habit constraints at a minimal amount of energy by using the list of locally available food in Malawi for the season of harvest and non-harvest [11]. The constraints used in the models were categorized into nutritional constraints and food consumption such as food and food groups' constraints. Moreover, Ref. [12] replaced the minimum cost with the minimum amount of calories. The content of nutrients and quantities for the proposed daily menu that normally consumed by a person was successfully discovered.

1.3 Integer Programming

Integer Programming (IP) is an optimization approach to build the menu planning model problem. 426 food variables were included during modeling. LPSolve IDE as an optimization tool is applied to compute the menu planning for Eczema patients [13]. Next, an Integer Programming model for meal planning of diabetic patients was proposed to meet their nutritional requirements [14]. Moreover, ref.[15] adopted LP and IP to generate the 1-day meal plan for high blood pressure patients of 80-year-old males and 54-year-old females.

2 Methodology

2.1 Data Collection

The food data were extracted from the Malaysian Food Composition Database (MyFCD) [16]. A total of 296 and 322 vegetarian food variables were adopted for Case 1 (Ovo) and Case 2 (Ovo-Lacto). The book, Recommended Nutrients Intake RNI by [17] was taken into consideration during developing the model of vegetarian menu planning for a particular age group of breast cancer patients. The nutrition information for breast cancer patients was verified by a dietician through an interview, named Madam Nik Mahani Binti Nik Mahmood from Hospital Pakar Sultanah Fatimah (HSPF) in Johor.

According to [18], breast cancer patients would require a higher intake amount of fiber to improve their illness. Since that the main source of fiber came from vegetables and fruits, thus, the number of requirements for vegetables and fruits were increased by one, respectively, to increase the amount of fiber intake. This came out with a total of 20 dishes needed to be consumed in a day as shown in Table 1.

Breakfast, morning tea, lunch, evening tea, and dinner were the five meals included in this study. These five meals were classified according to the food group in a day as shown in Table 1. A total of 20 dishes were included in the 1-week vegetarian menu planning in the study.

Table 1 The menu planning of food items per day

Meal	Type of food group	Amount
Breakfast	Beverage	1
	Cereal flour-based	1
	Fruits	1
Morning tea	Beverage	1
	Rice flour-based	1
Lunch	Beverage	2
	Cereal-based meal	1
	Vegetables	2
	Fruits	1
	Miscellaneous	1
Evening tea	Beverage	1
	Wheat flour-based	1
	Miscellaneous	1
Dinner	Beverage	1
	Cereal-based meal	1
	Vegetables	2
	Fruits	1
Total		20

2.2 Modeling

In this study, two types of models were developed, which are Small Data Model and Big Data Model. The small data model involved only 80 types of food from each food group for both Ovo and Ovo-Lacto vegetarian diets. Moreover, the big data model involved all food items excluding meat and seafood with a total of 296 and 322 types of food were adopted for Ovo and Ovo-Lacto vegetarian diet, respectively.

Objective Function and Constraints

The objective function, decision variables, and constraints are the three main components included in the mathematical model. The objective function in this study aimed to minimize the total food cost when developing the vegetarian menu plan. In addition, the constraints were divided into two parts: the general nutritional requirements and the food group requirements. The objective function was formulated as

$$\text{Minimize total cost} = \sum_{i=1}^{N}\sum_{j=1}^{P}\sum_{k=1}^{Q} c_i x_{ijk} \tag{1}$$

where x_{ijk} is the decision variable for food items *i*, food groups *j* and meals *k*, $c_i$is the cost for each food item *i*, *N* is the total number of food, *P* is the number of food groups and *Q* is the number of meal per day.

The constraints for the general nutritional requirements

$$LB_i \leq \sum_{i=1}^{N}\sum_{j=1}^{P}\sum_{k=1}^{Q} w_i x_{ijk} \leq UB_i \tag{2}$$

where LB_i is the lower bound for each nutrient, UB_i is the upper bound for each nutrient, and w_i is the weight of the nutrient for the food.

This study adopted seven constraints of nutrients with lower bound and upper bound values except for carbohydrate and fat where both of these two nutrients have only an upper boundary. The constraints were expressed as below:

$$\text{Carbohydrates: } \sum_{i=1}^{N}\sum_{j=1}^{8}\sum_{k=1}^{5} CHO_i x_{ijk} \leq UB_{CHO} \tag{3}$$

$$\text{Protein:} LB_{Pr} \leq \sum_{i=1}^{N}\sum_{j=1}^{8}\sum_{k=1}^{5} Pr_i x_{ijk} \leq UB_{Pr} \tag{4}$$

$$\text{Fats: } \sum_{i=1}^{N}\sum_{j=1}^{8}\sum_{k=1}^{5} Fat_i x_{ijk} \leq UB_{Fat} \tag{5}$$

$$\text{Fibre: } LB_{Fb} \leq \sum_{i=1}^{N}\sum_{j=1}^{8}\sum_{k=1}^{5} Fb_i x_{ijk} \leq UB_{Fb} \tag{6}$$

$$\text{Vitamin A: } LB_{A} \leq \sum_{i=1}^{N}\sum_{j=1}^{8}\sum_{k=1}^{5} A_i x_{ijk} \leq UB_{A} \tag{7}$$

$$\text{Vitamin C: } LB_{C} \leq \sum_{i=1}^{N}\sum_{j=1}^{8}\sum_{k=1}^{5} C_i x_{ijk} \leq UB_{C} \tag{8}$$

$$\text{Iron: } LB_{Fe} \leq \sum_{i=1}^{N}\sum_{j=1}^{8}\sum_{k=1}^{5} Fe_i x_{ijk} \leq UB_{Fe} \tag{9}$$

where CHO_i is the amount of carbohydrate in gram for food item *i*, Pr_i is the amount of protein in gram for food item *i*, Fat_i is the amount of fat in gram for food item *i*, Fb_i is the amount of fiber in gram for food item *i*, $A_i$is the amount of vitamin A in microgram for food item *i*, $C_i$is the amount of vitamin C in milligram for food item *i* and Fe_i is the amount of iron in milligram for food item *i*.

2.3 Definition of Decision Variable

The menu planning model involved numerous Malaysian styles of dishes as decision variables in this study. This combination of food will appear in the menu lists according to the five classifications of meals. The decision variables will be written as below:

$$f(x) = \begin{cases} 1 \text{ or } 2, \text{ if food } i \text{ appear in the menu} \\ 0, \text{ otherwise} \end{cases} \tag{10}$$

2.4 The Lower Bound and Upper Bound of Nutrient Contents

The diet between breast cancer patients and a normal person would have slight differences. As mentioned earlier, the diet for breast cancer patients should always be high in fiber, less sugar, and fewer fats. In this proposal, two types of vegetarian diets are being considered for breast cancer patients, which are Ovo and Ovo-Lacto vegetarian diets.

Table 2 Lower and upper boundaries of the nutrients for Case 1

Nutrients	Lower bound (LB)	Upper bound (UB)
Carbohydrate (g)	–	237.5
Protein (g)	71.25	95
Fat (g)	–	142.5
Fiber (g)	20	30
Vitamin A (μg)	600	3000
Vitamin C (mg)	70	2000
Iron (mg)	11	45

Table 3 Lower and upper boundaries of the nutrients for Case 2

Nutrients	Lower bound (LB)	Upper bound (UB)
Carbohydrate (g)	–	207.5
Protein (g)	62.25	83
Fat (g)	–	124.5
Fiber (g)	20	30
Vitamin A (μg)	600	3000
Vitamin C (mg)	70	2000
Iron (mg)	11	45

Case 1: A female with the age of 51 has a bodyweight of 60 kg and she is moderately active (PAL 1.6). She is allergic to all milk products and seldom eats meats. Therefore, she decides to practice an ovo-vegetarian diet after being diagnosed as a breast cancer patient as shown in Table 2.

Case 2: A 55-year-old lady with a bodyweight of 61 kg and is low active (PAL 1.4). She does not have any food allergies and she has the thought to become a vegetarian after being diagnosed as a breast cancer patient. Therefore, she chooses to practice Ovo-Lacto vegetarian diet during her breast cancer treatment (Table 3).

2.5 Assumptions in Preparing Menu

The food items for each menu are in standardized portions and whole units. The nutrients contained in each menu are assumed to be taken as a whole unit during the process of preparing, cooking, and serving the menus.

2.6 *One-Week Vegetarian Menu*

The food menus that have been optimized were reshuffled to produce the 7-day menu, which is based on the Delete-Reshuffle Algorithm introduced by Sufahani, [19]. In the food model, every food variable was assigned to have two occurrences. When the food items were selected for the particular day, the number of occurrences in the food model was reduced by one before the next computation. This approach excluded the two food items, which were plain rice and plain water. After a 1-week menu has been produced, the menus will be reshuffled.

In this study, the 7-day vegetarian food menu was computed for both case studies. The food menus using LP and IP approaches were computed from AMPL day by day. The best optimization method was concluded based on the economical factor of each menu scheduled.

3 Results and Discussion

Table 4 shows the 7-day food price comparison for Case 1 and Case 2 for both LP and IP approach.

Table 5 shows the total food cost for a 1-week menu for Case 1 and Case 2. The total food cost of the 1-week menu planning obtained using both approaches for Case 2 is relatively cheaper than Case 1. In other words, the total food cost of 1 week for the Ovo-Lacto vegetarian menu is cheaper than the Ovo-Vegetarian menu.

The nutrients and menu planning which computed through LP and IP were adequate for both Case 1 and Case 2. The food cost of the 1-week menu produced using IP would be slightly higher than the LP approach. Thus, it is concluded that LP is the most suitable method to be used in preparing vegetarian menus.

Table 4 Food price comparison by day for the case studies

Case study (approach)	Food price (RM)						
	Day 1	Day 2	Day 3	Day 4	Day 5	Day 6	Day 7
Case 1 (LP)	9.51	11.01	12.48	14.64	9.51	11.01	12.48
Case 2 (LP)	9.16	10.73	11.81	13.12	9.16	10.73	11.81
Case 1 (IP)	9.90	11.50	11.40	16.10	9.90	11.50	11.40
Case 2 (IP)	9.40	11.65	12.70	13.60	9.40	11.65	12.70

Table 5 The cost of the 1-week menu planning for the case studies

Menu planning		Case 1	Case 2
Approach	LP	RM80.64	RM76.52
	IP	RM81.70	RM81.10

Table 6 The values of optimal nutrient for 7 days, LB and UB for Case 1 using LP approach

Nutrient	LB	Optimal values							UB
		Day 1	Day 2	Day 3	Day 4	Day 5	Day 6	Day 7	
CHO (g)	–	237.50	237.50	237.50	237.50	237.50	237.50	237.50	237.50
Pr (g)	71.25	71.25	71.25	71.25	71.25	71.25	71.25	71.25	95.00
Fat (g)	–	94.81	56.01	34.39	63.25	94.81	56.01	34.39	142.50
Fb (g)	20.00	20.00	20.00	20.00	20.00	20.00	20.00	20.00	30.00
Vit A (mg)	600.00	2485.42	1000.84	600.00	600.00	2485.42	1000.84	600.00	3000.00
Vit C (mg)	70.00	262.47	297.07	106.12	616.88	262.47	297.07	106.12	2000.00
Fe (mg)	11.00	42.86	22.57	18.76	17.61	42.86	22.57	18.76	45.00

Table 7 The values of optimal nutrients for 7 days, LB and UB for Case 2 using LP approach

Nutrient	LB	Optimal values							UB
		Day 1	Day 2	Day 3	Day 4	Day 5	Day 6	Day 7	
CHO (g)	–	207.50	207.50	207.50	207.50	207.50	207.50	207.50	207.50
Pr (g)	62.25	71.25	71.25	71.25	71.25	71.25	71.25	71.25	83.00
Fat (g)	–	80.10	23.69	54.32	36.78	80.10	23.69	54.32	124.50
Fb (g)	20.00	20.00	20.00	20.00	20.00	20.00	20.00	20.00	30.00
Vit A (mg)	600.00	2532.65	982.02	600.00	743.75	2532.65	982.02	600.00	3000.00
Vit C (mg)	70.00	269.18	334.87	383.85	70.00	269.18	334.87	383.85	2000.00
Fe (mg)	11.00	43.39	11.27	11.01	18.34	43.39	11.27	11.01	45.00

In this section, Tables 6 and 7 show the sensitivity analysis of the best menu chosen for 7 days menu planning, which is the Linear Programming method.

The results from Tables 6 and 7 show that the nutrient values computed from AMPL were all within the nutrient boundaries for 1 week in each case study. The optimal values for all the seven nutrients understudy in 7 days varied slightly among each case study.

4 Conclusion

It is concluded that the Linear Programming method is suitable to prepare a vegetarian menu that achieved a lower price as compared with Integer Programming. This study has contributed a lot to society, especially to the medical sector. The limitations of the study were also pointed out. Moreover, recommendations were also discussed in detail about the improvement to be made during the further study. Overall, the optimization approaches of Linear Programming and Integer Programming could be applied during the menu planning and scheduling for particular consumers.

Acknowledgements Fundamental Research Grant Scheme (FRGS) Number K175 (FRGS/1/2019/STG06/UTHM/02/2) from the Ministry of Higher Education Malaysia (MOHE) and Universiti Tun Hussein Onn Malaysia (UTHM).

References

1. Indramalar, S.: The Star, pp. 1–2 (2019). https://www.thestar.com.my/lifestyle/family/2019/10/16/breast-cancer-2/. Accessed 20 June 2020
2. WCRF/AICR: Diet, Nutrition, Physical Activity and Breast Cancer Survivors. Continuous Update Project Expert Report (2018)
3. Sudin, A.M., Sufahani, S.: Mathematical approach for serving nutritious menu for secondary school student using "delete-reshuffle-reoptimize algorithm." J. Phys. Conf. Ser. **995**(1), 1–7 (2018)
4. Sufahani, S.F., Kamardan, M.G., Rusiman, M.S., Mohamad, M., Khalid, K., Ali, M., Khalid, K., Nawawi, M.K.M, Ahmad, A.: A mathematical study on "additive technique" versus "branch and bound technique" for solving binary programming problem. J. Phys. Conf. Ser. **995**(1), Article number 012001 (2018). https://doi.org/10.1088/1742-6596/995/1/012001
5. Latif, M.S.A., Fahmy-Abdullah, M., Sieng, L.W.: Determinants factor of technical efficiency in machinery manufacturing industry in Malaysia. Int. J. Supply Chain Manage. **8**, 917–928 (2019)
6. Soebagyo, D., Fahmy-Abdullah, M.O.H.D., Sieng, L. W., Panjawa, J.L.: Income Inequality and Convergence in Central Java under Regional Autonomy. Int. J. Econ. Manage. **13**(1) (2019)
7. Idris, A.I.M., Fahmy-Abdullah, M., Sieng, L.W.: Technical efficiency of soft drink manufacturing industry in Malaysia. Int. J. Supply Chain Manage. **8**, 908–916 (2019)
8. Harvie, M., Howell, A., Evans, D.G.: Can diet and lifestyle prevent breast cancer: what is the evidence? Am. Soc. Clin. Oncol. Educ. Book **35**, 66–73 (2015)
9. Jemal, A., Bray, F., Center, M.M., Ferlay, J., Ward, E., Forman, D.: Global cancer statistics. CA Cancer J. Clin. **61**, 69–90 (2011)
10. Key, T.J., Verkasalo, P.K., Banks, E.: Epidemiology of breast cancer. Lancet Oncol. **2**(3), 133–140 (2001)
11. BC Cancer Agency, & HealthLink BC.: A Nutrition Guide for Women with Breast Cancer (2012). https://www.bccancer.bc.ca. Accessed 20 June 2020
12. Darmon, N., Ferguson, E.: Linear and nonlinear programming to optimize the nutrient density of a population's diet: an example based on diets of preschool children in rural Malawi. Am. Soc. Clin. Nutr. **75**(2), 245–253 (2018)
13. Hreţcanu, C., Hreţcanu, C.: A linear programming model for a diet problem. Food Environ. Saf. J. **9**(1), 56–63 (2010)
14. Sheng, L.Z., Sufahani, S.: Optimal diet planning for eczema patient using integer programming. J. Phys: Conf. Ser. **995**(1), 1–9 (2018)
15. Sapri, N.S.M., Bedi, M.R.B.D.S., Abdul-Rahman, S., Benjamin, A.M.: A diet recommendation for diabetic patients using integer programming. AIP Conf. Proc. **2138**, 1–7 (2019)
16. Su Hui, L., Sufahani, S.F., Khalid, K., Wahab, M.H.A., Idrus, S.Z.S, Ahamd, A. & Subramaniam, T.S.: Menu scheduling for high blood pressure patient with optimization method through Integer Programming. J. Phys. Conf. Ser. **1874**(1), Article number 012088 (2021). https://doi.org/10.1088/1742-6596/1874/1/012088

17. IMR: Protocol for Sampling and Methods of Analysis for Malaysian Food Composition Database (2011). https://www.imr.gov.my/testlist/js/pdfjs/Protocol_Sampling_MY_FCD.pdf. Accessed 20 June 2020
18. Sieri, S., Krogh, V., Ferrari, P., Berrino, F., Pala, V., Thiébaut, A.C.M., Tjønneland, A., Olsen, A., Overvad, K., Jakobsen, M.U., Clavel-Chapelon, F., Chajes, V., Boutron-Ruault, M.C., Kaaks, R., Linseisen, J., Boeing, H., Nöthlings, U., Trichopoulou, A., Naska, A., Lagiou, P., Panico, S., Palli, D., Vineis, P., Tumino, R., Lund, E., Kumle, M., Skeie, G., González, C.A., Ardanaz, E., Amiano, P., Tormo, M.J., Martínez-García, C., Quirós, J.R., Berglund, G., Gullberg, B., Hallmans, G., Lenner, P., Bueno-de-Mesquita, H.B., Van Duijnhoven, F.J.B., Peeters, P.H.M., Van Gils, C.H., Key, T.J., Crowe, F.L., Bingham, S., Khaw, K.T., Rinaldi, S., Slimani, N., Jenab, M., Norat, T., Riboli, E.: Dietary fat and breast cancer risk in the European prospective investigation into cancer and nutrition. Am. J. Clin. Nutr. **88**(5), 1304–1312 (2008)
19. Ministry of Health Malaysia: Recommended Nutrient Intakes for Malaysia 2005. Ministry of Health Malaysia (2017)

Selection of Intrinsic Mode Function in Ensemble Empirical Mode Decomposition Based on Peak Frequency of PSD for EEG Data Analysis

Mohd Nurul Al Hafiz Sha'abani, Norfaiza Fuad, Norezmi Jamal, and Engku Mohd Nasri Engku Mat Nasir

Abstract Ensemble empirical mode decomposition (EEMD) is a powerful algorithm to decompose non-linear and non-stationary signals into several components called intrinsic mode function (IMF). EEMD has been used in EEG signal analysis, where the extracted IMFs need to be chosen properly to ensure the unwanted signal is effectively excluded. However, the method of selecting IMF has not been discussed widely. For that reason, this paper presents a method for selecting an appropriate IMF based on peak frequency of power spectral density for EEG application. The IMF is selected if the peak frequency lies on the brainwave signals of interest. Then, the selected IMFs are reconstructed as a clean EEG signal. A two-class problem of EEG cognitive task database was demonstrated to observe the feasibility of the selected IMFs. Shannon entropy and kurtosis were calculated on theta, alpha, and beta subbands of the reconstructed EEG signal. Several established classifiers were employed in the classification process. Results show that the classification achieved an acceptable accuracy, which varied between 70 and 86%.

Keywords Empirical mode decomposition · Intrinsic mode function · Electroencephalogram

M. N. A. H. Sha'abani (✉)
Electrical Engineering Department, Center for Diploma Studies, Universiti Tun Hussein Onn Malaysia, Pagoh, Malaysia
e-mail: nhafiz@uthm.edu.my

N. Fuad · N. Jamal · E. M. N. Engku Mat Nasir
Electronic Engineering Department, Universiti Tun Hussein Onn Malaysia, Parit Raja, Malaysia
e-mail: norfaiza@uthm.edu.my

N. Fuad
Computational, Signal, Imaging and Intelligent Focus Group (CSII), Parit Raja, Malaysia

M. N. A. H. Sha'abani
Microcontroller Technology for IoT Focus Group (MTIT), Pagoh, Malaysia

M. S. Kaiser et al. (eds.), *Proceedings of the Third International Conference on Trends in Computational and Cognitive Engineering*, Lecture Notes in Networks and Systems 348, https://doi.org/10.1007/978-981-16-7597-3_17

1 Introduction

Electroencephalograph (EEG) is one of the most widely used in groups of non-invasive techniques for analyzing neuronal activity of the brain. It is applied in clinical diagnosis, cognitive research, and human–machine interaction due to its advantages of portability, low cost, and non-invasive [1]. Principally, the electrical activity of the brain is recorded using EEG devices such as Emotiv Insight, InteraXon Muse, OpenBCI, and many more [2]. These devices consist of multiple electrodes that followed the standard 10–20 systems [3]. Several studies have stated that a fewer number of electrodes is feasible to be used instead of a higher number of electrodes to avoid the computational burden [4, 5]. For example, EEG-related studies in mental workload [6] and mind wandering [7] are using a minimal number of electrodes to capture the raw EEG signals on the frontal lobe area. Even though fewer electrodes are more ideal to be applied, the process of removing artifacts becomes challenging.

Recorded EEG signals are usually distorted by other physiological signals such as eye blink, heartbeat, and muscle movement signals. Conventional artifact removal method such as regression and adaptive filter requires additional reference channels [8]. Whereas decomposition method such as independent component analysis (ICA) requires a high number of EEG electrodes to be working properly. As an alternative, wavelet transform (WT) and empirical mode decomposition (EMD) can decompose a single source signal [9]. WT decomposes signal based on predefined basis function with a fixed set of high and low pass filters. Without knowing the pattern of the signal of interest, the chosen optimal basis function would be difficult [10]. Otherwise, EMD decomposes non-linear and non-stationary signals naturally without prior knowledge of the interest signal. The original EMD is facing a mode mixing problem. To overcome this issue, a noise-assisted EMD or ensemble EMD (EEMD) has been proposed [11]. EEMD decomposed a single EEG signal into a series of intrinsic mode function (IMF). However, not all IMF contained the valuable information of the targeted EEG application. Hence, it is important to select the significant IMF effectively.

Several works on selecting IMF have been found in the literature. Each method is using a different approach depending on specific applications. D. Boutana et al. proposed an IMF selection method based on Minkowski distance and Jensen Renyi divergence for heart sound signal application [12]. The method purposes to eliminate the redundancy of IMF. S. Cho et al. identify the significance of IMF based on energy and frequency peaks for vibration signal of an induction motor fault diagnosis [13]. IMF that has higher energy and contains fault-related peak was selected. In sleep disorder diagnosis, M. R. Islam et al. decomposed EEG signal using EMD and select the optimal IMF based on the classification result of entropy, standard deviation, skewness and kurtosis features. However, this method is not selecting IMF before the classification stage. O. Karabiber Cura et al. use a combination of energy, correlation, power spectral, and statistical to select IMF for epileptic seizure classification. The obtained IMFs are ranked according to the four selection criteria. Kotan et al. [14] compared three different IMF selection methods based on energy content [15], correlation [16], and power spectral density [17] for multiple sclerosis EEG data. By using three different classifiers, the author concluded that no method shows dominant of each other. Nevertheless, the highest classification result was obtained using k-nearest neighbor (kNN) when the power-based IMF selection method was used. Either energy, correlation or PSD method as the characteristic of selected IMF, a threshold value is required, which is left to be subjective to determine from one signal to another.

Brainwave is composed of five main frequency subbands; delta (0.5–4 Hz), theta (4–8 Hz), alpha (8–13 Hz), beta (13–30 Hz), and gamma (30–45 Hz). Meanwhile, the dyadic filter characteristic of EEMD makes the frequency content of the decomposed EEG signal may largely fall into the brainwave frequency subband naturally. For that reason, this paper proposed an IMF selection method based on peak frequency of power spectral density (PSD) to select the appropriate IMFs. By this method, it is assumed that any IMF which has peak frequency that falls within the brainwave subband of interest has more significant information of EEG signal. The detailed implementation of the proposed method is explained in the next section.

2 Methods

2.1 Dataset

The raw EEG dataset used to analyze the feasibility of the proposed IMF selection method was taken from a cognitive task experiment as in [18]. Thirty-one healthy subjects aged between 19.2 ± 0.7 years were voluntarily involved in the data collection. In that study, subjects were told to build a structure using a set of interlocking block toys within a limited time of 2 minutes. EEG signals and time of completion were recorded during the task execution. The recorded EEG signal was sampled at 128 Hz. A highpass filter at 0.1 Hz and a 50 Hz notch filter was applied. Only two frontal channels were utilized, specifically at AF3 and AF4. Based on the visual inspection of the recorded EEG signals, one subject was excluded from analysis due to poor EEG quality. So, the final sample size is 30 subjects. The subjects were grouped into two classes (i.e., Good and Bad) based on completion time by using a histogram. 10 samples were grouped in Good class and 20 samples were grouped in Bad class.

2.2 Ensemble Empirical Mode Decomposition

Ensemble empirical mode decomposition (EEMD) is a noise-assisted version of EMD. EMD is a data-driven decomposition method for non-linear and non-stationary signals. A non-stationary signal is decomposed into several IMFs through a sifting process according to two conditions: (i) the number of extreme points and number of zero-crossing should be equal or differ at most by one and (ii) the mean of local maxima envelope and local minima envelope should be zero. The operation of EEMD is similar to EMD, but the only difference is EEMD, which has an ensemble operation that added a series of white noise. The procedures of decomposition by EEMD are as follows:

i. Add a series of white noise, $w(t)$ to the original signal, $x(t)$ with noise amplitude a.
ii. Find local minima and local maxima and construct lower, e_{min} and upper, e_{max} envelopes using cubic spline interpolation.
iii. Calculate the mean of the envelopes, $m(t)$.
iv. Compute the difference, $d(t)$ between signal $x(t)$ and the mean envelopes, $m(t)$ such $d(t) = x(t) - m(t)$.
v. Repeat step (ii)–(iv) until $d(t)$ satisfies the IMFs basic conditions.
vi. Take $d(t)$ as IMF1 if $d(t)$ satisfying the IMF conditions.
vii. Repeat step (ii)–(vi) to find the next IMFs by replacing the $x(t)$ with residue $r(t)$, where $r(t) = x(t) - IMF1$.
viii. Stop the iteration if $r(t)$ exhibits a monotonic function.
ix. Repeat step (i)–(viii) with different series of $w(t)$ for n number of ensembles and find the ensemble mean of the corresponding IMFs as the final result.

The original signal, $x(t)$ can be represented as

$$x(t) = \sum_{i=1}^{K} IMF_i + r_K(t) \tag{1}$$

where K is the number of extracted IMF, IMF_i is the ith IMF and $r_K(t)$ is the final residual. In this paper, the noise amplitude, a was set to 0.4, 200 number of ensembles, n and 10 levels of IMFs were extracted. Figure 1 shows one of the EEG signal decompositions by using EEMD from the EEG database.

2.3 *IMF Selection Approach*

EEG is composed of several frequency subbands, where each subband is related to specific brain activities [19]. Hence, the frequency content of an IMF is important to determine its significance to the brainwave rhythm. Power spectrum can be used to describe the distribution of signal power over frequency. The power spectrum of a signal is estimated using Fourier transform method. In this paper, the Welch method [20] was used to estimate the power spectrum density.

Figure 2 illustrates the power spectral density of each IMFs from each sample in the EEG dataset used. It is clearly shown that the EEG signals were decomposed from higher to lower frequencies within a range of band frequency. Notice that the

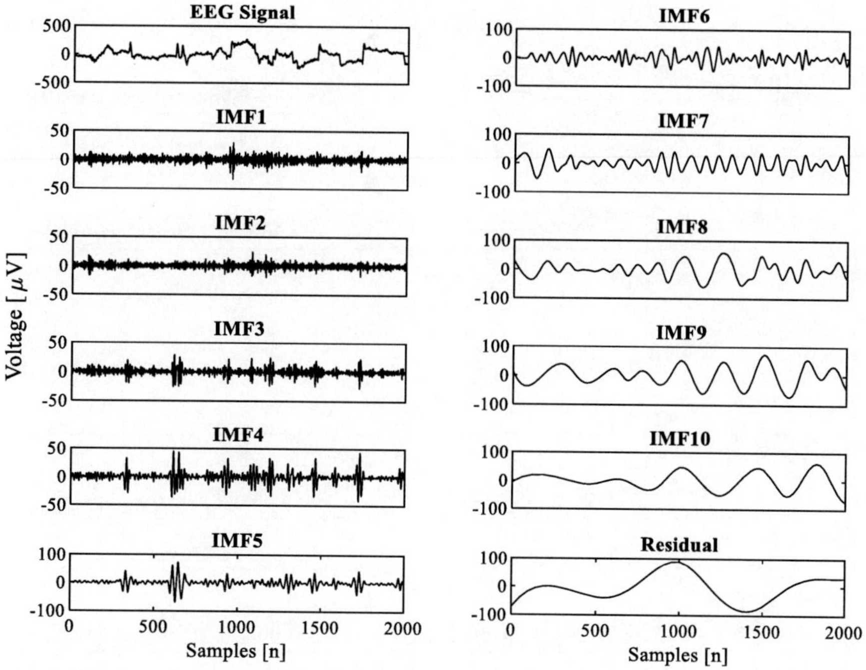

Fig. 1 EEG signal decomposition by EEMD method

peak frequency of IMFs lies within the brainwave subband such as IMF1 to IMF5 lies within gamma, beta, alpha, and theta subbands, respectively, whereas IMF6 to IMF10 lies within the delta subband. This indicates that the IMFs can be selected based on the peak frequency of power spectral density.

In this selection method, IMF is assumed not significant when the corresponding frequency of its peak power does not fall into the subband frequency of interest. The procedure of the IMF selection method is as follow:

i. Find peak amplitude, P_{max} of the power spectrum for each IMF.
ii. Obtain the frequency, f_p that corresponds to Pmax.
iii. Check whether f_p is lies within the interest EEG frequency subband, b.
iv. If $b_{low} \leq f_p \leq b_{high}$, then select IMF.
v. If several frequency bands are involved, for example, theta (4–8 Hz) and alpha (8–13 Hz), the b_{low} can be assigned as 4 Hz and b_{high} as 13 Hz.

In this study, the subbands of interest were theta, alpha, and beta since those subbands are dominant during cognitive activities [21]. After the selection process, the EEG signals were reconstructed by adding up all the selected IMFs.

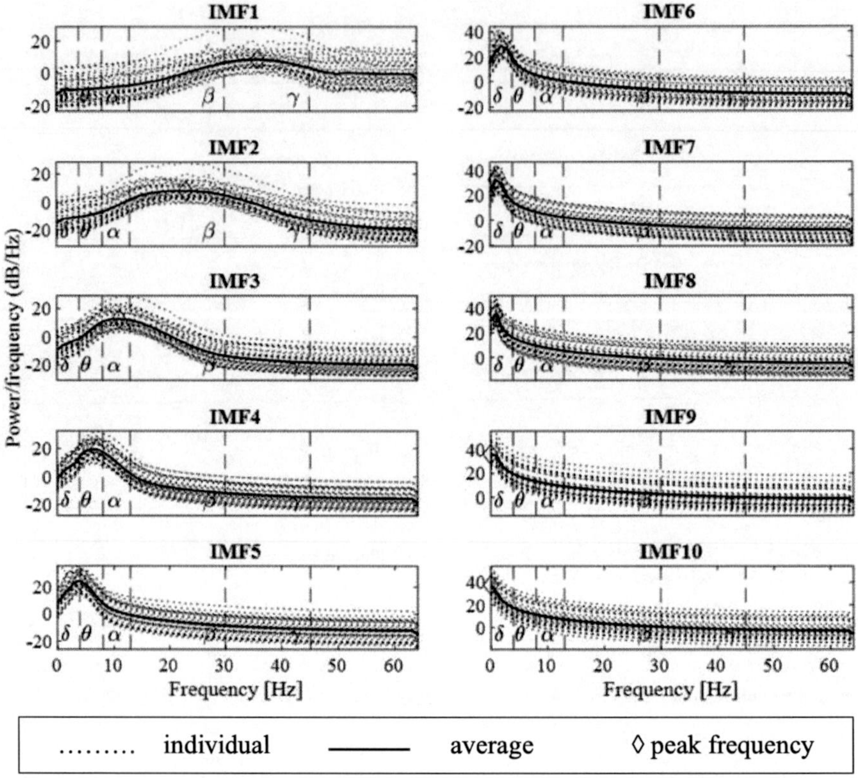

Fig. 2 Welch power spectral density of individual IMF by all subjects and its average

2.4 Performance

The performance of the selected IMFs was observed by classifying the sample of the used EEG dataset into two classes. Initially, theta, alpha, and beta subbands were obtained from the reconstructed EEG signal by using a bandpass filter. Then, features are extracted from each of those subbands. In this study, Shannon entropy and kurtosis were extracted as a feature set. Several classifiers as shown in Table 1 were used. The parameters' setting of each classifier is also provided. 10-fold cross-validation was utilized for all classification processes, except for the neural network, 60% of samples are used for training and 40% for testing.

3 Results and Analysis

Table 2 shows the range of selected IMFs associated with theta, alpha, and beta range

Table 1 Type of classifiers used for validation and their parameters setting

Classifiers	Type/model/learner	Parameters
Decision tree	Gini's diversity index	Maximum split number = 4
Discriminant analysis	Linear	–
Naïve Bayes	Gaussian kernel	–
SVM	Quadratic kernel	–
kNN	Cubic	Minkowski distance, k = 10
Ensemble	Subspace discriminant	Subspace dimension = 12
Neural network	Feed Forward	Hidden layers = 2, neuron = 10

Table 2 Selected IMFs at channels AF3 and AF4

Channel	Selected IMF	Number of samples
AF3	IMF1–IMF4	1
	IMF2–IMF4	5
	IMF2–IMF5	24
AF4	IMF1–IMF4	1
	IMF2–IMF4	4
	IMF2–IMF5	25

frequencies (i.e., 4–30 Hz). At both channel AF3 and AF4, the most selected IMF range was from IMF2–IMF5, which was found from 24 and 25 samples, respectively, whereas there are minor selections from IMF1–IMF4 and IMF2–IMF4. To be noted, the EEG signals recorded in the database were sampled at 128 Hz. The difference in sampling rate may return a different result of IMF selection.

Table 3 shows the classification results of two classes of the EEG database used. Most classifiers obtained more than 70% of accuracies. The highest accuracy was obtained by the kNN classifier with 86%, while the lowest was obtained by the

Table 3 Classification results for two classes

Classifiers	Accuracy (%)
Decision tree	70.0 ± 4.1
Discriminant analysis	76.6 ± 4.1
Naïve Bayes	60.0 ± 0.0
SVM	80.1 ± 1.5
kNN	86.0 ± 4.4
Ensemble	84.7 ± 1.9
Neural network	80.1 ± 14.5

Naïve Bayes classifier with only 60%. Since EEG signal is a non-linear problem, linear classifier such Naïve Bayes was not able to obtain a good result. In comparison, non-linear classifiers such as kNN, SVM, Ensemble, and Neural Network can provide higher accuracy.

4 Conclusion

In this study, a new IMF selection method of empirical mode decomposition-based technique is proposed using peak frequency of power spectral density. The method selected IMF according to the contribution of the IMF's frequency content to the preferred brainwave subband frequency. Several classifiers were used to study the feasibility of the selected IMFs by demonstrating two-class problem. The result shows a promising outcome with acceptable accuracies. To be noted, the order of selected IMF might be different depending on the sampling frequency of a recorded EEG signal.

Acknowledgements The author would like to thank the Universiti Tun Hussein Onn Malaysia for awarding the author a PhD scholarship. An appreciation also goes to the Center for Diploma Studies and Faculty of Electrical Engineering for supporting the author.

References

1. Xu, J., Zhong, B.: Review on portable EEG technology in educational research. Comput. Hum. Behav. **81**, 340–349 (2018). https://doi.org/10.1016/j.chb.2017.12.037
2. LaRocco, J., Le, M.D., Paeng, D.-G.: A systemic review of available low-cost EEG headsets used for drowsiness detection. Front. Neuroinformat. **14**, 553352 (2020). https://doi.org/10.3389/fninf.2020.553352
3. Homan, R.W., Herman, J., Purdy, P.: Cerebral location of international 10–20 system electrode placement. Electroencephalogr. Clin. Neurophysiol. **66**(4), 376–382 (1987). https://doi.org/10.1016/0013-4694(87)90206-9
4. James, C., Lowe, D.: Extracting information from multichannel versus single channel EEG data in epilepsy analysis. In: Hyder, A.K., Shahbazian, E., Waltz, E. (Eds.) Multisensor Fusion, vol. 1, pp. 889–895. Springer, Dordrecht (2002). https://doi.org/10.1007/978-94-010-0556-2_47
5. Troy M, L., Joseph T, G., Daniel P, F.: How many electrodes are really needed for EEG-based mobile brain imaging? J. Behav. Brain Sci. **2**(3), 387–393 (2012). https://doi.org/10.4236/jbbs.2012.23044
6. So, W.K., Wong, S.W., Mak, J.N., Chan, R.H.: An evaluation of mental workload with frontal EEG. PloS ONE **12**(4), e0174949 (2017). https://doi.org/10.1371/journal.pone.0174949
7. van Son, D., De Blasio, F.M., Fogarty, J.S., Angelidis, A., Barry, R.J., Putman, P.: Frontal EEG theta/beta ratio during mind wandering episodes. Biol. Psychol. **140**, 19–27 (2019). https://doi.org/10.1016/j.biopsycho.2018.11.003
8. Islam, M.K., Rastegarnia, A., Yang, Z.: Methods for artifact detection and removal from scalp EEG: a review. Neurophysiologie Clinique/Clin. Neurophysiol. **46**(4), 287–305 (2016). https://doi.org/10.1016/j.neucli.2016.07.002

9. Jiang, X., Bian, G.-B., Tian, Z.: Removal of artifacts from EEG signals: a review. Sensors **19**(5), 987 (2019). https://doi.org/10.3390/s19050987
10. Chen, X., Xu, X., Liu, A., McKeown, M.J., Wang, Z.J.: The use of multivariate EMD and CCA for denoising muscle artifacts from few-channel EEG recordings. IEEE Trans. Instrum. Meas. **67**(2), 359–370 (2017). https://doi.org/10.1109/TIM.2017.2759398
11. Wu, Z., Huang, N.E.: Ensemble empirical mode decomposition: a noise-assisted data analysis method. J. Adv. Adapt. Data Anal. **1**(01), 1–41 (2009). https://doi.org/10.1142/S1793536909000047
12. Boutana, D., Benidir, M., Barkat, B.: On the selection of intrinsic mode function in EMD method: application on heart sound signal. In: 3rd International Symposium on Applied Sciences in Biomedical and Communication Technologies, pp. 1–5. IEEE (2010). https://doi.org/10.1109/ISABEL.2010.5702895
13. Cho, S., Shahriar, M. R., Chong, U.: Identification of significant intrinsic mode functions for the diagnosis of induction motor fault. J. Acoust. Soc. Am. **136**(2), EL72-EL77 (2014). https://doi.org/10.1121/1.4885541
14. Kotan, S., Van Schependom, J., Nagels, G., Akan, A.: Comparison of IMF selection methods in classification of multiple sclerosis EEG data. In: Medical Technologies Congress, pp. 1–4. IEEE (2019). https://doi.org/10.1109/TIPTEKNO.2019.8895091
15. Junsheng, C., Dejie, Y., Yu, Y.: Research on the intrinsic mode function (IMF) criterion in EMD method. Mech. Syst. Signal Process. **20**(4), 817–824 (2006). https://doi.org/10.1016/j.ymssp.2005.09.011
16. Peng, Z., Peter, W.T., Chu, F.: A comparison study of improved Hilbert-Huang transform and wavelet transform: application to fault diagnosis for rolling bearing. Mech. Syst. Signal Process. **19**(5), 974–988 (2005). https://doi.org/10.1016/j.ymssp.2004.01.006
17. Kotan, S., Akan, A.: A new intrinsic mode function selection method based on power spectral density. In: Medical Technologies National Congress, pp. 1–4. IEEE (2018). https://doi.org/10.1109/TIPTEKNO.2018.8597127
18. Sha'abani, M.N.A.H., Fuad, N., Jamal, N., Marwan, M., Abd Wahab, M.H., Idrus, S. Z.S.: Development of cognitive and psychomotor task for EEG application with matlab-based GUI. In: IOP Conference Series: Materials Science and Engineering, pp. 012050. vol. 917, no. 1, IOP Publishing (2020). https://doi.org/10.1088/1757-899X/917/1/012050
19. Teplan, M.: Fundamentals of EEG measurement. Meas. Sci. Rev. **2**(2), 1–11 (2002). https://www.measurement.sk/2002/S2/p2.html
20. Kim, D.-W., Im, C.-H.: EEG spectral analysis. In: Im, C.-H. (Ed.) Computational EEG Analysis: Methods and Applications, vol. 1, pp. 35–53. Springer, Singapore (2018). https://doi.org/10.1007/978-981-13-0908-3
21. Plechawska-Wójcik, M., Tokovarov, M., Kaczorowska, M., Zapała, D.: A three-class classification of cognitive workload based on EEG spectral data. J. Appl. Sci. **9**(24), 5340 (2019). https://doi.org/10.3390/app9245340

A Brief Review of Computation Techniques for ECG Signal Analysis

Salleh Sh Hussain, Fuad Noman, Hadri Hussain, Chee-Ming Ting, Syed Rasul G. Syed bin Hamid, Hadrina Sh-Hussain, M. A. Jalil, Ahmad Zubaidi A.L., Syed Zuhaib Haider Rizvi, Kuryati Kipli, Kavikumar Jacob, Kanad Ray, M. Shamim Kaiser, Mufti Mahmud, and Jalil Ali

Abstract Automatic detection of life-threatening cardiac arrhythmias has been a subject of interest for many decades. The automatic ECG signal analysis methods are mainly aiming for the interpretation of long-term ECG recordings. In fact, the experienced cardiologists perform the ECG analysis using a strip of ECG graph paper in

---

S. S. Hussain · H. Hussain · H. Sh-Hussain
HealUltra PLT, Taman Pulai Utama, No-20 Jalan Pulai 18, 81300 Skudai, Johore, Malaysia
e-mail: sh.hussain@tutanota.com

F. Noman (✉) · C.-M. Ting
School of Information Technology, Monash University Malaysia, 47500 Bandar Sunway, Selangor, Malaysia
e-mail: fuad.noman@monash.edu

C.-M. Ting
e-mail: ting.cheeming@monash.edu

S. R. G. S. bin Hamid
Department of Cardiothoracic Surgery, Hospital Sultanah Aminah, 80100 Johor Bahru, Johor, Malaysia

M. A. Jalil
Department of Physics, Unviersiti Teknologi Malaysia, 81310 Skudai, Johor, Malaysia

A. Z. A.L.
Universiti Sultan Zainal Abidin, Medical Campus, 20400 Kuala Terengganu, Terengganu, Malaysia

S. Z. H. Rizvi · K. Jacob
Universiti Tun Hussein Onn Malaysia, Parit Raja, Malaysia
e-mail: syedzuhaib@uthm.edu.my

K. Kipli
Universiti Malaysia Sarawak (UNIMAS), Kota Samarahan, Malaysia
e-mail: kkuryati@unimas.my

K. Ray
Amity School of Applied Sciences, Amity University, Jaipur, India

M. Shamim Kaiser
Institute of Information Technology, Jahangirnagar University, Savar, Dhaka, Bangladesh
e-mail: mskaiser@juniv.edu

M. S. Kaiser et al. (eds.), *Proceedings of the Third International Conference on Trends in Computational and Cognitive Engineering*, Lecture Notes in Networks and Systems 348, https://doi.org/10.1007/978-981-16-7597-3_18

an event-by-event manner. This manual interpretation becomes more difficult, time-consuming, and more tedious when dealing with long-term ECG recordings. Rather, an automatic computerized ECG analysis system will provide valuable assistance to the cardiologists to deliver fast or remote medical advice and diagnosis to the patient. However, achieving accurate automated arrhythmia diagnosis is a challenging task that has to account for all the ECG characteristics and processing steps. Detecting the P wave, QRS complex, and T wave is crucial to perform automatic analysis of EEG signals. Most of the research in this area uses the QRS complex as it is the easiest symbol to detect in the first stage. The QRS complex represents ventricular depolarization and consists of three consequences waves. However, the main challenge in any algorithm design is the large variation of QRS, P, and T waveform, leading to failure for each method. The QRS complex may only occupy R waves QR (no R), QR (no S), S (no Q), or RSR, depending on the ECG lead. Variations from the normal electrical patterns can indicate damage to the heart, and these variations are manifested as heart attack or heart disease. This paper will discuss the most recent and relevant methods related to each sub-stage, maintaining the related literature to the scope of ECG research.

Keywords Classification · ECG · Feature extraction · Segmentation

1 Introduction

The heart has the ability to produce electrical pulses that travel through the myocardium cells to stimulate the heart chambers to contract in a cyclic manner. In fact, the brain is the main source of electrical pulses in the human body, but the heart has its own pacemakers that work independently of the brain.

The electrical conduction system of the heart is shown in Fig. 1a. The elements of the conduction system include the sinoatrial node (SA), atrioventricular node (AV), the bundle of His, left and right bundle branches, and Purkinje fibers [1]. The SA node is located at the superior wall (top-left corner) of the right atrium and acts as the primary pacemaker. The electrical pulses that originate from the SA node are propagated rapidly through the conduction fibers (or electrical bridges) and spread to the rest of the right and left atriums. Those pulses will stimulate both atriums to contract; synchronically, the AV node is another group of cells known as an ectopic pacemaker which acts as a hub to forward the pulses generated by the SA node down to the ventricular. If malfunction to the impulses generated by the SA node, the AV

M. Mahmud
Department of Computer Science, Nottingham Trent University, Clifton Lane, Nottingham NG11 8NS, UK
e-mail: mufti.mahmud@ntu.ac.uk

J. Ali
Asia Metropolitan University, Masai, Malaysia
e-mail: arifjalil@utm.my; jalilali@amu.edu.my

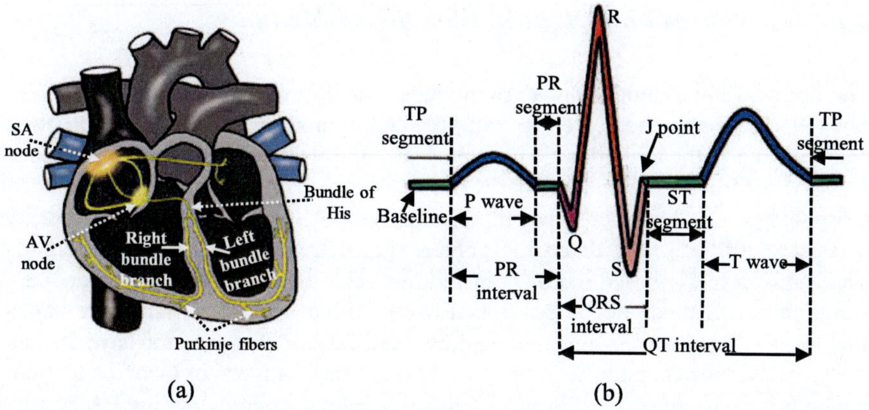

Fig. 1 **a** The electrical conduction system of the heart. **b** A typical ECG heartbeat, showing the ECG fundamental components, P wave, QRS segment, and T wave

node becomes the heart's pacemaker. Once the impulses of the SA node are received at the AV node, it delays the forwarding for about 0.12 s. This is important to ensure that the atriums had emptied their blood into the ventricles before the ventricles were stimulated for contraction.

The action potential of the heart's conduction system produces a small electrical current, which reflects the total electrical activity of the heart muscle. The heart electrical activity measurement is known as electrocardiography, which is still the most popular pre-screening tool in clinical settings. Figure 1b shows the basic components of a single ECG heartbeat.

2 Cardiovascular Signal Processing

Signal processing is the art and science of exploring time-series data, which involves a variety of simple and advanced operations and transformations to understand, analyze, or enhance the information contained in the data. Biomedical signal processing is the field of processing the signals produced by the human biological systems. Biosignals are the medium of communication between the machines or doctors, and the body organs, such as ECG and heart sound (PCG), are used to reflect the heart's health condition.

2.1 Derivatives and Digital Filter-Based Methods

The derivative and digital filters are the most widely used methods in the 80s and early 90s. These methods are less complex and do not require high computational resources due to the limitations in the processing and electronics technologies. Examples of time-domain methods are derivative methods, digital filters, and mathematical morphology. On the other hand, the second half of the 90s and the first decade of this century are considered the golden age of transformation-based methods. The transformation-based methods include wavelet transform (WT), phasor transform, empirical mode decomposition (EMD), Hilbert transform, and filter banks. Finally, machine learning methods are part of artificial intelligence and have increasingly gained more popularity in the past decade. Various types of machine learning methods have been involved in ECG fiducial point detection, including HMM, artificial neural network (ANN), k-nearest neighbor (KNN), K-means, support vector machine (SVM), and state-space modeling. This paper provides an overview of these algorithms and the recent developments linked to ECG fiducial point detection. The overview is focused on the description of the principles of the most but not all recent and famous methods reported in the literature. Algorithmic details can be found in the original work as per the citations.

2.2 Transformation-Based Methods

The Fourier transform is a technique used to transform the time-domain time-series signal into its global frequency spectrum. The short-term Fourier transform (STFT) solves the temporal resolution issue by using a windowed version of the Fourier transform. The wavelet transform provides multiple resolutions in different frequency bands, which saves both the time and frequency resolutions from the trade-off effect in STFT by using a set of wavelet functions called mother wavelet $\psi(t)$. The wavelet function can be obtained from (1),

$$\psi_{u,s}(t) = \frac{1}{\sqrt{|s|}}\psi\left(\frac{t-u}{s}\right) \tag{1}$$

where $s = 2^j$ is the wavelet frequency scale (dilation), which either contracts or expands the wavelet function. u is the translation parameter and is used to shift the location of the wavelet function. $\frac{1}{\sqrt{|s|}}$ is used to normalize the wavelet function energy level for different scales [2]. Yochum et al. [3] used a CWT-based method to delineate P, QRS, T waves since the CWT provides a constant frequency resolution over the wavelet scales, unlike DWT, which step-down (by half) the resolution at each scale. The author chooses the Daubechies-3 wavelet function, and the scale s was experimentally set to 38, in which the correlation with QRS complex is maximized. An automatic threshold and zero-crossing approach were used to find the R-peaks.

A temporal mask is applied to the transformed signal to restrain the P and T waves. Similarly, T waves were detected first by restraining the QRS complex from the original ECG signal (subtraction of ECG and masked ECG), then CWT was applied, and the maxima's or minima's points with zero-crossing were used to find T-peak. P wave was detected following the same procedure as T wave. However, the CWT keeps the frequency resolution through the lower scales; which makes the CWT contain much redundant data and requires more power and computational resources. According to [4], the CWT power consumption was four times higher compared with DWT. Therefore, the CWT is better to be used in offline ECG analysis.

Martínez et al. (2010) [5] proposed a phasor transform (PT)-based method for ECG (P, QRS, and T-wave) delineation method. The PT transfers the real part of the sinusoidal function: $x[t] = A\cos(\omega t \pm \Phi)$ from the time domain into the complex number domain which is also called the frequency domain. The phasor $y[t]$ was calculated at each time step t, $y[t] = R_v + jx[t]$, where R_v is a constant value. The ECG samples were used as the imaginary part of the phasor. The magnitude of the phasor was found by $M[t] = \sqrt{R_v^2 + x[t]^2}$, where the phase Φ was computed using the inverse-tangent $\Phi[t] = \tan^{-1}\left(\frac{x[n]}{R_v}\right)$. The R-peak is detected by applying a fixed threshold (0.003) to the phasor signal, the maxima point higher than the threshold is considered R-peak. The Q and S are detected around the candidate R-peak where the phasor signal is 25% lower than the max-value. The P and T waves are delineated in the same manner, using a search window around the detected QRS complex. The suggested approach is robust, mathematically simple and has low computational cost.

2.3 Machine Learning-Based Methods

Machine learning is an application of artificial intelligence (AI) that allows computers to be trained and automatically learns to assess humans on decision-making more accurately. Machine learning methods can be categorized into three groups: (1.) supervised learning methods that require hand-crafted labels for the observation or input data; (2.) the unsupervised learning methods, in contrast, are self-learning methods and do not require the labels for observation symbols; (3.) the semi-supervised methods are a hybrid system when the supervised techniques are combined with unsupervised. Here, some but not all of the recent machine learning-related previous work on ECG delineation is briefly discussed.

Akhbari et al. [6], in their recent work, proposed a trending approach in the ECG delineation research field. The approach is based on the switching state-space models known as SKF. In the SKF, at each time instance, the states are estimated by several Kalman filters (KFs). Furthermore, a hidden discrete state variable called switch was considered whose status changes over time according to a Markov model. The switch indicates the KF, which estimates the states better than others [6]. In the SKF model, each heartbeat was segmented into seven contiguous portions {B1, P,

PQ, QRS, ST, T, B2}. PQ and ST were considered to be baseline segments that were modeled with first-order autoregressive. Other segments {P, Q, R, S, T} were modeled with five Gaussian functions. The proposed idea of using SKF in ECG delineation is an interesting topic, however, the current approach still requires more human interactions, the testing phase only works on segmented heartbeats, the R-peak locations have to be provided with the test data.

2.4 *Other Methods for ECG Delineation*

Besides the above-discussed ECG wave detection and delineation approaches, various methods exist with different processing areas. A mathematical modeling-based method for T-wave peak and T-offset detection was proposed by Madeiro et al. [7]. The QRS detection is not in the scope of this method; therefore, an existed method based on Hilbert transform, Wavelet transform, the first derivative and thresholding are used to localize and delineate the QRS segments. Search blocks were defined for T-wave detection; the search segment starts at the offset of QRS to the onset of subsequent QRS. This interval was modeled with a single skewed Gaussian function. Tafreshi et al. [8] proposed a method capable of detecting nine types of QRS morphologies from 12-lead ECG data vis: QRS, QR, Q, RS, R, RSR, QRSR, RSRS, and QRSRS. Chen and Chuang [9] designed a real-time QRS detection algorithm for the application of wearable devices. The proposed method procedure is summarized as follows. First, the ECG segment (containing at least one heartbeat) was filtered using a bandpass filter of 0.5–17 Hz to remove the baseline drift noise. ECG enhancement was achieved by transforming the filtered signal using a moving window mask. This will emphasize the QRS segments and restrain the P- and T-waves. Define a searching window (0.3 s) to find all the peaks and valleys (crests and troughs). Then a set of heuristic rules were followed to detect and delineate the QRS segments. Four different templates of QRS morphologies were used for QRS recognition.

3 ECG Feature Extraction

In this sub-section, some of the recent methods for feature extraction will be briefly reviewed. The procedure of extracting features for the application of classification consists of two stages feature construction and feature selection [10]. Feature construction is a process of transforming the input data into mostly a lower dimension, which is assumed to be enough to represent the input data. The WPD can reveal all the frequency band decomposition of both detail and approximation coefficients, unlike the DWT, which only decompose the detail coefficients. Martis et al. [11] extracted the second, third, and fourth-order cumulants from the extracted QRS segment using the WPD coefficients. In total, 24 features from three cumulants at eight WPD scales were used. Furthermoe, the principal component analysis (PCA) is applied for feature selection; five features are selected out of 24 at each QRS complex. Chen and Yu [12] applied two stages; feature selection, where the redundant feature within a specific class is reduced. Feature-class selection then finalizes the process by selecting the optimum features between suggested classes. The features were extracted based on the HOS of sub-band components of DWT. The performance of the proposed feature selection methods is compared with the commonly used nonlinear Relief-F method and showed a noticeable improvement in the classification performance when using the proposed methods. The theory of extracting such nonlinear features [13] is discussed below. The ECG is a nonlinear and nonstationary signal, in which higher order statistics are required for ECG analysis. The first two-order statistics represent the Gaussian process of the signal. The HOS assumed that the ECG is non-Gaussian. In fact, the HOS of Gaussian process is identical to zero [14].

Let $\boldsymbol{x}(k) = [x(1), x(2), \ldots, x(K)]$ is a stationary time-series data, the statistical moments up to order n can be found by,

$$m_n^{\boldsymbol{x}}(\tau_1, \tau_2, \ldots, \tau_n) = E\big[\boldsymbol{x}(k)\boldsymbol{x}(k+\tau_1)\ldots\boldsymbol{x}(k+\tau_{n-1})\big] \quad (2)$$

where τ_i, is the time lag. For example, the second-order moment is $m_2^{\boldsymbol{x}}(\tau_1) = E[\boldsymbol{x}(k)\boldsymbol{x}(k+\tau_1)]$ and represents the autocorrelation of $\boldsymbol{x}(k)$ and its lagged version $\boldsymbol{x}(k+\tau_1)$. The nth order cumulant can be calculated as,

$$C_n^{\boldsymbol{x}}(\tau_1, \tau_2, \ldots, \tau_n) = m_n^{\boldsymbol{x}}(\tau_1, \tau_2, \ldots, \tau_n) - m_n^{\boldsymbol{G}}(\tau_1, \tau_2, \ldots, \tau_n) \quad (3)$$

where $m_n^{\boldsymbol{G}}(\tau_1, \tau_2, \ldots, \tau_n)$ is the Gaussian equivalent of signal $\boldsymbol{x}(k)$. Since $\boldsymbol{x}(k)$ is assumed to be non-Gaussian $\boldsymbol{x}(k)$ is a zero-mean signal, $m_n^{\boldsymbol{G}}(\tau_1, \tau_2, \ldots, \tau_n) = 0$ and the moment part of (3) is used to find the cumulants of $\boldsymbol{x}(k)$.

The DTW is a widely used method to measure the temporal similarity between two time-series. Unfortunately, the DTW needs to be modified to be used on higher dimension space; therefore, a dynamic spatiotemporal warping (DSW) was proposed by Yang et al. [15] to measure the dissimilarities between healthy control and MI heartbeats. The dissimilarity matrix was transformed into feature vectors and used as input to the classifiers. Reference [16] extracted a pool of features for the detection of myocardial scar using ECG/VCG signals. Four different sets of features were extracted. A total of 344 features were extracted, which is a large amount to represent a single heartbeat.

4 ECG Classification

The performance of ECG classification highly depends on the preprocessing and feature extraction stages. In previous sections, a wide variety of methods have been studied to fulfill these domains. So far, there is no universal standard for the best automatic methods that fit the needs of clinical monitoring applications. Although a huge amount of work exists on ECG classification, the performance of these methods is not stable with a vast of variations in ECG signals. However, once the ECG signal is enhanced and a set of features has been extracted, models can be trained and learnt the underlying dynamics from these feature data using artificial AI algorithms. Machine learning is one domain of AI that has the ability to learn and distinguish between the enrolled data classes.

4.1 Neural Network-Based Classifiers

Übeyli [17] used a recurrent neural network (RNN) for ECG heartbeat classification. The RNN is a multilayer neural network in which the output of some or all layers depends on the current input, and the previous output is looped back and reused as an extra input. RNNs can be configured in two designs, namely, fully connected or partially connected. The fully connected RNN appears to have high computational expenses, for that [17] used the partially connected RNN, which is simply a feed-forward neural network (NN) with a carefully selected set of feedback connections. Martis et al. [18] developed an ECG heartbeat automatic screening method based on an error backpropagation neural network (ERBNN). A simple three-layer ERBNN with 13 neurons in the input layer, six node in the hidden layer, and two neurons in the output layer to represent the binary classes of heartbeats. Chen and Yu [12] suggested a feedforward backpropagation neural network (FFBNN) that classifies seven different heartbeat classes. FFBNN is a three-layer network with a hidden layer that has 60 nodes. It is important to mention that the process at each node of the neural network occurs in the manner $output = activate(input \times weight + bias)$.

4.2 SVM-Based Classifiers

The SVM classifier is a powerful method and has been used for many applications in the literature. SVM simply is a single-layer nonlinear network that first transforms the data into a higher dimensional space using some specialized kernel transformation functions. Then uses the distance metric to create a boundary (hyperplane or support vector) between the data groups (classes), whereas this distance is simultaneously maximized [11]. Martis et al. [18] used a quadratic SVM to classify heartbeats into two binary classes. The SVM showed better performance compared with other classifiers (NN and GMM). However, the method does not detail the SVM hyper-parameters selection, which is the sensitive part of the SVM classifier. From the SVM-based reviewed studies in this subsection, it can be concluded that the SVM-based methods are highly dependent on the preprocessing methods, unlike the other machine learning algorithms.

4.3 HMM-Based Classifiers

Liang et al. [19] proposed an HMM-based heartbeat classifier. A four-state left-to-right HMM structure is to detect four different classes, viz, normal, premature ventricular contraction (PVC), premature atrial contraction (APC), and invalid class. The advantage of this method is that it is able to detect the invalid class, which means the input features to HMM are invalid (the feature extraction method has mislocated the ECG waves delineation points). Furthermore, it was designed to work on body sensor networks. The method only used four dynamic (interval-based) features to discriminate between the assigned classes, achieving classification accuracy above 95%.

4.4 GMM-Based Classifiers

GMM classifier is a basic supervised method that has the ability to automatically cluster the data into a limited set of overlapped clusters. The GMM method has been applied to ECG heartbeat classification in the literature. Chang et al. [20] proposed a four-dimensional future-based GMM classifier for the detection of myocardial infarction in 12-lead ECG signals. In training, two Gaussian mixtures were used to represent the normal and diseased datasets, the Gaussian parameters (mean, covariance, weights) were estimated iteratively using an expectation-maximization algorithm. In testing, the same feature vector from the test ECG heartbeats was used to find the fitted Gaussian parameters; the likelihood was calculated and compared with the already built GMM models. The main limitation of GMM-based methods is that the number of mixture models must be determined manually, which forces the GMM to cluster the data into a limited number of clusters, which is highly dependent on the correlation of the input data (feature vectors).

4.5 Other Selected Classifiers

Yang et al. [15] used a supervised self-organizing map (SOM) method to detect the myocardial infarction location. SOM is a type of NNs, which is able to transform the high-dimensional feature vectors of input into a two-dimensional space. The proposed SOM consisted of two layers, i.e., the input layer and a competitive layer, which follows an iterative procedure (updating the neighboring function) to learn the characteristics of the labeled input feature vectors. A set of hyper-parameters were required to be initialized or adjusted experimentally, i.e., map size, neighboring function, learning rate, and the number of neurons in the second layer. Bono et al. [21] proposed an automated implementation of the updated Selvester QRS scoring system (50 criteria/31-points) for myocardial scar detection using 12-lead ECG signals. The author follows a statistic confounders approach by implementing a series of heuristic rules comparing the QRS amplitude, duration, and axis with some reference normal values. The heuristic classifier provides good accuracies as long as the fiducial point extraction is valid. Any misdetection in the preprocessing step will be inherited and reflected as a false alarm in the classification.

5 Discussion and Summary

Literature has revealed several methods to detect and classify heart disease automatically. The developed approaches are based on three main steps: preprocessing, feature extraction, and classification. First, the elimination of noises and artifacts (i.e., baseline drift, power line interference, and muscle movement) enhances the ECG signals. This is followed by extraction or segmentation of the ECG waveforms. ECG waveforms are also known as PQRST, consisting mainly of P wave, QRS complex, and T wave. They are located using automatic fiducial point detection methods before being extracted by means of segmentation. Then several handcrafted features are calculated from these waveforms. Finally, these features are used for training, validation, and testing of the proposed machine learning classifier. For ECG filtering, Wavelet denoising is the most widely used method in the literature. Choosing the wavelet function is usually determined by the shape of the Wavelet, which should be closer to the ECG waveform shape. Most of the methods proposed in the literature rely on QRS complex detection and delineation as the first step, followed by defining search blocks around R-peaks to identify the P and T waves. In addition, the ECG signals from patients with cardiac abnormalities or signals with low signal-to-noise (SNR) ratios will significantly increase the chances of false detections of ECG components and become a more challenging problem.

6 Conclusion

Although a huge amount of work exists on ECG classification, the performance of these methods is not stable with a vast of variations in ECG signals. In addition to the feature extraction and the type of classifier, ECG classification performance measure is strongly relying on the evaluation scheme of the classification process. Two different classification evaluation schemes, namely, "class-oriented" and "subject-oriented", are used in the literature. In class-oriented, heartbeats of the same patient may partly appear in both training and testing subsets, whereas the subject-oriented partitions the whole ECG recordings into training and testing subsets. The class-oriented is usually adopted in most literature works, in which the classifiers usually produce over-optimistic results. However, testing such a scheme with any different database will decline the method's performance.

Acknowledgements This work was supported by the Ministry of Higher Education under Fundamental Research Grant Scheme with grant number FRGS/1/2019/STG06/UTM/02/14 (UTM vote number: R.J130000.7851.5F157) and associated facilities.

References

1. Kennedy, A., Finlay, D.D., Guldenring, D., Bond, R., Moran, K., McLaughlin, J.: The cardiac conduction system: generation and conduction of the cardiac impulse. Crit. Care Nurs. Clin. **28**(3), 269–279 (2016)
2. Gutiérrez-Gnecchi, J.A. et al.: DSP-based arrhythmia classification using wavelet transform and probabilistic neural network. Biomed. Signal Process. Control **32** (2017)
3. Yochum, M., Renaud, C., Jacquir, S.: Automatic detection of P, QRS and T patterns in 12 leads ECG signal based on CWT. Biomed. Signal Process. Control **25** (2016)
4. Merah, M., Abdelmalik, T.A., Larbi, B.H.: R-peaks detection based on stationary wavelet transform. Comput. Meth. Programs Biomed. **121**(3), 149–160 (2015)
5. Martínez, A., Alcaraz, R., Rieta, J.J.: Application of the phasor transform for automatic delineation of single-lead ECG fiducial points. Physiol. Meas. **31**(11), 1467 (2010)
6. Akhbari, M., Ghahjaverestan, N.M., Shamsollahi, M.B., Jutten, C.: ECG fiducial point extraction using switching Kalman filter. Comput. Meth. Programs Biomed. **157**, 129–136 (2018)
7. Madeiro, J.P.V., et al.: New approach for T-wave peak detection and T-wave end location in 12-lead paced ECG signals based on a mathematical model. Med. Eng. Phys. **35**(8), 1105–1115 (2013)
8. Tafreshi, R., Jaleel, A., Lim, J., Tafreshi, L.: Automated analysis of ECG waveforms with atypical QRS complex morphologies. Biomed. Signal Process. Control **10**, 41–49 (2014)
9. Chen, C.-L., Chuang, C.-T.: A QRS detection and R point recognition method for wearable single-lead ECG devices. Sensors **17**(9), 1969 (2017)
10. Chen, T.: Enhancing the diagnostic quality of ECGs in mobile environments. Diss. Univ. Southampt 225 (2015)
11. Martis, R.J., Acharya, U.R., Ray, A.K., Chakraborty, C.: Application of higher order cumulants to ECG signals for the cardiac health diagnosis. In: Engineering in Medicine and Biology Society, EMBC, 2011 Annual International Conference of the IEEE, 2011, pp. 1697–1700
12. Chen, Y.-H., Yu, S.-N.: Selection of effective features for ECG beat recognition based on nonlinear correlations. Artif. Intell. Med. **54**(1), 43–52 (2012)
13. da Luz, E.J.S., Nunes, T.M., de Albuquerque, V.H.C., Papa, J.P., Menotti, D.: ECG arrhythmia classification based on optimum-path forest. Expert Syst. Appl. **40**(9), 3561–3573 (2013)
14. Martis, R.J., Acharya, U.R., Adeli, H.: Current methods in electrocardiogram characterization. Comput. Biol. Med. **48**(1) (2014)
15. Yang, H., Kan, C., Liu, G., Chen, Y.: Spatiotemporal differentiation of myocardial infarctions. IEEE Trans. Autom. Sci. Eng. **10**(4), 938–947 (2013)
16. Dima, S.-M., et al.: On the detection of myocadial scar based on ECG/VCG analysis. IEEE Trans. Biomed. Eng. **60**(12), 3399–3409 (2013)
17. Derya Übeyli, E.: Recurrent neural networks employing Lyapunov exponents for analysis of ECG signals. Expert Syst. Appl. **37**(2), 1192–1199 (2010)
18. Martis, R.J., et al.: Automated screening of arrhythmia using wavelet based machine learning techniques. J. Med. Syst. **36**(2), 677–688 (2012)
19. Liang, W., Zhang, Y., Tan, J., Li, Y.: A novel approach to ECG classification based upon two-layered HMMs in body sensor networks. Sensors **14**(4), 5994–6011 (2014)

20. Chang, P.-C., Lin, J.-J., Hsieh, J.-C., Weng, J.: Myocardial infarction classification with multi-lead ECG using hidden Markov models and Gaussian mixture models. Appl. Soft Comput. **12**(10), 3165–3175 (2012)
21. Bono, V., et al.: Development of an automated updated selvester QRS scoring system using SWT-based QRS fractionation detection and classification. IEEE J. Biomed. Heal. Informat. **18**(1), 193–204 (2014)

Use of a Computational Tool for the Assessment of Attention of Medical Residents After a day on Duty

Argelia Pérez-Pacheco, José A. García-García, J. Eduardo Lugo, and Jocelyn Faubert

Abstract Medical residents usually spend long hours of sleep deprivation before graduation. Fatigue, physical wear due to long working hours result in sleep deprivation. Consequently, circadian cycles are altered increasing the probability of accidents and compromising patient care. Few studies have evaluated cognitive performance in physicians after a night shift. In this work, cognitive performance is compared, through a computerized system based on the paradigm of multiple objects tracking, between a group of young students without the effect of fatigue and a group of medical residents after a day on duty.

Keywords Multiple objects tracking · Fatigue · Sleep deprivation · Attention · Medical residents

1 Introduction

Medical residents (MR) are healthcare professionals who enter a medical service to continue their education and training in a specialized field of medicine. Medical residencies constitute intensive training where doctors put their knowledge into practice by developing skills to strengthen their training process using long working hours leading to sleep deprivation. This model is considered the best training system for specialist doctors, however, the negative effects of long working hours can have

A. Pérez-Pacheco (✉)
Research and Technological Development Unit, Hospital General de México "Dr. Eduardo Liceaga", C.P. 06726 Mexico City, Mexico
e-mail: argeliapp@ciencias.unam.mx

J. A. García-García
Directorate of Health Education and Training, Hospital General de México "Dr. Eduardo Liceaga", C.P. 06726 Mexico City, Mexico

J. E. Lugo · J. Faubert
Faubert Lab, École d'optométrie, Université de Montréal, 3744 Jean Brillant, Montréal, Québec 3T 1P1, Canada

M. S. Kaiser et al. (eds.), *Proceedings of the Third International Conference on Trends in Computational and Cognitive Engineering*, Lecture Notes in Networks and Systems 348, https://doi.org/10.1007/978-981-16-7597-3_19

repercussions on their physical and mental health, as well as the quality and safety of patient care [1].

In Mexico, the laws applicable to this topic establish that medical residents have a dual role: graduate students and health system workers [2]. In the academic field, there are very similar conceptualizations about what is a MR and a medical residency program, regardless of the country in question. There is also a certain consensus about what activities resident doctors should do; however, the maximum number of hours devoted to carrying out activities throughout the week is not homogeneous. The case of Libby Zion, an 18-year old woman, who died in 1984 while being cared for MR, was the original reason for reforming the number of working hours by resident physicians [3]. In 2003, the Accreditation Council for Graduate Medical Education (ACGME), introduced a set of rules for the reduction in the number of hours of resident doctors' activities [4]. During residency, physicians-in-training can work 60–130 h per week, depending on the residency training program [5]. Some countries have regulated a maximum of 40 h a week, but it is not a rule. The fact is that there are discrepancies between what is stipulated by law and what happens in situ in each hospital where a medical resident works [6–9].

The clinical work is an indispensable and irreplaceable element in the academic training of MR. However, there is controversy about the number of hours of clinical work that medical residents must comply with. Various deleterious situations have been documented in MR because of excess work hours: depression [10], acute stress and anxiety [11], increased risk of vehicular accidents [12, 13], headache with a higher prevalence than in the general population [14], drowsiness [15], sleep disorders [16], burnout [17], fatigue [18], global decrease in quality of life [19], deterioration of love life [20].

Some studies have been carried out to try to identify which are the pathophysiological mechanisms that underlie a situation inherent to activities during medical residencies: sleep deprivation [21], which is a triggering factor for negative effects on the health of MR. Sleep deprivation decreases cognitive and psychomotor skills [22]. An experimental study carried out with optical topography in MR reached the following conclusions: brain activity in the right prefrontal region decreased, measures of fatigue and daytime sleepiness were significantly strongly correlated with decreased brain activity. The results of that study suggest that the attention deficits due to the combined effects of fatigue and drowsiness could have a detrimental impact on the performance of MR after the night shift [23]

In this study, we analyze the impact that a day on duty has on the attention of MR through visual 3D multiple objects tracking (3D-MOT). The multiple objects tracking (MOT) task was developed by Pylyshyn and Storm (1988) as a laboratory analogue of the divided visual attention test [24, 25]. The 3D-MOT task consists of visually tracking multiple objects (targets) moving around a space while ignoring other physically indistinguishable objects (distractors). A thorough dissection of the MOT paradigm has suggested that this task is primarily concerned with the limits on divided or multifocal as well as attention and to understand processes of attention in typically and atypically developing individuals [26, 27]. 3D-MOT has been used to investigate a wide range of topics related to visual cognition due to the participant requires to: (i) selectively attend to targets while ignoring distractors, (ii) distribute attention across multiple objects, and (iii) maintain this effort throughout the test [24]. MOT has proven to be one of the most useful tools in the study of attention [28–30]. Few studies have evaluated cognitive performance in MR after a night shift. Night work is a requirement for doctors (medical residents) in their professional training as well as to guarantee constant access to patient care. However, long working hours are responsible for causing sleep deprivation and disruption of circadian cycles; in addition, it has been shown to have repercussions on worker health as well as negatively impacting patient safety (2, 10, 11).

2 Materials and Methods

Fifty-six subjects with no prior report of cognitive impairment voluntarily participated in the study. Forty-four physicians (33 females) from various medical and surgical specialties (age: 28.3 ± 2.3 years) from Hospital General de Mexico "Dr. Eduardo Liceaga" were recruited. Twelve subjects (6 females) (age: 23 ± 3.3 years) were students of different careers without the effect of fatigue due to not sleeping during the night prior and post to the intervention (Control group). In this population, the mean hours of sleep were 6 h 30 min, unlike the medical residents' group whose mean was 1 h 30 min (Table 1). Also, the medical residents' group was asked about the type of incident or accidents they suffered during or after a day on duty. Both groups agreed to participate in the study signing informed consent.

The computer program used to track multiple objects in 3D was NeuroTracker (CogniSensAthletics, Inc., Montreal, Quebec, Canada). The tracking task consists to identified four targets among eight spheres projected within a virtual cube space for 8 secs (one trial), controlling for a visual angle of 45°. Each sphere or ball contains

Table 1 Demographic and sleep characteristics on the medical residents and control group

Variables	Medical residents (n = 44)	Control group (n = 12)
Age (years) Sex Male Female Married (%)	28.3 ± 2.3 11 33 11.3	23 ± 3.3 6 6 0
Fortnightly guards	5.00 ± 0.61	–
Hours of pre-guard sleep/Hour of sleep prior to the 3D-MOT	5.68 ± 0.80	6.5 ± 1.4
Hours of sleep during the watch/Hours of sleep without duty	1.59 ± 1.38	6.6 ± 1.9
Previous illnesses Yes No	7 37	1 11

a number, and the observer must recall the spheres that were originally indexed. Then the balls move and when the movement is stopped after 8 s, the observer must identify the spheres' locations that were indexed initially. Dependent on correct or incorrect responses, the speed of the spheres increased or decreased with each trial. Each session lasted around 8 min and the subjects were not allowed to train for more than three sessions per day. A typical session based on a staircase procedure consisted of 20 trials.

The 3D-MOT was performed by using the CORE program of the NeuroTracker system with four tracking targets. Both the control group and the MR s' group carried out the 3D-MOT sessions early in the morning before starting their routine activities. All participants were sitting in a regular chair in front of the computer with a pair of 3D glasses for each trial. The speed threshold scores were obtained for each session. The first 3 sessions were before a 24 h call and the last 3 sessions were after a day on duty for the MR's group, while the first 3 sessions and last 3 sessions were both after a normal day of rest for the control group.

3 Results

As reported by MR about the incidents or accidents they had suffered during or after a day on duty, 75% said they had one related to (1) Sleeping in their hospital activities (31.8%), (2). None (25%), (3) Injuries during their hospital activities (13.6%), (4) Sleeping and making mistakes in their hospital activities, (5) Sleeping while driving without collisions, and (6) Vehicle crash (Fig. 1). Lockley et al. reported that MR who had worked 24 h or more were 2.3 times more likely to be in a car accident after their shift compared to those who worked less than 24 h, with a monthly risk of a

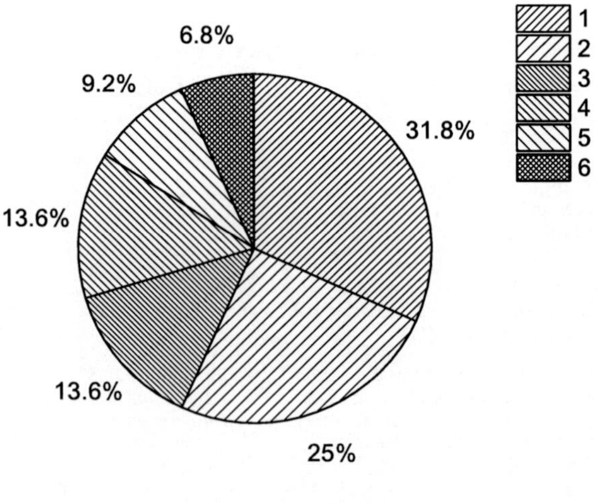

Fig. 1 Percentage of incidents and accidents post-guard of medical residents. 1. Dozing in their hospital activities, 2. None, 3. Injuries during their hospital activities, 4. Dozing and making mistakes in their activities, 5. Dozing while driving without collisions, 6. Vehicular collision

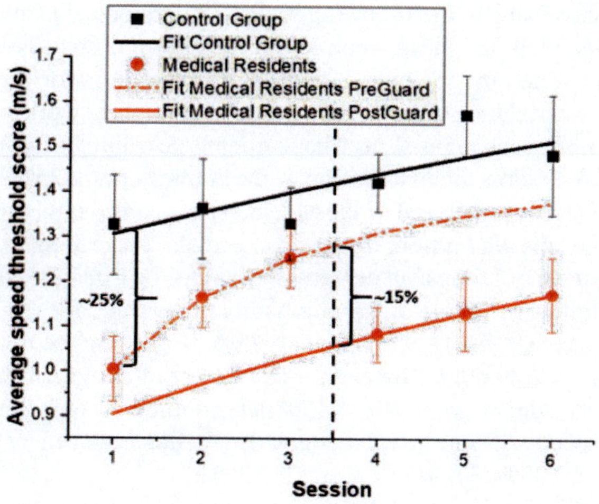

Fig. 2 Means and standard error (error bars) of 3D-MOT average speed threshold scores of the medical residents' group, pre-guard (first 3 sessions) and post-guard (last 3 sessions), and the control group

major accident in a year of 16.2% [31]. From their studies, they showed that eliminating long work shifts and reducing the number of hours worked allows significant improvements in interns' sleep and reductions in attentional failures.

Figure 2 shows average speed threshold scores, and standard error of 3D-MOT obtained each session of medical residents and control group. The number of sessions carried out by all the participants was 6. For the group of residents, it was divided into 3 sessions prior to the 24 h guard and the last 3 post-guard sessions (1.59 ± 1.38 h of sleep during the watch). In the case of the control group, the first 3 sessions and last 3 sessions were carried in two consecutively days both the next day of a normal rest day (6.5 ± 1.4 h of rest).

The 3 fittings curves of Fig. 2 correspond to the 3D-MOT performance obtained of: (1) the control group (upper curve), (2) the medical residents' group, pre-guard (middle curve), and (3) the medical residents' group, post-guard, (lower curve). Each curve is a learning curve, which has been perfectly described through a rational function of the type: $(a + bN)/(1 + cN)$, where N is the number session, a, b, and c are fitting parameters [32]. The fitting parameters for the pre-guard curve of the MR' group are a = 0.6257, b = 1.058, c = 0.6782, and post-guard are a = 0.8271, b = 0.1391, c = 0.06944, meanwhile for curve of the control group, are a = 1.253, b = 0.1045, c = 0.04018. This curve is related to the improvement of the cognitive performance and the learning capacity of the brain that culminates with the consolidation of the learning of a task represented as the curve plateau [28, 33]. As can be seen in Fig. 2, with only 6 sessions, learning has not yet been consolidated in any of the groups. Previous studies have found that reaching this plateau requires at least 15 sessions [30, 40].

The 3D-MOT speed threshold expected curve of a group without the effect of exhaustion would be like that of the control group (Fit Control Group), however, for the residents' group, 24 h before guard, the curve obtained was found below the control group curve (Fit Medical Residents PreGuard). Although this curve was increasing in the following pre-guard sessions, due to the brain's ability to continue learning, the mean value of the speed of the first session was well below the value obtained in the control group. One hypothesis for this behavior is that there is an accumulation of exhaustion due to their medical work even without having performed a previous night shift. The residents' learning curve begins with an average value well below the first session of the control group (25% lower). After the 2nd session, the response speed of the MR increases considerably (1.25 m/s), but after the guard, i.e., the 4th session, we would expect the curve to follow the trend of the fit to reach a speed of 1.3 m/s but decreases to a value of 1.08 m/s; comparing the trend that should follow the curve of a group without the effect of tiredness (Fit Control Group), this decrease would correspond to 25%. Regarding the trend that the medical residents' group should follow without the effect of the guard, this decrease was 15%. Bartel et al. demonstrated that a 15% deterioration in computerized tests to measure reaction speed after one guard is equivalent to that produced by a blood alcohol concentration of 0.05% [34, 41].

Fatigue is a complex phenomenon with repercussions on both physical and mental health and job performance. Establishing metrics to measure fatigue is a big challenge. Many studies of visualization-induced fatigue use subjective rating scales which are certainly helpful but need to be supplemented with more objective measures [35–37]. For instance, slow eye movements are correlated with poor performance in the psychomotor vigilance task and poorer neurobehavioral performance [38]. Other studies related to the 3D-MOT task have shown that physical fatigue has a detrimental impact on perceptual-cognitive processing and prior perceptual-cognitive training can dramatically reduce this negative consequence [39]. The execution of the 3D-MOT requires attention and peripheral vision to track moving objects, so slow eye movements will result in lower performance.

4 Conclusions

In this study, a computational tool based on visual 3D multiple objects tracking (3D-MOT) was used to analyze the effect that a day on duty has on the attention of a medical residents' group. The scores obtained regarding the speed of tracking multiple objects from the medical residents' group were lower than the control group, probably due to the cumulative effect of fatigue during their residency. The 3D-MOT results showed a decrease in the average speed threshold score of 25% in the medical residents' group after a day on duty regarding expected value in a group without the effect of fatigue. 75% of medical residents reported having a non-serious incident or accident after or during a 24 h call. Our results support the idea that computerized tests related to visual attention can be a tool for the attentional evaluation of medical residents, and therefore, in the quality of patient care.

References

1. Smith-Coggins, R., Rosekind, M.R., Hurd, S., Buccino, K.R.: Relationship of day versus night sleep to physician performance and mood. Ann. Emerg. Med. **24**, 928–934 (1994). https://doi.org/10.1016/S0196-0644(94)70209-8
2. SEGOB (2018) PROYECTO de Norma Oficial Mexicana PROY-NOM-001-SSA3-2018, Educación en salud, para la organización y funcionamiento de residencias médicas en establecimientos para la atención médica. Mexico
3. Brensilver, J.M., Smith, L., Lyttle, C.S.: Impact of the Libby Zion case on graduate medical education in internal medicine. Mt. Sinai J. Med. **65**, 296–300 (1998)
4. Simien, C., Holt, K.D., Richter, T.H., Whalen, T.V., Coburn, M., Havlik, R.J., Miller, R.S.: Resident operative experience in general surgery, plastic surgery, and urology 5 years after implementation of the ACGME duty hour policy. Ann. Surg. **252**, 383–389 (2010). https://doi.org/10.1097/SLA.0b013e3181e62299
5. Halbach, M.M., Spann, C.O., Egan, G.: Effect of sleep deprivation on medical resident and student cognitive function: a prospective study. Am. J. Obstet. Gynecol. **188**, 1198–1201 (2003). https://doi.org/10.1067/mob.2003.306
6. Temple, J.: Resident duty hours around the globe: where are we now? BMC Med. Educ. **14**, S8 (2014). https://doi.org/10.1186/1472-6920-14-S1-S8
7. Moonesinghe, S.R., Lowery, J., Shahi, N., Millen, A., Beard, J.D.: Impact of reduction in working hours for doctors in training on postgraduate medical education and patients' outcomes: systematic review. BMJ **342**, d1580–d1580 (2011). https://doi.org/10.1136/bmj.d1580
8. Han, E.-R., Chung, E.-K.: The perception of medical residents and faculty members on resident duty hour regulation. Korean J. Med. Educ. **32**, 67–72 (2020). https://doi.org/10.3946/kjme.2020.154
9. White, B.A.A., White, H.D., Arroliga, A.C.: Historical trajectory and implications of duty-hours in graduate medical education. J. Leg Med. **39**, 417–426 (2019). https://doi.org/10.1080/01947648.2019.1673264
10. Alshardi, A., Farahat, F.: Prevalence and predictors of depression among medical residents in western Saudi Arabia. J. Clin. Psychol. Med. Settings **27**, 746–752 (2020). https://doi.org/10.1007/s10880-019-09667-7
11. González-Cabrera, J., Fernández-Prada, M., Iribar, C., Molina-Ruano, R., Salinero-Bachiller, M., Peinado, J.: Acute stress and anxiety in medical residents on the emergency department duty. Int. J. Environ. Res. Public Health **15**, 506 (2018). https://doi.org/10.3390/ijerph15030506
12. McManus, B., Heaton, K., Mrug, S., Porterfield, J., Shall, M., Stavrinos, D.: The effect of poor sleep and occupational demands on driving safety in medical residents. Traffic Inj. Prev.. **19**, S137–S140 (2018). https://doi.org/10.1080/15389588.2018.1532202
13. Tornero, C., Ventura, A., Bourguet, M., Poquet, I.: Evaluation of driving ability among residents after the duty shift. Accid Anal. Prev. **47**, 182–183 (2012). https://doi.org/10.1016/j.aap.2012.01.007
14. Mélo Silva Júnior, M.L., Melo, T.S., Sousa Menezes, N.C., Valença, M.M., Sampaio Rocha-Filho, P.A.: Headache in medical residents: a cross-sectional web-based survey. Headache J. Head Face Pain **60**, 2320–2329 (2020). https://doi.org/10.1111/head.14000
15. Wada, K., Sakata, Y., Theriault, G., Narai, R., Yoshino, Y., Tanaka, K., Aizawa, Y.: Associations of excessive sleepiness on duty with sleeping hours and number of days of overnight work among medical residents in Japan. J. Occup. Health **49**, 523–527 (2007). https://doi.org/10.1539/joh.49.523
16. McManus, B., Galbraith, J.W., Heaton, K., Mrug, S., Ponce, B.A., Porterfield, J.R., Schall, M.C., Stavrinos, D.: Sleep and stress before and after duty across residency years under 2017 ACGME hours. Am. J. Surg. **220**, 83–89 (2020). https://doi.org/10.1016/j.amjsurg.2019.10.049
17. Lucas-Guerrero, V., Pascua-Solé, M., Ramos Rodríguez, J.L., Trinidad Borrás, A., González de Pedro, C., Jover Navalón, J.M., Rebasa, P., Targarona Soler, E.M., Serra-Aracil, X.: Desgaste

profesional o burnout en los residentes de Cirugía General. Encuesta de la Asociación Española de Cirujanos. Cir. Esp. **98**, 442–449 (2020). https://doi.org/10.1016/j.ciresp.2020.04.013

18. Schwartz, L.P., Hursh, S.R., Boyle, L., Davis, J.E., Smith, M., Fitzgibbons, S.C.: Fatigue in surgical residents an analysis of duty-hours and the effect of hypothetical naps on predicted performance. Am. J. Surg. **221**, 866–871 (2021). https://doi.org/10.1016/j.amjsurg.2020.08.015
19. Somsila, N., Chaiear, N., Boonjaraspinyo, S., Tiamkao, S.: Work-Related quality of life among medical residents at a university hospital in northeastern Thailand. J. Med. Assoc. Thail. **98**, 1244–1253 (2015)
20. Wang, Y.-J., Hsu, K.-L., Chang, C.-S., Wu, C.-H.: Interrelationships between romance, life quality, and medical training of female residents. J. Chin. Med. Assoc. **75**, 402–408 (2012). https://doi.org/10.1016/j.jcma.2012.06.009
21. Domínguez, P., Grosso, M.L., Pagotto, B., Taliercio, V., Allegri, R.: Effects of sleep deprivation on medical performance of pediatric residents. Arch. Argent Pediatr. **107**, 241–245 (2009). https://doi.org/10.1590/S0325-00752009000300012
22. Hamui-Sutton, L., Barragán-Pérez, V., Fuentes-García, R., Monsalvo-Obregón, E.C., Fouilloux-Morales, C.: Sleep deprivation effects on cognitive, psychomotor skills and its relationship with personal characteristics of resident doctors. Cir. Cir. **81**, 317–327 (2013)
23. Nishida, M., Kikuchi, S., Miwakeichi, F., Suda, S.: Night duty and decreased brain activity of medical residents: a wearable optical topography study. Med. Educ. Online **22**, 1379345 (2017). https://doi.org/10.1080/10872981.2017.1379345
24. Pylyshyn, Z.W., Storm, R.W.: Tracking multiple independent targets: evidence for a parallel tracking mechanism*. Spat. Vis. **3**, 179–197 (1988). https://doi.org/10.1163/156856888X00122
25. Doran, M.M., Hoffman, J.E.: The role of visual attention in multiple object tracking: evidence from ERPs. Attention, Perception, Psychophys. **72**, 33–52 (2010). https://doi.org/10.3758/APP.72.1.33
26. Koldewyn, K., Weigelt, S., Kanwisher, N., Jiang, Y.: Multiple object tracking in autism spectrum disorders. J. Autism Dev. Disord. **43**, 1394–1405 (2013). https://doi.org/10.1007/s10803-012-1694-6
27. O'Hearn, K., Hoffman, J.E., Landau, B.: Developmental profiles for multiple object tracking and spatial memory: typically developing preschoolers and people with Williams syndrome. Dev. Sci. **13**, 430–440 (2010). https://doi.org/10.1111/j.1467-7687.2009.00893.x
28. Faubert, J., Sidebottom, L.: Perceptual-cognitive training of athletes. J Clin Sport Psychol **6**, 85–102 (2012). https://doi.org/10.1123/jcsp.6.1.85
29. Faubert, J.: Professional athletes have extraordinary skills for rapidly learning complex and neutral dynamic visual scenes. Sci. Rep. **3**, 1154 (2013). https://doi.org/10.1038/srep01154
30. Tullo, D., Guy, J., Faubert, J., Bertone, A.: Training with a three-dimensional multiple object-tracking (3D-MOT) paradigm improves attention in students with a neurodevelopmental condition: a randomized controlled trial. Dev. Sci. **21**,(2018). https://doi.org/10.1111/desc.12670
31. Lockley, S.W., Cronin, J.W., Evans, E.E., Cade, B.E., Lee, C.J., Landrigan, C.P., Rothschild, J.M., Katz, J.T., Lilly, C.M., Stone, P.H., Aeschbach, D., Czeisler, C.A.: Effect of reducing interns' weekly work hours on sleep and attentional failures. N. Engl. J. Med. **351**, 1829–1837 (2004). https://doi.org/10.1056/NEJMoa041404
32. Wong-Loya, J.A., Andaverde, J.A., del Rio, J.A.: Improved method for estimating static formation temperatures in geothermal and petroleum wells. Geothermics **57**, 73–83 (2015). https://doi.org/10.1016/j.geothermics.2015.06.002
33. Oksama, L., Hyönä, J.: Is multiple object tracking carried out automatically by an early vision mechanism independent of higher-order cognition? An individual difference approach. Vis. Cogn. **11**, 631–671 (2004). https://doi.org/10.1080/13506280344000473
34. Bartel, P., Offermeier, W., Smith, F., Becker, P.: Attention and working memory in resident anaesthetists after night duty: group and individual effects. Occup. Environ. Med. **61**, 167–170 (2004). https://doi.org/10.1136/oem.2002.006197

35. Jaschinski, W., Bonacker, M., ALSHUTH E,: Accommodation, convergence, pupil diameter and eye blinks at a CRT display flickering near fusion limit. Ergonomics **39**, 152–164 (1996). https://doi.org/10.1080/00140139608964441
36. Kaneko, K., Sakamoto, K.: Spontaneous blinks as a criterion of visual fatigue during prolonged work on visual display terminals. Percept. Mot. Skills **92**, 234–250 (2001). https://doi.org/10.2466/pms.2001.92.1.234
37. Krupinski, E., Reiner, B.I.: Real-time occupational stress and fatigue measurement in medical imaging practice. J. Digit. Imaging **25**, 319–324 (2012). https://doi.org/10.1007/s10278-011-9439-1
38. Cajochen, C., Khalsa, S.B.S., Wyatt, J.K., Czeisler, C.A., Dijk, D.J.: EEG and ocular correlates of circadian melatonin phase and human performance decrements during sleep loss. Am. J. Physiol. Regul. Integr. Comp. Physiol. **277**,(1999). https://doi.org/10.1152/ajpregu.1999.277.3.r640
39. Faubert, J., Barthes, S.: Prior Perceptual-Cognitive Training Builds Mental Resistance During Acute Physical Fatigue in Professional Rugby Athletes. bioRxiv (2018). https://doi.org/10.1101/505313
40. Perceptual-Cognitive Training Can Improve Cognition in Older Adults with Subjective Cognitive Decline. (2009). Ageing Science & Mental Health Studies **3**(6). https://doi.org/10.31038/ASMHS.2019361
41. Williamson, A. M.: Moderate sleep deprivation produces impairments in cognitive and motor performance equivalent to legally prescribed levels of alcohol intoxication. Occupation. Environ. Med. **57**(10), 649–655, (2000). https://doi.org/10.1136/oem.57.10.649.

Internet of Things and Data Analytics

The Effectiveness of Public Transport System

Juwendra Tomas and Adlyn Nazurah Abdul Rahman

Abstract Malaysia is facing a challenge on how to improve the quality of the public transportation system along with the city's urbanization. An increase in air pollution in urban areas along with the number of vehicles in the city shows that the public transport system in the city areas needs to be improved. The objectives of the study were to investigate the current public transport and to improve the public transport systems in Batu Pahat. The study was conducted using a mixed-method. The data was collected using questionnaires, document reviews, and observation. The results showed that the affordability, safety, and comfort elements were recorded mean score at a high level along with the effectiveness attributes about the public transport system in Batu Pahat. Besides that, the elements such as ticket price requested by the bus operator, passenger safety while on public transport, and comfort of the seats in public transport were also recorded with the highest average mean score. The implication of this study shows that the number of users of public transport will increase due to the improvement that has been made. In addition, it will contribute to reducing the carbon emission from vehicles during traffic congestion in the study area.

Keywords Effectiveness · Public transport system · Accessibility · Reliability · Connectivity · Affordability · Safety · Comfort

1 Introduction

The process of development entails not only an economic phenomenon. It is unnecessarily assessed based on the achievement of national economic growth, but it has a broader perspective than merely an economic aspect [1]. The expansion of the economy also affects the growth in transportation. Transportation is one of the main human activities around the world. The transportation sector includes the movement of people and goods by cars, trucks, trains, ships, airplanes, and other vehicles [2]. In a country that is in the era of urbanization, there is always a problem that the country

J. Tomas · A. N. A. Rahman (✉)
Universiti Tun Hussein Onn Malaysia, 86400 Parit Raja Batu Pahat, Johor, Malaysia
e-mail: ap160260@siswa.uthm.edu.my

M. S. Kaiser et al. (eds.), *Proceedings of the Third International Conference on Trends in Computational and Cognitive Engineering*, Lecture Notes in Networks and Systems 348, https://doi.org/10.1007/978-981-16-7597-3_20

will face. Some of it was related to the transportation sector as traffic congestion leads to the rise of gas emissions that cause a lot of pollution in an urban city. The emission from motor vehicles is the most significant source of air pollution in many Malaysian urban areas [3].

To reduce the continuous increase in gas emissions and also reduce traffic congestion in the urban city area, there is a need to concern about the efficiency of public transport in the transport sector. This is because traffic congestion and the number of vehicles on the road will decrease if the public transport system is effective and reliable. The reliability of the public transport system that can accommodate passengers will influence the passenger's behavior in choosing to use public transport rather than private transport. But in Malaysia, even there is a growth in the use of public transport such as commuter train, buses, minibus, mass rapid transit (MRT), light rail transit (LRT), and taxi; people who use public transport less compared to the private vehicle [4]. This may be caused by the lack of quality services system provided in public transportation.

People are globally aware of the issues with climate change and global warming that significantly impact the environment, humans, and animals on the earth [5]. To prevent the gas emission from increasing and reducing traffic congestion in the urban city area, the effectiveness of public transport in the transportation sector needs to be a concern. This is because the number of vehicles on the road and traffic congestion may decrease if the public transport system is effective and efficient. Technical efficiency is crucial in order to determine the efficiency level in a firm and industry [6]. The concept of efficiency not only can be seen when a department or organization uses all available resources or inputs optimally to produce the maximum output [7], but also in the public transport system. The quality of the public transport system that can make passengers satisfied will affect the mindset of the passenger to keep using public transport rather than using private transport.

The authors of [8] defined public transport systems as transportation systems, either existing or envisaged for the future, that can be classified according to these components and their relations to the larger economic, social, and physical systems in which they occur. The effectiveness of the public transport system can be defined as the services provided by public transport that are successful in producing a desired or intended result that satisfies the demand from passengers and users. The public transport system's effectiveness can be described as public transport systems being efficient in achieving a desired or expected result that fulfilled passenger and user demand. The characteristics of an effective and efficient public transport system are providing safety, a transport that comfort to ride and on time just like the time scheduled that have been set, and making passengers of public transport satisfied. In order to keep and attract more passengers, public transport must have high service quality by satisfying and fulfilling a wide range of different customers' needs [4].

The effectiveness of the public transport system in the transportation sector can lead a city to a low carbon city as it can reduce traffic congestion. When the number of vehicles on the road decrease, it would reduce the pollutants released in the city at the same time. A low carbon city can be defined as a city consisting of communities using sustainable green technology, green, and low carbon practices to avoid adverse effects on climate change. The concept of a low carbon city (LCC) is closely aligned with sustainable development, which emphasizes the development that meets the needs of the present without jeopardizing the ability of future generations to meet their own needs. To achieve the goal of being a low carbon city, urban transportation is one of the reasons that need attention. It is because the increase of carbon emission that produces pollution in the cities is from transportation which is increasing, as in the past years motor and vehicle remain the major contributor of air pollution, especially in urban areas.

The public transport system should be improved to attract more people to ride public transport rather than a private car in the city because public transports such as city buses, rapid transit, passenger trains, electric buses and many more can reduce the population of vehicles in the city. Other than that, gas emissions can be reduced as electric buses or electric vehicles are much more environment-friendly, as their engines consume less energy, have greater efficiency as well as use the phenomenon of recovery.

In this study, investigation of the current public transport system in Batu Pahat Johor, Malaysia, can help to find out the effectiveness of the public transport system and make a suggestion to improve the public transport system. By improving the public transport system, we can attract people to ride a public transport that satisfied their demand rather than having a private car, and then it can reduce the carbon emission in the city as the number of vehicles decreases.

There are over 19 million registered vehicles in the country with cumulative production of over 1.4 million metric tons of pollution released in 2008 [9]. Besides that, the transportation sector is also the second-largest contributor to carbon dioxide (CO_2) emissions due to fossil fuel combustion [2]. This issue has led to increased carbon monoxide in urban cities.

Due to the increment of air pollution in urban areas along with the number of vehicles in the city, the public transport system in the city areas needs to be improved. It is because, to attract people's attention to use the public buses rather than a private vehicle, the public transport system must be efficient and effective to make passengers and users satisfied with the service provided such as safety, comfort, and can travel to the most common places that users tend to go in the city. In general, the need to improve the public transport system, providing safe and secure transportation is the main obstacle ahead [10].

Based on the problem and gaps of this study, this research, therefore, attempts to study the effectiveness of the public transport system in Batu Pahat Johor by investigating the current public transport system in the city and the improvisation needed in the public transport system in Batu Pahat Johor, Malaysia.

2 Research Methodology

2.1 Selection of Research Methodology

To complete the research, an in-depth study on research methodology is considered essential [11]. Because of the broad scope and the context of this study, a mixed-methods approach for data collection was used. Both quantitative and qualitative research techniques were adopted to achieve the research aim and objectives [12]. The reason for the choice is qualitative observation data provides an in-depth understanding of the state of the art in practice, while quantitative observation such as questionnaire survey and document review provides a wider view of the public transport system. Therefore, document review, observation, and questionnaire survey were used in this research study.

2.2 Sampling Technique

Population is a collection of individuals or objects that have similar characteristics [13]. Since it was impossible to administer surveys to all transportation system users, sampling was necessary to obtain a representative proportion of all users of the transportation system. Simple random sampling was used to ensure that each potential respondent within the target population stood an equal chance of being included in the sample. The respondents were asked to provide socioeconomic information, including gender, age range, education, occupational, and questions based on document reports.

2.3 Data Analysis Technique

Data analysis is the method by which the data is interpreted and evaluated using rational and critical thought. The data and information from the review and observation of the document will be analyzed by writing the review of the document and a report based on the research area observation. Both report results were compared and analyzed using Microsoft Word software. The questionnaire survey will be using IBM Statistical Package for Social Science (SPSS) version 25 to analyze the data accurately. Data analysis used in this study is a descriptive analysis. The descriptive analysis is used to present quantitative description in a proper form, which includes the calculation of percentage, statistic, mean, and frequency.

Table 1 Reliability test of the effectiveness attribute about the public transport system in Bandar Penggaram Batu Pahat

Variable	Cronbach's alpha	Number of items	Number of items deleted
Affordability	0.921	5	–
Safety	0.933	5	–
Comfort	0.848	5	–

2.4 Pilot Test

There are 30 respondents who took this pilot test. The reason the researcher conducted a pilot test for this research is to assess the feasibility, duration, cost, and improve the design of the study before a full-scale research project is completed. Cronbach's alpha has been used to determine the scales for the reliability of the questionnaire for this research. Cronbach's alpha is a metric used to assess a series of scale or test items' accuracy or internal consistency. In other words, any measurement's reliability refers to the extent to which it is a consistent measure of a concept, and Cronbach's alpha is one way to measure the strength of that consistency (Table 1).

3 Result and Discussion

3.1 Document Review

Analysis and findings of this study are described sequentially according to the question of the research and the objective of the study would like to achieve [14]. The document review is to guide how to make an observation of the case study and to conduct a questionnaire to formulate questions [15]. From the document review that has been conducted by the researcher, three phases of the public transport routes have been created in order to give benefits to residents in Batu Pahat. The public transport provided by GPS, CCTV, and Wi-Fi systems for passengers convenient.

Other than that, public transport also provides Kad Muafakat Johor (KMJ) to all users to enable users to board BMJ free of charge. In addition, for OKU, 60 or older, for 6 years and under, also free to ride the public transport. Cardholders are also covered in travel insurance for benefits to users. This card is also introduced to international students with the requirement registration, students currently studying at public colleges and universities throughout Johor, a copy of passport, and a copy of the student's matrix card when enrolling at the KMJ counter. Annual fee and registration for KMJ are only RM30 a year. Besides, public transport infrastructure is installed all over the public transport routes. The bus stop sign was built to provide information on the route and schedule of the Johor Muafakat Bus. In addition, Batu Pahat Municipal Council also has taken the initiative to upgrade the public bus route (bus stops) in P Town and adjacent residential areas in an effort to increase the flexibility of public transport services.

3.2 Observation

Accessibility. The researchers find that Batu Pahat public transport service needs to upgrade to make Batu Pahat more accessible. The result of the accessibility part is that the level and resistance footpath surface of the public transport station was not provided. Only the benches are provided but are not in good condition. Other than that, there is a provision for safe crossing in the waiting area for public transport and tickets for public transport, but there is no provision for disability accessibility for passengers.

In addition, there is no roof at the Jalan Pejabat bus stop. It makes users in the summer exposed to sunlight and in the rainy season exposed to rain while waiting for the bus. In addition, there is no proper bus stop at Jalan Ampuan (it only provided a signboard of a bus stop without a roof, bench, and level resistance). While at the Jalan Rahmat bus stop, there is a problem with public transportation to get in the passenger at the stop due to some other cars parking in front of the bus stop, which should be used by public transportation. There are no lamps at the bus stop as well. When waiting for public transport, it can be unsafe for passengers. Jalan Mohd Salleh was also not provided with a bus stop.

Reliability. Researchers on the quality of public transportation arriving on schedule or not find that the public transportation does not really arrive on time. The public transport arrives a few minutes late from the schedule from the researcher's observation made. Public transport always arrives at the bus stop for one hour each time. From some interviews with the people at the bus stop, public transport always arrives at the bus stop 5 min late from the scheduled but consistent. As for late at night public transportation, the city's public transportation is only available from 8.00 am to 7.00 pm.

Connectivity. The result of the observation for connectivity of the public transport is the public transport network is not comprehensive. The public transport network is not comprehensive from the observation because some routes that need a bus to through that way does not include in the public transport routes. For example, Jalan Masjid has residential areas and school. From the result of the questionnaire also there is a big amount of respondent says that the public transport is not comprehensive. As for public transport provides a customer service system, it was provided and passengers can take public transportation at each bus stop and passengers could access the time schedule for the public transport. Other than that, public transport Bas Muafakat Johor also provides passengers to be members using the public transports with a card membership. Public transport through each public transport stop was also consistent. Other than that, CCTV, WIFI, and security are provided to the public transport.

Table 2 Reliability test of the effectiveness attributes about the public transport system in Batu Pahat

Variable	Cronbach's alpha	No. of items
Affordability	0.916	5
Safety	0.913	5
Comfort	0.763	5

3.3 Descriptive Analysis

The reliability test of the effectiveness attributes of the public transport system in Bandar Penggaram Batu Pahat is shown in Table 2.

The value of alpha from 0.70 to 0.80 shows good internal consistency, acceptable instrumentation, while 0.80 to 0.90 show excellent internal consistency, considered highly reliable. At the same time, values less than 0.70 are considered poor consistency and, reliably, instrument items need to be modified. The result of the reliability test for affordability is 0.916 with five number of items without any deleted item, safety 0.913, and comfort is 0.763. The result shows that affordability and safety are excellent consistency, considered highly reliable, and for comfort shows good internal consistency and acceptable instruments.

Affordability dimension. Figure 1 shows the mean and standard deviation for the affordability dimension. It clearly indicates that the values of the mean ranged from 3.3960 to 3.5200, while the values of standard deviation range from 0.70112 to 0.89824. The results show the data points are closely distributed around the mean. We can conclude that the first rank of the question is the ticket price requested by the

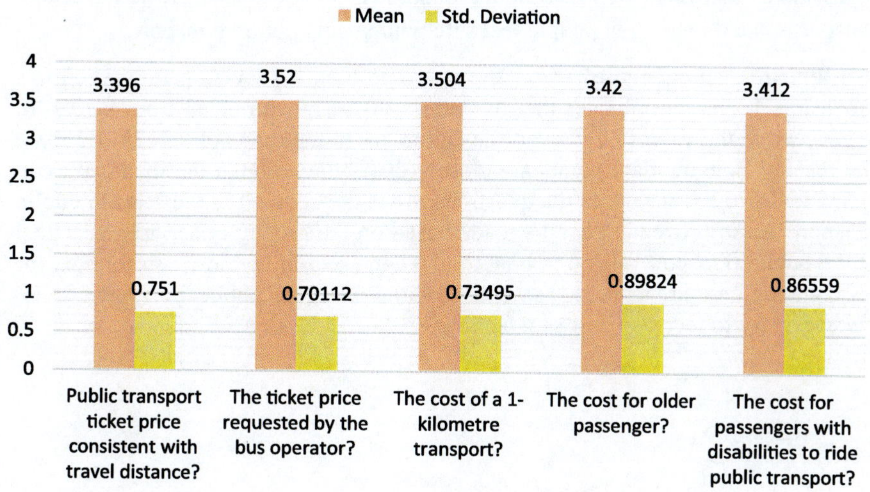

Fig. 1 Mean and standard deviation for the affordability dimension

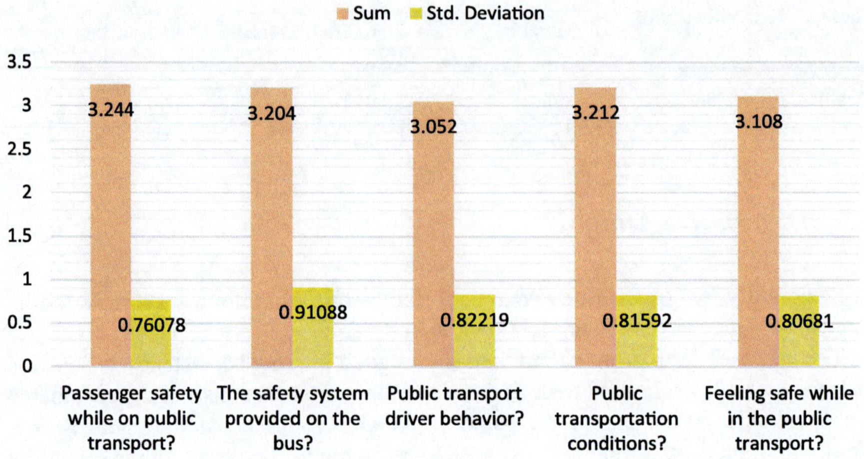

Fig. 2 Mean and standard deviation for the safety dimension

bus operator. An improvement can be made based on the quality of services provided by the public transport related to the ticket price that is given by the operator.

Safety dimension. Figure 2 shows the mean and standard deviation for the safety dimension. It clearly indicates that the values of the mean ranged from 3.0520 to 3.2440, while the values of standard deviation range from 0.76078 to 0.91088. The results show the data points are closely distributed around the mean. We can conclude that the first rank of the question is the feeling safe while using public transport. An improvement can be made based on the quality of services provided by the public transport related to making people feel safe while using public transport.

Comfort dimension. Figure 3 shows the mean and standard deviation for the comfort dimension. It clearly indicates that the values of the mean ranged from 2.7080 to 3.1760, while the values of standard deviation are in the range of 0.7179–1.39085. The results show the data points are closely distributed around the mean. We can conclude that the first rank in the questionnaire is the comfort of the seats in public transport. An improvement can be made based on the quality of services provided by the public transport related to how to make the seat to be more comfortable for passengers. Literature work [13] says that by improving the public transport system, it is necessary to increase user satisfaction.

4 Conclusion

Public transport system includes a variety of transit options such as buses, rail and subways. Generally, this system is available to the public, may require fare, and run with the schedule. Public transport system is implemented to support or ensure

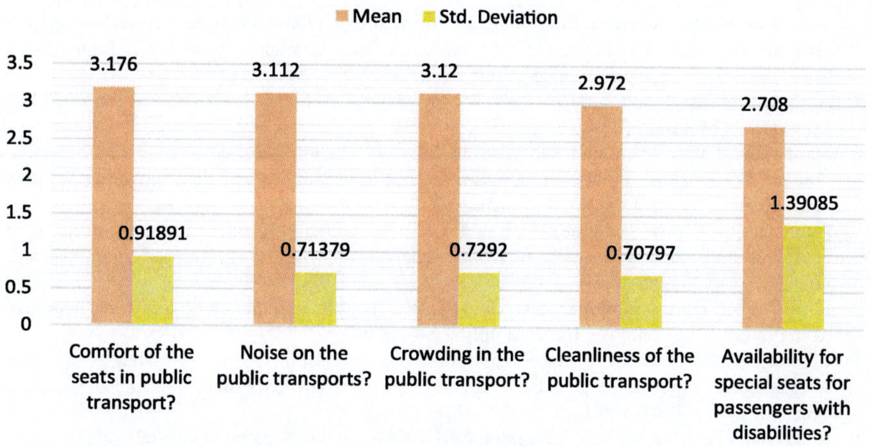

Fig. 3 Mean and standard deviation for the comfort dimension

people can reach every destination such as workplace, school, hospital, and places that often people to go safely and reliable. Public transport services play an important role for people who cannot drive, including those without access to personal vehicles, children, disabled people, and older adults.

In addition, motor vehicle air pollution that continues to contribute has an impact on health. Public transport systems are effective in reducing the number of health risk factors such as motor vehicle crashes, air pollution, and physical inactivity to reduce emissions from too many cars. The impacts of improving the public transport system capacity positively benefit, including savings in travel time, travel costs, and reductions in emissions.

Based on the result of analysis, the objective 1 and objective 2 that are discussed in Chap. 1 have been achieved. A conclusion can be made from the present study results. The result from document review, observation, and questionnaire shows that an improvement toward public transport system in the research is needed because accessibility for people to public transport in is poor. Some of the roads do not have a good bus stop, preventing people from getting into a public transport properly. Public transport service in Batu Pahat needs to be upgraded to create greater accessibility in Batu Pahat area, and by that, the provision of public transport systems available in the study area.

References

1. Soebagyo, D., Fahmy-Abdullah, M.O.H.D., Sieng, L.W., Panjawa, J.L.: income inequality and convergence in central java under regional autonomy. Int. J. Econ. Manage. **13**(1) (2019)
2. Munguia, N.E., Giannetti, B., Delakowitz, B., Zavala, A.G., Velazquez, L., Will, M., Verdugo, S.P.: Sustainable transportation strategies for decoupling road vehicle transport and carbon

dioxide emissions. Manage. Environ. Qual. Int. J. **26**(3), 373–388 (2015)

3. Shuhaili, A., Ihsan, S.I., Faris, W.F.: Air pollution study of vehicles emission in high volume traffic: Selangor, Malaysia as a case study. WSEAS Trans. Syst. **12**(2), 67–84 (2013)
4. Ambak, K., Tun, U., Onn, H., Ismail, R.: Passengers Preference and Satisfaction of Public Transport in Malaysia (2014)
5. Senin, S.N., Fahmy-Abdullah, M., Masrom, M.A.N.: The implementation of green transportation towards low carbon city. In: IOP Conference Series: Earth and Environmental Science, vol. 736, no. 1, p. 012063. IOP Publishing (2021)
6. Hamdan, H., Fahmy-Abdullah, M., Sieng, L.W.: Technical efficiency of Malaysian furniture manufacturing industry: a stochastic frontier analysis approach. Int. J. Supply Chain Manage. **8**(6), 929–37 (2019)
7. Idris, A.I.M., Fahmy-Abdullah, M., Sieng, L.W.: Technical efficiency of soft drink manufacturing industry in Malaysia. Int. J. Supply Chain Manage. **8**, 908–916 (2019)
8. Elkosantini, S., Darmoul, S.: Intelligent public transportation systems: a review of architectures and enabling technologies. In: 2013 International Conference on Advanced Logistics and Transport, ICALT 2013, (May), pp. 233–238 (2013)
9. Malaysian Department of Environment. (2010). http://www.doe.gov.my/portal/
10. Liyanage, C., Dias, N., Amaratunga, D., Haigh, R.: Current context of transport sector in South Asia: recommendations towards a sustainable transportation system. Built Environment Project and Asset Management (2017)
11. Norazlan, M.A.E., Fahmy-Abdullah, M., Masrom, M.A.N.: Promoting carpooling and vanpooling program to reduce the use of private motorised transportation. In: IOP Conference Series: Earth and Environmental Science, vol. 736, no. 1, p. 012050. IOP Publishing (2021
12. Sabli, M.A.N., Fahmy-Abdullah, M., Sieng, L.W.: Application of two-stage data envelopment analysis (DEA) in identifying the technical efficiency and determinants in the plastic manufacturing industry in Malaysia. Int. J. Supply Chain Manage. **8**, 899–907 (2019)
13. Mohamad, N.F.N., Fahmy-Abdullah, M., Masrom, M.A.N.: Transit Oriented Development (TOD) typology. In: IOP Conference Series: Earth and Environmental Science, vol. 736, no. 1, p. 012037. IOP Publishing (2021
14. Latif, M.S.A., Fahmy-Abdullah, M., Sieng, L.W.: Determinants factor of technical efficiency in machinery manufacturing industry in Malaysia. Int. J. Supply Chain Manage. **8**, 917–928 (2019)
15. Ahadi, A.I., Fahmy-Abdullah, M., Rasi, R.Z.R.M., Masrom, M.A.N.: Integrating walking and cycling facilities towards green mobility in Bandar Penggaram Batu Pahat. In: IOP Conference Series: Earth and Environmental Science, vol. 736, no. 1, p. 012003. IOP Publishing (2021)

A Study of Smart People Toward Smart Cities Development

Choo Mun Chye, Mohd Fahmy-Abdullah, Suliadi Firdaus Sufahani, and Mohammad Kamarulzzaman Bin Ali

Abstract Smart people are very important to our future because only humans can utilize technology and improve economic and political efficiency, and play a role in social, cultural and urban progress. However, low moral intelligence, low skilled manpower and conflicts between multi-ethnic are major problems that often lead to social issues. The objectives of the study were to identify smart people concept for a smart city and examine the elements of smart people in Pengerang. The study was conducted using a mixed-method. The data were collected using questionnaires, document reviews and observations. The results showed that the agreeableness, conscientiousness, emotional stability, extraversion and experience to openness elements recorded mean score at a high level. Besides that, the element of agreeableness recorded as the highest average mean score with 3.78 rather than the other four elements, conscientiousness, extraversion, emotional stability and experience to openness. The implications of this study show that the local authorities and government need to draw the strategies and policies for build-up smart people in order to develop and promote a smart city in Malaysia.

Keywords Smart people · Smart cities · Data analytic · Statistical analysis

C. Mun Chye · M. Fahmy-Abdullah (✉) · M. K. Bin Ali
Faculty of Technology Management and Business, Universiti Tun Hussein Onn Malaysia, 86400 Parit Raja Batu PahatJohor, Malaysia
e-mail: mohdfahmy@uthm.edu.my

M. Fahmy-Abdullah · S. F. Sufahani
Oasis Integrated Group, Institutefor IntegratedEngineering, Universiti Tun Hussein Onn Malaysia, 86400 Parit Raja, Batu Pahat, Johor, Malaysia

S. F. Sufahani
Faculty of Applied Science and Technology, Universiti Tun Hussein Onn Malaysia, 84600 Pagoh CampusParit RajaPagoh, Johor, Malaysia

M. S. Kaiser et al. (eds.), *Proceedings of the Third International Conference on Trends in Computational and Cognitive Engineering*, Lecture Notes in Networks and Systems 348, https://doi.org/10.1007/978-981-16-7597-3_21

1 Introduction

The development of smart city becomes more and more important in our future lives. This is because more than half of the global population now lives in cities [1] and the prediction is by the year 2050 the population of urban areas will increase to 70%. When cities become larger and more populous, the resources of urban become limited and we need to make full use of resources urban in the future and serve a more urban population. The smart city concept is coming up and has become important to develop a city for the future.

According to [2], the elements of smart people can be defined in agreeableness, conscientiousness, emotional stability, extraversion and experience to openness. According to [3], making good decisions involved in public life, learning, flexibility and creativity are the necessary temperaments that make people brilliant in order to improve their work performance. Agreeableness can be defined as extremely agreeable people who are warm, happy, courteous, cooperative, organizational team members satisfied and enjoy in their jobs. Superior organizational ability can bring greater efficiency in a student, a teacher, a business owner, an employee, an artist or a celebrity. People with agreeableness can participate in various city activities to establish a happier atmosphere harmony in the city and cooperate toward development. This is one of the essential factors for building smart people.

Extraversion can be defined as the comfort level of the relationship between humans. Extroverts can show confidence in social, have a lot of positive energy, like to get along with people, and perform well in all types of work in their job, willing to make changes in order to get more success in work and actively participate in every decision processes. Extroverts should be in smart cities instead of introverts, who are usually cowardly, shy, quiet and reserved. Next, conscientiousness can be defined as a measure of dependability. A conscientious person is more organized, responsible, reliable and interested in learning new things. These types of people usually live longer because they will take better care of themselves. In a smart city, paying attention to changes in the surrounding environment and taking corresponding actions are crucial for the achievement of smart people. This is because in a smart city it will build a crime-free, corruption-free, safe and healthy ecosystem.

Low moral intelligence is one of the problems that often lead to social issues [4]. The number of social issues is increasing such as vandalism, drugs, terror attacks, business scandals, military interventions or the development of new technologies or means such as biotechnology, cloning and research of stem cells. These activities are recognized as a moral hazard or moral crisis by the public [5]. Moral intelligence is the ability to understand right from wrong, to have a powerful righteous belief and to act in them to behave in the right and way of honorable [5]. Hope humans can try hard to use moral principles and self-regulation skills to bring benefits to society. Therefore, moral intelligence is the most important to build a harmony social.

Besides that, skilled manpower is important for a developed country. Skilled manpower has the professional knowledge, experience and training to conduct more complex mental or physical tasks than conventional job functions. Due to the

Malaysian rapid pace of development and the need for skilled labor is undoubtedly higher. Therefore, low skilled workforce and talented human capital have become one of the issues faced by Malaysia [4]. Shamsuddin worries about if no more skilled workers, our economy will be more difficult to escalate the value chain and will not have the ability to attract more large capital to investments. Based on Department of Statistics Malaysia (DOSM) shows that in 2017 the percentage of employment by skill level in Malaysia is 24%, even though increased by 0.5% as compared to 23.5% in 2015. However, compared with other countries such as Finland, German and Switzerland the rate of skilled workers in Malaysia is still low.

In addition, in the 2019 World Happiness Report, Malaysia ranks the 80th happiest nation [6]. However, compared with 2018 at 35th it has shown a massive drop of 45 places. In this situation, we obviously observe that the country is facing a lack of harmonious public interaction [4]. The entire world knows Malaysia as a multi-ethnic and pluralistic country. However, it also means that it has conflicts between multi-ethnic. Therefore, harmonious public interactions become one of the important factors for a country to be a harmonious society. According to [7], "Harmonious Society" is all the people will do their best, each individual has their own proper place and everybody will get along in harmony with each other. Thus, smart people can help to solve the lack of harmonious public interaction. Furthermore, people are less clear about the concept of smart people. This is likely to have to impact on issues that are stated before and change a city's profile and urban quality.

The elements of smart people have become a critical part to achieve. Based on the research [2], they use the five-factor model (FFM) to explain the needed elements of smart people. The five-factor model, also known as the big-five personality traits, includes all of an individual's characteristics and outlines the major various critical aspects of human personality. The five-factor model includes agreeableness, conscientiousness, emotional stability, extraversion and experience to openness. These personality traits can effectively predict people's behavior and use the five-factor model to lead how people in a smart city should behave and work smartly to achieve high-quality and sumptuous life [2].

Previous research has shown that agreeableness is one of the elements needed for smart people [2]. Agreeableness can refer to people who are highly cooperative, trusting, tender-minded and good-natured, and they are very concerned with the welfare of others [8]. On the opposite, disagreeable people tend to be more self-centered and come across as cynic, tough-minded, domineering and are between these extremes. Studies have found that agreeableness has a moderate positive correlation with tests of emotional intelligence (EQ) [9]. This means that people who are agreeableness achieve more easily to success. This is because emotional intelligence is essential for success [10]. Generally, people who are agreeableness tend to be loved by others. In a smart city, they play the role in cooperative development and create a happier, harmonious and more positive atmosphere in the city.

The personality trait of conscientiousness is the second element needed for smart people [2]. Based on the previous study, conscientiousness refers to the attitude of a person toward missions or duties. In the research of personality development, conscientiousness is known as being cautious, deliberate and careful, as well as in need of achievement [11]. Such people like to learn new knowledge and find some enjoyment in their work. People who are conscientious achieve more success in their job because they are more organized, self-controlled, have less behavioral issues [12] and always comply with a plan to meet their special purpose in order to improve organization efficiency. In addition, they usually live longer because they will take better care of themselves [13]. In a smart city, paying attention to changes in the surrounding environment and taking corresponding actions are crucial for the achievement of smart people. This is because this personality trait can lead to less chaos, processes and activities are better organized, and always go according to the plan, and the city will grow successfully [2].

The personality trait of extraversion is the third element needed for smart people [2]. Extraversion can refer to active people who are sociable, talkative and self-confident [14]. Extroverts can show confidence in social, have more positive energy, like to get along with people and perform well in all types of work. In their job, they are willing to make changes in order to get more success in work and actively participate in every decision process and motivate others to take up a new project. People who are extroverts have leadership skills and can lead a team to work effectively because they can create a good and harmonious relationship with the team members. Smart people should have this personality trait. This is because, in a smart city, information and communications technology (ICT) will become a pillar to bring cities smarter [15]. Thus, smart cities need extroverts to utilize better and smarter new ICT infrastructure and services [2].

The personality trait of emotion stability (neuroticism) is the fourth element needed for smart people [2]. Emotional stability refers to the ability of an individual to keep stable and equilibrated [14]. The emotional stability factor is ordinarily associated with personal life satisfaction, job satisfaction and levels of stress. In professional tests, people with high emotional stability normally have fewer emotional reactions and are less to feel depressed, not susceptible to the stress of physical and psychological effects. They tend to maintain emotional stability, are cool-headed and do not frequently experience negative emotions [16]. Emotionally stable people have better ways to overcome their barriers and find solutions to compensate for their weaknesses. Smart people require this personality trait. This is because the smart city will easily deal with various obstacles because of the optimistic, calm and stress-free characteristics of the smart people [2].

Lastly, the personality trait of openness to experience is the fifth element needed for smart people [2]. Openness to experience can be defined as "person who is a rich life of fantasy, awareness of inner feelings, aesthetic sensibility, knowledge curiosity, need for variety in behavior, and liberal value systems" [14]. People with high openness are keen to explore new experiences and environment more, in order to increase their knowledge [17]. Such people can learn from their mistakes to prevent negative impacts on their lives. Smart people always look at problems from a different perspective to find the most effective solving method. These types of people do not lose the opportunity to expand their experience more successfully. In a smart city, they can find out new opportunities and development of technologies and make smart cities more sustainable [2].

Based on previous research, studies on smart people are still lacking compared with other research. Studies such as [18, 19] and [20] only discussed the smart economy, smart living and smart environment. In fact, a recent study like Perbadanan Putrajaya (2019) only focused on big cities like Putrajaya. Hence, based on the problems and gaps of this study, an effort is made to study the concept of smart people and the way to examine the elements of smart people in Pengerang.

2 Research Methodology

2.1 *Selection of Research Methodology*

In order to satisfy the objectives of this research, the mixed-method is selected. This is because the mixed-method can provide data that has credibility and validity of data because this study is seen in various directions where it uses quantitative methods as well as qualitative methods at the same time. The document review was used to identify the smart people concept for a smart city but also conducted questionnaires to examine elements of smart people in Pengerang.

In this research, the researcher used primary data and secondary data as the source of getting the data. For the secondary data, the researcher would use various forms to conduct document review; one of the forms was governmental sources because they come from transparent and trustworthy governmental agencies [21]. The researcher would also conduct primary data for obtaining the research result of the second objective. Thus, the survey questionnaire was used as the instrument to collect the data. The researcher distributed the questionnaire by online Google form to the respondents who live in Pengerang.

2.2 *Sampling and Population Techniques*

The target population for this study is the residents of Pengerang. The total population in Pengerang is about 15,169 people. The sample size for this research was determined by referring to the Krejcie and Morgan table. For this research, the non-random sampling technique was used due to easy to find the respondents and there are no specific conditions. Our sample size was 370 people.

2.3 *Data Analysis Technique*

Data analysis was done using Statistical Package for The Social Sciences (SPSS) Version 22 to process and analyze quantitative data. It is often used as a data collection

Table 1 Result reliability statistics for the pilot test

	Cronbach's Alpha	Cronbach's Alpha Based on Standardized Items	No. of Items
Agreeableness	0.767	0.780	5
Conscientiousness	0.756	0.765	5
Extraversion	0.706	0.704	5
Emotional stability	0.786	0.787	5
Openness to experience	0.711	0.711	5

tool by other researchers, especially to analyze the results of the questionnaire. SPSS used to run the statistical tests because it has a wide variety of functions for creating and recoding variables.

2.4 Reliability of Pilot Test

Table 1 shows the Cronbach's alpha of the variables in this research. The total measured item is 25 items for elements of smart people. It shows the Cronbach's alpha value of agreeableness as 0.767, conscientiousness 0.756, extraversion 0.706, emotional stability 0.786 and openness to experience as 0.711, which in the interval consistency is acceptable. Thus, the questionnaire can be continuously distributed.

3 Result and Discussion

3.1 Document Review

The researcher [22] believes that smart cities need to begin with human development because only human beings can use technology to improve economic and political efficiency and play a role in social, cultural and urban progress. Same with Hollands' view, [23] also stated public participation in smart city initiatives is the key factor that the probability of the project will succeed or fail. Thus, smart people initiatives become an important part to achieve smart cities.

The development of smart people can be implemented by several strategies to achieve it. The first strategy is to increase the public willingness to adapt to emerging technologies [4]. Technology is changing at such a rapid pace that some of us are finding it hard to blend in. Therefore, if a city is to be "smart", in terms of changing mindset and willingness to accept the changes, this approach turns citizens from end-users to begin-users as well as to give initial exposure to people of what will happen to their city in the future. The second strategy is enhancing public participation and community empowerment [4]. Public participation can be any process that directly engages the public in decision-making and consider public input in making that decision. Public participation involves seeking public opinions at specific points and specific issues in the decision-making process, and these opinions may indeed help make decisions or take actions.

The third strategy is to increase skilled and talented human capital at every level [4]. The Department of Statistics Malaysia (DOSM) stated that the percentage of skilled and talented human capital in Malaysia is relatively low at 24.4% in 2019 compared to advanced countries such as Finland (100%) and Germany (80%). Another main concern for Malaysia is the lack of talent in information technology, computer manufacturing industry and digital industry sector. Therefore, this strategy aims to emphasize and enhance the existing initiatives with additional support targeting to increase talented human capital and lead to the generation of smart people. Lastly, the strategy of build-up smart people is to improve moral education in school [4]. Moral values require conscious knowledge, guided by the positive effect that is carried out in righteous action. Moral intelligence involves a combination of knowledge, desire and willpower. It involves the way people think, feel and act. The knowledge of right and wrong alone may not change our feelings, skills or will to act.

3.2 *Reliability of Real Test*

Table 2 shows the Cronbach's alpha of the variables in this research. The total measured items are 25 items for elements of smart people. It shows the agreeableness Cronbach's alpha is 0.738, conscientiousness is 0.708, extraversion is 0.690, emotional stability is 0.734 and openness to experience is 0.717, which is referred to in Table 2. The interval consistency is acceptable. It is because mostly the range is higher than 0.70.

Table 2 Reliability statistics result for real test

	Cronbach's Alpha	Cronbach's Alpha Based on Standardized Items	No. of Items
Agreeableness	0.738	0.738	5
Conscientiousness	0.708	0.709	5
Extraversion	0.690	0.690	5
Emotional stability	0.734	0.733	5
Openness to experience	0.717	0.717	5

3.3 Demographic Analysis

The researcher discussed in detail about the background of the respondents collected from questionnaires in Sect. 1. The background of the respondents in this research are gender, race, age group, employment status, educational level, residential and non-residential origin, and frequency go to Pengerang. Tables 3, 4, 5, 6, 7 and 8 show the respondent demographic analysis.

Table 3 Respondent gender

	Frequency	Percentage
Female	86	40.2
Male	128	59.8
Total	214	100

Table 4 Respondent race

	Frequency	Percentage
Malay	72	33.6
Indian	53	24.8
Chinese	89	41.6
Total	214	100

Table 5 Respondent age

	Frequency	Percentage
20 years old and below	20	9.3
21–30 years old	55	25.7
31–40 years old	59	27.6
41 years old and above	80	37.4
Total	214	100

Table 6 Respondent employment status

	Frequency	Percentage
Full time	105	49.1
Part time	23	10.7
Unemployed/Retired	21	9.8
Student	39	18.2
Other	26	12.1
Total	214	100

Table 7 Respondent educational level

	Frequency	Percentage
No Education	11	5.1
Primary School	30	14.0
Secondary School	79	36.9
Bachelor's Degree	75	35.0
Higher	19	8.9
Total	214	100

Table 8 Respondent residential status

	Frequency	Percentage
Residential	170	79.4
Non Residential	44	20.6
Total	214	100

3.4 Central Tendencies and Standard Deviation

Based on the results in Table 9, each of the items for the agreeableness was recorded with a high mean score according to the central tendency level. Meanwhile, the standard deviation for each item is in the range of 0.82395 to 0.86474. For A5, "People with agreeableness can treat all people equally" is the highest level compared with other items with a mean score of 3.8364. Based on the results, the researcher agrees with the past researcher [24] and stated agreeable people experience emotional concern for others' well-being, treat others with regard for their personal rights and preferences, hold generally positive beliefs about others and can treat people equally

Table 9 Results of agreeableness

	Question	N	Mean	Std. Deviation	Level	Rank
A1	Agreeableness is important in your life at Pengerang	214	3.8271	0.82395	High	2
A2	Agreeableness can help you to decrease contradiction with others	214	3.8131	0.82940	High	3
A3	People with agreeableness will sympathize with others' feelings	214	3.7290	0.85107	High	4
A4	Agreeableness can increase trust that others have good intentions	214	3.7009	0.86361	High	5
A5	People with agreeableness can treat all people equally	214	3.8364	0.86474	High	1
	Average	214	3.7813	0.59167	High	–

Table 10 Results of conscientiousness

	Question	N	Mean	Std. Deviation	Level	Rank
C1	Conscientiousness is important in your life at Pengerang	214	3.9579	0.81252	High	1
C2	Conscientiousness have a strong ability to control impulses	214	3.6963	0.91738	High	3
C3	Conscientiousness can help you to complete your tasks	214	3.8131	0.90518	High	2
C4	Conscientiousness can help you more success in your academic or job	214	3.6963	0.91225	High	3
C5	Conscientiousness can help you pay more attention in your eating healthily and taking exercise	214	3.6589	0.99788	High	4
	Average	214	3.7645	0.61884	High	–

without any prejudice. The results showed an overall mean value of 3.7813 where it remained at a high level with a standard deviation of 0.59167.

Based on Table 10, each item recorded a high mean score. The highest value of the mean score is 3.9579 from C1, "Conscientiousness is important in your life at Pengerang". In this result, the researcher agrees with the past researchers [25] and stated that conscientiousness plays a role in most of the major domains of life and positive aging. This is because it can influence an individual's health, longevity, and success in love, academic and work. C2 and C4 have the same mean score of 3.6963. Meanwhile, the standard deviation for each item is in the range of 0.81252–0.99788. These results showed an overall mean value of 3.7645 where it remained at a high level with a standard deviation of 0.61884.

Based on Table 11, the results show that all the items have a high mean score between 3.6729 and 3.8832. Meanwhile, the standard deviation for each item is in the range from 0.85052 until 0.96510. E1, "Extraversion is important in your life at Pengerang", has the highest score level compared to other items with a mean score of 0.85052. The result parallels with past researchers [26, 28] that extraversion is positively correlated with life satisfaction judgments. This is because extraversion increases the amount of positive affect, leading to a more positive hedonic balance, and a positive hedonic balance enhances life satisfaction. The results showed an overall mean value of 3.7495 and the standard deviation is 0.60613.

Based on the results in Table 12, each of the items for emotional stability was recorded with a high mean score according to the central tendency level. Meanwhile, the standard deviation for each item is in the range of 0.84446–0.96814. N1, "Emotional stability is important in your life at Pengerang", has the highest mean score with 3.8178 in this section compared to other items. This is because emotional stability has a relationship between job satisfaction and life satisfaction [29]. Stable emotion indicates that people can respond and adapt to circumstances effectively, and their needs can be satisfied. It also represents good emotional adjustment that is

Table 11 Result for extraversion

	Question	N	Mean	Std. Deviation	Level	Rank
E1	Extraversion is important in your life at Pengerang	214	3.8832	0.85052	High	1
E2	Extraversion can help you to get energy that is more positive in life	214	3.7336	0.88232	High	3
E3	Extraversion can help you more easily to being with others	214	3.7757	0.90702	High	2
E4	Extraversion can help you good at social conversation	214	3.6822	0.96510	High	4
E5	Extraversion can help get good performance in your academic or job	214	3.6729	0.92717	High	5
	Average	214	3.7495	0.60613	High	–

Table 12 Results of emotional stability

	Question	N	Mean	Std. Deviation	Level	Rank
N1	Emotional stability is important in your life at Pengerang	214	3.8178	0.84446	High	1
N2	Emotional stability can help you control your emotion in every moment	214	3.6916	0.96814	High	3
N3	Emotional stability can help you to reduce the stress-producing effects of difficult life situations	214	3.5841	0.91412	High	4
N4	Emotional stability can help you in employment contexts	214	3.5093	0.95799	High	5
N5	Emotional stability can help you with the ability for good judgment, trust and minimize conflict	214	3.7430	0.96625	High	2
	Average	214	3.6673	0.64681	High	–

likely to facilitate job satisfaction and life satisfaction. The results showed an overall mean value of 3.6692 where it remained at a high level with a standard deviation of 0.64835.

Based on the results shown in Table 13, the openness to experience was recorded with a high mean score according to the central tendency model. Meanwhile, the standard deviation for each item is in the range of 0.84419–0.93129. The highest value of the mean score 3.8598 is from O1, "Openness to experience is important in your life at Pengerang". Based on the result, openness to experience is important in life because it can encourage people to try new things, explore and reflect on new developments, and thus has the potential to increase knowledge [30]. They are mostly

Table 13 Results of openness to experience

	Question	N	Mean	Std. Deviation	Level	Rank
01	Openness to experience is important in your life at Pengerang	214	3.8598	0.84419	High	1
02	Openness to experience have a good imagination	214	3.7243	0.93129	High	4
03	Openness to experience can motivate you to try something new	214	3.7710	0.89280	High	2
04	Openness to experience can lend you mostly easier to find work in a complex environment	214	3.6262	0.91440	High	5
05	Openness to experience can help you learn from your mistakes to prevent negative impacts on lives	214	3.7570	0.88642	High	3
	Average	214	3.7477	0.61275	High	–

easier to find work in a complex environment [29] because they have the ability to adapt to different cultures [31]. The results showed an overall mean value of 3.7477 with a standard deviation of 0.61275.

Based on Table 14, all the elements were recorded with the mean score at a high level between 3.6673 and 3.7813. Meanwhile, the standard deviation value is in the range of 0.59167–0.64681. This shows that the data points are closely grouped around the mean. The highest value of the mean score of 3.7813 is from agreeableness. However, the result has shown that all the elements listed above consist of some degree of importance concerning to become smart people toward a smart city. This is because the means of the computed items are above 3.50 which is 3.7421.

Table 14 Summary of the means of computed items

Element	N	Mean	Std. Deviation	Level	Rank
Agreeableness	214	3.7813	0.59167	High	1
Conscientiousness	214	3.7645	0.61884	High	2
Extraversion	214	3.7495	0.60613	High	3
Emotional stability	214	3.6673	0.64681	High	5
Openness to experience	214	3.7477	0.61275	High	4
Average	214	3.7421	0.52094	High	–

4 Conclusion

This study is aimed to identify smart people concept for a smart city and examine the elements of smart people in Pengerang. This research used the mixed method, including both qualitative and quantitative methods. Based on the analysis for document review, the concept of smart people is they have a lifelong zeal to learn and have the flexibility to adapt to changes in the environment, as well as the creativity to contribute to education. Smart people possess a democratic nature and they like to participate in public life and use their knowledge, creativity as well as continuous interaction to develop a smart city. Due to the smart people existing in a smart city, thus smart city can be divided into four types of cities, which are the creative city, humane city, learning city and knowledge city.

The second objective of this research is to examine elements of smart people in Pengerang. Based on the result, all elements are on a high level. The element of agreeableness was recorded as the highest average mean score rather than the other four elements, which are conscientiousness, extraversion, emotional stability and experience to openness. This indicated that respondents look at it as the most obvious and important element in Pengerang. People with agreeableness are socially nimble or deft, they have the self-censorship ability and only say something that is suitable for a decent and civilized to avoid conflict with others and create a harmonious atmosphere.

Agreeableness is an essential characteristic of leadership for building an atmosphere of justice because they are trustworthy, compassionate and give response to others' goals and needs. As we all know, social harmony can promote the development of a smart city. Therefore, people with agreeableness have the ability to lead others to create a happier atmosphere of harmony in the city and bring higher efficiency toward the development of a smart city.

References

1. Brockerhoff, M., & Nations, U: World Urbanization Prospects: The 1996 Revision. Popul. Dev. Rev. **24**(4), 883 (1998)
2. Gupta, S., Mustafa, S. Z., & Kumar, H. (2017). Smart People for Smart Cities: A Behavioral Framework for Personality and Roles. In Advances in Smart Cities (pp. 23–30). Chapman and Hall/CRC.
3. Giffinger, R., Fertner, C., Kramar, H., Kalasek, R., Pichler-Milanovic, N., and Meijers, E. 2007. Smart cities. Ranking of European medium-sized cities, Final Report, Centre of Regional Science, Vienna UT.
4. Kementerian Perumahan dan Kerajaan Tempatan. (2018). Final Report Malaysia Smart City Framework. Malaysia Smart City Framework, 251.
5. Tanner, C., & Christen, M. (2014). Moral Intelligence – A Framework for Understanding Moral Competences. (January 2014), 119–136.
6. Walker, A., Councillor, M. L., Woodward, A., Hales, S., de Wet, N., WWF., Sachs, J. D. (2017). World Happness Report. Oecd, (March), 20.
7. Rothman, A.: Harmonious society. China Business Review **35**(2), 24–28 (2008)

8. Levesque, R. J. R. (2011). Encyclopedia of Adolescence (Agreeableness). In Encyclopedia of Adolescence
9. Mayer, D.M., Aquino, K., Greenbaum, R.L., Kuenzi, M.: Who displays ethical leadership, and why does it matter? An examination of antecedents and consequences of ethical leadership. Acad. Manag. J. **55**(1), 151–171 (2012)
10. Bhootrani, M.L., Junejo, J.: Emotional intelligence is a key to success. Journal of the Liaquat University of Medical and Health Sciences **15**(3), 108–109 (2016)
11. Levesque, R.J.R.: Encyclopedia of Adolescence (Conscientiousness). Encyclopedia of Adolescence (2016). https://doi.org/10.1007/978-3-319-32132-5
12. Abe, J.A.A.: The predictive validity of the Five-Factor Model of personality with preschool age children: A nine year follow-up study. J. Res. Pers. **39**(4), 423–442 (2005). https://doi.org/10.1016/j.jrp.2004.05.002
13. Martin, L.R., Friedman, H.S., Schwartz, J.E.: Personality and Mortality Risk Across the Life Span: The Importance of Conscientiousness as a Biopsychosocial Attribute. Health Psychol. **26**(4), 428–436 (2007)
14. McCrae, R.R.: In the public domain Creativity, Divergent Thinking, and Openness to Experience. J. Pers. Soc. Psychol. **52**(6), 1258–1265 (1987)
15. Baghdadi, Youcef, Marta Musso, A. H. (2020). ICT for an Inclusive World : Industry 4.0-Towards the Smart Enterprise.
16. Hills, P., Argyle, M.: Emotional stability as a major dimension of happiness. Personality Individ. Differ. **31**(8), 1357–1364 (2001)
17. Furnham, A., Swami, V., Arteche, A., Chamorro-Premuzic, T.: Cognitive ability, learning approaches and personality correlates of general knowledge. Educ. Psychol. **28**(4), 427–437 (2008)
18. Mwaniki, D., Kinyanjui, M., & Opiyo, R. (2017). Chapter 27: Towards Smart Economic Development in Nairobi In Smart Economy in Smart Cities. https://doi.org/10.1007/978-981-10-1610-3
19. Lai, P. C. (2016). SMART LIVING for SMART CITIES @ the palm of your hand. (February).
20. Pradhan, M. A., Patankar, S., Shinde, A., Shivarkar, V., & Phadatare, P. (2018). IoT for smart city: Improvising smart environment. 2017 International Conference on Energy, Communication, Data Analytics and Soft Computing, ICECDS 2017, (August 2018), 2003–2006. https://doi.org/10.1109/ICECDS.2017.8389800
21. Martins, F.S., da Cunha, J.A.C., Serra, F.A.R.: Secondary Data in Research – Uses and Opportunities. Revista Ibero-Americana de Estratégia **17**(04), 01–04 (2018). https://doi.org/10.5585/ijsm.v17i4.2723
22. Hollands, R.G.: Will the real smart city please stand up? Intelligent, progressive or entrepreneurial? City **12**(3), 303–320 (2008). https://doi.org/10.1080/13604810802479126
23. Chourabi, H., Nam, T., Walker, S., Gil-Garcia, J. R., Mellouli, S., Nahon, K., … Scholl, H. J. (2012). Understanding smart cities: An integrative framework. Proceedings of the Annual Hawaii International Conference on System Sciences, (January), 2289–2297. https://doi.org/10.1109/HICSS.2012.615
24. Bornstein, M. H. (2018). Big Five Personality Traits. The SAGE Encyclopedia of Lifespan Human Development, (March). https://doi.org/10.4135/9781506307633.n93
25. Roberts, B.W., Lejuez, C., Krueger, R.F., Richards, J.M., Hill, P.L.: What is conscientiousness and how can it be assessed? Dev. Psychol. **50**(5), 1315–1330 (2014). https://doi.org/10.1037/a0031109
26. Steel, P., Schmidt, J., Shultz, J.: Refining the Relationship Between Personality and Subjective Well-Being. Psychol. Bull. **134**(1), 138–161 (2008). https://doi.org/10.1037/0033-2909.134.1.138
27. Schimmack, U., Radhakrishnan, P., Oishi, S., Dzokoto, V., Ahadi, S.: Culture, personality, and subjective well-being: Integrating process models of life satisfaction. Jour- nal of Personality and Social Psychology **82**(4), 582–593 (2002). https://doi.org/10.1037/0022-3514.82.4.582
28. Rode, J.C.: Job satisfaction and life satisfaction revisited: A longitudinal test of an integrated model. Human Relations **57**(9), 1205–1230 (2004). https://doi.org/10.1177/0018726704047143

29. Nekljudova, S. (2019). Six aspects of openness to experience. Journal of Psychology and Clinical Psychiatry Review, 10(2), 78–81. https://doi.org/10.15406/jpcpy.2019.10.00632
30. Nieß, C., Zacher, H.: Openness to experience as a predictor and outcome of upward job changes into managerial and professional positions. PLoS ONE **10**(6), 1–22 (2015). https://doi.org/10.1371/journal.pone.0131115
31. Ones, D.S., Viswesvaran, C.: Relative Importance of Personality Dimensions for Expatriate Selection: A Policy Capturing Study. Hum. Perform. **12**(3–4), 275–294 (1999). https://doi.org/10.1207/s15327043hup1203&4_4

Smart Cities with Smart Environment

Muhamad Syazreen Md Salleh, Mohd Fahmy-Abdullah, Suliadi Firdaus Sufahani, and Mohammad Kamarulzzaman Bin Ali

Abstract Smart environment in a smart city is the changes of a city and shape the pure environment with the element to achieve a human settlement. This change is achieved by extensive and persuasive infrastructure and building which give a big impact on the environment. However, the lack of a smart environment regulatory framework was increasing the level of difficulties for the implementation of smart environmental practices. The objectives of the study were to identify the smart environment concept for a smart city and examine the elements of the smart environment in Pengerang. The study was conducted by using a mixed method. The data was collected by using questionnaires, document reviews and observations. The results showed that the quality of life, facilities and automated system element recorded mean score at a high level. Besides that, the element of quality of life recorded the highest average mean score with 4.48 rather than the other four elements, facilities, awareness, safety and automated system. The implications of this study show that the implementation needs deep planning by local authorities and government to realize the changes toward a smart environment.

Keywords Smart environment · Smart cities · Data analytic · Statistical analysis

---

M. S. M. Salleh · M. Fahmy-Abdullah (✉) · M. K. Bin Ali
Faculty of Technology Management and Business, Universiti Tun Hussein Onn Malaysia, RajaBatuPahat, 86400 Johor, Malaysia
e-mail: mohdfahmy@uthm.edu.my

M. Fahmy-Abdullah · S. F. Sufahani
Oasis Integrated Group, Institutefor IntegratedEngineering, Universiti Tun Hussein Onn Malaysia, 86400 Parit Raja, Batu Pahat, Johor, Malaysia

S. F. Sufahani
Faculty of Applied Science and Technology, Universiti Tun Hussein Onn Malaysia, 84600 Pagoh CampusParit RajaPagoh, Johor, Malaysia

M. S. Kaiser et al. (eds.), *Proceedings of the Third International Conference on Trends in Computational and Cognitive Engineering*, Lecture Notes in Networks and Systems 348, https://doi.org/10.1007/978-981-16-7597-3_22

1 Introduction

Smart environment in a smart city is the changes of a city and shape the pure environment with the element to achieve a human settlement. This change is achieved by extensive and persuasive infrastructure and building which give a big impact on the environment [1]. Based on KPKT there has been an indicator to do the evaluation to achieve a smart environment. The elements are preserving green areas, enhancement of trees in public areas, strengthening the integrated and sustainable solid waste management, improving air quality and water quality and its monitoring system. The smart environment is the important element that needs to be successfully transformed before creating a smart city because it needs supporting ecosystem facilities and has to invest into system and skills to achieve smart city in an efficient way [2].

Smart environment is one of the elements in the implementation of a smart city. Based on Ministry Housing and Local Government (KPKT) there are have five elements to develop a smart environment. The high excellent of life in housing areas has been cited as the first component in developing a smart environment. Then, environmental safety is the second element to encouraging a clever environment. The next issue is a clean environment, sustainable aid management and lastly, readiness toward disaster. Furthermore, smart surroundings are to overcome all the challenges that show up nowadays. The challenges are loss of inexperienced area, enhanced stable waste management, pollution of air and water, flash flood, landslide, a high charge of non-revenue water, excessive utilization of non-renewable strength and lastly, excessive carbon footprint [3].

The lack of smart environment regulatory framework increased the level of difficulties for the implementation of smart environment practices [4]. The legislation and policy of smart environment need to be defined by the government. The legislation and policy will help all sectors in implementing or practicing a smart environment. There are so many challenges to implementing a smart environment such as legislation and policy framework. Then, traffic forms an environmental sustainability aspect because the number of vehicles has increased due to the growth of technology. The increasing number of vehicles would affect the environment [1].

Besides that, source management is the second obligation in imposing a smart environment to implement supply administration to aid the smart city. The historic gadget shape wants to be substituted to allow the gadget in a position to control all the monitoring and supply management [5]. Source management is very important to reduce waste in energy or source usage in the city. There must be a new system that can control all the usage of energy and resources [6]. For example, the usage of electricity recorded very high in the city. It has become the biggest challenge to reduce the consumption of electricity in a smart city.

Moreover, water pollution also challenges toward smart environment. Water can be considered polluted if there are some substances or conditions present that show the degree of water cannot be used for a specific purpose [7]. Based on previous research, studies on the smart environment are still lacking compared with other research works. Smart environment only discusses the framework and does no demonstrate at all. In fact, the smart environment is a new idea to develop, protect and maintain a sustainable environment for the better life of the next generation.

Hence, based on the problems and gaps of this study, an effort was made to study the concept of a smart environment and the way to examine the elements of a smart environment in Pengerang.

2 Research Methodology

2.1 Selection of Research Methodology

To fulfill the destinations of this exploration, the blended technique was chosen. This is because the blended technique can give information that has believability and legitimacy of information since this examination is seen in different ways where it utilizes quantitative strategies just as subjective strategies simultaneously. The archive survey would be utilized to recognize the smart people idea for a smart city yet in addition leaded polls to look at components of smart people in Pengerang.

In this research, the researcher used primary data and secondary data as the source of getting the data. For the secondary data, the researcher would use various forms to conduct document review; one of the forms was governmental sources because they come from transparent and trustworthy governmental agencies [8]. The researcher also conducts primary data collection for obtaining the research result of the second objective. For that, the survey questionnaire was used as an instrument to collect the data. The researcher distributed the questionnaire by online Google form to the respondents who live in Pengerang.

2.2 Sampling and Population Techniques

The objective population for this investigation are the residents of Pengerang. The overall population in Pengerang is around 15,169 people. The sample size for this exploration is determined by referring to the Krejcie and Morgan table. For this examination, the non-arbitrary inspecting strategy was being utilized because it was simple to discover the respondents and there are no specific conditions. Our sample size was 370 individuals (Fig. 1).

2.3 Data Analysis Technique

Data analysis was done using (SPSS) version 22 for handling and examining the quantitative information. It is normally utilized as an information assortment device

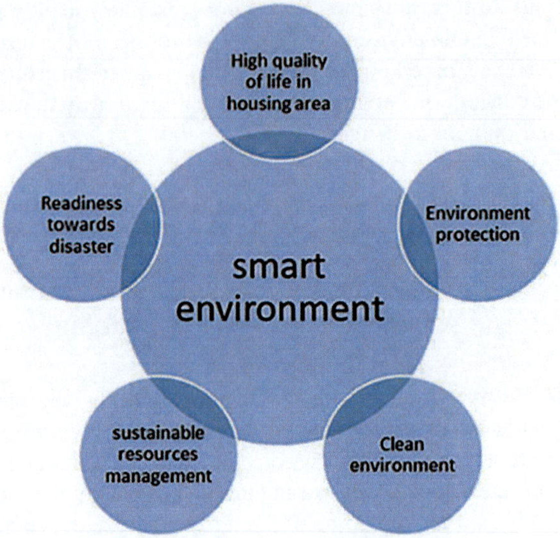

Fig. 1 Smart environment elements

by different analysts, particularly to examine the result from the questionnaire. SPSS was utilized to run the factual tests since it has a wide assortment of functions for making and recoding factors.

2.4 Reliability of the Pilot Test

Table 1 shows the Cronbach's alpha of the independent variable for this research. A total of 25 items were measured for smart environment elements. Based on the table the Cronbach's alpha value for the quality of life was 0.912, facilities were 0.924, awareness 0.912, safety 0.890 and the automated system was 0.920. The result of the five elements shows excellent interval consistency. Therefore, the questionnaire is continuously being distributed.

Table 1 Result reliability statistics for the pilot test

	Cronbach's Alpha	Cronbach's Alpha Based on Standardized Items	No. of Items
Quality of life	0.912	0.916	5
Facilities	0.924	0.931	5
Awareness	0.912	0.928	5
Safety	0.890	0.900	5
Automated system	0.920	0.919	5

3 Result and Discussion

3.1 Document Review

The concept of a smart environment was started when the progress of urbanization grew rapidly. The process had caused a significant effect on the environment such as air quality, water pollution and radiation pollution [9]. Besides, the meaning of a smart environment concept was illustrated in combination with artificial intelligence (AI) and the Internet of Things (IoT) [10]. It means the environment had been connected to sensors that can collect and analyze the data and propose the best solution in maintaining the diversity of environments.

The changes on the environment to the smart environment can be achieved by several strategies. The first strategy was implementing high quality of life in the housing area. The quality of the environment in housing was very important because the high quality of the environment can maintain physical and mental health. The quality of the environment can change due to the behavior of the community in the area and the pollution from the factory, etc. in a way to maintain the high quality of life in the housing area, they must be followed by waste management. Every housing will create many types of waste. Poor handling of waste management can cause pollution to the environment. For example, the local government increases the capacity for waste transportation for collecting waste from the housing area.

The second strategy was environmental protection. Environmental protection was very important to protect and maintain the quality of the environment. The environmental protection strategy is to ensure that there is no people or organization that can easily ignore the importance of the environment. This strategy leads to enhancing the laws toward the environment to anybody who is trying to make something that can reduce the quality of the environment. For example, every manufacturing process will produce smoke and waste. The waste must handle with care and the liquid waste cannot be easily thrown into the river. The departure of liquid waste will cause water and air pollution. Local government must always do an inspection and take some action laws for the parties who make it.

Lastly, a clean environment was the strategy in implementing a smart environment. A clean environment means the degree of pollution was very low. For this strategy, all the communities in the area must take part to perform this strategy. Local governments must play their role to give awareness to the community toward the importance of a clean environment. Every community must know the minimum level of cleanliness to ensure the quality of the environment. The high quality of the environment shows the cleanliness of the city or environment.

3.2 Reliability of Real Test

Table 2 shows the Cronbach's alpha for this research independent variable. A total

Table 2 Result reliability statistics for real test

	Cronbach's Alpha	Cronbach's Alpha Based on Standardized Items	No. of Items
Quality of life	0.817	0.829	5
Facilities	0.809	0.820	5
Awareness	0.778	0.817	5
Safety	0.692	0.698	5
Automated system	0.836	0.841	5

of 25 items were measured for the elements of a smart environment. The result statistic shows the quality of life was 0.817, facilities 0.809, awareness 0.778, safety 0.692 and the automated system is 0.836. The value of Cronbach's alpha shows good interval consistency.

3.3 Central Tendencies and Standard Deviation

Based on the results in Table 3, the result of each item for quality of life elements shows a high mean based on the central tendency table. Meanwhile, the standard deviation for each item was in the range of 0.546–0.837. The highest value for mean had been scored on (Qol) with the mean score of 4.74. Based on the result, it shows that the respondent in Pengerang, Johor was concerned about the quality of life. Based on research by [11], it shows that the quality of life was affected by the quality of the environment. The higher degree of environment will produce the convenience surrounding.

Based on Table 4, the result of each item for facilities elements shows a high mean based on the central tendency table. Meanwhile, the standard deviation for each item was in the range of 0.751–1.093. The highest value for mean had been scored on (F1) with the mean score of 4.04. Based on the result, it shows that the

Table 3 Result of quality of life

Question	N	Mean	Std. Deviation	Level
Do you think that quality of life is important	116	4.74	0.546	High
Did the changes of environment affect the quality of life	116	4.55	0.702	High
Did the sustainable environment will lead to high quality of life	116	4.59	0.620	High
The quality of life can be measured by the quality of	116	4.34	0.814	High
environment	116	4.44	0.837	High
Average	116	4.478	0.7038	High

Table 4 Results of facilities

Question	N	Mean	Std. Deviation	Level
How was your experience when using urban facilities?	116	4.04	1.093	High
According to you, did the great facilities give an impact toward environment?	116	4.41	0.875	High
How important the facilities to community?	116	4.47	0.751	High
Did the facilities need to be upgraded?	116	4.52	0.774	High
Do you think the good facilities can save our environmental?	116	4.53	0.751	High
Average	116	4.35	0.8488	High

respondent in Pengerang, Johor were having a better experience toward the provided facilities in the area. Based on [12], facilities can be divided into many types, such as water supply, electricity, public transportation and communication. All the facilities were to provide for the basic needs of the community. The proper facilities must be developed to ensure the information were received to their target. However, based on the result, Bandar Pengerang had provided proper basic facilities to their community.

Based on Table 5, the result of each item for awareness elements shows a high mean based on the central tendency table. Meanwhile, the standard deviation for each item was in the range of 0.614–1.108. The highest value for mean had been scored on (A2) with the mean score of 3.87. Based on the result, it shows that the respondents in Pengerang, Johor did not receive enough awareness toward the importance of the environment. In order to develop a smart environment, all parties such as government, non-government and community should play their role to maintain the smart environment. However, the level of awareness toward the environment should be implemented among all the people in the area. Based on [13], Malaysia had been placed as the top country with the highest level of air pollution globally. To prevent the entire situation, all the community must have a high level of awareness to avoid pollution in all aspects.

Table 5 Results of awareness

Question	N	Mean	Std. Deviation	Level
Did awareness increase the environment quality?	116	4.62	0.614	High
Lack of environment awareness campaign	116	3.87	1.108	High
Lack of awareness will destroy the quality of environment	116	4.46	0.838	High
Is awareness an important thing needed to be applied among community?	116	4.59	0.632	High
Should government and non-government do more awareness campaign?	116	4.47	0.796	High
Average	116	4.402	0.7976	High

Table 6 Results of safety

Question	N	Mean	Std. Deviation	Level
Do you feel safe with nowadays environment?	116	3.59	1.112	High
How satisfied you are with the environment safety nowadays?	116	3.77	1.041	High
Are the safety elements important in an environment?	116	4.34	0.844	High
Do you think that safety can sustain the environment quality?	116	4.44	0.714	High
According to you, should safety be highlighted in creating smart environment?	116	4.31	1.025	High
Average	116	4.09	0.9472	High

Based on the results in Table 6, the result of each item for safety elements shows a high mean based on the central tendency table. Meanwhile, the standard deviation for each item was in the range of 0.714–1.112. The highest value for mean had been scored on (S4) with the mean score of 4.44. Based on the result, it shows that the respondents in Pengerang, Johor agreed that the safety elements such as enhancing laws can maintain a sustainable environment degree. Safety refers to a plan or framework to reduce the effect of disaster on the environment. This is because Bandar Pengerang was known as an industrial place. Based on an explosive incident in the Diesel Hydro Treater Unit at Pengerang that had happened, the explosive had killed five workers. However, the explosive of chemicals did not affect the environment. This is because the implementation of the plan or the action of safety elements has been done to ensure the quality of the environment.

Based on the results shown in Table 7, the result of each item for automated system elements shows a high mean based on the central tendency table. Meanwhile, the standard deviation for each item was in the range of 0.866–1.130. The highest value for mean had been scored on (As2) with the mean score of 4.16. Based on the result, it

Table 7 Results of an automated system

Question	N	Mean	Std. Deviation	Level
Do you ever hear an automated system in the environment?	116	3.96	1.130	High
Did automated system help in controlling environment quality?	116	4.16	0.910	High
Do you agree our environment is monitored by an automated system?	116	3.97	1.000	High
Did an automated system increase the efficiency of energy usage in urban?	116	4.12	0.925	High
Implementation of the automated system will stop the wrong-doing that can decrease the quality of the environment	116	4.12	0.866	High
Average	116	4.066	0.9662	High

Table 8 Summary of the means of computed items

Element	Mean	Std. Deviation	Level
Quality of life	4.478	0.7038	High
Facilities	4.350	0.8488	High
Awareness	4.402	0.7976	High
Safety	4.090	0.9472	High
Automated system	4.066	0.9662	High
Average	4.2772	0.85272	High

shows that the respondents in Pengerang, Johor realized that the implementation of an automated system such as a monitoring system using a sensor or artificial intelligence and the internet of things can increase the efficiency in controlling the environment quality. Based on the research [14], an automated system was used in increasing the efficiency of manufacturing monitoring. It can show that the automated system will provide the actual data and automatically make a decision to reduce defects in production. Thus, an automated system can be used in monitoring the quality of an environment.

Based on Table 8, all the elements were recorded as the mean score at a high level. The score is between 4.066 and 4.478. Meanwhile, the standard deviation value is in the range of 0.7038–0.9662. This shows the data points are closely grouped around the mean. The highest value of the mean score, 4.478 is from the quality of life. However, the result has shown that all the elements listed above consist of some degree of importance concerning to become a smart environment toward a smart city.

4 Conclusion

This study aimed to identify the smart environment concept for a smart city and examine the elements of a smart environment in Pengerang. This research used the mixed method, including both qualitative and quantitative methods. Based on the analysis for document review, the smart environment is a concept that has a variety of environmental aspects such as hazardous environment, greenhouse, automated control system and controlling resources. Generally, the concept had been realized in other countries. The development of a smart environment is to support the desire of the government to control and maintain the diversity in urbanization. Urban is the place where most people live. Environment is an important thing for a community to stay healthy and keep the productivity for enhancing the economy. The way to keep the community is by changing the environment. Environment plays a very important role for all the community and living things who are still alive.

In addition, the quality of the environment was assisted by the growth of urban. The rapid development of urbanization had decreased the green plant used to maintain the diversity of the environment. Smart environment is an effective way to support rapid urbanization while ensuring the stability of the environment quality through the automated system. The implementation of a smart environment will monitor the environment through a sensor. The sensor will detect all the wrongdoing toward the environment. This can ensure the stability of the environment for the community. Besides, the implementation of a smart environment will enclose all aspects in maintaining and improving the diversity of the environment for future uses.

The second objective of this research is to examine the elements of smart environment in Pengerang. Based on the result, all the smart environment elements show a high level of central tendency. From all the elements, quality of life element had shown the highest result of mean. This shows that the quality of life is an important thing for the respondents who have experience in Bandar Pengerang environment.

References

1. Aletà, N.B., Alonso, C.M., Ruiz, R.M.A.: Smart Mobility and Smart Environ- ment in the Spanish cities. Transportation Research Procedia **24**, 163–170 (2017). https://doi.org/10.1016/j.trpro.2017.05.084
2. Stone, M., Knapper, J., Evans, G., Aravopoulou, E.: Information management in the smart city. Bottom Line **31**(3–4), 234–249 (2018). https://doi.org/10.1108/BL-07-2018-0033
3. Ministry of Housing Smart City Framework. In Malaysia Smart City Framework (2018)
4. Weber, M., Žarko, I. P.: A regulatory view on smart city services. Sensors (Swit- zerland), 19(2), 1–18 (2018). https://doi.org/10.3390/s19020415
5. Nugent, C D., McClean, S.I., Cleland, I., Burns, W.: Sensor Technology for a Safe and Smart Living Environment for the Aged and Infirm at Home. In Comprehensive Materials Processing (Vol. 13) (2014). https://doi.org/10.1016/B978-0-08-096532-1.01319-4
6. Albino, V., Berardi, U., Dangelico, R.M.: Smart cities: Definitions, dimensions, performance, and initiatives. J. Urban Technol. **22**(1), 3–21 (2015). https://doi.org/10.1080/10630732.2014.942092
7. Owa, F.D.: Water pollution: Sources, effects, control and management. Mediterr. J. Soc. Sci. **4**(8), 65–68 (2013). https://doi.org/10.5901/mjss.2013.v4n8p65
8. Martins, F.S., da Cunha, J.A.C., Serra, F.A.R.: Secondary Data in Research – Uses and Opportunities. Revista Ibero-Americana de Estratégia **17**(04), 01–04 (2018). https://doi.org/10.5585/ijsm.v17i4.2723
9. Ullo SL, Sinha GR. Advances in smart environment monitoring systems using iot and sensors. Sensors (Switzerland). 20(11) (2020). https://doi.org/10.3390/s20113113
10. Mohey El-Din, Doaa & Hassanein, Aboul & Hassanien, Ehab & Hussein, Walaa. (2020). E-Quarantine: A Smart Health System for Monitoring Coronavirus Patients for Remotely Quarantine.
11. Miguel, L., Ram, J.M.: The impact of quality of life on the health of older. Jounal of Anging Research **2018**, 1–9 (2018)
12. Ng, S.L., Zhang, Y., Ng, K.H., Wong, H., Lee, J.W.Y.: Living environment and quality of life in Hong Kong. Asian Geogr. **35**(1), 35–51 (2018). https://doi.org/10.1080/10225706.2017.1406863

13. Huda, M.: Analysis of Factors that affect the Risk of Implementation of Underpass Project Construction in Mayjen Sungkono Surabaya. Int J Civ Eng Technol. **10**(06), 483–493 (2019)
14. Kulkarni, A. A., Dhanush, P., Chetan, B. S., Thamme Gowda, C. S., & Shrivastava, P. K. (2019). Recent development of automation in vehicle manufacturing industries. International Journal of Innovative Technology and Exploring Engineering, 8(6 Special Issue 4), 410–413. https://doi.org/10.35940/ijitee.F1083.0486S419

Smart Economy Through Smart Cities

Nurlatifah Diana Binti Pajilani, Mohd Fahmy-Abdullah, Suliadi Firdaus Sufahani, and Mohammad Kamarulzzaman Bin Ali

Abstract Smart economy has emerged as part of the smart city framework to encourage urban growth which the urban population currently lives in a digital society. The smart economy has emerged as part of the smart city framework to encourage urban growth. However, with technological and economic shifts brought about by globalization, cities are now facing the challenges of simultaneously sustaining productivity and sustainable urban development. The objectives of the study were to identify the smart economy concept for a smart city and examine the elements of smart economy in Pengerang. The study has conducted by using a mixed method. The data have collected by using questionnaires, document reviews and observations. The results showed that respondents fully understood the concept of smart economy that enables, encourages and stimulates economic activity in Pengerang. Besides, there will be future strategies and initiatives in order to encourage people to implement the smart economy. The implications of this study show that people need to pay attention to the issues and strategies that have been proposed by the government to implement and promote the smart economy toward a smart city.

Keywords Smart Economy · Smart Cities · Data Analytic · Statistical Analysis

---

N. D. B. Pajilani · M. Fahmy-Abdullah · M. K. B. Ali
Faculty of Technology Management and Business, Universiti Tun Hussein Onn Malaysia, 86400 Parit Raja, Batu Pahat, Johor, Malaysia

M. Fahmy-Abdullah (✉) · S. F. Sufahani
Oasis Integrated Group, Institutefor IntegratedEngineering, Universiti Tun Hussein Onn Malaysia, 86400 Parit Raja, Batu Pahat, Johor, Malaysia
e-mail: mohdfahmy@uthm.edu.my

S. F. Sufahani
Faculty of Applied Science and Technology, Universiti Tun Hussein Onn Malaysia, 84600 Pagoh CampusPagoh, Johor, Malaysia

M. S. Kaiser et al. (eds.), *Proceedings of the Third International Conference on Trends in Computational and Cognitive Engineering*, Lecture Notes in Networks and Systems 348, https://doi.org/10.1007/978-981-16-7597-3_23

1 Introduction

In recent decades, the world has been increasingly urbanizing. In 1950, just 30% of the global population lived in urban areas, while the number expanded to 55% in 2018 [1]. Despite the rapid increase in the global population, cities around the world face a wide variety of emerging challenges and issues in delivering public services. Another strategy popularly suggested to overcome this challenge is to build smart cities for governments and city planners [2]. Smart city concept based on the ideas of urban development followed the current industrial revolution 4.0, the idea which is more advanced due to increasingly urban problems [3]. When investments in human and modern transport, social capital, high quality of life and communication networks drive sustainable economic development with wise environmental management, cities are considered intelligent [4].

According to Zuraida Kamaruddin, Minister of Housing and Local Government, in the next five years, Pengerang in Johor is planned to become a smart city that includes an area of 121 hectares south of the district that has been identified as a smart city area, which must comply with the Malaysian Smart City Framework guidelines launched by the government [5]. Malaysian cities need to be more linked both physically and economically to reach smooth interaction between people, jobs, educational institutions and facilities, which is an important part of a creative city and also contributes to the quality of life [6].

According to Prime Minister Tun Dr Mahathir Mohamad, when Malaysia Smart City Framework was introduced, it was stated that the importance of making Malaysia a smart city is one of the priorities in the Rancangan Malaysia Ke-12 (RMKe-12) concept for smart cities is integrated with sustainable urban services technology, which will enhance public safety and the quality of life of the city including the fifth-generation (5G) connectivity, cashless communities, efficient public transport, energy-efficient buildings, water treatment and smart waste management [5]. Figure 1 shows that analysts from the World Bank have stated that without building smart cities, Malaysia will not achieve the objective. Statisticians predict that 90% of the population of the world will live in city regions by 2050. The fact that Malaysia wants to invest in its smart cities is important to ensure that economic growth and competitiveness do not slow [7].

Smart economy relies on a new version of cooperation between growth, distribution and consumption [3]. Technological and economic shifts brought about by globalization, cities are now facing the challenges of simultaneously sustaining productivity and sustainable urban development. Such problems would have an impact on various aspects such as economic, socio-cultural, housing environmental and living conditions on issues relating to the quality of urban lifestyle [8]. While the focus of urban appears to be on regional metropolises, most urbanites are located in medium-sized cities facing the challenge of dealing with tremendous urban center rivalry on similar issues [8].

On the other hand, one of the problems that occur about the smart city information is data not available either due to no authoritative entity or due to the organization

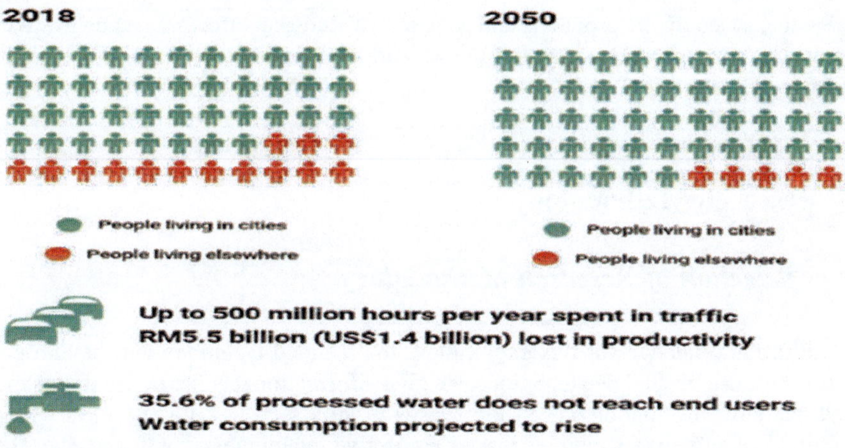

Fig. 1 The need for smart cities in Malaysia

responsible for collecting such information because of confidential and private information that cannot be obtained. As discovered in this report, the information restriction severely limits the usefulness of the conventional approach in evaluating cities for quick comparison, even when ports are within the same areas [9]. Therefore, in most developed countries, including Malaysia there has been a lack of literature, and this includes a local smart city study to address local issues that article particularly transforms the local concerns [10]. The solution to the problem is to resolve the issues, problems and goals found in technology opportunities that are essential to understanding local needs. Therefore, it helps to spark future research directions and to make future innovations and policy decisions while accelerating the implementation of government policies at the same time [10].

Based on previous research, studies on the smart economy focusing on e-commerce or e-business, ICT is used in business such as to increase productivity, innovation on new services or products and delivery of services [11]. Besides that, the previous study also discussed smart economy principles that aim for competitiveness and economic development which link with the smart city in terms of creativity and ideas, opportunities, growth of entrepreneurship, and regional or global competitiveness [3]. However, there are still no fixed indicators and variables used in the implementation of the concept of the smart economy in smart cities, based on some information and literature on the dimensions of the smart economy in smart cities. It is important for research to obtain indicators and variables related to the dimensions of the smart economy in the implementation of the smart city structure [12].

In the future, to get more information more case studies must be included, more informants and improve preparedness for further testing [13]. Other than that, it remains uncertain how the system and infrastructure can be managed by policy to allow the city area, which the smart economy can be conducted in a city [12].

Hence, based on the problems and gaps of this study, an effort was made to study the concept of smart economy and the way to examine the elements of the smart economy in Pengerang.

2 Research Methodology

2.1 Selection of Research Methodology

To fulfill the objectives of this exploration, the blended technique will be chosen. This is because the blended technique can give information that has believability and legitimacy of information since this investigation is seen in different ways where it utilizes quantitative strategies just as subjective techniques simultaneously. The document review would be used to identify the smart economy concept for a smart city but also conducted questionnaires to examine elements of smart economy in Pengerang.

In this examination, the researcher utilized the primary and secondary information as the wellspring of getting the information. For the secondary information, the researcher would utilize different structures to direct report audit; one of the structures was administrative sources because they come from straightforward and reliable legislative offices [8]. The researcher would likewise lead essential information for getting the exploration consequence of the subsequent target. Along these lines, the review survey was utilized as the instrument to collect the information. The researcher conveyed the poll by online Google structure to the respondents living in Pengerang.

2.2 Sampling and Population Techniques

The target population for this study is the residents of Pengerang. The total population in Pengerang is about 15,169 people. The Krejcie and Morgan table determined the sample size for this research. For this research, the non-random sampling technique is used due to easy to find the respondents and there are no specific conditions. Our sample size was 370 people.

2.3 Data Analysis Technique

Information investigation was finished by utilizing the Statistical Package for the Social Sciences (SPSS) Version 22 to prepare and break down quantitative information. It is normally utilized as an information assortment apparatus by different scientists, particularly to examine the result of the survey. SPSS utilized to run the

Table 1 Mean and standard deviation for the concept of smart economy

	Concept of Smart Economy	Mean	Std. Deviation	N
C1	Smart economy can evolve economic growth for digital technology, financial or information purposes, distribution and processing of goods through economic activities	4.375	0.570	136
C2	Smart economy enables a productive and competitive market environment, encourages and stimulates creativity application of technologies, connectivity and information (Internet of Things)	4.338	0.559	136
C3	Smart economy can improve the attractiveness of E-commerce, productivity, information and communications technology (ICT), innovation	4.338	0.520	136

factual tests since it has a wide assortment of capacities for making and recording factors.

3 Result and Discussion

3.1 Document Review

Table 1 demonstrates the central tendencies measurement of the concept of smart economy. The mean value for the concept of smart economy is between 4.338 and 4.375. The result shows that most respondents choose to answer agree. The table shows that C1 scores the highest standard deviation of 0.570, whereas C3 scores the lowest standard deviation of 0.520.

C1 with 4.375 mean stated that smart economy can evolve economic growth for financial, digital technology, information purposes, distribution and processing of goods through economic activities. Smart economy will strengthen economic activities from a few aspects that will increase the quality and standard of life. In addition, it is enabling a productive and competitive market environment, which irrespective of results encourages and stimulates creativity. It will ensure stability in a competitive labor market and the ability to adapt, if necessary, and make changes. Moreover, market conflicts are limited not just to the distribution of profit resources but also to creativity, competition, including the application of technologies, connectivity and information (IoT) [9]. Moreover, it can guarantee a city's prosperity and economic growth, as well as its residents' livelihoods [10].

3.2 *Reliability of Real Test*

Table 2 shows the Cronbach's alpha values for the independent variable in this research. For the E-commerce variable, it consists of six items, which are: "'How was your experience using E-commerce?', 'E-commerce helpful to the consumer in the E-business domain?', 'Are you satisfied using E-commerce as commercial advantages over the traditional commercial methods?', 'Do you agree online purchasing is more convenient rather than visiting a physical store?', 'Do you agree online payment are more convenient rather than using cash money?', and 'Do you agree that E-commerce can provide an alternative marketing channel by eliminating middleman?'". The E-commerce Cronbach's alpha is 0.843 and indicates good internal consistency. The variable Productivity consists of six items, which are: "'How efficient the productivity of economy in Pengerang?', 'Productivity growth constitutes an important element for modelling the productive capacity of economics?', 'Productivity growth defines quality of life, promote digitalization can accelerate productivity?', 'If the productivity level is low, the economy will turn to become slow?' and 'Do you receive any goods / services within the stipulated time?'". Productivity Cronbach's alpha value is 0.773 and indicates acceptable internal consistency.

Next, ICT consists of six items, which are: "'The improvement of internet connectivity helps to succeed in the implementation of smart economy?', 'Do you have a stable internet connectivity?', 'Do you agree that Pengerang area has fully utilized using ICT (i.e. Internet, wireless network, cell phones, software)?', 'ICT help to reach new customer/user beyond your area?', 'ICT enables users to access, store and manipulate information in digital form?' and 'ICT could help business to present a creative economy?'". ICT Cronbach's alpha value is 0.713 and indicates acceptable internal consistency.

Lastly, Innovation consists of six items, which are: "'Do you think a product/services should be under innovation development to sustain in business industry?', 'Innovation can increase quality of life?', 'do you think further innovation can increase the value added of a product?', 'Implementation of innovation is a must in all sectors of the economy (businesses, services and enterprises)?', 'Innovation induces technological changes by adopting new technology?', and 'Provide incentives can encourage innovation'" Innovation Cronbach's alpha is 0.902 and indicates excellent internal consistency. Among these variables, innovation has the highest Cronbach's alpha with the value of 0.902, whereas ICT has the lowest Cronbach's alpha with the value of 0.713.

Table 2 Cronbach's alpha for each variable

Variable	Cronbach's Alpha	No. of Items	Deleted Item	Internal Consistency
E-commerce	0.843	6	No	Good
Productivity	0.773	6	No	Acceptable
Information and Communication Technology (ICT)	0.713	6	No	Acceptable
Innovation	0.902	6	No	Excellent

Table 3 Result of E-commerce

	Question	Mean	Std Deviation	N
E1	How was your experience using E-commerce?	4.404	0.682	136
E2	Is E-commerce helpful to the consumer in the E-business domain?	4.368	0.528	136
E3	Are you satisfied using E-commerce as commercial advantages over the traditional commercial methods?	4.294	0.610	136
E4	Do you agree online purchasing is more convenient rather than visiting a physical store?	4.368	0.665	136
E5	Do you agree online payment is more convenient rather than using cash money?	4.360	0.605	136
E6	Do you agree that E-commerce can provide an alternative marketing channel by eliminating middleman?	4.463	0.729	136

3.3 Central Tendencies and Standard Deviation

The results in Table 3 demonstrate the central tendencies measurement of E-commerce. The mean value for E-commerce is between 4.294 and 4.463. The result shows that most respondents choose to answer agree. The table shows that E6 scores the highest standard deviation of 0.729, whereas E2 scores the lowest standard deviation of 0.528. Statement E6 got a higher mean of 4.463 because of the eliminating intermediaries for E-commerce used. Today's global E-commerce network connects all sizes and systems of nations, individuals and organizations, opening up locally, regionally and nationally. Internet technology creates the nature of the internet market where its national and international technological architecture constantly redisplays and reinterprets location, duration and individuality [17]. As some of these natural advantages like the "no-middleman" have evidently been neutralized by technological developments, many have predicted the eventual extinction of such "intermediaries", leading to a coming period of disintermediation of the supply chain. Intermediaries, however, flourished and even thrived in a ruthlessly efficient environment that left little space for intermediaries. The internet is meant to put manufacturers into direct communication with end customers [18]. For some, E-commerce is all about using technology to change business operations in which disintermediation by cutting out the intermediary and re-intermediation, meaning which the company becomes the intermediary, are strategic instruments to make the supply chain more effective [19].

Table 4 demonstrates the central tendencies measurement of Productivity. The mean value for Productivity is between 4.316 and 4.500. The result shows that most respondents choose to answer agree. The table shows that P1 scores the highest standard deviation of 0.772, whereas P3 scores the lowest standard deviation of 0.559. For item P4, it stated that promoting digitalization can accelerate productivity, which got higher mean with 4.500, supported by statement said that the improvement of productivity, output and growth of city functions through the use of technology to

Table 4 Results of Productivity

	Question	Mean	Std Deviation	N
P1	How efficient the productivity of economy in Pengerang?	4.338	0.772	136
P2	Productivity growth constitutes an important element for modelling the productive capacity of economics	4.397	0.681	136
P3	Productivity growth defines quality of life	4.338	0.559	136
P4	Promote digitalization can accelerate productivity	4.500	0.571	136
P5	If the productivity level is low, the economy will turn to become slow	4.419	0.591	136
P6	Do you receive any goods/services within the stipulated time?	4.316	0.707	136

efficiently track, analyze and disseminate information in real time, make better use of existing resources and actually be beneficial to economic growth and standard of living [20].

Table 5 demonstrates the central tendencies measurement of Information and Communication Technology (ICT). The mean value for Information and Communication Technology (ICT) is between 4.147 and 4.647. The result shows that most respondents choose to answer agree. The table shows that ICT3 scores the highest standard deviation of 0.792, whereas ICT1 scores the lowest standard deviation of 0.510.

The statement stated in ICT1 is supported by the previous study that today's capital is a high-speed connectivity hub that offers stable ICT infrastructure that connects the city in real time to cities around the world [16]. Therefore, the concept of a city that can be smart and digital is ancient by using technology and specifically ICT to enhance the quality of living in the urban landscape [21]. Through the successful use of ICTs, smart city aims to change every part of its economic activities that show high potential [22] by showing that ICTs are important ingredients for improving

Table 5 Results of ICT

	Question	Mean	Std Deviation	N
ICT1	The improvement of internet connectivity helps to succeed in the implementation of smart economy	4.647	0.510	136
ICT2	Do you have a stable internet connectivity?	4.147	0.756	136
ICT3	Do you agree that Pengerang area has been fully utilized using ICT (i.e. Internet, wireless network, cell phones, software)	4.309	0.794	136
ICT4	ICT helps to reach new customer/user beyond your area	4.478	0.655	136
ICT5	ICT enables users to access, store and manipulate information in digital form	4.507	0.584	136
ICT6	ICT could help business to present a creative economy	4.493	0.558	136

Table 6 Results of Safety

	Question	Mean	Std Deviation	N
I1	Do you think a product/services should be under innovation development to sustain in the business industry?	4.610	0.533	136
I2	Innovation can increase quality of life	4.434	0.641	136
I3	Do you think further innovation can increase the value added of a product?	4.434	0.540	136
I4	Implementation of innovation is a must in all sectors of the economy (businesses, services and enterprises)	4.316	0.641	136
I5	Innovation induces technological changes by adopting new technology	4.493	0.596	136
I6	Provide incentives can encourage innovation	4.331	0.621	136

efficiency and competitiveness as a means of achieving economic development and improving the quality of life.

The results in Table 6 demonstrate the central tendencies measurement of Innovation. The mean value for Innovation is between 4.316 and 4.610. The result shows that most respondents choose to answer agree. The table shows that I2 and I4 score the highest standard deviation of 0.641, whereas I1 scores the lowest standard deviation of 0.533.

Mean for I1 is 4.610, which states that the products or services should be under innovation to sustain in the business industry supported by the statement that innovation plays a vital role in gaining competitive advantage in a highly dynamic market environment because versatile innovation capabilities allow a company to find and use opportunities to develop new services and products [23]. Innovation assumes the potential to predict the demands of the market. At the same time, it appears to help the enterprise to increase the quality of either its services or products it provides. It has organized more effectively and kept the operating budget under control given that the entire economic environment is unsure. At the same time, it may also be violent in some cases; it seems that innovation and creativity can help entrepreneurs keep their business reliable, flexible and efficient in meeting the needs of their customers [24]. Nevertheless, the most efficient invention is special, and research and development (R&D) is always the need or the action of safety elements, which has been done to ensure the quality of the environment.

Table 7 Results of automated system

Variable	Mean	Std. Deviation	N
E-commerce	4.376	0.350	136
Productivity	4.385	0.365	136
Information Communication and Technology (ICT)	4.430	0.346	136
Innovation	4.436	0.366	136

Table 7 demonstrates the average mean measurement of the entire variable. The mean value for the entire variable is between 4.376 and 4.436. The result shows that the higher average of mean is for innovation, while the rest is lower than 4.436. The table shows that innovation scores the highest standard deviation of 0.366, whereas ICT scores the lowest standard deviation of 0.346.

4 Conclusion

This study aimed to identify the smart economy concept for a smart city and examine the elements of a smart economy in Pengerang. This research used the mixed method, including both qualitative and quantitative methods.

Based on the analysis for document review, the smart economy can evolve economic growth for digital technology, financial or information purposes, distribution and processing of goods through economic activities. It is important for this entire element to grow on the right path in order to have stability in the market activities. From the data, the review shows that smart economy implementation will encourage the economic stability for the growth of a city as well as the welfare of the people in those residential areas. Other than that, a smart economy enables a productive and competitive market environment, encourages and stimulates the creative application of technologies, connectivity and information (Internet of Things). During this technological era in which the urban population actually lives in a digital society by strengthening their economic activity through technologies and the Internet of Things. The transition from the old economy style to the new style will encourage the level of readiness to transition high value-added industries. Furthermore, based on the observation it shows that the number of respondents mostly agree with the statement that a smart economy can improve the attractiveness of E-commerce, productivity, Information and Communications Technology (ICT), and innovation. It is because this entire element mostly has been discussed from the previous research.

The second objective of this research is to examine the elements of a smart economy in Pengerang. From the data analysis, there are a few elements that can be observed in Pengerang for the element of smart economy. There are E-commerce, productivity, ICT and innovation that have been observed in this questionnaire survey. This element gives a view for the people living in Pengerang on what smart economy will give those benefits in terms of economic growth. Based on the data analysis, it shows the acceptance toward smart economy is high where most of the respondents agree with the characteristics for the element of smart economy. Thus, with the high acceptance from the respondents toward the element of the smart economy, fast development must be implemented in order to attract more people to change their environment to achieve a smart city framework.

The first element for a smart economy is E-commerce; in this technological era, most people will use E-commerce platform services to get their daily basic needs. The experience of using this platform may lead to a change in purchasing methods. For example, back then people usually buy their needs when visiting a physical store but nowadays people just do online shopping.

Other than that, productivity in one area will show how their environment needs to be improved in order to make sure they have efficient economic growth. It shows how important productivity growth will constitute in modeling the productive economic capacity. Promotion of digitalization will boost in terms of productivity in order to become more productive in SMEs, manufacturing or even entrepreneurship.

Moreover, ICT will become the most expected element that will help to perform smart economy. It is because to perform a smart economy application, people need to have gadgets, internet connectivity and software that helps them to connect with one another. Stable connectivity is important to analyze how ICT helps people reach their friends or customer beyond their area. Furthermore, ICT will enable users to access, manipulate and store information in a digital form such as using a software application, Google Drive and Gmail. It will also help in terms of marketing of a product, back then people hard to promote their product especially for those who have a business, but nowadays with the help of ICT, the marketing channel can be easily shared with others.

In addition, to achieve the transition from low-level readiness to high value-added industries, innovation plays a crucial role in order to increase the value added of the product. A transformation of how cities function is underway through innovations emerging from IoT, cloud computing, big data analytics, mobility and empowered citizens [25]. Implementation of innovation is a must in all sectors of the economy to make sure the growth will accelerate and be efficient. Inducing technological changes by having new technology will increase the value of the product by undergoing R&D. Incentives from the government or local authorities will encourage SMEs to improve in terms of their product. Innovation centre can provide a smart business resource platform for projects and products and a hub for interaction, partnership or collaboration between participating businesses, governments of R&D institutes and the citizens of the region [25].

This entire element is very helpful to be an indicator of the smart economy toward a smart city in Pengerang. A few point of views can be improved to implement the smart economy in Pengerang. First, Pengerang local authorities need to provide information on what is actually smart economy will look like in Pengerang. Lack of information provided to residents who live in Pengerang about the framework of smart city may lead to misunderstanding of information.

References

1. United Nations: the speed of urbanization around the world. Population Bulletin of the United Nations **8**, 16–33 (2018)
2. Inn, T. L. (2019). Penang: Becoming A Smart State. Penang Institute Issues
3. Tyas, W. P., Nugroho, P., Sariffuddin, S., Purba, N. G., Riswandha, Y., & Sitorus, G. H. I. (2019). Applying Smart Economy of Smart Cities in Developing World: Learnt from In- donesia's Home Based Enterprises. IOP Conference Series: Earth and Environmental Sci- ence, 248(1).

4. Lekamge, S., & Marasinghe, A. (2013). Developing a smart city model that ensures the optimum utilization of existing resources in cities of all sizes. Proceedings - 2013 International Conference on Biometrics and Kansei Engineering, ICBAKE 2013, 202–207.
5. Metro, H. (2020). Pengerang jadi bandar pintar. Mymetro, 10–11. Retrieved from https://www.hmetro.com.my/mutakhir/2020/01/536979/pengerang-jadi-bandar-pintar
6. Worldbank. (2011). Malaysia economic monitor: smart cities. 1, 1–108. Retrieved from http://documents.worldbank.org/curated/en/639301468052737567/Malaysia-economic-monitor-smart-cities
7. Aseantoday. (2018). the link between smart cities and Malaysia's economic growth. 1–6. Retrieved from https://www.aseantoday.com/2018/03/the-link-between-smart-cities-and-malaysias-economic-growth/
8. Tahir, Z., & Malek, J. A. (2016). Main criteria in the development of smart cities deter- mined using analytical method. Planning Malaysia, 14, 1–14. https://doi.org/10.21837/pmjournal.v14.i5.179
9. Mohd Adnan, Y., Hamzah, H., Md Dali, M., Nasir Daud, M., & Anuar Alias. (2016). An initiatives-based framework for assessing smart city. Planning Malaysia, (5), 13–22. https://doi.org/10.21837/pmjournal.v14.i5.189
10. Yau, K. L. A., Lau, S. L., Chua, H. N., Ling, M. H., Iranmanesh, V., & Kwan, S. C. C. (2016). Greater Kuala Lumpur as a smart city: A case study on technology opportunities. 2016 8th International Conference on Knowledge and Smart Technology, KST 2016, 96–101. https://doi.org/10.1109/KST.2016.7440496
11. Soe, R. M., & Mikheeva, O. (2017). Combined model of smart cities and electronic pay- ments. Proceedings of the 7th International Conference for E-Democracy and Open Gov- ernment, CeDEM 2017, 194–205. https://doi.org/10.1109/CeDEM.2017.11
12. Indrawati, Azkalhaq, N., & Amani, H. (2018). Indicators to measure smart economy: An Indonesian perspective. ACM International Conference Proceeding Series, 173–179. https://doi.org/10.1145/3278252.3278278
13. Jnr, B.A., Majid, M.A., Romli, A.: A Trivial Approach for Achieving Smart City: A Way Forward towards a Sustainable Society. 21st Saudi Computer Society National Computer Conference. NCC **2018**, 1–6 (2018). https://doi.org/10.1109/NCG.2018.8592999
14. Martins, F.S., da Cunha, J.A.C., Serra, F.A.R.: Secondary Data in Research-Uses and Opportunities. Revista Ibero-Americana de Estratégia **17**(04), 01–04 (2018). https://doi.org/10.5585/ijsm.v17i4.2723
15. Yudono, A., Satria, D., & Erlando, A. (2019). Toward Inclusive Development Through Smart Economy in Malang Regency. IOP Conference Series: Earth and Environmental Science, 328(1). https://doi.org/10.1088/1755-1315/328/1/012008
16. Stawasz, D.: The Concept of a Smart City in Urban Management. Business, Man- agement and Education **14**(1), 34–49 (2016). https://doi.org/10.3846/bme.2016.319
17. Javalgi, R., Ramsey, R.: Strategic issues of e-commerce as an alternative global distribution system. **18**(4), 376–391 (2000)
18. Popkova, E. G., & Bruno, S. S. (2020). The 21st Century from the Positions of Modern Science : Intellectual , Digital and Innovative Aspects.
19. Simpson, M., Docherty, A.J.: E-commerce adoption support and advice for UK SMEs Determine the barriers facing SMEs adopting. **11**(3), 315–328 (2004)
20. .Govada, S., Sprujit, W., & Rodgers, T. (2017). Smart city concept and framework. Smart Economy in Smart Cities, 749–790. https://doi.org/10.1007/978-981-10-1610-3
21. Dameri, R. P. (2017). Smart City Implementation.
22. Kumar, V., D. B.: Smart Economy is smart cities. In Smart Economy in Smart Cities (2017). https://doi.org/10.1007/978-981-10-1610-3
23. Ho, T.C.F., Ahmad, N.H., Ramayah, T.: Competitive Capabilities and Business Performance among Manufacturing SMEs: Evidence from an Emerging Economy. Malay- sia. Journal of Asia-Pacific Business **17**(1), 37–58 (2016)

24. Belias, D., Trivellas, P., & Koustelios, A. (2017). Entrepreneurship and Innovation: Current Aspects. 189–205. https://doi.org/10.1007/978-3-319-47732-9
25. MCMC. (2016). Smart city: The new frontier of innovation. (13), 63. Retrieved from http://www.skmm.gov.my/Resources/Publications/MyConvergence.aspx

Exploring the Factor of Public Transport Ridership

Zaid Al-Muzzammil Abdul Halim and Adlyn Nazurah Abdul Rahman

Abstract Malaysia is facing a challenge on how to improve factor public transport ridership along with the urbanization in the city. With the air pollution increment in urban areas along with the numbers of vehicles in the city, the public transport in the city areas need to improve. The research study, therefore, attempts to study the effectiveness of public transport ridership by investigating the factor of public transport ridership and improve the public transport ridership. The study has been conducted by using a triangulation method. The results showed that the elements of convenience, consistency/continuity, and comfortability are factors of the public transport feeder. Besides that, the factor of the attitude of the crew; the physical condition of public transport; environmental-friendly service; availability of seats, safety, and security; and punctuality of service, service frequency, and bus connections between bus stops recorded mean score at a high level under the limitation for public transport ridership. The implication of this study shows that the public transport operators and government must do some initiative to improve the public transport ridership.

Keywords Factor · Public transport ridership · Improvement

1 Introduction

Public transport is defined as buses, trains, and other forms of transport that are available to the public, charge set fares, and run-on fixed routes [1]. While ridership means the number of passengers using a particular form of public transport [2]. In public transport ridership, patrons or passengers refer to the number of people using transit services [3]. Public transport also known as those who ride public transport from one place to another regardless of any kind of vehicles such as buses, trains,

Z. A.-M. Abdul Halim (✉) · A. N. Abdul Rahman
Faculty of Technology Management and Business, Universiti Tun Hussein Onn Malaysia, 86400 Parit Raja Batu Pahat, Johor, Malaysia
e-mail: ap160260@siswa.uthm.edu.my

M. S. Kaiser et al. (eds.), *Proceedings of the Third International Conference on Trends in Computational and Cognitive Engineering*, Lecture Notes in Networks and Systems 348, https://doi.org/10.1007/978-981-16-7597-3_24

taxis, or bicycles [4]. According to [5], users who use public transport can enjoy the facilities provided to go to one destination regardless of distance or proximity.

The concept of public transport is seen as a medium that enables users to go to a place without having to think about vehicle servicing. Some people will be servicing their vehicles before going to a remote place but for users who use public transport do not have to think about it. They only need to pay a courtesy charge. They only need to state where they want to go to public transport drivers and wait to reach the destination. While public vehicles can be concluded as a field for drivers to gain profit from the purchase of tickets or charges imposed on public transport users. While public transport ridership can be concluded as those who use public transport no matter what kind of vehicles like bus, train, carpooling, and so forth.

Vehicles are one of today's human needs to ease travel or movement of humans from one point to another and accelerate their journey to do their daily tasks. Most of the vehicles today use a gasoline or easy to call petrol and diesel used as fuel which is known to cause air pollution and are also blamed for contributing to climate change and global warming. People are globally aware regarding the issues with climate change and global warming, which significantly affect the environment, human, and animal on earth [6]. The better known pollutants emitted by air fleets have been a major research question for authorities responsible for site management and managing the impact of vehicle emissions on air quality especially in urban areas. Because of the affordable and more reliable than other transport available, private cars have become one of the major means of transport in Malaysia. In addition, it is easier for folk to move from one side to another side and at once can save their time and money than exhausted to public transport charges [7]. In general, due to the conveniences, speed, comfort, and individual freedom, private cars are the most attractive mode of transportation, which has often promoted the commercial car.

While private vehicles are capable to being owned by those who are capable, they will still use public transport as a medium to travel to a distant destination. It cannot be denied that public transport is one of the vehicles that is easier to use for anyone who need it. Private vehicles are also capable of providing a satisfaction to the users. However, behind the scenes there is a disadvantage in relation to public transport. It can produce large amounts of carbon and contributes to air pollution and global warming. The Malaysian transport ministry has addressed the problem of private vehicle use and temperature rise in various ways. One of them is by providing and demanding the use of public transport in Malaysia. According to [8], SPAD general public transport passenger volume increased in 2017 up by 3.7%, reaching 1.2 million. This shows that more and more people are aware about the status and importance of public transport. Therefore, to increase the use of public transport every year, this scenario should be maintained and improved.

Another reason that some of the studies have been found to be important in explaining transit passengers is the arrival of private vehicles. Larger home segments with vehicles, all of which are similar, may result in less public transport demand. This will affect the technical efficiency of public transport. Technical efficiency is crucial in order to determine the efficiency level in a firm and industry [9]. The concept of efficiency can be seen when a department or organization uses all available resources or inputs optimally to produce maximum output [10]. Efficiency can also be seen in public transport system.

The part of the car has decreased over the last few decades, and the number of vehicles available per household has increased [11]. Public transport passengers have been affected by other competitive or complimentary transportation options, in particular, the cost of owning and operating a private vehicle [5].

Parking trouble, traffic congestion, lack of public transport, and pedestrian entrance interconnected. This problem needs to be overcome by controlling the influx of vehicles to reduced public transport problems and traffic congestion by creating an efficient, easy, and systematic transport system. Hence, the proposed park-and-ride concept was introduced to support this study, in which park-and-ride encourages people to use private transport, park their vehicles in designated areas, and have to use bus services operating at a shuttle around the Bandar Penggaram [12]. Cities require a multi-modal transportation system that includes walking, cycling, public transportation, car trips, and transmission and telecommunications services. Excess car reduces the efficiency of urban transport systems by creating traffic and parking congestion, accident risk, and pollution, negatively influencing the pattern of land and city use density [13]. There is some policy that has been made by Johor Public Transport Masterplan (2015–2045) [14]. Intensive and relevant implementation strategy is necessary to achieve the vision of Johor PA, namely, "Creating a Comprehensive, Effective and Efficient Transport Network System in Johore 2020". The Implementation Indicator was created as a guide to designing implementation strategies for each transport sector comprising private transport (PT) and transit transport (TT).

The public transportation sector is a challenging thing in the present revolution. Various efforts are made to encourage consumers to use public transport and facilitate users to go to a place while reducing the release of unhealthy smoke. Nowadays, Malaysian urban areas, such as Kuala Lumpur, Penang, Johor Bahru, Kuching, and Kota Kinabalu, have begun to decrease along the way of dependency of the car as it happens in many western countries cities. Therefore, many other cities in Malaysia are also facing the same trend, probably more because those towns and cities are even in lower supply and transportation compared to the Kuala Lumpur metropolitan area [15]. Awareness of public transport is one of the challenges to implement public transport in the cities. Because it has limited capability to influence the policy process since government has their authorities in making decision toward economic growth. Firms cannot make their own decision to implement sustainable transport without permission of government, since government controls the economic growth [16].

Besides, the main problem lacks facilities for public transport and the use of low-quality mood of public transport. Public transports like bus, taxi, train, and more are certainly having their own place and not covered at a certain part in cities and villages [17]. Therefore, few places have a public transport and some of them are lacking with good facilities. Presently, most Malaysians are dependent on their own private cars to commute from one place to another because it is more comfortable and does not waste time to wait for public transport. This is evident from the percentage of car ownership in Malaysia is constantly increasing every year [18]. The public transport system is taken seriously especially in KL. This includes various vehicle services designed to transport customers in local and regional routes. Public transport is therefore made available to the public by car, buses, vans, and trains, either privately or publicly [3].

The first determinant factor comprises suitable vehicles as public transport feeder. Through the provision of a suitable vehicle as a public transport feeder, appropriate vehicle use can be identified. Various types of vehicles are suitable for feeder vehicles. According to [19], through the study that feeder vehicles are one of today's public transports. According to [20], the feeder's gear must follow the set requirements for a public vehicle. In addition, the importance in determining feeder vehicles that are compatible as a medium of public transport such as the use of buses that redirect from one point to another and facilitate the user to ride it. [21] state that feeder are also the transit carriers, private landowners, and developers. These entities are the one that may control the station access services, and the development and traffic. While [22] tell that among public transport modes other than heavy rails, franchise bus services have a high capacity and can use more flexibility for feeder vehicles and [23] said that feeder vehicles could reduce traffic congestion while saving on land use. As well as accommodating many passengers to take to the destination. Traveling by public transport uses less energy and produces less pollution than traveling in private vehicle. To make progress in reducing our reliance abroad impact of oil and climate impacts, public transportation must be part of settlement [5]. By reducing the growth in motor vehicle travel, reducing congestion, and supporting the use of land patterns efficiently, public transport can reduce harmful carbon production of 37 million metric tons per year. This savings represents the start of potential public transport contribution to national efforts to reduce greenhouse gases release and promote energy conservation. By providing suitable vehicles with low carbon emissions as feeder transport for passenger traveling to public transport is one of the ideas on how to reduce the carbon emission in Bandar Penggaram. [24] tell that feeder service is important for any public transport system to work effectively because integration between several different modes of transportation can contribute to a more efficient and highly patronage public transport services. The advantage of riding a feeder vehicle can alter the lifestyle of the community to reduce the use of private vehicles [25].

The second determining factor is the aspect of providing park and ride. Changing the park and riding on bike trips and trips can create an important environment, energy conservation, and public health benefits [26]. Based on [27] park-and-ride facilities range from multi-story parking garages with customer amenities to simple surface parking lots and they may vary in purpose from serving a major intermodal transportation center to facilitate carpooling. From [28], parking is a significant factor influencing transit access and ridership. Many communities and transit agencies have been revising their parking policies to encourage the use of transit and to minimize resources expended on parking. [29] said public transportation can have a variety of benefits in parklands. Air pollution, noise, wildlife impacts, and automobile congestion are reduced, while more visitors can be accommodated in fewer vehicles and historic roadways and views capes can be better preserved. The definition of Park and Ride (P & R) is a combination of car and public transport [30] which means parks use a private transport in one place and use the public transport to the new destination. Park and Ride (P & R) is a step internationally used as a 1 minute to handle traffic congestion and pollution. P & R is used for persuading drivers to transfer to public transport mode by offering prices or save time on driving alternatives for the entire trip and parking. To reduce the parking space requests and the environmental impact of private cars, universities around the world are implementing strategies to reduce dependence on private vehicles and improvements in the use of alternative transport modes [7]. Taman Putrajaya Sentral and riding facilities are registering low occupancy rates in multi-story car parks (below 50%) but a high occupancy rate in surface parking (85%). In addition, all the park stations and boarding used by long-term marketers [31]. Reference [32] give a statement that in their studies on Park-and-Ride (P&R) facilities in Putrajaya showed few factors that influence respondents to use P&R. It is found that 42% of the respondents agreed that they will switch to P&R facility if they can save their travel cost, meanwhile more than half (58%) of the respondents will switch to use P&R if they can save their travel time.

The third determinant factor is to provide ample and secure parking. According to [33], car-parking policy is significant in influencing transport, since almost all car trips start and end in a parking space. Even at an airport, bus stations, and shopping centers, problems with parking are an everyday occurrence. Lack of accessible parking can hurt local business and decrease the quality of life residents. Parking lots should be taken into consideration by the government and agencies. This is because, with the convenient and secure parking facilities, users do not have to worry about the vehicles that left before boarding public transport. In fact, with the availability of parking facilities in public transport areas, it also facilitates the public transport system without being disturbed by traffic congestion. Good on-street parking management can end on-street parking chaos. It enables streets to function more efficiently and to be better places. It makes them safer. Most of the smart parking systems (SPS) proposed in literature over the past few years provide solution to the design of parking availability information system, parking reservation system, occupancy detection, and management of parking lot real-time navigation within the parking facility and others [34]. In Malaysia, special in Terengganu, automatically rotating parking system also known as Automatic Rotary Type (ART) is the first parking system introduced in Malaysia and commence operations in early 2011 within the trading area Kuala Terengganu City [35]. Not only can it inspire other agencies, but it can also save space for vehicles to park their vehicles.

The fourth determinant factor is the facility of public transport. The nation's growth and the need to meet mobility, environmental, and energy objectives place demands on public transit systems. Current systems, some of which are old and in need have upgrading, must expand service area, increase service frequency, and improve efficiency to serve these demands [21]. Plant shrubs, trees, and flowers, then make sure of well-maintenance and think about attractive lighting. It is as important for safety and enhancing the character of the community. Based on empirical studies, it is important to maintain and care the facilities well. Good facility care can attract consumers to visit regularly and use public transport. With the complete facilities in the transportation sector, it can attract users to use the facilities provided and will make it a start to sustainability in that place. Next, [32] stated that some respondents from the interview indicated that they were dissatisfied with the CSAs' (customer service assistant) attitude that does not have eye contact and lack of smile when communicating. One respondent noticed that some staff may be very helpful during morning hours but due to long working hours, they seemed to be not "helpful" or "friendly" in the late afternoon due to fatigue. Apart from the visible facilities, such as sustainability, hygiene, and other facilities, customer service assistant plays an important role in maintaining the quality of public transport. The facilities of transportation generally lead to economic growth for a country, especially developing countries and investment in facilities or infrastructure can increase demand for goods and services [36].

Based on the previous study, public transport placement cannot cover all the places because of some of the inevitable factors. Among them are non-residents or no public facilities such as shopping complexes, networking facilities, facilities, and other facilities. This is because there is no high demand in some places and it is difficult for public transport to provide the facility other than to the detriment. Additionally, poor facilities result in less public transport users and have an effect on the public transport sector. The absence of organized systems also has adverse effects on the use of public transport in Malaysia. Users must know the bus schedules, fare prices, and estimated time of arrival to the destination so they did not get confused with the tickets they want to buy. Unsatisfactory infrastructure facilities such as no maintenance of surau, toilets, and seats also have some effect on the use of public use.

Hence, this research aims for the use of ridership that used public transport and what is needed for upgrading the public transport and the facilities at Batu Pahat by investigating the current reason of public transport ridership. Based on [37], in park-and-ride system, people can easily park their own transportation and use public transport to move from one point to another point. The facilities of park and ride are suggest to hire security to ensure that their transportation in a safe condition. Lack of facilities need to be improved and is held extensively to public transport like [7] and [38]. Therefore, with the availability of parks and rides available, it can be a bit easier for users to use public transport without worrying about parking their vehicles.

2 Research Methodology

2.1 Selection of Research Methodology

In this study, the researcher makes an observation on 11 roads at Batu Pahat. An in-depth study on research methodology is considered essential in order to complete the research [39]. The observation has been done based on the checklist of public transport ridership. This study is done by using quantitative method to collect the primary data for this study. After that, the collected data are analyzed through SPSS software.

The quantitative approach is being practiced by identifying the existence of walking and cycling facilities at the 11 roads which are Jalan Sultanah, Jalan Rahmat, Jalan Mohd Salleh, Jalan Pejabat, Jalan Ampuan, Jalan Zaharah, Jalan Tanjong Laboh, Jalan Masjid, Jalan Syahbandar, Jalan Muhammad Akil, and Jalan Zabedah. The observation is made and the checklist is used to identify the public transport ridership.

The quantitative approach is being practiced by interviewing the selected respondents at Majlis Perbandaran Batu Pahat (MPBP). The respondent has given the instruction on how to answer the questions before starting the interview. The respondent needs to answer all the questions and the interview was recorded to analyze the data. Lastly, the information obtained through the interview sessions are tabulated in the table for data analysis. A mixed methods approach for data collection using both quantitative and qualitative research techniques was adopted to achieve the research aim and objectives [40].

2.2 Sampling Technique

Sampling is the selection of a subset, which is a statistical sample of individual from within a statistical population to estimate characteristic of the whole population. Population is a collection of individuals or objects that have similar characteristics [41]. The targeted respondents are comprised of different age bracket, races, professions, and background. There are two types of sampling, which is random sampling and non-random sampling. Random sampling is the likelihood of any member of the population being selected known while non-random sampling is the likelihood of any member of the population selected is unknown. In this research, the non-random sampling is chosen.

2.3 Data Analysis Technique

Analysis and findings of this study has described sequentially according to the question of the research and the objective of the study would like to achieve [42]. Analysis of data is the process of interpreting and analyzing this data using logical and analytical thinking. The data and information obtained from interviews and observations were analyzed, and all these data were subsequently analyzed after being copied. All conversations have been recorded through the interview, and translated into an easy-to-understand form. Similarly, observations are made to enable researchers to obtain results to compare secondary data and to achieve the study's objective. Using Microsoft Word software, these data were analyzed. The document review is to

Table 1 Reliability analysis in the pilot study and the actual study

	Cronbach's Alpha	Number of item	Number of respondents
Pilot test	0.810	20	30
Actual study	0.817	20	250

guide how to make an observation of the case study and to conduct a questionnaire to formulate questions [43].

2.4 *Pilot Test*

Table 1 shows the reliability analysis for both pilot test and the actual in this research. The analysis shown for Cronbach's Alpha coefficient in the pilot test study was 0.810 while the analysis for the actual data was 0.810. It demonstrates that the validity of the questionnaire used in the study is good and therefore acceptable.

3 Result and Discussion

3.1 *Document Review*

Public Transport. By reviewing Johor's Public Transportation master plan (PAJ), it consists of the policies that include guidelines for the action plan development. There are 2,244 bus huts identified for the entire state of Johor. Because of the observation, almost all (97.4%) bus huts are still in good condition and safe to use by the public (Johor Public Transport Masterplan). In an effort to improve reliability in the use of public bus services, Batu Pahat Municipal Council has taken the initiative to upgrade the public bus route (bus stops) in Batu Pahat Town and nearby residential areas. The public transport path through the Batu Pahat Municipal Council includes three main phases to be implemented by approved public transport operators.

Ample and Secure Parking. Batu Pahat Municipal Council aims to provide a new infrastructure that will improve parking to reduce land use and conserve parking space. Proposed parking lot is supported to promote the concept of "Pedestrianization" in the inner regions of the road closure proposal.

Proposed Parking Level Road (Jalan Syahbandar). Among the five roads planned by the municipality, one road to focus on is Jalan Syahbandar. This is because the path was included in the 11 pathways selected in this study. Proposed parking on this site is to accommodate the potential for leisure activities and to apply the principle of "pedestrianization" in urban areas.

3.2 Result Analysis

Park and Ride. Based on Fig. 1, 178 respondents (71.2%) know about definition of park and ride while 72 respondents did not know what is park and ride. Most of the respondents (60.8%) have used park-and-ride service in Malaysia.

On the other perspective, 91 (36.4%) of the respondents suggest that park-and-ride systems are needed at Batu Pahat while most of respondents said no which is 159 respondents (63.6%).

Ample and Secure Parking. Based on Fig. 2, 158 respondents (63.2%) park their own transport on parking space provided along the road at Batu Pahat. Most of the respondents (76.0%) agreed that parking slots at Batu Pahat are not enough.. On the other perspective, 145 (58.0%) of the respondents worried about their own transport because parking slots are not secured.

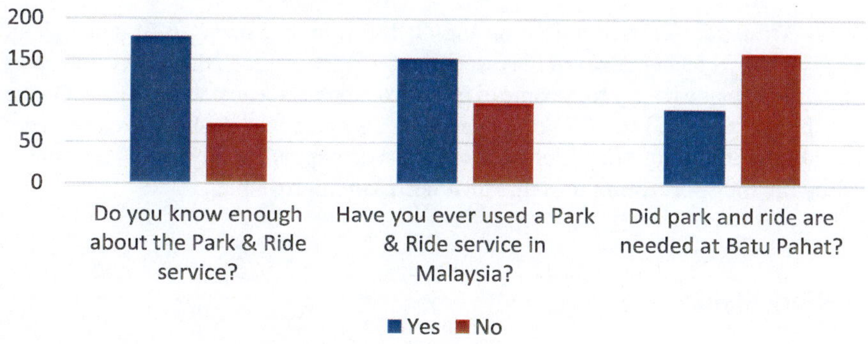

Fig. 1 Summary of the knowledge about park and ride

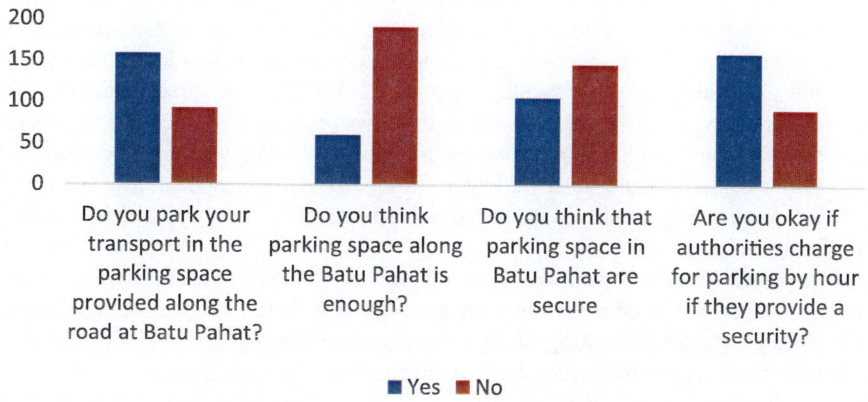

Fig. 2 Summary of the knowledge about ample and secure parking

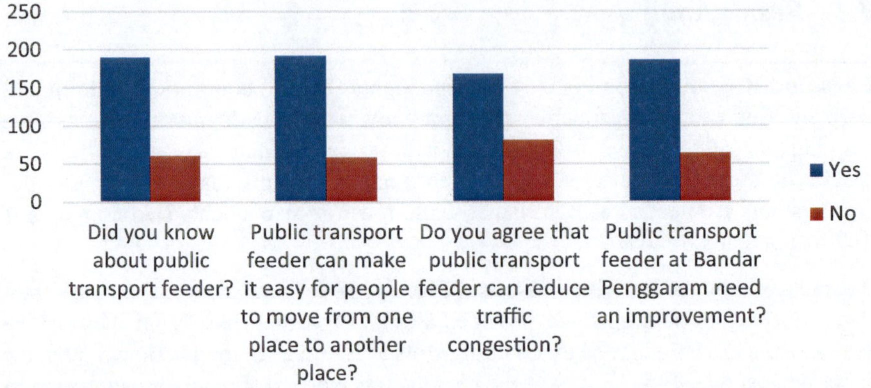

Fig. 3 Summary of the knowledge about public transport feeder

Public Transport Feeder. Based on Fig. 3, 189 respondents (75.6%) know about public transport feeder. Most of the respondents (67.2%) agreed that public transport feeder could make it easy for people to move from one place to another place. On the other perspective, 168 (67.2%) of the respondents agree that public transport feeder can reduce traffic congestion. Furthermore, most of respondents, 186 (74.4%), agreed that public transport feeder at Batu Pahat needs an improvement.

4 Conclusion

Based on document review and observation, the data have been compared to identify what is the factor that influence public transport ridership. MPBP has been planning to make Batu Pahat as one of the cities with low carbon emission. One of the plans is to improve their public bus as a feeder transport in Batu Pahat and build a more parking lot. With ample parking facilities, people can easily leave their vehicles and take public transport without worry. Batu Pahat Municipal Council aims to provide a new infrastructure that will improve parking to reduce land use and conserve parking space. Proposed parking lot is supported to promote the concept of "Pedestrianization" in the inner regions of the road closure proposal. Site selection location for parking levels shall be determined based on the "walkable distance" capability radius.

Among the five roads planned by the municipality, one road to focus on is Syahbandar Road. This is because the path was included in the 11 pathways selected in this study. Proposed parking on this site accommodates the potential for leisure activities and to apply the principle of "predestination" in urban areas.

From the data analysis, there are few programs that can be made to introduce and promote the park and ride in Batu Pahat. It is because some of the residents in the city do not know anything about park and ride and benefit of using public transport. Based on the data analysis, it has been shown that the bus connections between bus stops are poor where most of the respondents agree with the characteristics. Thus, with the high acceptance from the respondents toward bus connections, the programmed must be implementing in order to make it clear.

In addition, the other incentives that can be proposed are the attitude of worker and the buses physical. In this part, it is necessary to keep it up positive vibes between crew and customer. The users need to ensure that they know their bus route so that they can enjoy the accommodation provided by the parties involved (Batu Pahat/Majlis Perbandaran, 2016).

This study is important to know the public transport ridership factor in Batu Pahat. Previous studies are very important to be taken as references for the future researcher to do a research and analyze whether the research that is being done fulfill the theory and criteria needed in this chapter. Criteria and theory stated by the previous researchers are full of facts and the information needed are to be fulfilled. Theory and the conclusion that are being stated by many researchers maybe different but it got the same idea and criteria.

References

1. Saif, M. A., Zefreh, M.M., Torok, A.: Public Transport Accessibility : A Literature RevieWw (2018)
2. Cervero, R.: Forthcoming: Environment And Planning A (2020)
3. Rahman, A., Ashikin, N., Ayu, A.Y.: Theorizing The Concept Of Urban Public Transportation Institutional Framework In Malaysia **00043**, 1–6 (2016)
4. Golub, A.: Informal Public Transport, A Global Perspective (2011)
5. Holmgren, J.: A Strategy For Increased Public Transport Usage : The Effects Of Implementing A Welfare Maximizing Policy, (48), 22–1226 (2016).
6. Senin, S.N., Fahmy-Abdullah, M., Masrom, M.A.N.: The implementation of green transportation towards low carbon city. In IOP Conference Series: Earth and Environmental Science (Vol. 736, No. 1, p. 012063). IOP Publishing (2016)
7. Mohammed, A.L.I.A., Shakir, A.A.: Factors That Affect Transport Mode Preference For Graduate Students In The National University Of Malaysia By Logit Method 2. Study Methodology And Model Structure **8**(3), 351–363 (2013)
8. Spad. (2017). Trasportation Statistic Malaysia.
9. Hamdan, H., Fahmy-Abdullah, M., & Sieng, L. W. (2019). Technical Efficiency of Malaysian Furniture Manufacturing Industry: A Stochastic Frontier Analysis Approach.". International Journal of Supply Chain Management8, (6), 929–37.
10. Idris, A.I.M., Fahmy, M., Sieng, L.W.: Technical Efficiency of Soft Drink Manufacturing Industry in Malaysia". International Journal of Supply Chain Management **8**, 908–916 (2019)
11. Mallett, W. J. (2018). Trends In Public Transportation Ridership : Implications For Federal Policy.
12. Johan, N. (2012). Potential Concept Of 'Park And Ride' At Universiti Teknologi Malaysia At Skudai Campus, 28–33.
13. Frederik Strompen, T. L., & Bongardt, D. (2012). Reducing Carbon Emissions Through Transport Demand Management Strategies.
14. Pengangkutan, P., Johor, A., & Utama, T. T. (2016). Johor Public Transport Masterplan.
15. Raja Noriza Raja Ariffina, And R. K. Z. (2006). Urban Transport Growth: The Challenges Ahead – The New Realism And Institutional Changes Dr. Lawrence Tseu, 1–23.

16. Valley, K. (2015). The Challenges Of Implementing Urban Transport Policy In The Klang Valley , The Challenges Of Implementing Urban Transport Policy In The. Procedia Environmental Sciences, 17(December 2013), 469–477.
17. Pojani, D., & Stead, D. (2015). Sustainable Urban Transport In The Developing World: Beyond Megacities, (2), 7784–7805.
18. Borhan, M. N., Nazrul, A., & Ibrahim, H. (2019). Why Public Bus Is A Less Attractive Mode Of Transport : A Case Study Of.
19. Kwarteng, A.C., Iyer-Raniga, U., Guillermo, A.: Transport And Accessibility Challenges Facing The Rural People Living Along Feeder Roads In Ghana **6**(5), 257–267 (2018)
20. Yim, Y. B. (2006). Smart Feeder / Shuttle Bus Service : Consumer Research And Design, 19–43
21. Coffel, K., Parks, J., Semler, C., & Ryus, P. (2012). Guidelines For Providing Access To Public Transportation Stations
22. Bureau, H. (2017). Public Transport Strategy Study, (June)
23. Nielsen, G. (2005). Public Transport - Planning The Networks
24. Bachok, S., S. H. M. M. Z.: Feeder Mode Choice Selection Behavioural Modelling: The Case Of Ktm Komuter. Kuala Lumpur **15**(1), 65–80 (2017)
25. Shafii, H., Meryam, S., & Musa, S. (2009). Urban Transportation : Issue And Solution, 31–46
26. Cervero, R., Caldwell, B., & Cuellar, J. (2013). Bike-And-Ride : Build It And They Will Come, 83–105
27. Pool, P., Barker, J. B., Antion, K., Barnes, R. L., Garber, C., Gilliam, F. M., … Rosenberg, J. M. (2004). Secretary
28. Jacobson, L. (2016). Transit Cooperative Research Program Tcrp Synthesis 122 Transit Supportive Parking Policies And Programs.
29. Holly, F. (2009). Incentives And Disincentives For Day Visitors To Park And Ride Public Transportation At Acadia National
30. Norhisham, S., Sidek, L. M., Beddu, S., Usman, F., & Basri, H. (2015). Awareness And Level Of Usage For Park And Ride Facilities In Putrajaya , Malaysia, (August).
31. Hamsa, A.A.K., Adnan, S.A.A.S., Khalid, U.A.: Analysis Of Parking Usage At The Park And Ride Facility In Klang Valley. Malaysia **138**, 179–193 (2014)
32. Dahalan, D., Silva, J.L.D., Abdullah, H., Ismail, I.A., Ahmad, N.: Youth Confidence In The Quality Of Public Transport Services : The Case Of Greater Kl. Malaysia **9**(9), 12–22 (2015)
33. Christiansen, P., Engebretsen, Ø., Fearnley, N., Hanssen, J.U.: Parking Facilities And The Built Environment : Impacts On Travel Behaviour Transportation Research Part A Parking Facilities And The Built Environment : Impacts On Travel Behaviour. Transp. Res. Part A **95**(March), 198–206 (2017)
34. Sadhukhan, P. (2017). An Iot-Based E-Parking System For Smart Cities, (September).
35. Rafhanah, N. U. R., Ibrahim, B., & Jamal, M. (2017). Automatic Rotating Parking System
36. Umar, S., Shariff, B., Kadir, A.: Infrastructure Impact Of Land Transportation To Economic Growth: Malaysia Case Study **3**(2013), 1536–1545 (2013)
37. Parkhurst, G., Meek, S., Rail, S., Road, B., & Se, L. (2014). The Effectiveness Of Park-And-Ride As A Policy Measure For More Sustainable Mobility.
38. Kang, A. S., Jayaraman, K., Soh, K. L., Wong, W. P., & Kang, A. S. (2019). Social Predictors And Implementation Intention Of Drivers To Use Public Bus Transport.
39. Norazlan, M. A. E., Fahmy-Abdullah, M., & Masrom, M. A. N. (2021, April). Promoting carpooling and vanpooling program to reduce the use of private motorised transportation. In IOP Conference Series: Earth and Environmental Science (Vol. 736, No. 1, p. 012050). IOP Publishing.
40. Sabli, M.A.N., Fahmy, M., Sieng, L.W.: Application of Two-Stage Data Envelopment Analysis (DEA) in Identifying the Technical Efficiency and Determinants in the Plastic Manufacturing Industry in Malaysia. International Journal of Supply Chain Management **8**, 899–907 (2019)
41. Mohamad, N. F. N., Fahmy-Abdullah, M., & Masrom, M. A. N. (2021, April). Transit Oriented Development (TOD) typology. In IOP Conference Series: Earth and Environmental Science (Vol. 736, No. 1, p. 012037). IOP Publishing.

42. Latif, M.S.A., Fahmy, M., Sieng, L.W.: Determinants factor of technical efficiency in machinery manufacturing industry in Malaysia. International Journal of Supply Chain Management **8**, 917–928 (2019)
43. Ahadi, A. I., Fahmy-Abdullah, M., Rasi, R. Z. R. M., & Masrom, M. A. N. (2021, April). Integrating walking and cycling facilities towards green mobility in Bandar Penggaram Batu Pahat. In IOP Conference Series: Earth and Environmental Science (Vol. 736, No. 1, p. 012003). IOP Publishing.

Analyze Scheduling Problem on Operation Theatre in Malaysian Public Hospital Using Integer Linear Programming Method

Low Qiau Han, Suliadi F. Sufahani, and Mohd Fahmy-Abdullah

Abstract Operation theatre is one of the largest revenue centres of hospitals but also required high costs, therefore, the schedule of operation theatre should be effective and maximize the usage of operation theatre to improve their performance and meet patients' satisfaction levels. This study is aimed to construct a one-month schedule of an operation theatre at the strategic level by using the Integer Linear Programming method, maximize the usage of operation theatres and compare the existing schedule with the constructed schedule to validate the effectiveness of the proposed model. The coding is programmed into AMPL software and the schedule is solved weekly since the demand of operating hours of each department is updated each week by adding the unfulfilled demand in the previous week to the original demand of the same department in the current week. The constructed schedule has allocated a total of 54 operation theatres in a month. It has also achieved 96.52% of the overall allocated operating hours compared to the existing schedule. The usage of operation theatre has been maximized and the effectiveness of the proposed model is validated. However, this study can be improved by using various optimization methods and taking more variables and parameters into account.

Keywords Scheduling Problem · Operation theatre · Strategic level

L. Q. Han
Faculty of Applied Sciences and Technology, Universiti Tun Hussein Onn Malaysia, 84600 Pagoh, Johor, Malaysia

S. F. Sufahani (✉) · M. Fahmy-Abdullah
Faculty of Technology Management and Business, Universiti Tun Hussein Onn Malaysia, 86400 Parit Raja, Johor, Malaysia
e-mail: suliadi@uthm.edu.my

M. Fahmy-Abdullah
Oasis Integrated Group, Universiti Tun Hussein Onn Malaysia, 86400 Parit RajaBatu Pahat, Johor, Malaysia

M. S. Kaiser et al. (eds.), *Proceedings of the Third International Conference on Trends in Computational and Cognitive Engineering*, Lecture Notes in Networks and Systems 348, https://doi.org/10.1007/978-981-16-7597-3_25

1 Introduction

Operation theatre is an office or a spot in medical clinics where a few kinds of operations can be done to treat the patients. With the quick improvement in science and innovation, careful activity has turned into a typical clinical therapy with cutting-edge clinical hardware [1]. In this way, it required a lot of cash for clinical equipment. Hence, activity theatre required a significant expense, yet it is likewise the biggest income focus of emergency clinics [2–4].

Planning of activity theatre is fundamental for clinics to work on their presentation and meet the level of patients' fulfillments [5, 6]. Be that as it may, the planning of an operation theatre is anything but a simple matter. This is because of certain factors that should be considered like the accessibility of working rooms, the number of specialists and attendants, accessibility of patients, recuperation beds, and emergency medical procedures [7–9]. For the medical procedure technique, the Peri-operative (PERIOP) measure is comprised of three Pre-operative phases (preop) stage, the Intra-operative (intraop) stage, and the Post-operative (postop) stage. In the preop stage, the data of the patients were gathered and the medical procedures were ready. Then, at that point, the patients go through medical procedures in the intraop stage, while the postop stage was for the patient recuperation [4, 10, 18].

Besides, there are three levels or stages associated with the booking of operation theatre, which are key level, strategic level, and functional level. For a vital level, it normally included drawn-out arranging and it likewise had been called case blend arranging because the accessible season of activity theatre was distributed among the various offices or claims to fame. For the strategic level, it involved medium-term arranging and the Master Surgical Schedule (MSS) was built in this level. The MSS will allow the number of activity rooms to every one of the medical procedure gatherings. For the functional level, it required a momentary arrangement and the timetable that was built was comprised of the subtleties, for example, the beginning time and finish a season of activity [11–13]. The process was divided into two stages which are progressed planning (initial step) and allocation booking (second step). The development booking required of the meeting of the activity rooms and dates for every activity though the portion planning required of the assurance of the grouping of patients and tasks [14–16]. The three levels in the planning of activity theatre had an association with one another as the yields of the past level will be the contributions of a higher level [4] (Table 1).

Optimization methods are valuable in tracking down a most extreme or least number for a particular target work with a few imperatives. The planning of activity theatre is one of the model issues that can be tackled by utilizing enhancement techniques. The problem (operation theatre planning) can also be solved using Integer Linear Programming (ILP), Mixed Integer Linear Programming (MILP), Heuristic calculation, etc. [2]. Because of various genuine situations and goals, the strategies used to tackle the issues are likewise unique. In ILP, the model that is being utilized is straight. ILP and Straight Programming (LP) are very comparable, however, one of the contrasts between these two strategies is that ILP consists of entire numbers

Table 1 Week 1 schedule

Operation Theatre 1						
Department	Weekday					
	Monday	Tuesday	Wednesday	Thursday	Friday	Saturday
Ophthalmology	1	0	0	0	0	–
Surgery	0	1	1	0	0	-
Orthopaedic	0	0	0	0	1	–
Otorhinolaryngology	0	0	0	0	0	–
Oral Surgery and Orthodontic	0	0	0	1	0	–
Obstetrics and Gynaecology	0	0	0	0	0	–
Operation Theatre 2						
Ophthalmology	0	0	0	0	1	0
Surgery	0	0	0	0	0	0
Orthopaedic	1	0	0	0	0	0
Otorhinolaryngology	0	1	0	0	0	0
Oral Surgery and Orthodontic	0	0	0	0	0	0
Obstetrics and Gynaecology	0	0	1	1	0	0
Operation Theatre 3						
Ophthalmology	0	0	0	0	0	–
Surgery	0	0	0	0	0	–
Orthopaedic	0	1	0	0	0	–
Otorhinolaryngology	0	0	0	0	0	–
Oral Surgery and Orthodontic	0	0	0	0	0	–
Obstetrics and Gynaecology	1	0	0	0	0	–

just in the ideal arrangement, while LP comprises numbers with decimal spots. Consequently, ILP is more reasonable to be utilized to tackle the planning issue instead of the LP technique in this examination.

This investigation is a cross-sectional examination where the activity theatre information is gathered from various offices at a similar point on the schedule. The goals in this investigation are to build an agenda theatre by utilizing ILP strategy, amplify the us-period of activity theatres dependent on the accessible use time through the proposed model of ILP technique, and contrast the current timetable and the built timetable through the illustrative examination. Be that as it may, this examination zeroed in on an essential level just as it is drawn-out arranging and worry about the general heading of the planning of activity theatre [17]. To approve the viability of the proposed model, the spellbinding investigation that engaged with contrasting the timetables are from the parts of the complete number of activity theatres being assigned and the by and large dispensed working hours all through the one-month time frame (Table 2).

Table 2 Week 2 schedule

Weekday	Department					
	Monday	Tuesday	Wednesday	Thursday	Friday	Saturday
Operation Theatre 1						
Ophthalmology	1	0	0	0	0	–
Surgery	0	1	1	0	0	–
Orthopaedic	0	0	0	0	1	–
Otorhinolaryngology	0	0	0	0	0	–
Oral Surgery and Orthodontic	0	0	0	1	0	–
Obstetrics and Gynaecology	0	0	0	0	0	–
Operation Theatre 2						
Ophthalmology	0	0	0	0	1	0
Surgery	0	0	0	0	0	0
Orthopaedic	1	0	0	0	0	0
Otorhinolaryngology	0	1	0	0	0	0
Oral Surgery and Orthodontic	0	0	0	0	0	0
Obstetrics and Gynaecology	0	0	1	1	0	0
Operation Theatre 3						
Ophthalmology	0	0	0	0	0	–
Surgery	0	0	0	0	0	–
Orthopaedic	0	1	1	1	0	–
Otorhinolaryngology	0	0	0	0	0	–
Oral Surgery and Orthodontic	0	0	0	0	0	–
Obstetrics and Gynaecology	1	0	0	0	0	–

2 Methodology

2.1 *Flowchart*

To conduct a study, a lot of activities are involved to solve the proposed problem. A flow chart that shows the procedure of activities in the study is shown in Fig. 1.

2.2 *Data Collection*

Data used in the study was obtained from a hospital located in Muar, Johor. It was divided into two categories, which were primary and secondary data. The primary data, such as types of departments, numbers of operating rooms, availability of operation theatre every day, availability of human resources, weekly demand of operation

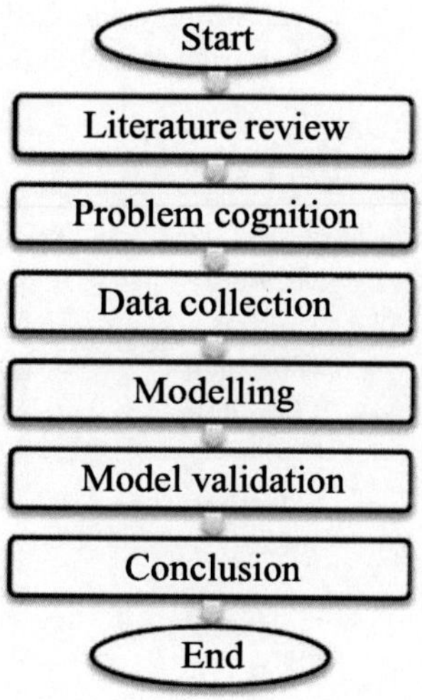

Fig. 1 Flow chart of the study

for each department, the maximum and the minimum number of operating rooms that required for each department per week, were obtained through hospital staff. The existing schedule was considered as the secondary data and was taken from hospital staff. The duration of all the data taken from the hospital was a one-month period, which was from 6 January 2020 until 1 February 2020. Due to the privacy and confidentiality the data from the hospital will not be revealed in this study.

2.3 Modelling

The mathematical formulation that was used in this study is shown as follows:

- Types of operation theatre, i:
  $i = 1, 2, 3$
  where

  1 Operation Theatre 1;
  2 Operation Theatre 2;
  3 Operation Theatre 4.

- Types of departments, j:
  $j = 1,2,3,4,5,6$
  where

  1 Ophthalmology Department;
  2 Surgery Department;
  3 Orthopaedic Department;
  4 Otorhinolaryngology Department;

5 Oral Surgery and Orthodontic Department;
6 Obstetrics and Gynaecology Department.

- Day, k:
  $k = 1,2,3,4,5,6$
  where

1 Monday;
2 Tuesday;
3 Wednesday;
4 Thursday;
5 Friday;
6 Saturday.

- Week, l:
  $l = 1,2,3,4$
  where

1 Week 1;
2 Week 2;
3 Week 3;
4 Week 4.

- x_{ijkl} = Number of operation theatre i that assigned to department j on day k for week l
- h_{ikl} = Available usage hours of operation theatre i on day k for week l
- r_{ikl} = Number of available operation theatre i on day k for week l
- d_{jl} = Weekly demand of operating hours in department j for week l
- t_{jkl} = Number of surgical teams in department j on day k for week l
- a_{jl} = Minimum number of operation theatre required by department j for week l
- b_{jl} = Maximum number of operation theatre required by department j for week l

The objective function of this model was to maximize the sum of the weekly percentage of demand of operating hours allocated in each department. The constraints of the model were the ideal from Sufahani and Ismail [6] and Dios et al. [8] (Table 3).

Objective function: Maximize

$$\sum_{i=1}^{3}\sum_{j=1}^{6}\sum_{k=1}^{6}\sum_{l=1}^{4}\frac{(h_{ikl} \times x_{ijkl})}{d_{jl}} \tag{1}$$

Subject to

$$\sum_{i=1}^{3}\sum_{j=1}^{6}\sum_{k=1}^{6}\sum_{l=1}^{4}(h_{ikl} \cdot x_{ijkl}) \le d_{jl} \tag{2}$$

Table 3 Week 3 schedule

Operation Theatre 1						
Weekday	Department					
	Monday	Tuesday	Wednesday	Thursday	Friday	Saturday
Ophthalmology	0	0	0	0	0	–
Surgery	0	0	0	0	0	–
Orthopaedic	1	1	0	0	0	–
Otorhinolaryngology	0	0	0	0	1	–
Oral Surgery and Orthodontic	0	0	0	0	0	–
Obstetrics and Gynaecology	0	0	0	1	0	–
Operation Theatre 2						
Ophthalmology	0	0	1	0	0	0
Surgery	1	0	0	0	0	0
Orthopaedic	0	0	0	1	1	0
Otorhinolaryngology	0	0	0	0	0	1
Oral Surgery and Orthodontic	0	0	0	0	0	0
Obstetrics and Gynaecology	0	1	0	0	0	0
Operation Theatre 3						
Ophthalmology	0	0	0	0	1	–
Surgery	0	1	1	0	0	–
Orthopaedic	0	0	0	0	0	–
Otorhinolaryngology	0	0	0	0	0	–
Oral Surgery and Orthodontic	0	0	0	1	0	–
Obstetrics and Gynaecology	1	0	0	0	0	–

$$\sum_{i=1}^{3}\sum_{j=1}^{6}\sum_{k=1}^{6}\sum_{l=1}^{4}(x_{ijkl}) \le t_{jkl} \tag{3}$$

$$\sum_{i=1}^{3}\sum_{j=1}^{6}\sum_{k=1}^{6}\sum_{l=1}^{4}(x_{ijkl}) \ge a_{jl} \tag{4}$$

$$\sum_{i=1}^{3}\sum_{j=1}^{6}\sum_{k=1}^{6}\sum_{l=1}^{4}(x_{ijkl}) \le b_{jl} \tag{5}$$

$$\sum_{i=1}^{3}\sum_{j=1}^{6}\sum_{k=1}^{6}\sum_{l=1}^{4}(x_{ijkl}) \le r_{ikl} \tag{6}$$

$$h_{ikl}, x_{ijkl}, d_{jl}, t_{jkl}, a_{ijkl}, b_{ijkl}, r_{ikl} \ge 0 \tag{7}$$

Description of the constraints:

- (2): The sum of the available usage hours of operation theatre that multiply with the number of operation theatre that is assigned to each department on a particular day for a particular week has to be less or equal to the weekly demand of operating hours of a department for that week.
- (3): The total number of operation theatres assigned to each department on a particular day of a particular week has to be less than or equal to the number of surgical teams in the same department of the day for that week.
- (4): The total number of operation theatres assigned to each department on a particular day of a particular week has to be greater than or equal to the minimum number of operation theatres required by that department of that week.
- (5): The total number of operation theatres assigned to each department on a particular day of a particular week has to be less than or equal to the maximum number of operation theatres required by that department of that week.
- (6): The total number of operation theatres that are assigned to each department on a particular day of a particular week has to be less than or equal to the number of operation theatres that are available on that day of the particular week.
- (7): All the values of the variables are positive.

In developing the proposed model, all the coding was programmed into AMPL (A Mathematical Programming Language) software to obtain an optimal solution for the scheduling problem through the ILP method. Since the scheduling problem was solved based every week, so the steps of constructing a 6-day schedule were repeated four times. Hence, four weekly schedules were constructed and then combined to become a one-month schedule. The weekly demand of operating hours for each department was updated every week, except on the first week, before solving the scheduling problem of operation theatre for that particular week. The weekly demand was updated by adding the unfulfilled demand in the schedule on the previous week to the original weekly demand in the current week.

2.4 *Model Validation*

To validate the effectiveness of the proposed model, the constructed schedule produced by using the proposed model was compared to the existing schedule given by the hospital. Besides, the feasibility of the optimal solutions obtained from AMPL software was checked manually. If the optimal solutions followed all the restrictions, then it could be said that it is also feasible solutions.

3 Results and Discussion

3.1 Constructed Schedules

A one-month schedule that starts from 6 January 2020 until 1 February 2020 is constructed using the hospital data and it is the combination of four weekly schedules. All the results obtained are optimal and feasible. In the constructed schedules, the operation theatres are available from Monday until Saturday each week. Based on Week 1 schedule, the departments assigned into Operation Theatre 1 on Monday until Friday are Ophthalmology, Surgery, Surgery, Oral Surgery and Orthodontic, and Orthopaedic departments, respectively. For Operation Theatre 2, the departments assigned on each weekday are Orthopaedic, Otorhinolaryngology, Obstetrics and Gynaecology, Obstetrics and Gynaecology, and Ophthalmology departments, respectively. For Operation Theatre 4, there are only 2 departments assigned throughout the week, which are Obstetrics and Gynaecology department that was assigned on Monday, and the Orthopaedic department that was assigned on Tuesday. From the aspect of the percentage of departmental allocations, only the Ophthalmology department has achieved 100% of weekly demand. The rest departments such as Surgery, Orthopaedic, Otorhinolaryngology, Oral Surgery, and Orthodontic, and Obstetrics and Gynaecology departments have achieved 95%, 77%, 69%, 86%, and 90%, respectively. So, the unfulfilled demands of these five departments are 1 h (5%), 8 h (23%), 4 h (31%), 1 h (14%), and 3 h (10%), respectively.

In Week 2 schedule, the departments assigned into Operation Theatre 1 on Monday, Tuesday, Wednesday, Thursday, and Friday are Ophthalmology, Surgery, Surgery, Oral Surgery, and Orthodontic, and Orthopaedic departments, respectively. For Operation Theatre 2, the Orthopaedic department is assigned on Monday, the Otorhinolaryngology department is assigned on Tuesday, Obstetrics and Gynaecology department is assigned on both Wednesday and Thursday, and the Ophthalmology department is assigned on Friday. Last but not least, for Operation Theatre 4, the Obstetrics and Gynaecology department is assigned on Monday, while the Orthopaedic department is assigned on the next 3 days, which is from Tuesday until Thursday. From the aspect of departmental allocations, both Ophthalmology and Obstetrics and Gynaecology departments are fully met their weekly demand of operating hours. However, for Surgery, Orthopaedic, Otorhinolaryngology, and Oral Surgery and Orthodontic departments, their percentage of departmental allocations are 69%, 92%, 53%, and 75%, respectively. Therefore, the unfulfilled demand of operating hours for the corresponding departments are 8 h (31%), 4 h (8%), 2 h (18%), and 2 h (25%), respectively.

Then in Week 3 schedule, the Orthopaedic department is assigned into Operation Theatre 1 on both Monday and Tuesday, Obstetrics and Gynaecology department is assigned on Thursday, and the Otorhinolaryngology department is assigned on Friday. For Operation Theatre 2, the departments assigned from Monday until Friday are Surgery, Obstetrics and Gynaecology, Ophthalmology, Orthopaedic, and Obstetrics and Gynaecology departments, respectively. Then for Operation Theatre

4, the departments assigned on Monday, Thursday, and Friday are Obstetrics and Gynaecology, Oral Surgery and Orthodontic, and Ophthalmology, respectively. On both Tuesday and Wednesday, Operation Theatre 4 is assigned to the Surgery department. From this schedule, Ophthalmology, Otorhinolaryngology, Oral Surgery and Orthodontic, and Obstetrics and Gynaecology departments have reached 100% of departmental allocations or fully met their weekly demand of operating hours. For the Surgery and Orthopaedic departments, the percentage of departmental allocation that they have reached is 82% and 92%, respectively. So, the unfulfilled demand for these two corresponding departments is 6 h (18%) and 3 h (8%), respectively.

Based on Week 4 schedule, the Ophthalmology department is assigned on Monday, the Surgery department is assigned on both Tuesday and Wednesday, Oral Surgery and Orthodontic department is assigned on Thursday, while the Orthopaedic department is assigned on Friday for Operation Theatre 1. For Operation Theatre 2, Oral Surgery and Orthopaedic department are assigned on both Monday and Tuesday, Obstetrics and Gynaecology are assigned on both Wednesday and Thursday, while the Ophthalmology department is assigned on Friday. In Operation Theatre 4, the Surgery department is assigned on both Monday and Thursday, while the Orthopaedic department is assigned on Wednesday. From the aspect of departmental allocations, only the Ophthalmology department has reached 100% of departmental allocation which means they have fully met their weekly demand. The rest departments such as Surgery, Orthopaedic, Otorhinolaryngology, Oral Surgery, and Orthodontic, and Obstetrics and Gynaecology, have reached 97%, 95%, 0%, 86%, and 75% of departmental allocations, respectively. Hence, their corresponding unfulfilled demand of operating hours are 1 h (3%), 2 h (5%), 7 h (100%), 1 h (14%), and 6 h (25%), respectively.

Within one month, the total number of operation theatres assigned to each department is 54 rooms, and the overall operating hours that being allocated is 471 h. Within the 54 operation theatres, 12 operation theatres in Week 1, 14 operation theatres in Week 2, 15 operation theatres in Week 3, and 13 operation theatres in Week 4. In addition, the objective function of the proposed model in Week 1, Week 2, Week 3, and Week 4 are 5.2, 5.2, 5.7, and 4.5, respectively. The objective function in Week 4 is the lowest because the unfulfilled demand of operating hours in that particular week is the highest compared with others. However, these schedules have maximized the usage of the operation theatre each week through the proposed model.

3.2 Comparison Between Constructed Schedule an Existing Schedule

Based on the existing schedule, the total number of operation theatres allocated to the departments each week is the same, which is 13 operation theatres. Therefore, there are a total of 52 operation theatres that have been assigned to the departments throughout the month. When compared to the constructed schedule, the existing schedule has allocated a higher total amount of operation theatre within the period with the difference of 2 operation theatres. During Week 1, the total number of operation theatres allocated to each department in constructed schedule is lower than that in the existing schedule. But, the total number of operation theatres becomes higher than that in the existing schedule during the next two weeks. Then, the number of operation theatres assigned to the departments is the same during Week 4.

Besides, the overall operating hours that are allocated in the constructed schedule is 471 h, whereas in the existing schedule, it is 488 h. This indicates that the constructed schedule has achieved about 96.52% of the overall allocated operating hours in the existing schedule in one month. Hence, the effectiveness of the proposed model is validated.

3.3 Sensitivity Analysis

The sensitivity analysis is done by assuming that the operation theatres are available on Sundays from 8 a.m. to 2 p.m. Hence, each of the three operation theatres is assumed to be available for 6 h on Sundays and each department also has available surgical teams on that particular day. The other parameters and variables are the same as constructing the previously constructed schedule. The results of sensitivity analysis are then compared with the previously constructed schedule to check whether the sensitivity analysis can achieve higher demand of operating hours. Since the objective function of the sensitivity analysis in each week is always higher than that in the constructed schedules, so the sensitivity analysis has better results than the constructed schedules. The availability of operation theatres on Sundays provides an opportunity for each department to increase their chances of meeting higher weekly demand of operating hours throughout the week. The objective function of the sensitivity analysis in Week 3 is 6.0, which indicates that each department has fully met its weekly demand of operating hours. In other words, the departmental allocations of each department in Week 3 are 100%.

The sensitivity analysis has allocated four additional rooms within the one-month period, where two of them are in Week 1, one in Week 2, and another one in Week 4. Hence, the total number of operation theatres assigned to each department in Week 1, Week 2, Week 3, and Week 4 are 14, 15, 15, and 14 operation theatres, respectively. From the aspect of each department, the availability of operation theatre on Sundays has increased the number of operation theatres assigned to three departments which include Orthopaedic, Otorhinolaryngology, and Obstetrics and Gynaecology departments. This means that these three departments have achieved higher demand for operating hours compared to the previously constructed schedules. In the sensitivity analysis, the overall operating hours that are being allocated is 480 h, where the allocated operating hours in each week are 114 h, 123 h, 129 h, and 114 h, respectively. The overall allocated operating hours in sensitivity analysis is 9 h more than that in the one-month constructed schedule.

The results show that the sensitivity analysis has achieved about 98.36% of the overall allocated operating hours in the existing schedule. It means that the total demand for operating hours achieved by all departments within one month can increase about 1.84% (from 96.52% to 98.36%) if the operation theatres are available on Sundays from 8 a.m. to 2 p.m. So, it is recommended for the operation theatres to be available on every Sunday and assigned to the departments that have a high demand.

4 Conclusion

In conclusion, the first objective of this study, which is to construct a schedule of operation theatre by using the ILP method is achieved successfully in Chapter 3, subtopic 3.1. The constructed schedules in Chapter 3, subtopic 3.1 are shown on weekly basis. Based on the departmental allocations of the constructed schedules, the unfulfilled demand of operating hours of each department is determined. The unfulfilled demand for operating hours indicates that the department does not fully reach its target for operations in that particular week. Therefore, those unfulfilled demands are added to the original demand of the same department in the following week. When came to the last week of the one-month period, which is Week 4, only the Ophthalmology department has fully met their demand of operating hours. The rest department which include Surgery, Orthopaedic, Otorhinolaryngology, Oral Surgery, and Orthodontic, and Obstetrics and Gynaecology departments, have 1 h, 2 h, 7 h, 1 h, and 6 h of unfulfilled demand of operating hours, respectively. Since the method used to construct the schedules is the ILP method, so the optimal solutions are integers. All the optimal solutions obtained from the AMPL software are fall within the restrictions, which means that the results are also feasible (Table 4).

Next, the second objective of this study that focussed on maximizing the usage of operation theatres based on the available usage time succeeds to achieve in both Chapter 2, subtopic 2.3 and Chapter 3, subtopic 3.1. The model proposed in Chapter 2, subtopic 2.3 is a maximization model. Hence, the results obtained through this proposed model have maximized the usage of operation theatres based on their available usage time of each operation theatre per day. The Week 4 constructed schedule has the lowest objective function, which is 4.5, is due to the restriction of the constraint (2.2) that restricted the allocated hours of an operation theatre not exceeding the demand of operating hours of each department. There are a total of 54 operation theatres allocated by the departments throughout the one-month schedule.

Furthermore, the third objective is fulfilled by comparing the existing schedule with the constructed schedules in Chapter 3, subtopic 3.2. The effectiveness of the proposed model is validated. The constructed schedule has allocated 2 more operation theatres than the existing schedule from the aspect of the total number of operation theatres allocated by the departments throughout the month. Besides, the comparison result shows that the constructed schedule has achieved about 96.52% of the overall allocated operating hours in the existing schedule in one month. Hence,

Table 4 Week 4 schedule

Operation Theatre 1

Department	Weekday					
	Monday	Tuesday	Wednesday	Thursday	Friday	Saturday
Ophthalmology	1	0	0	0	0	–
Surgery	0	1	1	0	0	–
Orthopaedic	0	0	0	0	1	–
Otorhinolaryngology	0	0	0	0	0	–
Oral Surgery and Orthodontic	0	0	0	1	0	–
Obstetrics and Gynaecology	0	0	0	0	0	–

Operation Theatre 2

Weekday Department	Monday	Tuesday	Wednesday	Thursday	Friday	Saturday
Ophthalmology	0	0	0	0	1	0
Surgery	0	0	0	0	0	0
Orthopaedic	1	1	0	0	0	0
Otorhinolaryngology	0	0	0	0	0	0
Oral Surgery and Orthodontic	0	0	0	0	0	0
Obstetrics and Gynaecology	0	0	1	1	0	0

Operation Theatre 3

Weekday Department	Monday	Tuesday	Wednesday	Thursday	Friday	Saturday
Ophthalmology	0	0	0	0	0	–
Surgery	1	0	0	1	0	–
Orthopaedic	0	0	1	0	0	–
Otorhinolaryngology	0	0	0	0	0	–
Oral Surgery and Orthodontic	0	0	0	0	0	–
Obstetrics and Gynaecology	0	0	0	0	0	–

the constructed schedule can be said as effective. However, the usage of operation theatres on Sundays is recommended since the overall operating hours allocated throughout the one month can be increased from 471 to 480 h. The increment in the percentage of overall allocated operating hours is about 1.84%.

For future study, the objective of the study is suggested to not only focus on the maximizing of the usage of operation theatre, but also minimizing the costs such as staff overtime cost, utility costs, purchases of medical equipment, and maintenance costs. It is important to control the operation theatre costs to maximize the revenue generated from operation theatres and allow for long-term performance. Besides, the future study can use more than one optimization method in analyzing and constructing the operation theatre schedule as each optimization method has its advantages and disadvantages. Moreover, the variables such as the availability of recovery beds, arrivals of emergency cases, and cleaning time of operation theatre are suggested to be included in solving the scheduling problem of operation theatre in the future study. These variables can be set as the constraints in the model to obtain a better schedule of operation theatre that fulfilled all the restrictions.

Acknowledgements Fundamental Research Grant Scheme (FRGS) No. K175 (FRGS/1/2019/STG06/UTHM/02/2) from the Ministry of Higher Education Malaysia (MOHE) and Universiti Tun Hussein Onn Malaysia (UTHM)

References

TG Weiser Ampleur et répartition du volume mondial d'interventions chirurgicales en 2012 Bull. World Health Organ. 94 3 201 209F https://doi.org/10.2471/BLT.15.159293

B Cardoen E Demeulemeester J Beliën 2010 Operating room planning and scheduling: A literature review Eur. J. Oper. Res. 201 3 921 932 https://doi.org/10.1016/j.ejor.2009.04.011

H Saadouli B Jerbi A Dammak L Masmoudi A Bouaziz 2015 A stochastic optimization and simulation approach for scheduling operating rooms and recovery beds in an orthopedic surgery department Comput. Ind. Eng. 80 72 79 https://doi.org/10.1016/j.cie.2014.11.021

A Abedini W Li H Ye 2017 An Optimization Model for Operating Room Scheduling to Reduce Blocking Across the Perioperative Process Procedia Manuf. 10 859 60 70 https://doi.org/10.1016/j.promfg.2017.07.022

S. Ghazalbash, M. M. Sepehri, P. Shadpour, and A. Atighehchian, "Operating room scheduling in teaching hospitals," *Adv. Oper. Res.*, vol. 2012, 2012, DOI: https://doi.org/10.1155/2012/548493.

S Sufahani Z Ismail 2014 A real scheduling problem for hospital operation room Appl. Math. Sci. 8 113–116 5681 5688 https://doi.org/10.12988/ams.2014.46413

PC Kuo RA Schroeder S Mahaffey RR Bollinger 2003 Optimization of operating room allocation using linear programming techniques J. Am. Coll. Surg. 197 6 889 895 https://doi.org/10.1016/j.jamcollsurg.2003.07.006

M Dios JM Molina-Pariente V Fernandez-Viagas JL Andrade-Pineda JM Framinan 2015 A Decision Support System for Operating Room scheduling Comput. Ind. Eng. 88 430 443 https://doi.org/10.1016/j.cie.2015.08.001

G Latorre-Núñez A Lüer-Villagra V Marianov C Obreque F Ramis L Neriz 2016 Scheduling operating rooms with consideration of all resources, post anesthesia beds and emergency surgeries Comput. Ind. Eng. 97 248 257 https://doi.org/10.1016/j.cie.2016.05.016

M. Belkhamsa, B. Jarboui, and M. Masmoudi, "Two metaheuristics for solving no-wait operating room surgery scheduling problem under various resource constraints," *Comput. Ind. Eng.*, vol. 126, no. December 2017, pp. 494–506, 2018, DOI: https://doi.org/10.1016/j.cie.2018.10.017.

E. W. Hans and P. T. Vanberkel, "Operating Theatre Planning and Scheduling," 2014, DOI: https://doi.org/10.1007/978-1-4614-1734-7_5.

S Sufahani I Zuhaimy 2014 A real scheduling problem for hospital operation room Applied Mathematical Sciences 8 113–116 5681 5688 https://doi.org/10.12988/ams.2014.46413

L Koppka L Wiesche M Schacht B Werners 2018 Optimal distribution of operating hours over operating rooms using probabilities Eur. J. Oper. Res. 267 3 1156 1171 https://doi.org/10.1016/j.ejor.2017.12.025

A Kumar AM Costa M Fackrell PG Taylor 2018 A sequential stochastic mixed integer programming model for tactical master surgery scheduling Eur. J. Oper. Res. 270 2 734 746 https://doi.org/10.1016/j.ejor.2018.04.007

PM Castro I Marques 2015 Operating room scheduling with Generalized Disjunctive Programming Comput. Oper. Res. 64 262 273 https://doi.org/10.1016/j.cor.2015.06.002

JM Molina-Pariente V Fernandez-Viagas JM Framinan 2015 Integrated operating room planning and scheduling problem with assistant surgeon dependent surgery durations Comput. Ind. Eng. 82 8 20 https://doi.org/10.1016/j.cie.2015.01.006

M Vandenberghe S Vuyst De EH Aghezzaf H Bruneel 2019 Surgery sequencing to minimize the expected maximum waiting time of emergent patients Eur. J. Oper. Res. 275 3 971 982 https://doi.org/10.1016/j.ejor.2018.11.073

JJ Pandit M Pandit JM Reynard 2010 Understanding waiting lists as the matching of surgical capacity to demand : are we wasting enough surgical time ? J. Assoc. Anaesth. Gt. Britain Irel. 65 625 640 https://doi.org/10.1111/j.1365-2044.2010.06278.x

Enhancing the Queuing Management System in Malaysian Public Hospital

Ng Rou Ting, Suliadi F. Sufahani, and Mohd Fahmy-Abdullah

Abstract In a public hospital, patients always have to wait for a long time at the registration desk before being served by physicians. Hence, it is necessary to overcome this problem by improving its queuing system to increase effectiveness. In this research, the methods used are queuing theory and simulation. Based on the data collected, most of the patients waited for 1 h 20 min to receive treatment. So, to reduce patients' waiting time, it is necessary to increase the number of servers in both the registration and consultation phase especially during the peak hours, and allocate resources in the best way as well.

Keywords Public hospital · Queuing system · Waiting time

1 Introduction

The problem of overcrowding in a public hospital is often happening which causes a negative effect which is a long waiting time. Longer waiting time in a public hospital will influence the satisfaction of patients, which includes both satisfaction with the provider and specific perception on the care quality and physicians' abilities [1]. Patients who visit public hospitals need treatment immediately, but they need to wait a very long time to be served, so this consequently will cause dissatisfaction. The public hospital faces the overcrowding problem due to the limited capacity [2]. So, to overcome this problem, better resource utilization can increase the efficiency of a hospital and reduce patient's waiting time [3]. To evaluate and increase the efficiency

Ng R. Ting
Universiti Tun Hussein Onn Malaysia, 84600 Pagoh, Johor, Malaysia

S. F. Sufahani (✉) · M Fahmy-Abdullah
Universiti Tun Hussein Onn Malaysia, 86400 Parit Raja, Johor, Malaysia
e-mail: suliadi@uthm.edu.my

M Fahmy-Abdullah
Oasis Integrated Group, Universiti Tun Hussein Onn Malaysia, 86400 Parit Raja, Batu Pahat, Johor, Malaysia

M. S. Kaiser et al. (eds.), *Proceedings of the Third International Conference on Trends in Computational and Cognitive Engineering*, Lecture Notes in Networks and Systems 348, https://doi.org/10.1007/978-981-16-7597-3_26

of queuing system, queuing theory gives important performance to measure the number of servers and queues [4].

Queuing theory is a mathematical approach to study the waiting line which can use in healthcare to deal with the patient flow [5]. Queuing systems consist of characteristics such as arrival pattern, service system, queue discipline, and arrival behavior [5]. In the outpatients' department, the size of arrivals is unlimited and they arrive randomly. Since the population size is unlimited, the number of arrivals at any given time is only a small portion compared with all of the potential arrival [6]. For queuing problems, the random arrivals generally follow Poison distribution [7]. Service systems consist of several channels and several phases [8]. The service system for the hospital can be classified as a multichannel, multiphase system due to the system consist of several servers and waiting lines. To evaluate the service system, the measurements required are an average number of patients waiting, average patients' waiting time, capacity utilization, capacity cost, and probability of patient will have to wait until get the service [9]. In most healthcare institutions, the queue disciplines used are first-in-first-out (FIFO) or based on categories of patients to decide priorities if an appointment system is not provided, as in the emergency department, patients with life-threatening injuries will be treated first before other patients [8]. The assumption for most of the queuing models is arrival will be patient to wait until being served. However, some arrivals have been known as balking or reneging. Balking refers to arrivals who reject to join in queue to get the service since the waiting time is too long while reneging refers to arrivals who join the waiting line but are impatient to wait so they leave without getting any service [7].

Other than queuing theory, simulations also can be applied to find out the reality of queues and get the results [4]. The simulation model is a set of assumptions that relates to the operations of a system in which the assumption are expressed in the mathematical, logical, and symbolic relationship between entities [10]. To prevent the risk of degradation in the level of healthcare and spending a huge amount of money, simulation can be applied which allows significant exploration of multiple choices [11]. Hence, the simulation of the queuing system can be used in real-world applications which imitate the reality that exists or is contemplated [12]. The main problem of long spending time at Jitra Health Centre is due to the uneven between the number of patients and the number of doctors which occurs especially in the morning [13]. So re-schedules the activities are suggested to make sure that enough number of doctors at the health center in that time.

Hospitals should treat their patients promptly; however, it is not be achieved in public hospitals due to high demand and limited resources [14]. Even though many patients choose private healthcare centers to get treatment, but the public hospital still has the problem of long waiting times. The Malaysian health ministry has set the waiting time for each patient only 60 min and this guideline is already included in the customer charter [15]. The average waiting time in a public hospital to get the prescription slip during registration is more than two hours while the average consultation time is only 15 min [16]. The factors that cause the longer waiting time are physician behavior, management problem, and insufficient facilities [16]. This problem consequently results in poor access to the services provided, difficulties to make an appointment, long waiting times, and dissatisfaction of patients [17, 18].

Hence, based on these problems, the objectives of this study are (i) to analyze the information of the queuing system in the outpatient department, (ii) to simulate and validate outpatient queuing model, (iii) to simulate queuing models which can reduce patients' waiting time, and (iv) to propose an effective queuing model which can improve the efficiency of services. This research focuses on queuing management system for a public hospital in Johor, Malaysia. One of the public hospitals located in Johor, Malaysia was selected to understand the queuing system in the outpatient department for improving the efficiency.

2 Materials and Methods

In this study, the data was collected from one of the outpatient department of a public hospital located in Johor, Malaysia. By using the data collected, more effective queuing models were simulated to reduce patient's waiting time in the outpatient department.

2.1 *Data Collection*

In this study, the data required is patients' waiting time in the outpatient department until they receive a consultation. Before getting the required data, this research has been registered on National Medical Research Register (NMRR) website for approval. After that, one of the public hospitals located in Johor, Malaysia was contacted to get the required data. The data used in this research was collected by hospital staff in the outpatient department on February 3, 2020. The variables for the data are arrival time at the outpatient department, registration time, and service time. The data collection was started from 7.40 am until 1.30 pm which was 5 h 50 min. During the data collection, there were 136 patients arrived in the outpatient department to receive a consultation.

2.2 *Queuing Theory*

Arrival Pattern of Patients

Due to the highly variable arrivals and service patterns, waiting lines will occur which will cause the system overload temporarily [9]. When the arrivals are independent of one another and cannot be predicted, arrivals are considered random and the size of arrival is unlimited or infinite when the number of arrivals is just a small portion of all potential arrivals [6]. Hence, the patients involved in this study were considered as random and unlimited arrivals which follow the Poisson distribution.

Service Characteristics

The capacity of each server and the number of servers used can determine the capacity of the queuing system [9]. However, the time needed for each patient to be served is different as the nature of the illness and patients' conditions are varying. Service systems are classified by the number of channels and number of phases [6]. They are a single-channel queuing system, multichannel queuing system, single-phase system, and multiphase system. In this study, the service time of the outpatient department was follow a negative exponential probability distribution and the queuing system was a multichannel, multiphase system as multiple servers existed and patients had to go through more than one server.

Queue Discipline

There are two types of the queue which are structured and unstructured queues. Structured queue form when people that are included in the queue are predictable and in a fixed location while unstructured queue form when people that included in the queue are unpredictable and varying locations [19]. In addition, one of the common rules in the queuing system is first-in, first-out (FIFO). FIFO behavior is the most fairly as people leave the queue in the order, which they arrive [19]. However, priority queues can be used to avoid congestions which protect the rights of priority patients such as old patients and severely ill patients [4]. In this research, the queue discipline was assumed as FIFO, although there were some priorities in the system for the emergency case.

Patients' Behaviors

The assumption for most of the queuing models is arrival will be patient to wait until being served. However, some arrivals have been known as balking or reneging. Balking refers to arrivals who reject to join in the queue to get the service since the waiting time is too long while reneging refers to arrivals who join the waiting line but are impatient to wait so they leave without getting any service [7]. So, these behaviors exist in a strong argument with the uses of queuing theory.

Operating Characteristics

Patients who arrived randomly will follow Poisson distribution while the service time follows Exponential distribution. There is some assumption for this model:

1. Arrivals are served by using a FIFO basis.
2. No balking or reneging exist in the queue.
3. Arrivals are independent of each other but the rate is constant over time.
4. Arrivals follow a Poisson distribution.
5. Service times are variable and independent but the average is known.
6. Service times follow a negative exponential distribution.
7. The average service rate is greater than the average arrival rate.

Table 1 Symbols for queuing model and calculation

Symbol	Definition	Calculation in M/M/1 model	Calculation in M/M/m model (m = number of channels open)
P_0	The probability that there are zero customers in the system	$P_0 = 1 - \frac{\lambda}{\mu}$	$P_0 = \frac{1}{[\sum_{n=0}^{n=m-1} \frac{1}{n!}(\frac{\lambda}{\mu})^n] + \frac{1}{m!}(\frac{\lambda}{\mu})^m \frac{m\mu}{m\mu - \lambda}}$ for $m\mu > \lambda$
L	The average number of customers in the system	$L = \frac{\lambda}{\mu - \lambda}$	$L = \frac{\lambda\mu(\frac{\lambda}{\mu})^m}{(m-1)!(m\mu-\lambda)^2} P_0 + \frac{\lambda}{\mu}$
W	The average time a unit spends in the waiting line or being served, in the system	$W = \frac{1}{\mu - \lambda}$	$W = \frac{\mu(\frac{\lambda}{\mu})^m}{(m-1)!(m\mu-\lambda)^2} P_0 + \frac{1}{\mu}$ $= \frac{L}{\lambda}$
L_q	The average number of customers or units in line waiting for service	$L_q = \frac{\lambda^2}{\mu(\mu - \lambda)}$	$L_q = L - \frac{\lambda}{\mu}$
W_q	The average time a customer spends waiting in the queue	$W_q = \frac{\lambda}{\mu(\mu - \lambda)}$	$W_q = W - \frac{1}{\mu}$ $= \frac{L_q}{\lambda}$
P	The probability that a service facility is being used	$P = \frac{\lambda}{\mu}$	$P = \frac{\lambda}{m\mu}$

Before computing the operating characteristics for the queuing system, average arrival rate, λ and average service rate at each channel, μ was determined. The average arrival rate is the average number of patients who arrived at the outpatient department during a given period while the average service rate is the average number of patients served per period.

Operating characterizes for a queuing system include the probability that there are zero customers in the system (P_0), the average number of customers in the system (L), the average time a unit spends in the waiting line or being served in the system (W), the average number of customers or units in line waiting for service (L_q), the average time a customer spends waiting in the queue (W_q), and the probability that the service facility is being used (P). In this research, the operating characteristics for the outpatient department of a public hospital were computed by using the software Excel QM. Table 1 shows the symbols for queuing model and the calculations.

2.3 Simulation

In this research, the student version of Arena software was used for simulation. Each of the simulations of queuing models was run in 10 replications to investigate the average waiting time at registration counters and consultation rooms. The average and maximum waiting times were recorded from the reports output generated by SIMAN.

Firstly, the outpatient queuing system was simulated, then the model was validated by comparing actual and simulated waiting times. The commonly used validation tolerance is 10% [20]. It means that the difference between actual and simulated average waiting times cannot exceed 10%. Then, 3 queuing models which can reduce patients' waiting time were simulated. For these simulation models, the number of servers provided in both registration and consultation phase increase, so that more patients can be served and at the same time reduce their waiting time. Next, these queuing models were compared with the outpatient queuing model by calculating the differences in percentage (%). Hence, the percentage of improvement was computed.

3 Results and Discussion

3.1 Descriptive Analysis

The basic features of the collected data from the outpatient department located in Johor, Malaysia are described by using the graph. In this research, there are 136 patients involved in this study. Figure 1 shows the patients' arrival time at the outpatient department.

During the data collection period, the first patient arrived at 8:00 am while the last patient arrived at 12.30 pm. Many patients arrived at 8.00 am then the number of arrivals decreased after 8.30 am. Most of the patients arrived at 9.30 am and 10.00 am, which were 24 patients. Then, 22 patients arrived at 10.30 am which was the second-highest number of arrivals. The maximum inter-arrival gap was 30 min while the minimum inter-arrival gap was 0 s. The first patient registered at 8.10 am while the last patient registered at 12.40 pm. Most of the patients registered at 9.40 am

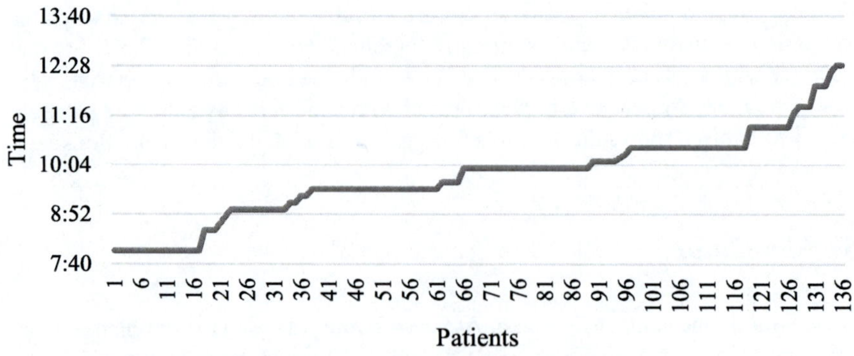

Fig. 1 Patients' arrival time

Table 2 Operating characteristics for the queuing system in the registration and consultation phase

Operating characteristics	Registration phase	Consultation phase
P_0	0.0138	0.0011
L	17.3175	19.7204
W	0.7428	0.8459
L_q	14.4915	14.0616
W_q	0.6216	0.6031
p	0.9420	0.9431

and 10.10 am, which were 24 patients. Then, the longest waiting time spent in the queue was 2 h 20 min while the shortest waiting time spent in the queue was 20 min and most of the patients waited for 1 h 20 min. Moreover, the first patients received treatment at 9 am while the last patients received treatment at 1:15 pm and the average service time was about 15 min.

3.2 Operating Characteristics

The data collection was started from 7.40 am until 1.30 pm, which was 5 h 50 min. During this period, there are 136 patients arrived, hence patients arrived at an average rate of 23.3143 per hour (Poison distributed). There are 3 registration counters and 10 consultation rooms in the outpatient department. However, only 3 registration counters and 6 consultation rooms were used during the data collection period.

The service rate for registration was 8.25 patients per hour while the service rate for consultation was 4.12 patients per hour. Table 2 shows the operating characteristics of the queuing system in the registration and consultation phase.

In the registration phase, the average waiting time a patient spends in the system is 0.7428 h while the average waiting time a patient spends in the queue is 0.6216 h. On the other hand, in the consultation phase, the average waiting time a patient spends in the system is 0.8459 h while the average waiting time a patient spends in the queue is 0.6031 h.

3.3 Simulation of Outpatient Queuing System

During the data collection period, the average inter-arrival time was 2.5735 min. The estimated time spent for registration is between 5 and 15 min. On the other hand, the estimated duration of consultation time is between 10 and 45 min, which most likely is 15 min. Figure 2 shows the simulation of the outpatient queuing system. There are 3 registration counters and 6 consultation rooms provided. After patients received consultation, they will leave the system.

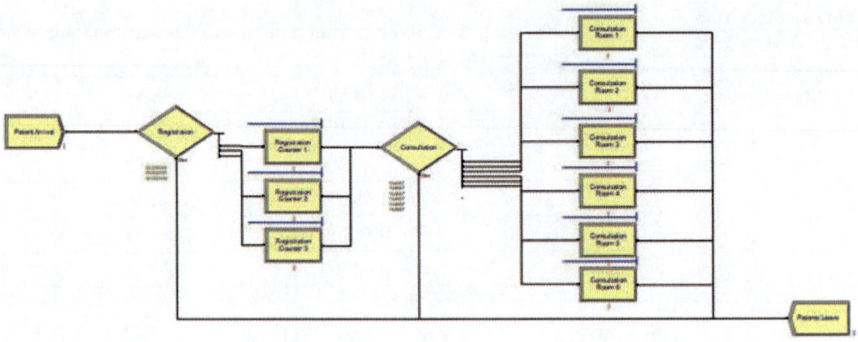

Fig. 2 Simulation of the outpatient queuing system

Table 3 Comparison between actual and simulated average waiting time

Phase	Average waiting time (h)		Differences (%)
	Actual	Simulation	
System	1.5887	1.6036	0.9379
Registration	0.6216	0.6810	9.5560
Consultation	0.6031	0.5475	9.2190

The actual and simulated average waiting times are compared to validate the simulation model. Table 3 shows the comparison between actual and simulated average waiting times.

Based on the table, the maximum difference is only 9.5560% which does not exceed 10%, so the simulation queuing model is considered valid.

3.4 Sensitivity Analysis

3 queuing models were simulated in which the patients' waiting time can be reduced at the same time more patients can be served. In these simulation queuing models, there is no fixed arrival quantity. Table 4 shows the number of servers provided

Table 4 The number of servers provided for each simulation model

Phase	Number of servers			
	Outpatient queuing model	Simulated queuing model 1	Simulated queuing model 2	Simulated queuing model 3
Registration	3	3	4	3
Consultation	6	10	10	11

Table 5 Average waiting time for each simulated queuing model

Phase	Average waiting time (h)			
	Outpatient queuing model	Simulated queuing model 1	Simulated queuing model 2	Simulated queuing model 3
System	1.6036	1.4719	1.1800	1.4119
Registration	0.6810	0.8420	0.3615	0.7857
Consultation	0.5475	0.1984	0.2832	0.1876

in each simulation model. These queuing models were simulated by adding more servers for the registration or consultation phase.

Table 5 shows the average waiting time for each simulated queuing model. Each simulation queuing model was compared with the outpatient queuing model.

For simulated queuing model 1, the average waiting time in the system decrease 8.2128% when compared with the outpatient queuing model. Then, the average waiting time in queue for registration increased by 23.6417% while the average waiting time in queue for consultation decreased by 63.7626%.

For simulated queuing model 2, the average waiting time in the system decreased by 26.4156% when compared with the outpatient queuing model. Then, the average waiting time in queue for registration and consultation decreased by 46.9163% and 48.2740%, respectively.

For simulated queuing model 3, the average waiting time in the system decreased by 11.9544% when compared with the outpatient queuing model. On the other hand, the average waiting time in queue for registration increased by 15.3744% while the average waiting time in queue for consultation decreased by 65.7352%.

Based on the table, the average waiting time in the system for simulation model 2 is greatly reduced compared with other simulation models. This is because registration counter and consultation rooms were added for patients. For simulation model 1 and simulation model 3, only consultation rooms were added, hence this caused overcrowded at the registration phase which ultimately influence the average waiting time in the system. Hence, simulated queuing model 2 is the most effective simulation model since the average waiting time for both registration and consultation phases was reduced.

4 Conclusion

Many patients spent a very long time waiting and most of them waited for 1 h 20 min. This is considered a long waiting time as the Malaysian health ministry set the waiting time for each patient to be only 60 min. In addition, there were a lot of patients waiting in the queue in which the average number of patients in the queue was 14 patients for both the registration and consultation phase. So, to reduce patients' waiting time, it is necessary to improve the queuing system by increasing the number of servers in both the registration and consultation phase especially during the peak hours as proposed in Simulated Queuing Model 2. Here is the contribution that can be highlighted through this paper:

- Analyzing the information of the queuing system in the outpatient department,
- Simulate and validate the outpatient queuing model,
- Simulated queuing models which reduce patients' waiting time by increasing the number of servers and counter
- Proposed an effective queuing model which can improve the efficiency of services.

For future research, Software Arena Full Version can be used to simulate a more complex queuing model as there are some limitations in the student version. Next, the researcher can test the applicability of the simulated queuing model to make sure it can be used by any outpatient department of the public hospital to reduce patients' waiting time. Then, the simulated queuing model can be manipulated to become more appropriate. For example, adding a registration counter and consultation room for senior citizens so that they can receive treatment faster. Not only that, a more effective queuing model which consists of an appointment system can be simulated, so that patients can make an appointment at home first before they arrived at the outpatient department. This is very useful to reduce their waiting time and avoid the overcrowded problem. In addition, future research also can study the doctors' and nurses' idle time during the operation hour to allocate resources in the best way.

Acknowledgements Fundamental Research Grant Scheme (FRGS) No. K175 (FRGS/1/2019/STG06/UTHM/02/2) from the Ministry of Higher Education Malaysia (MOHE) and Universiti Tun Hussein Onn Malaysia (UTHM).

References

1. Bleustein, C., Rothschild, D.B., Valen, A., Valaitis, E., Schweitzer, L., Jones, R.: Wait times, patient satisfaction scores, and the perception of care. Am. J. Manag. Care **20**(5), 393–400 (2014)
2. Chandra, D.: Reducing waiting time of outdoor patients in hospitals using different types of models : a systematic survey. Int. J. Adv. Res. Innov. **3**(1), 81–87 (2015)
3. Dellaert, N., Cayiroglu, E., Jeunet, J.: Assessing and controlling the impact of hospital capacity planning on the waiting time. Int. J. Prod. Res. **54**(8), 2203–2214 (2016). https://doi.org/10.1080/00207543.2015.1051668
4. Aslan: Applications of queues in hospitals in Istanbul. J. Soc. Sci. **4**(2), 770–794 (2015). https://doi.org/10.25255/jss.2015.4.2.770.794
5. Dhar, S.K.: Case study for bank ATM queuing model. IOSR J. Math. **7**(1), 01–05 (2013). https://doi.org/10.9790/5728-0710105
6. Dilrukshi, P.A.D., Nirmanamali, H.D.I.M., Lanel, G.H.J., Samarakoon, M.A.S.C.: A strategy to reduce the waiting time at the outpatient department of the national hospital in Sri Lanka. Int. J. Sci. Res. Publ. **6**(2), 281–2250 (2016)
7. Heizer, J.H., Render, B.: Module D: waiting-line models. Pearson Oper. Manage. (2013)
8. Fomundam, S., Herrmann, J.: A survey of queuing theory applications in healthcare. ISR Tech. Rep. **24**, 1–22 (2007)

9. Vass, H., Szabo, Z.K.: Application of queuing model to patient flow in emergency department. case study. Proc. Econ. Financ. **32**(15), 479–487 (2015). https://doi.org/10.1016/s2212-5671(15)01421-5
10. Kocaleva, M., Stojanova, A., Zlatanovska, B.: Simulation of M/M/n/m queuing system (2016)
11. Garza, N., Station, T.A.E.: Proceedings of the 1997 Juniper symposium. J. Range Manage. **50**(3), 330 (1997). https://doi.org/10.2307/4003741
12. Lade, I.P., Choriwar, S., Sawaitul, P.B.: Simulation of queuing analysis. Int. J. Mech. Eng. Robot. Res. **2**(3), 122–128 (2013)
13. Najib, M., Yusop, N., Ali, H.H.: Using simulation to reduce out-patient waiting time : a case study at Jitra Health Center, p. 7 (2002)
14. O. S. A, A. R. A, O. T. O: Queueing analysis of patient flow in hospital. IOSR J. Math. **10**(4), 47–53 (2014). https://doi.org/10.9790/5728-10464753
15. Aziati, A.H.N., Hamdan, N.S.N.S.B.: Application of queuing theory model and simulation to patient flow at the outpatient department. Proc. Int. Conf. Ind. Eng. Oper. Manage. (2013), 3016–3028 (2018)
16. Sufahani, S., Zuhaimy, I.: A real scheduling problem for hospital operation room. Appl. Math. Sci. **8**(113–116), 5681–5688 (2014). https://doi.org/10.12988/ams.2014.46413
17. Pillay, D., et al.: Hospital waiting time: The forgotten premise of healthcare service delivery? Int. J. Health Care Qual. Assur. **24**(7), 506–522 (2011). https://doi.org/10.1108/09526861111160553
18. Najmuddin, A.F., Ibrahim, I.M., Ismail, S.R.: Simulation modeling and analysis of multiphase patient flow in obstetrics and gynecology department (O&G Department) in specialist centre. Int. Conf. Appl. Math. Simul. Model. Proc. **9**(10), 125–130 (2010)
19. Bahadori, M., Teymourzadeh, E., Ravangard, R., Raadabadi, M.: Factors affecting the overcrowding in outpatient healthcare. J. Educ. Health Promot. **6**(1), 21 (2017). https://doi.org/10.4103/2277-9531.204742
20. Ahmed, S.Z.: Automated Queue Management System (2016)

Shooting and Discretization Method in Settling the Royalty Payment Problem

W. N. A. Wan Ahmad, S. F. Sufahani, and M. A. I. Kamarudin

Abstract Royalty payment has become one of the sources of income today. This research involved a non-classical Optimal Control problem (OCP), which is an economic application of the royalty problem. The first condition applied when the final state variable was unknown. The main goal was to maximize the functional performance index. However, the performance index was in terms of the unknown final state variable. Moreover, the unknown final state value produced a necessary boundary condition of the nonzero final shadow value. In this study, the three-stage royalty function was used and approximated into the continuous approximation of the hyperbolic tangent (tanh) procedure. This paper exhibits the output through shooting and discretization methods by manipulating the C++ and AMPL program language, respectively. The shooting method was constructed by combining the Newton's and Golden Section Search methods. At the end of the study, a validation process was conducted. This was done by comparing the shooting result with the discretization methods such as Euler, Runge–Kutta, Trapezoidal, and Hermite–Simpson methods. It is expected that the shooting method yields a more accurate optimal solution.

Keywords Discretization method · Non-classical optimal control · Royalty problem · Shooting method · Two-point boundary value problem

W. N. A. Wan Ahmad · S. F. Sufahani (✉)
Universiti Tun Hussein Onn Malaysia, 84600 Pagoh, Johor, Malaysia
e-mail: suliadi@uthm.edu.my

M. A. I. Kamarudin
School of Business Management, Universiti Utara Malaysia, 06010 Sintok, Kedah, Malaysia

S. F. Sufahani
Oasis Integrated Group, Universiti Tun Hussein Onn Malaysia, 86400 Parit Raja, Batu Pahat, Johor, Malaysia

M. S. Kaiser et al. (eds.), *Proceedings of the Third International Conference on Trends in Computational and Cognitive Engineering*, Lecture Notes in Networks and Systems 348, https://doi.org/10.1007/978-981-16-7597-3_27

1 Introduction

Optimal Control (OC) is an extension of Calculus of Variations (CoV). It is defined as a process of looking over among all permissible control variables $u(t)$, the one that takes the dynamical framework from some initial state $y(t_0)$ at time t_0 to a few terminal states $y(T)$ at some terminal time T to accomplish a maximum or minimum of a certain objective function or performance index [1, 2].

Other than the economics application [3, 4], the study of OC can also be applied in other various areas. For example, the OC was applied in the hazing study by [5].

Referring to the classical setting, the final value of the state variable $y(T)$ that is unknown will create an essential boundary condition of the final costate value that is equal to zero, $p(T) = 0$. In addition, the integrand of the performance index or objective function does not rely on the unknown final value of the state variable $y(T)$.

However, the problem in this research considers the objective function that relies on the final state value $y(T)$ that is unknown. Furthermore, the final costate value $p(T)$ is not equal to zero and this case can be classified as a non-classical OC problem. Therefore, we referred to Malinowska and Torres [6] to solve the matter regarding the nonzero $p(T)$ in the non-classical OC problem. This is to prove the new boundary condition. Thusly, the problem is unsolvable by using Pontryagin's Minimum Principle with the final value of the classical boundary conditions. As the integrand is the component of the final state value $y(T)$, a new necessary condition for $y(T)$ is set up to be equal to a certain continuous system which is z.

This research resolves the non-classical OC problem with the execution of a royalty function ρ that is equal to a three-stage piecewise function. The piecewise function is in the discrete form. Thus, the difficulty arises in the calculation process since the discrete royalty function is non-differentiable at a certain process. Hence, the discrete function will be converted into a form of continuous approximation of the hyperbolic tangent (tanh) function. This is to overcome the difficulty.

The shooting method is used to solve the problem, and it is constructed in the C++ programming language. The algorithm is a combination of Newton's method with the minimization techniques (Golden Section Search method) in the shooting method. After that, the result will undergo a validation process. In this process, the discretization methods such as Euler, Runge–Kutta, Trapezoidal, and Hermite–Simpson methods are applied to validate the result obtained.

This paper is organized into six sections. The next section will explain the shooting method applied in solving the royalty problem. After that, the non-classical OC problem will be explained briefly, together with an illustrative example of the royalty problem. Finally, a brief conclusion will be made at the end of the study. Some other researches involve business and management as well [7–9].

2 Shooting Method

This study uses the combination of Newton's method with the Golden Section Search method as a part of the shooting method in the tackling of the non-classical OC problem. The constructed algorithm utilizes C++ programming language with Numerical Recipes library routine (NR3) [10] which is profoundly precise. The flowchart of the method is well illustrated in the following figure (Fig. 1).

3 Non-Classical Optimal Control Problem

This study, for the most part, is centered on the non-classical OC issue. In this study, the objective function depends on the final value of the state which is free and priori unknown. Regardless of that, the final costate value is not equivalent to zero and this ought to be identical to another boundary condition, which varies from the classical optimal control theory. Malinowska and Torres [6] and Cruz et al. [2] have demonstrated this new boundary condition for CoV on time scales.

In light of the information from the above discourse, let us proceed with the numerical problem which was examined by [11, 12]. Let us consider the following ODE system [13]

$$\dot{y}(t) = u(t) \tag{1}$$

in optimizing (maximizing) the following objective function [11, 12]

$$J[u(t)] = \int_{t_0}^{T} \left(a(t)u^{1-\alpha} - \left(\rho + m_0 + c_0 e^{-\lambda y}\right)u(t)\right)e^{-rt}dt \tag{2}$$

As mentioned previously, the J system depends on the state variable $y(t)$ and the royalty function ρ is a three-stage piecewise function with the $y(T)$ expression in this study. Here, the proposed settings are $t_0 = 0$ and $T = 10$ while the underlying known state is $y(0) = 0$ and $y(T)$ is unknown.

There are a few fundamental conditions that should be satisfied; the state equation and the costate equation along with the stationary condition. In addition, the underlying condition of $y(0) = 0$ is specified and a guessed initial value for $p(0)$ is chosen. The boundary condition of the integral likewise needs to be satisfied at the final time T. Additionally, the iterated value of z utilized in the state equation needs to guarantee that it will be equivalent to the value $y(T)$ at the final time so that the system will be close to zero. This is fulfilled in our calculation only when the costate equation converges. At that point, the ideal solution will be attained. The following section will illustrate the problem and the results obtained.

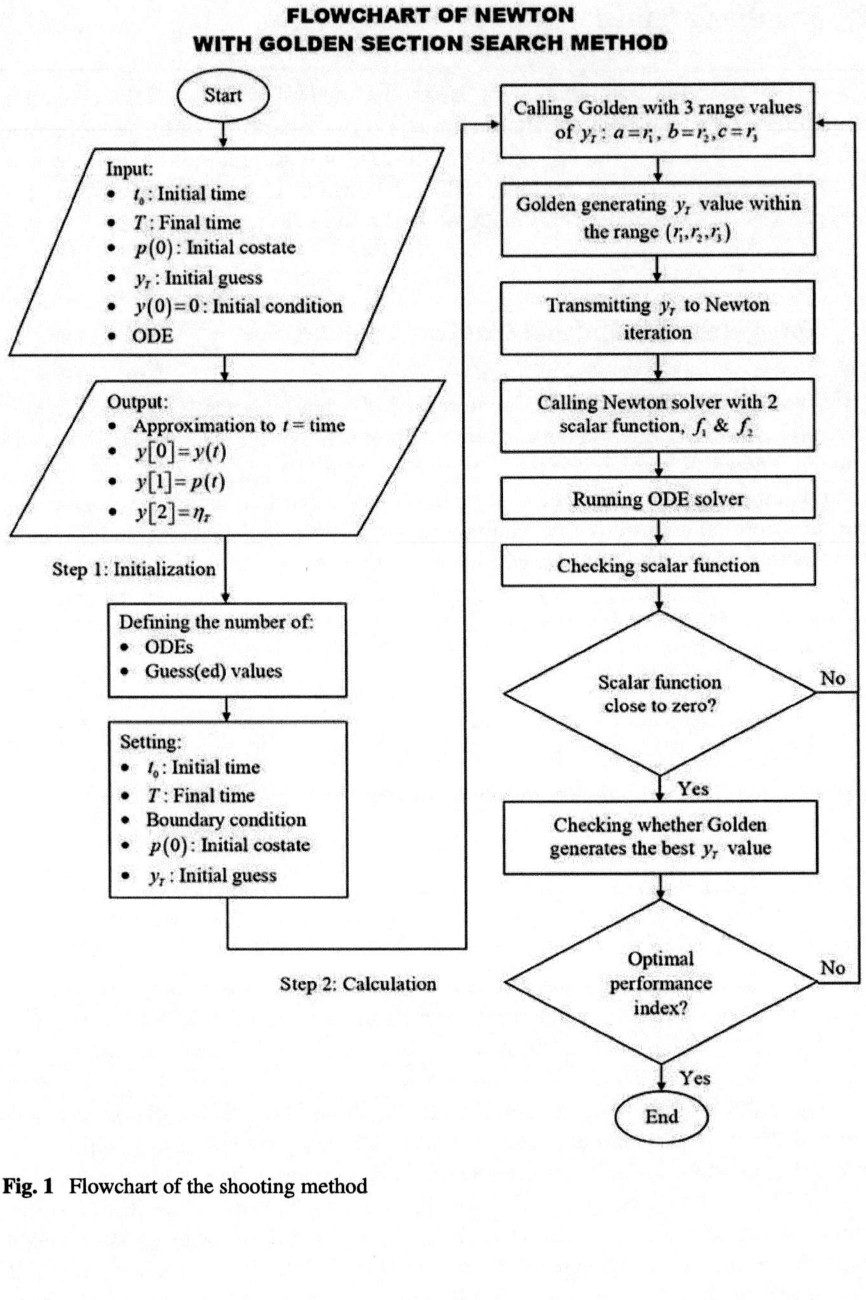

Fig. 1 Flowchart of the shooting method

4 An Illustrative Example

The proposed problem will use the following three-stage royalty function

$$\rho(y(t)) = \begin{cases} 0 & \text{for } 0 \le y(t) \le 0.2z \\ 0.6 & \text{for } 0.2z < y(t) \le 0.6z \\ 1 & \text{for } 0.6z < y(t) \le z \end{cases} \quad (3)$$

The royalty function can be converted into a hyperbolic tangent (tanh) function

$$\rho(y(t)) = 0.5 + 0.3\tanh(k(y - 0.2z)) + 0.2\tanh(k(y - 0.6z)) \quad (4)$$

and this study will consider smoothing value $k = 250$ to approximate Eq. (4). The greater the k is, the smoother the ρ will be.

The result for the royalty function is equal to the three-stage piecewise with the smoothing value $k = 250$ for the shooting method and the discretization validations are now will be presented.

5 Result and Discussion

The results obtained from the discretization and the shooting technique are an ideal solution with high accuracy. The results are tabled as shown in Table 1.

Based on Table 1, the values of $y(T)$, $p(0)$, and $J(T)$ obtained are slightly different for the shooting method when compared with the discretization techniques. The Trapezoidal approximations give the result for $J(T)$ only comparing up to two decimal places when contrasted with the nonlinear shooting procedure. In the meantime, Euler's, Runge–Kutta's, and Hermite–Simpson's methods produce the answer with comparatively up to one decimal place when compared with the shooting results.

Table 1 Results for the shooting and discretization methods

Methods	Results				
	$y(T)$	$p(0)$	$p(T)$	$\eta(T)$	$J(T)$
Shooting method					
Newton and Golden	0.527102	−0.420910	0.216728	0.216728	0.782687
Discretization method					
Euler	0.534359	−0.503092	–	–	0.789992
Runge–Kutta	0.533420	−0.428101	–	–	0.790565
Trapezoidal	0.523972	−0.446858	–	–	0.788824
Hermite–Simpson	0.542751	−0.420502	–	–	0.790716

The initial costate value gives the answer that is comparatively up to two decimal places for the Runge–Kutta and Hermite–Simpson results when compared with the shooting calculation. At the same time, the underlying costate value for the Trapezoidal approximation is similar to up to one decimal place with the shooting value. However, the Euler approximation tends to be slightly different from the shooting result for the initial costate value. Likely, there are a few errors that occur in the calculation due to the discretization inaccuracy as the step size increments and this affects the costate values.

The $J(T)$ values give the solution with similar up to two decimal places for Euler's and Trapezoidal methods when contrasted with the shooting result. The Runge–Kutta and Hermite–Simpson approximations managed to obtain a comparable answer up to one decimal place with the shooting technique for the objective function or performance index value. In light of the shooting results, the performance index can be transformed into a graph. Based on the results also, the optimal curve for the state, costate, and control values will be figured in the following graphical form.

Figure 2 figures the optimal curve for the performance index and shows the optimal curves for the state, costate, and control values, which are identical for both the shooting technique and the discretization methods. The demonstrated plots for

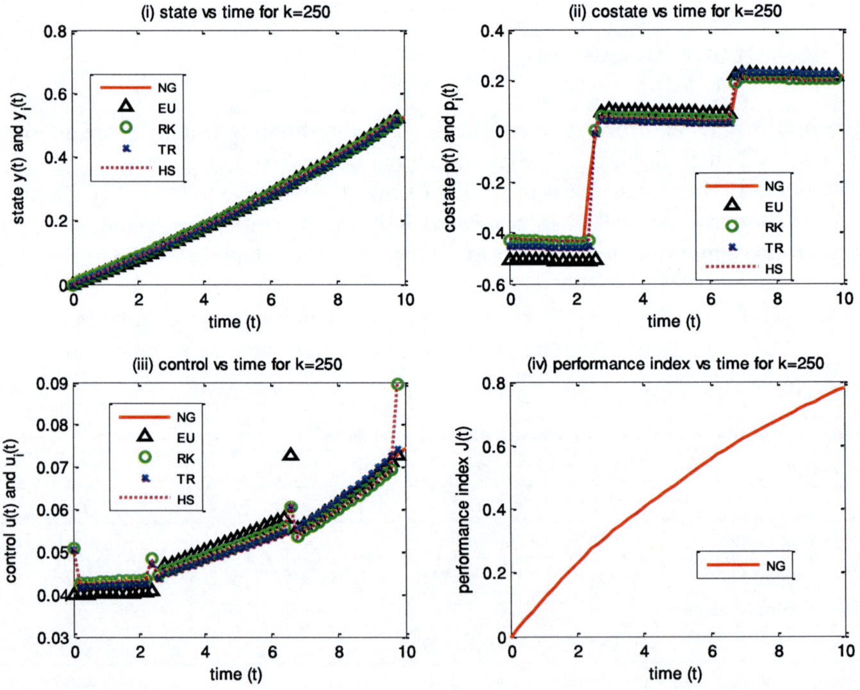

Fig. 2 The plots were generated from the shooting and discretization methods (NG = Newton and Golden; EU = Euler; RK = Runge–Kutta; TR = Trapezoidal; HS = Hermite–Simpson)

the costate and control values are slightly different for the discretized values when compared with the shooting results. Despite that, the curve for the shooting values is smoother compared to the discretized values.

6 Conclusion

This paper has shown the non-classical OC problem with the results by applying the shooting technique. The problem applied shooting method that involved a combination of the Newton with the one-dimensional minimization technique, which was the Golden Section Search method. Then, the results obtained have been compared with the discretization approaches, which were the Euler, Runge–Kutta, Trapezoidal, and Hermite–Simpson approximations. This was to validate the answers. This paper additionally showed the usage of necessary boundary conditions and numerical techniques for tackling non-classical OC problems to get an optimal solution. The shooting methods used C++ programming language, while the AMPL program language applied the discretization techniques. The result obtained is accurate and in the optimal solution. In conclusion, this research can be a stepping-stone to other ambitious researchers in exploring new mathematical methods to solve a real-world economic problem. This is to make sure that the methods used are up-to-date as time goes by. Usually, Runge–Kutta's and Newton's methods are applied in the shooting method [14]. However, this research applied new methods such as Newton's and Golden Section Search in the shooting method.

Acknowledgements The authors thank the referees for their supportive comments on improving the script. Thank you to Universiti Tun Hussein Onn Malaysia (UTHM) and Research Management Centre (RMC) for kindly proving us with the internal funding through GPPS H418.

References

1. Bryson, A.E.: Applied Optimal Control: Optimization. Routledge, Estimation and Control (2018)
2. Zinober, A.S.I.: Optimal Control Theory Lecture Notes. The University of Sheffield (2010)
3. Cruz, P.A.F., Torres, D.F.M., Zinober, A.S.I.: A non-classical class of variational problems. Int. J. Math. Model. Numer. Opt. **1**(3), 227–236. Inderscience Publishers
4. Zinober, A.S.I., Sufahani, S.: A non-standard optimal control problem arising in an economics application. Pesquisa Oper. **33**(1), 63–71 (2013). Brazilian Operations Research Society
5. Ahmad, A., Quegan, S.: The effects of haze on the accuracy of maximum likelihood classification. Appl. Math. Sci. Rousse **10**(39), 1935–1944 (2016)
6. Malinowska, A.B., Torres, D.F.M.: Natural boundary conditions in the calculus of variations. Math. Meth. Appl. Sci. **33**(14), 1712–1722. Wiley Online Library
7. Latif, M.S.A., Fahmy-Abdullah, M., Sieng, L.W.: Determinants factor of technical efficiency in machinery manufacturing industry in Malaysia. Int. J. Supply Chain Manage. **8**, 917–928 (2019)

8. Soebagyo, D., Fahmy-Abdullah, M.O.H.D., Sieng, L.W., Panjawa, J.L.: income inequality and convergence in central java under regional autonomy. Int. J. Econ. Manage. **13**(1) (2019)
9. Idris, A.I.M., Fahmy-Abdullah, M., Sieng, L.W.: Technical efficiency of soft drink manufacturing industry in Malaysia". Int. J. Supply Chain Manage. **8**, 908–916 (2019)
10. Press, W.H., Teukolsky, S.A., Vetterling, W.T., Flannery, B.P.: Numerical Recipes: The Art of Scientific Computing, 3rd edn. Cambridge University Press, Cambridge (2007)
11. Zinober, A.S.I., Kaivanto, K.: Optimal Production Subject To Piecewise Continuous Royalty Payment Obligations. University of Sheffield (2008)
12. Spence, A.M.: The learning curves and competition. Bell J. Econ. JSTOR 49–70 (1981)
13. Fourer, R., Gay, D.M., Kernighan, B.W.: A modelling language for mathematical programming. Manage. Sci. **36**(5), 519–554 (1990)
14. Betts, J.T.: Practical Methods for Optimal Control Using Nonlinear Programming. Advance in Design & Control, Society for Industrial & Applied Mathematics, Philadelphia, PA (2010)

Low-Cost Stand-Alone Smart Irrigation System: A Case Study

Farzana Haque Chowdhury, Roksana Akter Raisa, Md. Sharif Uddin Azad, M Shamim Kaiser, and Mufti Mahmud

Abstract Bangladesh is heavily dependent on agriculture for its crop production, food supply, and crop rotation. About 50% of the population in Bangladesh is working in the agriculture sector; agriculture occupies 70% of the country's territory. To ensure a bountiful harvest, a soil condition suitable for cultivation and the judicious use of irrigation is essential. A fuzzy neural network-controlled irrigation controller system was developed using the research presented here. The system comprises a feedback Fuzzy Neural Network (FNN) controller that keeps track of important system measurements using sensors. The controller bases its findings on crop production, which guides it in determining when it is appropriate to irrigate. MATLAB may assign triangular and trapezoidal membership functions to every input variable. This inference engine uses Max-Min methods, which serve to derive the optimum answer for every case. Also, water consumption is lessened, and freshwater supplies are thereby protected. the system is created and tested for plant growth that reduces water usage by about 50–60% and reduces energy generating costs by the same amount. Improved irrigation management can be achieved when FNN is combined with data logging. By implementing this strategy, the overall energy use, water demand, total energy use, battery, and power control unit expenses can be reduced.

Keywords Fuzzy Logic · Smart pump · Irrigation · Sensors · Micro-controller

---

F. H. Chowdhury · R. A. Raisa
Department of Information and Communication Technology,
Bangladesh University of Professionals, Mirpur, Dhaka-1216, Bangladesh

Md. S. U. Azad · M. S. Kaiser
Institute of Information Technology, Jahangirnagar University, Savar, Dhaka-1342, Bangladesh

M. S. Kaiser (✉)
Applied Intelligence and Informatics (AII), Wazed Miah Science Research Centre (WMSRC),
Jahangirngar University, Savar, Dhaka-1342, Bangladesh
e-mail: mskaiser@juniv.edu

M. Mahmud
Department of Computer Science, Nottingham Trent University, Nottingham, NG1 4FQ, UK

M. S. Kaiser et al. (eds.), *Proceedings of the Third International Conference on Trends in Computational and Cognitive Engineering*, Lecture Notes in Networks and Systems 348, https://doi.org/10.1007/978-981-16-7597-3_28

1 Introduction

Intending to make human lives easier, the field of technology is undergoing a huge evolution. With the advancement of technology, the use of automation is increasing and having a great impact on human life. Automation refers to a process where machines conduct the assigned task according to given instructions, which plays a significant role in reducing human labor, maintaining accuracy, minimizing execution time, and higher work efficiency. Moving toward economic development, agriculture is one of the biggest concerns of the masses. Though the world is rapidly advancing, there are still sides that need to be furnished. The irrigation system is quite challenging when it comes to the proper utilization of water. Irrigation is a technique for transporting water to crops to increase crop yield [9, 13].

Roughly 72% of water is used for irrigation purposes worldwide. Many of the existing irrigation schemes are inefficient in their use of water. This results in the use of more water than is needed or in a lack of water to ensure healthy crop growth. Again, excess water or lack of water may lead to Transpiration is a necessary evil in the case of crops, which leads to the necessity of the proper use of water in the irrigation process.

The irrigation system is carried out conventionally, which includes the Check Basin Method; Furrow Irrigation Method; Strip Irrigation Method; Basin Irrigation Method, etc. These are traditional methods that require a lot of time as well as do not ensure precise use of water. Automation in irrigation systems can reduce human labor, cut off extra costs since water will be supplied when needed and to specific places only, and keep the sideways clean too.

With the availability of low-cost micro-controllers (such as Arduino, Raspberry Pi, etc.) and sensing devices (such as soil moisture, humidity, temperature, rainfall, etc.), researchers are working on automation (on-board and off-board) of irrigation systems [1, 2]. On-board automation systems use stand-alone micro-controllers which are not connected to the cloud or any other platform via Internet/GSM connection, whereas off-board automation systems send data to any processing unit/storage unit via Internet connection [3, 8, 18].

Fuzzy logic (FL) uses the cognition process of the human brain to deal with uncertainty. As a result, an FL-based controller can function in a manner comparable to the human perception process, generating action based on inputs that are incomplete, imprecise, and uncertain [10, 11]. Fuzzy neural networks (FNN), on the other hand, combine the qualities of both neural networks and FL, resulting in an extremely strong hybrid tool that combines the best of both domains. They make it possible to incorporate expert information into the system, and they are regarded to be intrinsically more intelligible because of the utilization of human-like fuzzy inference in the system [12, 14].

In this paper, we proposed a low-cost stand-alone smart irrigation system. The contribution in this paper is given below:

- We studied various types of stand-alone smart irrigation systems.

- We proposed a FNN-based controller for fussing sensor data collected using distributed sensors such as soil moisture, humidity, temperature, rainfall. The action generated using FNN can be used to control the flow of the pump.
- The performance of the proposed system is evaluated and compared with the similar systems.

The remaining sections are Sect. 2 discussed the literature review of smart irrigation systems. The proposed FNN model is outlined in Sect. 3. The results discussion is presented in Sect. 4. In the end, the work concluded in Sect. 5.

2 Literature Review

Several research studies have been published in which the authors proposed the use of a smart irrigation system. Sahu and Behera presented a prototype of a low-cost smart irrigation system that was designed to be environmentally friendly. In this study, the authors employed a variety of sensor nodes to collect data from the environment, which they then passed to an ATMEGA-328 micro-controller for processing, after which the processed data was delivered to a smartphone via the RASPBERRY-Pi [15]. The authors have not considered artificial, or bio-inspired algorithms to control the irrigation system. Salazar and Rodolfo proposed a fuzzy-based multi-agent system for monitoring and controlling irrigation that makes use of a soil moisture sensor. The prototyping system made use of JADE, NetBeans, the Arduino IDE, and the ARDUINO MEGA [2, 16]. A single sensor, on the other hand, cannot provide an accurate irrigation system control. In their paper [4, 6], Priya et al. developed an irrigation system comprised of a humidity and moisture sensor that is placed near the crop's root zone, and watering is performed depending on the readings produced by a micro-controller. This technique, on the other hand, precludes the farmer from gaining a thorough understanding of the field.

Anushree and Krishna came up with an AI or bio-inspired algorithm that uses Arduino to design smart irrigation systems. Several sensors were employed to acquire data and a comparison was then made with a weather forecast repository to obtain a real-time weather forecast. The authors employed three distinct forms of control systems that could turn into a difficult system and an expense that is out of proportion to the benefit[5]. Sneha Angal designed an inexpensive automatic pumping system that saves water. The author combined a Raspberry Pi, Arduino micro-controllers, and the Zigbee protocol to create a smart drip irrigation system that keeps consumers informed. Artificial intelligence or bio-inspired algorithms was not considered in this case [4].

Thakare and Bhagat [3, 17] and Benyezza and Bouhedda [7] proposed the Internet of Things based smart irrigation system using ESP-MODULE 8266. These systems sent data to a cloud server and can be monitored and controlled from anywhere. The running cost of the system is high and these are not user friendly from the perspective of local farmers.

However, the FNN-based stand-alone smart irrigation system is preferred over the existing system as it can be trained using a small dataset and expert opinion; it saves the running cost of the systems and reduces greenhouse gas emission.

3 Proposed System Model

3.1 Data collection

The proposed model aims to make the traditional irrigation system more convenient and low cost through automation. Arduino Uno R3 is the micro-controller of this model and the main work of this Arduino is to compare received values with predefined values and reach a decision regarding supplying water or not. We used two DHT11 temperature and relative humidity sensors at two edges of our model, rainfall sensor and four YL-69 hygrometer humidity and soil moisture sensors at the intersecting points of the ditches. The weather condition is notified via DHT11 and the moisture level of the soil is notified via YL-69. Thus based on data, the micro-controller takes the right decision after analyzing the values.

Arduino Uno R3: Arduino is a single-board micro-controller designed to make interactive objects and their environments more usable to users. It is an open-source hardware board based on ATmega328 with 14 digital (input or output) pins. This can be directly powered using a USB port which makes it easier for development purposes.

Mini Water Pump: 5v DC Mini Water Pump Micro Submersible Motor Pump 2.5-6V 120L/H is used to pump water under the instructions of the micro-controller. It has a flow rate of 80 to 120 L/H with a diameter of approximately 24mm.

Soil Moisture Sensor: This sensor has two forks that are inserted into the soil. It uses the resistance between the two metallic probes of the forks to determine the moisture level of the soil. The sensor gives high analog voltage when the resistance is high and low analog voltage when the resistance is low which means the soil is dry and wet, respectively.

Rainfall Sensor: This rain sensor module is constructed of high-quality double-sided plastic. It is attached to the irrigation system; And serves as a water conservation mechanism, shutting down the system in the event of rain. The voltage required is 5V. This sensor has a 5cm x 4cm surface area and may be constructed with a nickel plate on the side.

Temperature and Humidity Sensor: This sensor reports both temperature (ranging from 0° to 50°) and humidity (ranging from 20% to 90%) using a one-wire interface with an accuracy of +/-2° and +/-5%, respectively. It monitors the airflow with a capacitive humidity sensor and a thermistor and outputs a digital signal on the data pin.

Relay Board Module: This module works as an electromagnetic switch that uses a small amount of current to regulate a much larger amount of current.

A 5V DC motor pump is used for the proper water supply and the connection has been given directly to the laptop because of the value taken from the serial monitor of Arduino software. The code has been generated by Python language which is also known as Arduino code. Void setup() is used for taking all the inputs of the sensors and for the output void loop() is used. All the soil moisture sensor is connected with the analog pin (A1, A2, A3, A4) of the Arduino, pin 13 is selected for LED and the digital pin 7,8 is connected with the DHT11 sensors. Temperature and humidity are taken individually from each DHT11 sensor simultaneously.

3.2 *FNN-based Irrigation System*

Neural networks are regarded as universal approximators because they can provide a generalized framework for a wide range of issues. If properly built, it is capable of approximating any continuous multivariate function arbitrarily well. While training a neural network necessitates the use of a large training data set, whereas FNN can reduce the size of the training data set by mixing expert knowledge with the fuzzy idea. In this case, the data collected from the selected location using temperature, humidity, and soil sensors distributed land, as well as expert opinion, were used to build the model. The FNNs of local and overall feelings are both four-layer feed-forward backpropagation networks. The detail of the FNN model is not discussed in the paper which can be found in detail in [10] A similar number of input units as input variables are used; the first hidden layer represents fuzzy subsets of the input variables, and the second hidden layer represents fuzzy subsets of the output variables; the fourth layer is an output layer with only one unit related to a single output; and the fifth layer is a hidden layer with a single unit related to a single output. The sigmoid functions are used as input activation functions, and linear functions are used for the rest of the layering structure. As a result, the complex non-linear map is created, and the output range is virtually limitless. The FNN-based smart irrigation system is presented in Fig. 1. The input-output membership values of soil moisture, rainfall, humidity, and temperature, as well as the input-output relationship, and hours of the waterfall are given in Fig. 2.

4 Results and Discussion

The proposed FNN-based smart irrigation system is tested using test the model. The seasonal variations and abnormal environment conditions were not considered in this case. The data collected using temperature, humidity, and soil moisture sensors are presented using a violin plot in Fig. 3a.

Figure 3b depicts the total amount of water consumed by the proposed and manual irrigation system for 3 days. As shown in this comparison, the suggested FNN-based smart irrigation system pumps less water than the manual system.

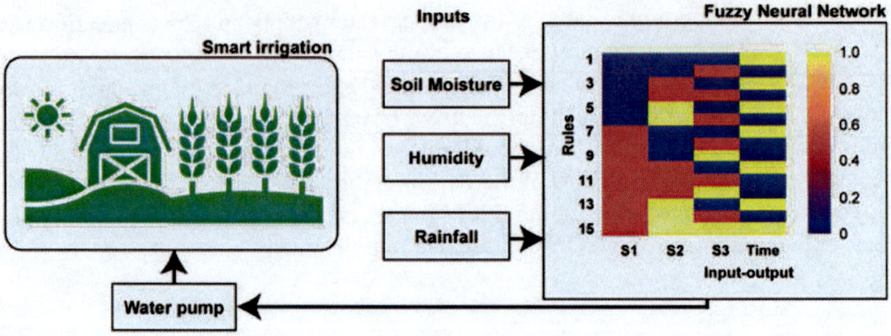

Fig. 1 FNN-based smart irrigation system. In this case, soil moisture, humidity, and rainfall sensors are employed to collect data from the field and environment. Then the FNN model supply required water to the field

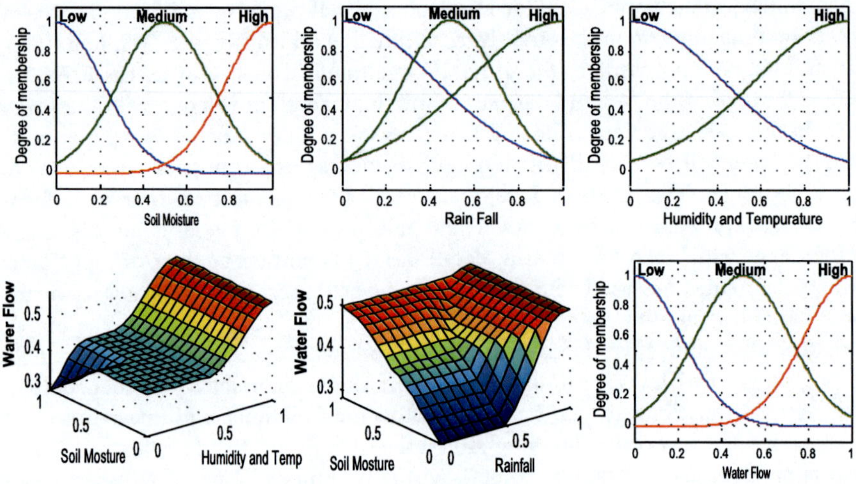

Fig. 2 Input/Output membership functions for soil moisture, rainfall, and humidity sensors. moisture, rainfall has three linguistic terms (Low, Medium, High) whereas humidity has two linguistic terms (Low, High). The input-output relationship is presented by the surf plot

Figure 3c shows the box plot of the unit running cost of the proposed and manual irrigation system for 3 days. As shown in this comparison, the suggested FNN-based smart irrigation system pumps require less running cost than the manual system.

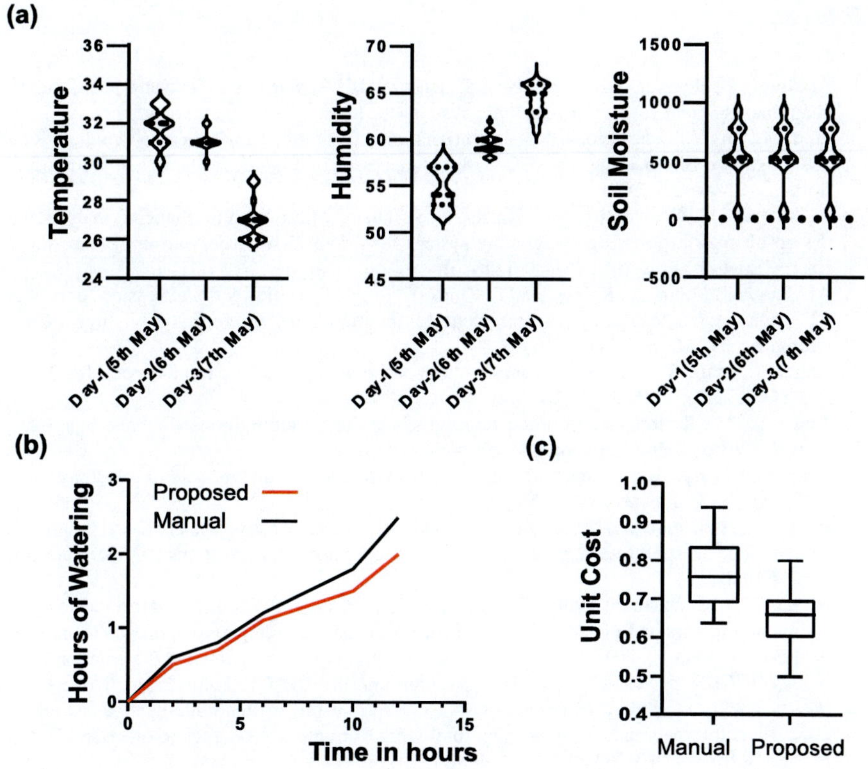

Fig. 3 Performance evaluate of the proposed FNN model used in low-cost stand-alone Irrigation System. **a** Temperature, humidity, and soil moisture values are shown using violin plot; **b** shows the hour of supplied water as a function of time which shows the proposed model supplies less water in the field; **c** the unit cost of the proposed FNN system is less than that of the manual system as it saves water and the running cost

5 Conclusion

In this paper, the authors proposed and developed a prototype low-cost stand-alone smart irrigation system. The FNN has been used for controlling the irrigation system. The system collects soil conditions (using soil moisture sensors) and environmental conditions (using humidity and rainfall sensor) via sensors placed at a different part of the field. It has been found that the cost of the system is low and it is effective in saving water usage. An additional Wi-Fi module, Android application, or cloud service can allow the system to be observed and controlled from a distance, as well as to offer visualization and alerts to app users.

References

1. Al-Amin, S., Sharkar, S.R., Kaiser, M.S., Biswas, M.: Towards a blockchain-based supply chain management for E-Agro business system. In: Kaiser, M.S., Bandyopadhyay, A., Mahmud, M., Ray, K. (eds.) Proceedings of International Conference on Trends in Computational and Cognitive Engineering. pp. 329–339. Advances in Intelligent Systems and Computing, Springer, Singapore (2021)
2. Al Mamun, S., Chowdhury, Z.I., Kaiser, M.S., Islam, M.S.: Techno-financial analysis and design of on-board intelligent-assisting system for a hybrid solar-deg-powered boat. Int. J. Energy Environ. Eng. **7**(4), 361–376 (2016)
3. Al Mamun, S., Lam, A., Kobayashi, Y., Kuno, Y.: Single laser bidirectional sensing for robotic wheelchair step detection and measurement. In: International Conference on Intelligent Computing. pp. 37–47. Springer (2017)
4. Angal, S., Mali, R.: Raspberry pi and arduino based automated irrigation system. Int. J. Sci. Res. (IJSR) **5**(7), 1145–1148 (2016)
5. Anushree, M., Krishna, R.: A smart farming system using arduino based technology. Int. J. Adv. Res. Ideas Innov. Technol **4**(4), 850–856 (2018)
6. Archana, P., Priya, R.: Design and implementation of automatic plant watering system. Int. J. Adv. Eng. Glob. Technol. **4**(1), 1567–1570 (2016)
7. Benyezza, H., Bouhedda, M., Djellout, K., Saidi, A.: Smart irrigation system based thingspeak and arduino. In: 2018 International Conference on Applied Smart Systems (ICASS). pp. 1–4. IEEE (2018)
8. Biswas, S., Anisuzzaman, Akhter, T., Kaiser, M.S., Mamun, S.A.: Cloud based healthcare application architecture and electronic medical record mining: an integrated approach to improve healthcare system. In: 2014 17th International Conference on Computer and Information Technology (ICCIT). pp. 286–291 (2014). https://doi.org/10.1109/ICCITechn.2014.7073139
9. Blessy, J.A., et al.: Smart irrigation system techniques using artificial intelligence and iot. In: 2021 Third International Conference on Intelligent Communication Technologies and Virtual Mobile Networks (ICICV). pp. 1355–1359. IEEE (2021)
10. Kaiser, M.S., Al Mamun, S., Mahmud, M., Tania, M.H.: Healthcare robots to combat COVID-19, pp. 83–97. Springer Singapore, Singapore (2021)
11. Kaiser, M.S., et al.: iworksafe: towards healthy workplaces during covid-19 with an intelligent phealth app for industrial settings. IEEE Access **9**, 13814–13828 (2021). https://doi.org/10.1109/ACCESS.2021.3050193
12. Mahmud, M., Kaiser, M.S., Rahman, M.M., Rahman, M.A., Shabut, A., Al-Mamun, S., Hussain, A.: A brain-inspired trust management model to assure security in a cloud based iot framework for neuroscience applications. Cogn. Computation **10**(5), 864–873 (2018)
13. Mousavi, S.K., Ghaffari, A.: Data cryptography in the internet of things using the artificial bee colony algorithm in a smart irrigation system. J. Inf. Sec. Appl. **61**, 102945 (2021)
14. Rahman, S., Kaiser, M.S., Ahmmed, M.U.: Fnn based adaptive route selection support system. Int. J. Adv. Comput. Sci. Appl. (IJACSA) **7**(10), 356–358 (2016)
15. Sahu, C.K., Behera, P.: A low cost smart irrigation control system. In: 2015 2nd International Conference on Electronics and Communication Systems (ICECS). pp. 1146–1152. IEEE (2015)
16. Salazar, R., Rangel, J.C., Pinzón, C., Rodríguez, A.: Irrigation system through intelligent agents implemented with arduino technology. Adv. Distrib. Comput. Artif. Intell. J. **2** (2013)
17. Thakare, S., Bhagat, P.: Arduino-based smart irrigation using sensors and esp8266 wifi module. In: 2018 Second International Conference on intelligent computing and control systems (ICICCS). pp. 1–5. IEEE (2018)
18. Zaman, S., Kaiser, M.S., Tasin Khan, R., Mahmud, M.: Towards SDN and blockchain based IoT countermeasures: a survey. In: 2020 2nd International Conference on Sustainable Technologies for Industry 4.0 (STI). pp. 1–6 (Dec 2020)

Network and Security

Improving the Life Span of IoT Sensor Devices Using Smart Packet Filtration Algorithm

Wan Suhaimzan Wan Zaki, Ashok Vajravelu, Mohd Helmy Abd Wahab, G. Murugesan, M. Joseph Auxilus Jude, and N. Nishrhutha

Abstract The Internet of Things (IoT) refers to connecting electronic devices to the Internet. It refers to a network comprised of physical objects capable of collecting and sharing information. The physical object includes sensors, aggregators, communication channels, etc. In wireless sensor networks (WSNs), the long lifetime requirement of various applications and limited energy storage capability of sensor nodes had led us to find out new horizon for reducing power consumption upon nodes. The sensors used are of delay tolerant application such as agricultural field, which are mostly battery dependent. Hence conserving the energy of this sensor may increase its durability for a longer period of time. This project is a mixture of hardware and software technology that the existing architecture varies with the domain of design, quality of service factor, interoperability, etc. Here we are majorly focused on the perception layer and the processing layer of the sensor device. The problem focused in the processing layer is storage management and heterogeneous task scheduling which increases complexity. Thus we design our algorithm for the efficient storage of data, load distribution, and communication efficiency by altering the existing algorithm

W. S. W. Zaki (✉) · A. Vajravelu
Biomedical Engineering and Measurement System Research Group, Faculty of Electrical and Electronic Engineering, Universiti Tun Hussein Onn Malaysia, Johor 86400, Malaysia
e-mail: mizan@uthm.edu.my

A. Vajravelu
e-mail: ashok@uthm.edu.my

M. H. A. Wahab
Faculty of Electrical and Electronic Engineering, Universiti Tun Hussein Onn Malaysia, Johor 86400, Malaysia
e-mail: helmy@uthm.edu.my

G. Murugesan
Department of Electronics and Communication, Kongu Engineering College, Perundurai-638060, India

M. J. A. Jude · N. Nishrhutha
Vinton Network Lab, Department of Electronics and Communication, Kongu Engineering College, Perundurai-638060, India

M. S. Kaiser et al. (eds.), *Proceedings of the Third International Conference on Trends in Computational and Cognitive Engineering*, Lecture Notes in Networks and Systems 348, https://doi.org/10.1007/978-981-16-7597-3_29

for better durability. The simulation is done using SENSEnuts software and the algorithm is implemented in the microcontroller which is embedded in the sensor. The result shows that there is a markable improvement in conserving the energy of the sensor.

Keywords Quality of service · Heterogeneous task scheduling · Storage management

1 Introduction

The Internet of Things (IoT) is an upcoming technique that connects the hardware package with software via the Internet/Web. The IoT is a network of physical devices embedded with software, sensor, and connectivity which enables to connect, collect, and exchange information. IoT consist of physical object which includes sensors, aggregators, communication channel, decision trigger, and electronic-utility.

The superimposed design aims to attain presence and generality in IoT systems by sensing, analyzing, and processing of enormous information. Communication protocols and different process technologies square measure utility in different layers. It conjointly emphasizes on the foremost power hungry nodes, protocols, middleware parts, and application. This design is split into 5 layers which are Perception, Transport, Processing, Networking, and Application layers as shown in Fig. 1.

The elements of the IoT are segregated into power hungry and non-power hungry elements [1]. Power hungry elements of perception layer are sensors, actuators, and the gateway, and the non-power hungry components of perception layer are accelerometers, bootstrap, ADC, as shown in Fig. 2.

Wireless sensor network (WSN) refers to a bunch of sensors for monitoring and recording the physical conditions of the environment and organizing the collected information at a central location. WSNs measure environmental conditions like

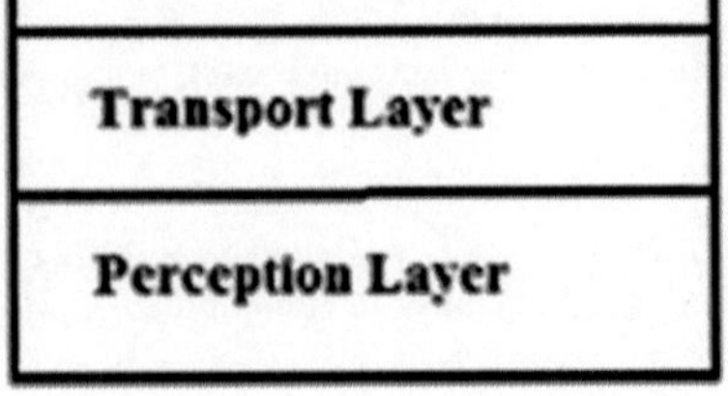

Fig. 1 Internal module of sensor

Fig. 2 Power consumption and non-consumption of perception layer

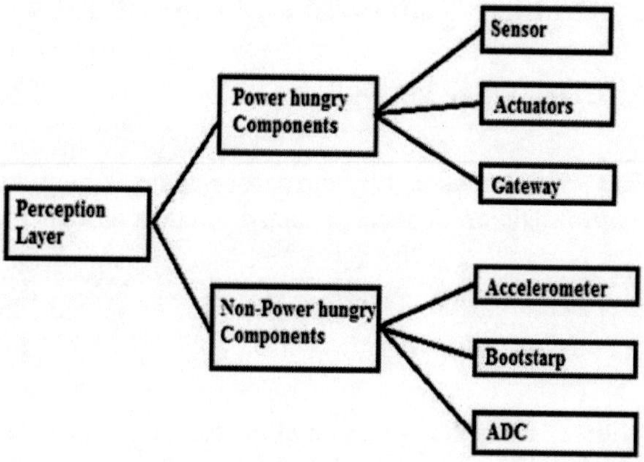

temperature, sound, pollution levels, humidity, wind speed and direction, pressure, etc. The architecture of WSN consists of Sensing Fields where there are numerous sensors, a sink node known as a gateway. Internet and the user, from the Internet the user can access the information gathered by the sensor nodes. WSN may be outlined as a network of devices which will communicate the information gathered from a monitored/sensed field through wireless links [2]. This information is forwarded through multiple nodes, and with a gateway, the information is connected to other networks like wireless Ethernet.

A sensing element which is in the sensing field of the WSN architecture is an electronic device which measures physical properties such as temperature, acceleration, weight, sounds, location, presence, identity, and so on. Sensor's sensitivity indicates how much the sensor's output changes when the input quantity being measured changes. All sensors employ mechanical, electrical, chemical, optical, or other effects at an interface to a controlled process or open environment. The modules of the sensor device are shown in Fig. 3. Sensor module consists of a microcontroller along with memory, transceiver, sensing module, and a battery [3]. The sensing module which is active all the time will sense the information and send the sensed data to the microcontroller where the information are converted into packets and these generated packets are transmitted to the successive sensing element through the transceiver [4].

Fig. 3 Internal module of sensor

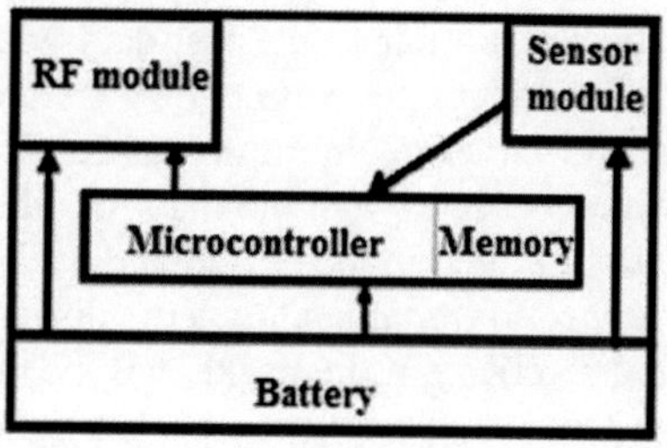

2 Problem Statement and Contribution

2.1 Problem Statement

The power consumed by the sensor nodes is high due to needless generation and transmission of identical packets from the consequent duty cycle.

2.2 Contribution

With regard to the limitation of the existing system, we have proposed an algorithm which will avoid the needless packet generation and transmission by comparing the current data sensed with the data for which the packet has been formed. Here we have chosen the delay tolerant application which is a temperature sensor used in the agricultural field. If the temperature of the soil is high, automatic irrigation takes place. So that if there is any delay in processing the sensed data won't cause any serious problems.

3 Packet Filtration Algorithm

The proposed work was incorporated with the alteration of storage management to the packet filtration algorithm. This algorithm will increase the life span of the IoT sensor devices. By avoiding the generation of redundant packets, which will reduce the power consumption of the IoT sensor devices [5].

This algorithm is the altered version of the existing storage management algorithm. The main objective of this algorithm is to avoid the storage and generation of the redundant packet which consumes lot of power. Thus in this algorithm we are avoiding the storage of the redundant packet in the buffer of the microcontroller. The alteration starts from storage of the sensed data, which was sensed at present, and this sensed data is then compared with the data for which the packet was generated before, if there is no change in the data sensed by the sensor and the data for which the last packet was generated [6]. Those data which is sensed by the sensor module which is similar to the last data will not be stored in the buffer of the microcontroller and packet will not be generated for that data. The flow chart of the proposed packet filtration algorithm has be shown in Fig. 4. If there is any change in the data which is sensed and the data for which the last packet was generated then the data will be stored in the buffer, and packet will be generated. If the packet generation is stopped for redundant data the power consumption for the generation, transmission, and receiving of the packet will be reduced [7, 8].

Thus by filtering the packet we can conserve the power consumption thereby increasing its lifespan. The number of packets generated will vary according to the

```
Start
    Turn ON Sensor Module
    Sense the Required parameter
    Store the parameter, send the parameter to the Gateway
    Turn OFF
    After a specified Time Turn ON sensor module
    Sense the Required parameter
    IF(sensed parameter value = stored parameter value)
        ABORT
    ELSE
        Replace the stored old parameter value by new parameter value
        Send the Stored value to Gateway
    Turn OFF sensor module
    End
End
```

Fig. 4 Packet filtration algorithm

application of the sensor and where it is used. Since we have focused on the delay tolerant application we choose agro-based application, if the sensed temperature is above the threshold then the operation mentioned will be performed. The limitation of these algorithms is it not suitable for non-delay tolerant application.

4 Implementation

4.1 Software Simulation

The sensor is made to run and the data are displayed in the SENSEnuts. By using the already existing algorithm the sensor is made to sense in real time for 1000 s, the data sensed are displayed in the software thus by using those data, we have calculated the number of packets generated and the power consumed for the generation of packets [9].

Power consumed for generating single packet is equal to 1370.88 μW.

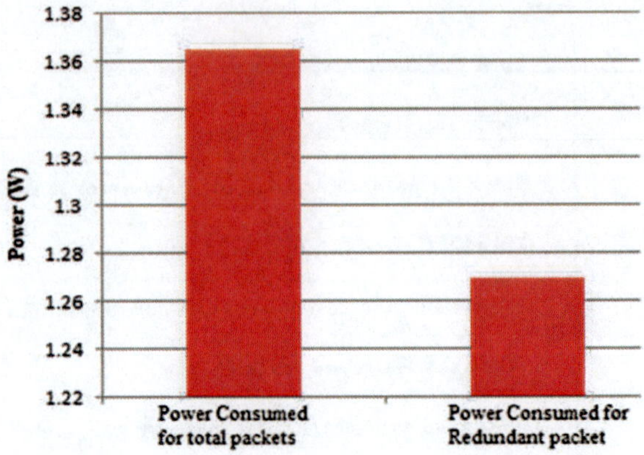

Fig. 5 Power comparing between the existing algorithm for redundant and total packet

Total number of packets generated by existing algorithm is 996 packets.

Power consumed for generating 996 packets = 1.365 W.

Number of redundant packets generated = 972 packets.

Power consumed for generating 972 packets = 1.270 W.

The power consumption between the total packets generated and the power generated by the redundant packets (data packets which are same as value as the previous data packet) and is shown in Fig. 5. The aim of our packet filtration algorithm is to reduce the power consumed by that redundant packet [10, 11].

Then the existing algorithm is replaced by our proposed packet filtration algorithm using SENSnuts software and the altered code has been dumped into the microcontroller of the sensor module. Then sensor is made to sense in real time for 1000 s by using the packet filtration algorithm and we have calculated the number of packets generated and the power consumed for the generation of packets.

Total number of packets generated by packet filtration algorithm = 200 packets.

Power consumed for generating 200 packets = 0.274 W.

Number of packets generated using both the algorithm has been compared and plotted which is shown in Fig. 6.

Then power consumed for the generation of packets using the existing algorithm and the power consumed due to the packet filtration algorithm was calculated and compared as shown in Fig. 7.

Fig. 6 Number of packets generated

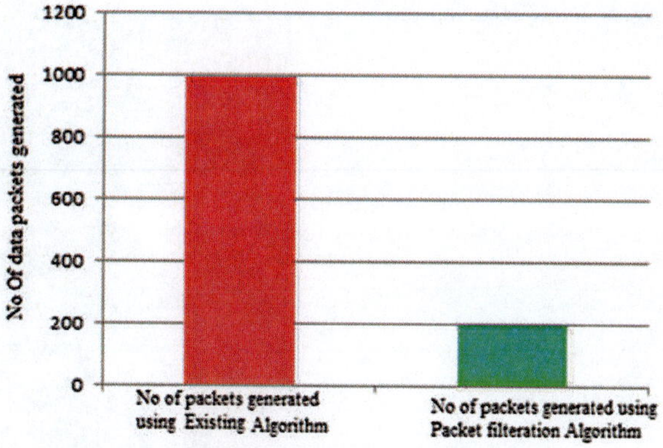

Fig. 7 Power consumption comparison

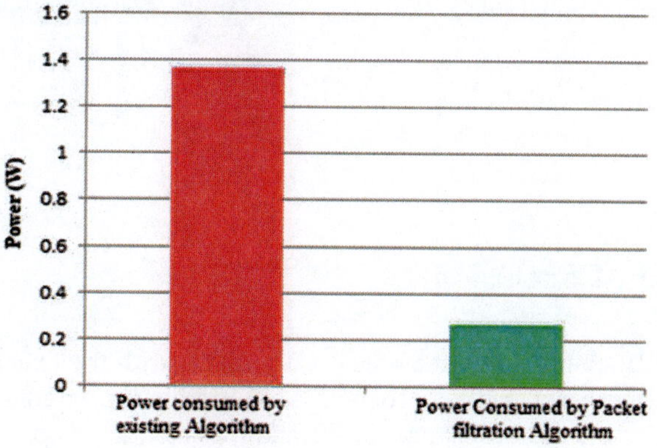

4.2 Hardware Setup

The implementation of the proposed working model portion is as shown in Fig. 8. The system was fabricated with three modules as control box, mobile application, and web-based application. In which the control box helps to keep the electronic devices to monitor the events as event manager with centralized processing unit.

The control box does the another important function as to locate a place wherever in the field, i.e., in the farm as well as near to the farm, also it will monitor soil moisture with the sensors, a solenoid valve is provided to act as a switch, then the sensor DHT22 also employed for monitoring the humidity and moisture of the surroundings to the control unit. In this study, IoTs with the networking technology are applied to the agriculture application to sense the soil moisture measurements, and the measurements of the humidity of the soil as well as operate the control switch for water sprinklers in an automatic manner. The next portion of the setup is web-based one which will collect the information about the agriculture from different Node MCU. It will access Internet through Wi-Fi.

Fig. 8 Experimental setup

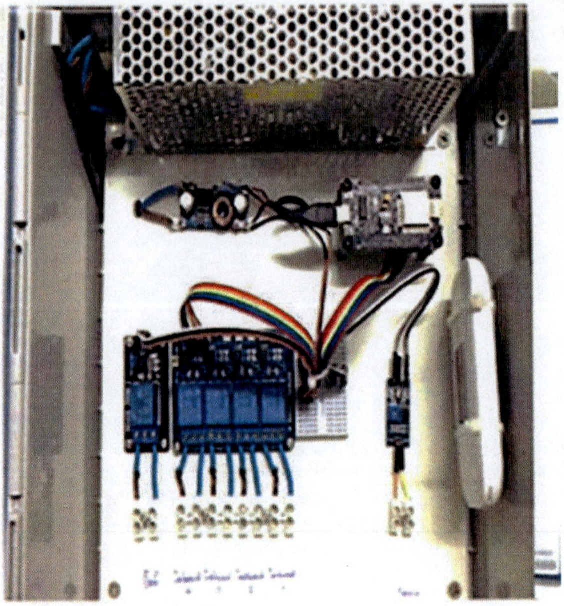

5 Conclusion

By comparing our proposed algorithm with the existing algorithm by making the sensor run for 1000 s for both the algorithm and we analyzed that the power consumed by our proposed packet filtration algorithm is approximately 79% less than the existing system algorithm.

References

1. Froytlog, A., Foss, T., Bakker, O., Jevne, G., Haglund, M.A., Li, F.Y., Oller, L., Li, G.Y.: Ultra-low power wake-up radio for 5G IoT. IEEE Commun. Mag. **57**(3), 111–117 (2019)
2. Tahiliani, V., Dizalwar, M.: Green IoT systems: an energy efficient perspective. In 2018 Eleventh International Conference on Contemporary Computing (IC3), pp. 1–6. IEEE (2018)
3. Steinfeld, L., Silveira, F., Ritt, M., Carro, L.: A new memory banking system for energy-efficient wireless sensor networks. In 2013 IEEE International Conference on Distributed Computing in Sensor Systems, pp. 215–222. IEEE (2013)
4. Ali, A., Abo-Zahhad, M., Farrag, M.: Modeling of wireless sensor networks with minimum energy consumption. Arab. J. Sci. Eng. **42**(7), 2631–2639 (2017)
5. Abo-Zahhad, M., Farrag, M., Ali, A., Amin, O.: An energy consumption model for wireless sensor networks. In 5th International Conference on Energy Aware Computing Systems & Applications, pp. 1–4. IEEE (2015)
6. Khriji, S., Chéour, R., Goetz, M., El Houssaini, D., Kammoun, I., Kanoun, O.: Measuring energy consumption of a wireless sensor node during transmission: Panstamp. In 2018 IEEE

32nd International Conference on advanced information networking and applications (AINA), pp. 274–280. IEEE (2018)

7. Li, K., Ni, W., Duan, L., Abolhasan, M., Niu, J.: Wireless power transfer and data collection in wireless sensor networks. IEEE Trans. Veh. Technol. **67**(3), 2686–2697 (2018)
8. Kamyabpour, N., Hoang, D. B.: Modeling overall energy consumption in Wireless Sensor Networks. arXiv, 1112.5800. (2011)
9. Calle, Maria.: Energy consumption in wireless sensor networks using GSP. Sch. Inform. Sci. (2006)
10. JN516x Microcontroller data sheet. https://www.nxp.com/docs/en/data-sheet/JN516X.pdf. Accessed 4 Jul 2021
11. IEEE Standard for Local and metropolitan area networks-Part 15.4: Low-Rate Wireless Personal Area Networks (LR-WPANs). In: IEEE Std 802.15.4–2011 (Revision of IEEE Std 802.15.4–2006), pp.1–314. IEEE (2011)

Modal Share Concept for Mobility

Zumrotul Jannah Supaat and Adlyn Nazurah Abdul Rahman

Abstract Modal share plays an important role in transportation and conditions in Malaysia. However, the previous study on modal share for mobility was still underdeveloped to introduce the concept of public transport routes, bicycle lanes, and pedestrian routes. Therefore, this research is to identify the concept of modal share for mobility and to examine the concept of modal share for mobility. The data have been collected by using questionnaires, document reviews, and observations of the 11 roads focused in this study. The results showed that several of the 11 roads lacked development in upgrading transport routes such as public transport routes, pedestrians, and bicycles. Besides that, the elements of the awareness, safety, facilities, cost, and carbon dioxide reduction in modal share and environment impact are significantly moving the modal share acceptance and benefit of the environment. The implications of this study show that the local authorities and government need to better access to real modal share and improve their organization capability as well.

Keywords Concept · Modal share · Mobility · Acceptance · Knowledge

1 Introduction

Modal share concept is an efficient car sharing where it can accommodate a large number of passengers at five passengers at a time [1]. Modal share is a transportation system with different road routes, such as walking routes, bicycle routes, and public and private transportation routes [1]. Based on previous studies, [1] explained that all the proposed routes were aimed at raising public awareness of the importance of walking, bicycling, and the benefits of using public transport.

The modal share also has important dependence on the transport sector. The transportation sector is one of the sectors that contribute to CO_2 emissions globally around 27% of energy users [2]. However, transportation issues also harm the release

Z. J. Supaat · A. N. A. Rahman (✉)
Faculty of Technology Management and Business, Universiti Tun Hussein Onn Malaysia, 86400 Parit Raja Batu Pahat, Johor, Malaysia
e-mail: Ap160260@siswa.uthm.edu.my

M. S. Kaiser et al. (eds.), *Proceedings of the Third International Conference on Trends in Computational and Cognitive Engineering*, Lecture Notes in Networks and Systems 348, https://doi.org/10.1007/978-981-16-7597-3_30

of carbon dioxide. The scale-up of carbon emissions has a direct impact on the environment, especially against the social. The impact of social transport is closely related to the lack of a healthy lifestyle, increased accidents, and traffic congestion [3]. Road jams cause road users to spend on roads and waste petrol. All their social activities are affected and there is no time to do healthy activities.

With the modal share, it is possible to create awareness among the public about the importance of using better available routes, also enhancing a healthy lifestyle. The model share can increase the use of public transport to reduce the use of private transportation and reduce carbon emissions [1]. Someone spends as much as 1.1 h a day for travel and little money for travel. Smaller and steady use can measure the level of mobility and appropriate modes of transport [4]. The use of model share is a serious and effective measure to reduce the environmental impact of carbon. People globally are aware of the issues with climate change and global warming that significantly impact the environment, humans, and animals on earth [5].

Low carbon cities are defined as cities comprising societies that consume sustainable green technology and green practice to prevent adverse impact on climate change compared to nowadays. The concept of low carbon cities is close with the development environment. By adopting the principle of sustainability, carbon emissions can be reduced and developing cities and consuming resources [1]. Low carbon cities are essentially cities that take serious and effective measures to reduce their impact on the environment. The process of development entails not only an economic phenomenon. It is unnecessarily assessed based on the achievement of national economic growth, but it has a broader perspective than merely an economic aspect [6].

In Asian countries, low carbon societies are one way of reducing pollution as well as representing sustainable development [7]. To identify the changes, effects, causes, and problem solving, many articles, papers, and workshops have been conducted for [7]. However, some countries that are still lagging behind the environment, its effects; carbon reduction cannot be implemented and no sustainable development and strategy. Malaysia is a middle country that is also no exception. Therefore, this study has been conducted for the purpose of selecting and adapting recovery options for Malaysia within 95 years since 2010 [8].

In Malaysia, the increased number of cars and users on the road occurred over the last two decades. The increase was fourfold from 5 million in 1991 to 21.4 million in 2011. The average growth rate was 7.5% [6]. This causes an increase in the number of private cars and registered consumer accidents, making the safer mode of transit mode very important to improve road safety [9]. Most model share transports such as public transport are combined under one authority. They plan, construct, develop, monitor, and operate effectively so that the communities can easily move to their destination as well as their travels smoothly [10].

According to [11], another of the public transport missions in Johor is to create a more effective, efficient, and comprehensive transportation system to reduce dependency on private mode transportation, create an integrated and sustainable transit transportation system, adopt the use of green transportation, and the logistics and transport system of goods efficiently and smoothly. Technical efficiency is crucial in order to determine the efficiency level in a firm and industry [12]. The concept of efficiency not only can be seen when a department or organization uses all available resources or inputs optimally to produce the maximum output [13], but also can be seen in public transport. Besides, goals in the state of Johor can be achieved by reducing the dependence on private transport by providing public transportation and reducing traffic congestion problems. In addition, it can create an integrated and efficient transit transportation system and in line with land use planning. The process of development entails not only an economic phenomenon [14].

Finally, yet importantly, the efficient use of model share mobilitycan increase the efficiency of the car sharing or the use of the public transport services easily and comfortably also can achieve the stated goals. According to [6], with advantages such as low per capita users and environmental protection of green, urban public transport is one way to reduce carbon. Vehicle model sharing can also reduce the amount of traffic congestion, reduce vehicle parking costs, reduce traffic accidents, reduce air pollution and noise, and support land use strategies [15].

2 Research Methodology

2.1 *Selection of Research Methodology*

To complete the research, an in-depth study on research methodology are considered essential [16]. Selection of research methods, various criteria that are emphasized, and psychometric and legal criteria must be fulfilled, including standardization, validity, fairness, and reliability [17]. The questionnaire and observation are the methods used to obtain accurate data and information in carrying out this research. The researcher chose the triangulation to analyze the independent variable, which is (i) awareness, (ii) safety, (iii) facilities, (iv) cost, and (v) carbon dioxide reduction in modal share and environmental effect toward the understanding of modal share concept (dependent variable).

2.2 *Sampling Technique*

Because of the broad scope and the context of this study, a mixed method approach for data collection was used as both quantitative and qualitative research techniques that were adopted to achieve the research aim and objectives [18]. The population is a collection of individuals or objects that have similar characteristics [19]. Since it was impossible to administer surveys to all transportation system users, sampling was necessary to obtain a representative proportion of all users of the transportation system. Simple random sampling was used to ensure that each potential respondent within the target population stood an equal chance of being included in the sample. The researcher will be concerned about the user's profile. The respondents were asked to provide socioeconomic information including gender, age range, education, occupation, and question based on document reports.

2.3 Data Analysis Technique

For this study, the sample that was used by the researcher is the technique of probability sampling. This is planned for a decrease of sampling error in collected data and will guide to improved correctness and steadiness of the result generated. The sampling technique also selected a large number of units from the population [20]. Analysis and findings of this study are described sequentially according to the question of the research and the objective of the study would like to achieve [21]. The sample size of this research is 370 samples based on Morgan's table. The probability sampling with the simple random in this study is an appropriate method for the large resident in Batu Pahat. Population is a collection of elements that has some or other characters in the sample. However, the population might be difficult to research as not all elements are easy to access. The researcher finds the population as something might be more manageable and this is called the target population. Probability sampling provides an alternative technique to select the samples. The majority are including an element of subjective judgement.

The total of population is accretion of entity or person which have applicable info and who has the in order necessary by researcher in undertake their research. In this case study, random sampling which were concentrating on modal share chose the target population [22].

2.4 Pilot Test

The purpose of the pilot test is to examine the usefulness of the strategy that was intended to calculate the items and to validate the instrument. According to [23], the aim of the pilot test is to enhance the questionnaire so that the participants will have no issues answering the questions and no issues recording the information. For [24], it began that there should be at least 50 individuals to attend. It is reasonable for pre-testing questionnaires for a sample of 30 participants by [25]. The first main of the pilot test was to measure the understanding of respondents about the modal share concept for mobility in Batu Pahat. The data collection process for the pilot test began on 13 September 2019. The pilot test was conducted by 30 samples of questionnaire. For the result, there was a 100% response rate from the respondents.

The content validity was established by evaluating the acceptance: (i) awareness, (ii) safety, (iii) facilities, (iv) cost, and (v) carbon dioxide reduction in modal share and environment impact (independent variable) toward understanding modal share concept (dependents variable). The information gathered from the pilot research was evaluated using the software version 20 of the Statistical Package for Social

Table 1 The reliability statistic for the pilot test

No	Acceptance	Cronbach's alpha	Number of items
1	Awareness	0.822	6
2	Safety	0.941	5
3	Facilities	0.728	4
4	Cost	0.815	4
5	Carbon dioxide reduction in modal share and environment impact	0.809	5
6	Understanding of modal share concept	0.747	4

Sciences (SPSS). To assess the internal consistency and reliability of the instrument, the reliability analysis was carried out using Cronbach's Alpha, which ranges from 0 to 1. When Cronbach's Alpha is above 0.7 [23, 26], they are considered acceptable.

The reliability of the questionnaire concerns the robustness of the questionnaire [23]. Reliability findings for this pilot test research were also regarded to be good, as Cronbach's Alpha was more than 0.7. However, some corrections have been made as the consequence of checking the validity of the questionnaire by the specialist and making some modifications before continuing with the final allocation of the questionnaire. The findings of the pilot research are shown in Table 1 below.

3 Result and Discussion

3.1 Reliability Analysis

Based on the Table 2, the highest value among all the independent variables is carbon dioxide reduction in modal share and environment impact. This is because the use of the transportation of modal share can increase the health of environment. This is related to [27]; the public transport also can reduce the greenhouse gas emissions and improve the health of the community by promoting walking.

Table 2 The analysis of reliability

No	Modal share acceptance	Cronbach's alpha	No. of items
1	Awareness	0.876	6
2	Safety	0.842	5
3	Facilities	0.788	4
4	Cost	0.865	4
5	Carbon dioxide reduction in modal share and environment impact	0.895	5
6	Understanding of modal share concept	0.858	4

3.2 *Frequency Analysis for Acknowledgement Among Respondents Toward the Modal Share for Mobility*

The Biggest Issue in the Modal Share. The dominant answer that respondents chose is the commitment from the government and all the parties. This is because the commitment from all parties is important to increase the understanding of the society using the modal share. This is also related to previous research [28]; with commitment from all parties, the cities are increasingly developing sustainable urban transport systems with efforts to improve quality and access to public transportation and non-motorized modes (Table 3).

Organization That Can Increase the Society Use on Transportation of Modal Share. The dominant of respondents chose the answer MPBP. This is related to previous research [29]; the Dewan Bandaraya Kuala Lumpur (DBKL) had made efforts to make the Tuanku Abdul Rahman Road as a pedestrian area starting April 1 for a trial period. This research related to Majlis Perbandaran Batu Pahat also had their responsibility to make the Batu Pahat a low carbon city by 2030 and also increase the use of public transport in that area and also in Batu Pahat (Table 4).

Table 3 The biggest issue in the modal share

(A) The biggest issue in the modal share	Frequency
Lack of awareness among the residents in Bandar Penggaram	140
Commitment from the government and all the parties	161
Lack of the campaign in reducing the use of the private vehicle	114

Table 4 Organization that can increase the society use on transportation of modal share

(B) Organization that can increase the society use on transportation of modal share	Frequency
Federal government	93
State government	94
MPBP	206
Government agency	113
Development and private agency	79
NGO/Resident association	89

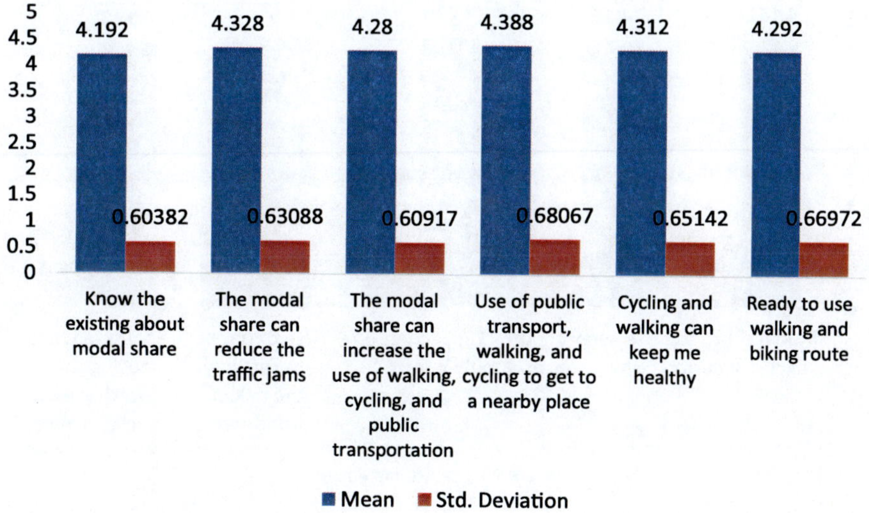

Fig. 1 Descriptive statistic for awareness

3.3 Descriptive Analysis

Awareness. The highest mean score was 4.388. This finding in line with research by [30] proved that the programs to reduce private transportcould greatly affect the use of roads for sustainable transport and, thus, reduce the use of private vehicles (Fig. 1).

Safety. The highest mean score was 4.2320. This is because the use of public transport has a positive significant effect on society, which can reduce accidents while using public transport. These findings supported by [30–32] also said use the public transport high risk safety since had some happen rather than use the one car (Fig. 2).

Facilities. The highest mean score was 4.1960. This finding is supported by [33] because the modal share of public transport and non-motorized transport routes can reduce 40% carbon emission by private transport by 2050 if, respectively, use the public transport, use the walking route, and also cycling in cities (Fig. 3).

Cost. The highest mean score was 4.1640. This finding was supported by [34]; it can reduce the gas usage to engage in economic and minimize the negative impacts on the environment (Fig. 4).

Carbon Dioxide Reduction in Modal Share and Environment Impact. The result above shows the most important items of these factors are modal share which can reduce carbon dioxide (item CD3), with the highest mean score was 4.1720. This finding supported by [35], that can improve the active transportation such as walking

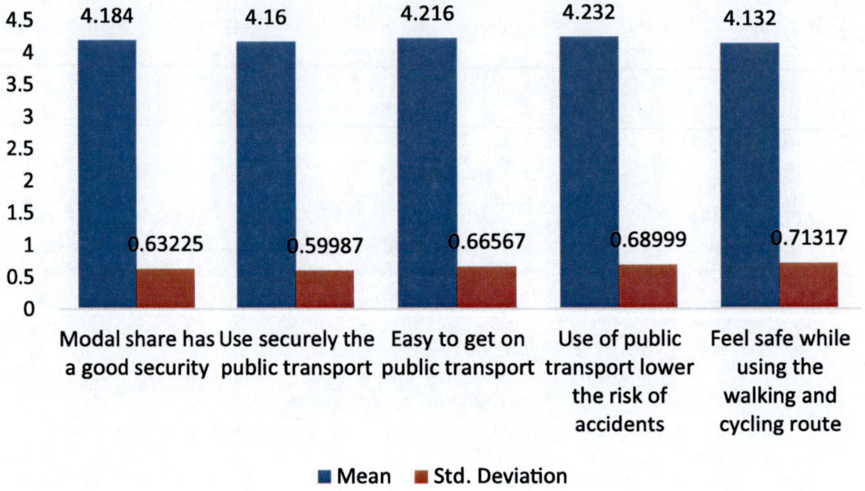

Fig. 2 Descriptive statistics of safety

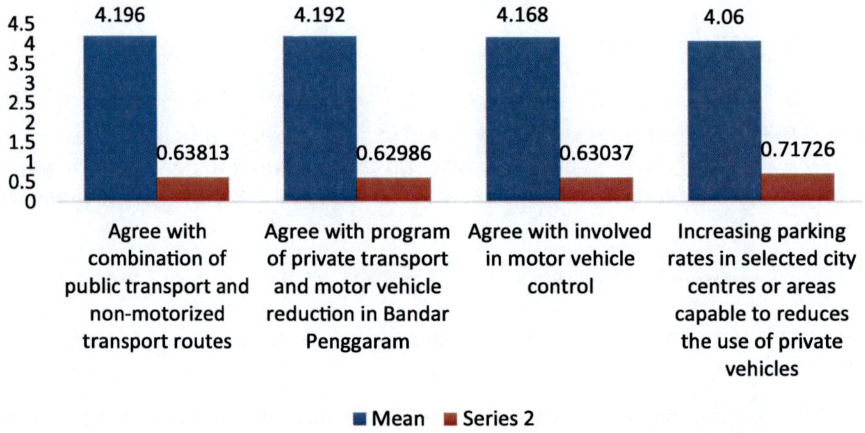

Fig. 3 Descriptive statistics of facilities

trips and enhance walking is safe mobility, include need increase the healthy environment. Willis et al. [36] also said the use of public transportation can reduce greenhouse gas emissions and improve the health of the community by promoting walking (Fig. 5).

Understanding of Modal Share Concept. The result showed the most important times of factor is all parties need to work together to drive community engagement in the modal share (item U2) and also walking routes, buses, and biking trails are effective ways of reducing traffic congestion (item U4), with the highest mean

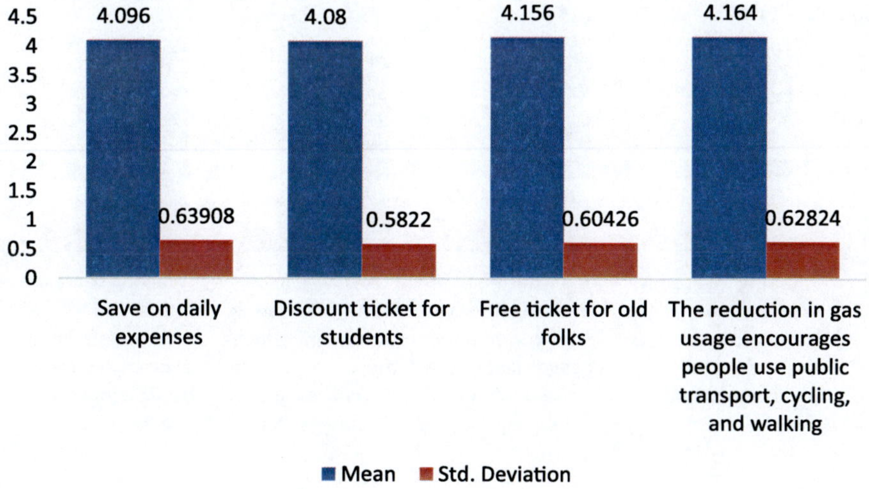

Fig. 4 Descriptive statistics of cost

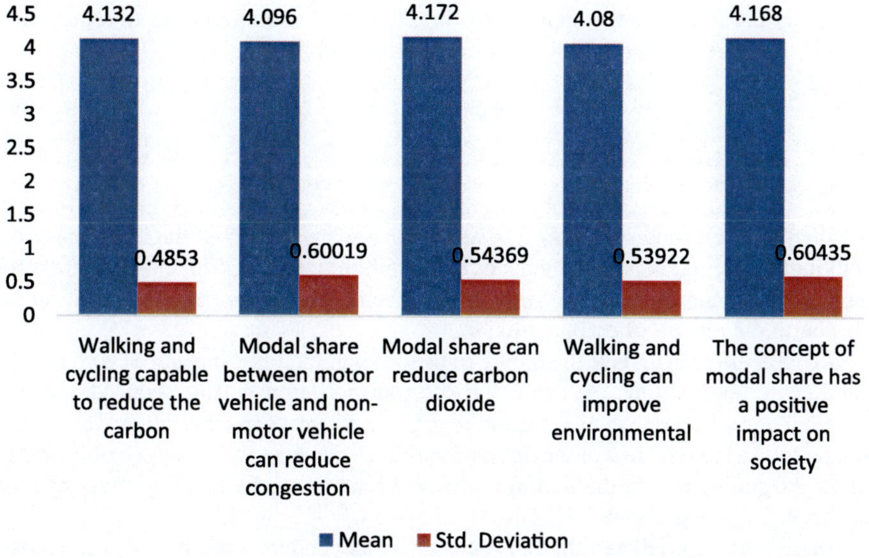

Fig. 5 Descriptive statistics of carbon dioxide reduction in modal share and environment impact

score of 4.2000, respectively. In the understanding of modal share is the desire of the individual to take the opportunity, involve, and engage the individual by understanding and knowing the modal share is important to give a good impact in the future. According to [36], it was found about the perception related to the service

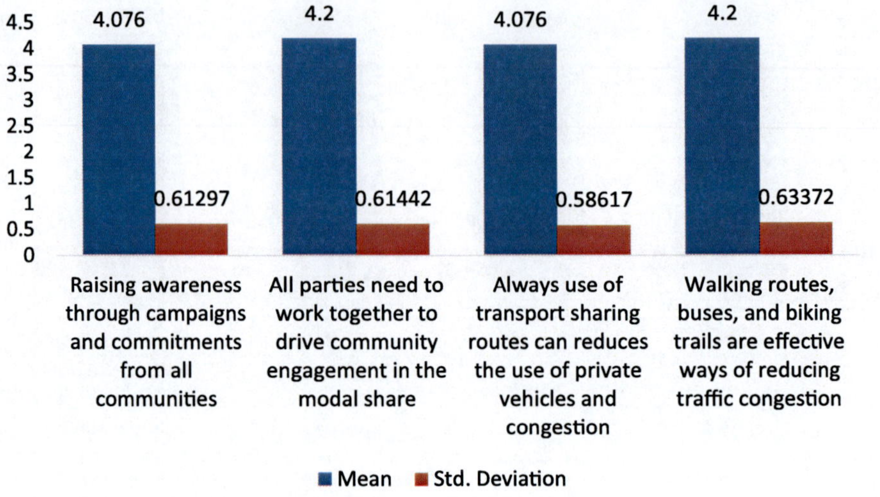

Fig. 6 Descriptive statistics of understanding of modal share concept

of quality urban public transportation with all parties and alternative to reduce the private motorize in the road using the car sharing and a bike sharing (Fig. 6).

4 Conclusion

All the proposed routes are aimed at raising public awareness of the importance of walking, bicycling, and the benefits of using public transport. The modal share also has important dependence on the transport sector. However, transportation issues also harm the release of carbon dioxide.

So, with the plan to have the public transportation facilities in parallel with developed countries and thus the use of private transportation will be reduced and the development of modal share concept be more efficient [37]. This study was represented following the view of the discussion of the answers to the research objectives. Thus, the summaries of the findings will be the focus on the two objectives of this research.

These studies predicted that all elements (independent variables) of the modal share concept are positively significant toward the understanding of the modal share concept (dependent variable). These all elements are meeting the answer with a positive result, that can benefit the environment and society around the Batu Pahat. Thus, all variable indicators with a high degree of understanding of the modal share concept were unrestrictedly verified (confirmed with unconstrained).

References

1. Energy, M.O.F., Technology, G.: Low carbon (2017)
2. Batty, P., Palacin, R., González-Gil, A.: Challenges and opportunities in developing urban modal shift. Travel Behav. Soc. **2**(2), 109–123 (2015)
3. Miralles-Guasch, C., Domene, E.: Sustainable transport challenges in a suburban university: the case of the autonomous university of Barcelona. Transp. Policy **17**(6), 454–463 (2010)
4. Schafer, A., Victor, D.G.: The future mobility of the world population. Transp. Res. Part A: Policy Pract. **34**(3), 171–205 (2000)
5. Senin, S. N., Fahmy-Abdullah, M., Masrom, M.A.N.: The implementation of green transportation towards low carbon city. In: IOP Conference Series: Earth and Environmental Science, vol. 736, No. 1, p. 012063). IOP Publishing (2021)
6. Soebagyo, D., Fahmy-Abdullah, M.O.H.D., Sieng, L.W., Panjawa, J.L.: Income inequality and convergence in Central Java under regional autonomy. Int. J. Econ. Manage. **13**(1) (2019)
7. Ali, G., Abbas, S., Qamer, F.M.: How effectively low carbon society development models contribute to climate change mitigation and adaptation action plans in Asia. Renew. Sustain. Energy Rev. **26**, 632–638 (2013)
8. Al-Amin, A.Q., Rasiah, R., Chenayah, S.: Prioritizing climate change mitigation: an assessment using Malaysia to reduce carbon emissions in future. Environ. Sci. Policy **50**, 24–33 (2015)
9. Chiu Chuen, O., Karim, M. R., Yusoff, S.: Mode choice between private and public transport in Klang Valley, Malaysia. Sci. World J. **2014**(Figure 1), 1–14 (2014)
10. Nurdden, A., Rahmat, R.A.O.K., Ismail, A.: Effect of transportation policies on modal shift from private car to public transport in Malaysia (2007)
11. Mohamed, M., Ponrahono, Z., Osman, M.M., Ibrahim, M., Bachok, S., Jaafar, S.: Apreliminary study of sustainable transport indicators in Malaysia: the case study of Klang Valley public transportation. Procedia Environ. Sci. **28**(sustain 2014), 464–473 (2015)
12. Hamdan, H., Fahmy-Abdullah, M., Sieng, L.W.: Technical efficiency of Malaysian furniture manufacturing industry: a stochastic frontier analysis approach. Int. J. Supply Chain Manage. **8**(6), 929–37 (2019)
13. Idris, A.I.M., Fahmy-Abdullah, M., Sieng, L.W.: Technical efficiency of soft drink manufacturing industry in Malaysia. Int. J. Supply Chain Manage. **8**, 908–916 (2019)
14. Johor, P.P.A.: Pelan Induk Pengangkutan Awam (2015–2045) (2016)
15. District, H., District, H.: Prediction model of public transport vehicle allocation based on multiple regression analysis, 69–79 (1824)
16. Litman, T.: Quantifying bicycling benefits for achieving TDM objectives. Transp. Res. Rec. **1441**, 134–140 (1994)
17. Norazlan, M.A.E., Fahmy-Abdullah, M., Masrom, M.A.N.: Promoting carpooling and vanpooling program to reduce the use of private motorised transportation. In: IOP Conference Series: Earth and Environmental Science, vol. 736, No. 1, p. 012050. IOP Publishing (2021)
18. Patterson, F., Baron, H., Carr, V., Plint, S., Lane, P.: Selection evaluation of three short-listing methodologies for selection into postgraduate training in general practice, 50–57 (2009)
19. Sabli, M.A.N., Fahmy-Abdullah, M., Sieng, L.W.: Application of two-stage data envelopment analysis (DEA) in identifying the technical efficiency and determinants in the plastic manufacturing industry in Malaysia. Int. J. Supply Chain Manage. **8**, 899–907 (2019)
20. Mohamad, N.F.N., Fahmy-Abdullah, M., Masrom, M.A.N.: Transit oriented development (TOD) typology. In: IOP Conference Series: Earth and Environmental Science, vol. 736, No. 1, p. 012037. IOP Publishing (2021)
21. Teddlie, C., Yu, F.: Mixed methods sampling. J. Mixed Methods Res. **1**(1), 77–100 (2007)
22. Saunders, M., Lewis, P., Thornhill, A.: Research methods for business students. Pearson Education (2009)
23. Latif, M.S.A., Fahmy-Abdullah, M., Sieng, L.W.: Determinants factor of technical efficiency in machinery manufacturing industry in Malaysia. Int. J. Supply Chain Manage. **8**, 917–928 (2019)

24. Krosnick, J.A.: Questionnaire Design. In the palgrave handbook of survey research, pp. 439–455 (2017)
25. Kumar, A., Schmeler, M.R., Karmarkar, A.M., Collins, D.M., Cooper, R., Cooper, R.A., Holm, M.B.: Test-retest reliability of the functional mobility assessment (FMA): a pilot study. Disabil. Rehabil. Assistive Technol. **8**, 213–219 (2013)
26. Perneger, T.V., Courvoisier, D.S., Ange, P.M.H.: Sample size for pre-tests of questionnaires. Qual Life Res **24**, 147–151 (2015)
27. O'Sullivan, E., Rassel, G., Berber, M.: Research methods for public administrators, 6th edn (2017)
28. Taylor, B.D., Morris, E.A.: Public transportation objectives and rider demographics: are transit's priorities poor public policy? Transportation **42**(2), 347–367 (2015)
29. Basu, R., Araldo, A., Akkinepally, A.P., Basak, K., Sehadri, R., Biran, B.H.N., Ben-Akiva, M.: Implementation & policy applications of AMOD in multi-modal activity-driven agent-based urban simulator simmobility. Trans. Res. Rec. J. Transp. Res. Board 1–20 (2018)
30. Lim, J.: Jln TAR to close for one month from April 1 (2019)
31. Nasrudin, N., Rostam, K., Rose, R.A.C.: Persepsi penduduk Shah Alam terhadap dasar pengangkutan dan kesediaan mengguna pengangkutan mampan Transportation policies and the readiness of local public for sustainable transportation - A perception study of Shah Alam, Malaysia. Malays. J. Soc. Space **2**(2), 133–142 (2014)
32. Joewono, T.B.: Safety and security improvement in public transportation based on public perception. Int. Assoc. Traffic Saf. Sci. **30**(1), 86–100 (2006)
33. Bernama (2019) Masyarakat disaran beralih kepada pengangkutan awam
34. Mouwen, A.: Drivers of customer satisfaction with public transport services. Transp. Res. Part A **78**, 1–20 (2015)
35. Sietchiping, R., Permezel, M.J., Ngomsi, C.: Transport and mobility in Sub-Saharan African cities: an overview of practices, lessons and options for improvements. Cities **29**(3), 183–189 (2012)
36. Willis, D.P., Manaugh, K., El-geneidy, A.: Cycling under influence: summarizing the influence of perceptions, attitudes, habits, and social environments on cycling for transportation. Int. J. Sustain. Transp. **9**(2010), 565–579 (2015)
37. Guglielmetti, R., Toni, M., Raharjo, H., Di, L., Petros, S.: Does the service quality of urban public transport enhance sustainable mobility ? J. Clean. Prod. **174**, 1566–1587 (2018)

Dynamic Traffic Flow to Promote Sustainable Mobility

Adlyn Nazurah Abdul Rahman and Suliadi Firdaus Sufahani

Abstract Dynamic traffic flow is one of the main challenges for a state or country after a city is created. Since the population is increasing, it also affects the number of vehicles and problems occurring in the area. Increasing road capacity by constructing new roads and widening existing roads do not, in the end, resolve the situation but simply postpone the problem until more roads need to be built. The objectives of the study were to identify the dynamic traffic flow and ways to improve it. The study was conducted by using a mixed-method. The data was collected by using questionnaires, document reviews, and observation. The results showed that the traffic light was recorded as mean score at a high-level cause of traffic congestion compare with the junction, two-way road, roundabout, and road structure. Besides that, the elements lessening traffic light, upgrading road structure, upgrading junction, and adding two ways may be taken to reduce traffic congestion. The implications of this study show that the local authorities should have strategic planning such as lessening the traffic light or upgrading the road structure to ensure dynamic traffic flow to promote sustainable mobility.

Keywords Dynamic traffic flow · Traffic congestion · Low-carbon · Road · Carbon emission

---

A. N. A. Rahman (✉)
Faculty of Technology Management and Business, Universiti Tun Hussein Onn Malaysia, 86400 Parit Raja Batu Pahat, Johor, Malaysia
e-mail: ap160260@siswa.uthm.edu.my

S. F. Sufahani
Oasis Integrated Group, Institute for Integrated Engineering, Universiti Tun Hussein Onn Malaysia, 86400 Parit Raja, Batu Pahat, Johor, Malaysia

Faculty of Applied Science and Technology, Universiti Tun Hussein Onn Malaysia, 86400 Parit Raja, Batu Pahat, Johor, Malaysia

M. S. Kaiser et al. (eds.), *Proceedings of the Third International Conference on Trends in Computational and Cognitive Engineering*, Lecture Notes in Networks and Systems 348, https://doi.org/10.1007/978-981-16-7597-3_31

1 Introduction

Traffic dynamics describes the interplay of many vehicles and drivers. Moreover, the interaction of the vehicles and drivers, technically termed driver vehicle units, leads to new collective effects that do not depend on the details of individuals. Traffic flow models can be described as the dynamics of vehicles and drivers in terms of mathematical equations. The predictions can be obtained by running the model simulation. For the simulation that produces the best fit with the data (model calibration), it is necessary to choose the values of the model's parameter. Once calibrated, the model may be used for the traffic flow prediction and other applications [1]. A condition where traffic demands exceed traffic capacity is defined as traffic congestion which is also known as a global problem that occurs all around the world especially in the metropolitan cities. One of the methods that is capable of dispersing traffic congestion efficiently has been recognized as dynamic traffic routing. An alternative to this is to increase the highway capacity by improving the efficiency and management of traffic operation and demand. An Intelligent Transport System (ITS) emerged as a means to implement such initiatives for efficient operation and better traffic management [2].

Climate change is a subject that generates global risks and uncertainties [3, 4] due to a series of extreme weather events that have occurred in recent years [5, 6]. As was indicated by [7], the largest Green House Gas (GHG) producers are from road transportation in the Organization for Economic Cooperation and Development (OECD) countries, while road freight typically accounts for slightly less than half of the road transportation. In general, as a dominant mode of freight movement, road transportation accounts for the largest share of the freight-related emissions [8].

Road is the most important route for a city because nowadays the number of vehicles is increasing whether private or public transport. This increment causes road upgrades to be frequent in urban areas. A traffic congestion at the junctions or highway leads to time delays and waiting queues, resulting in more vehicular pollutants released in the air, specifically greenhouse gases. People are aware of the issues with climate change and global warming that significantly impact the environment, humans, and animals on earth [9].

Road surfacing has known as a crucial component of highway development and maintenance. An efficient running of the road network has the success of national and local economies as well as the quality of the dependency of public life [10]. It is important to reduce the amount of carbon in the atmosphere because this carbon has caused global warming, also the depletion of the ozone layer. Due to the increase in the number of vehicles now, the amount of carbon is increasing especially during peak hours of traffic congestion. The process of development entails not only an economic phenomenon. It is unnecessarily assessed based on the achievement of national economic growth, but it has a broader perspective than merely an economic aspect [11].

The car registrations in Malaysia recorded an increment to 114 600 in January from 96 201 in December of 2018. Car registrations in Malaysia averaged 68 111.33

from 1 988 until 2019, reaching an all-time high of 138 727 in March of 2015 and a record low of 9 732 in May of 1988 [12]. This shows that the increment in the number of vehicles has contributed to the problem of traffic congestion especially at peak hours and increasing carbon emissions in the air.

The increment in carbon is alarming because of the previous implementation of this problem, still inadequate. One of the ways that has been in practice in Malaysia to overcome the traffic congestion problem, which invites the carbon improvement is the Smart Traffic Management system in Cyberjaya [13]. Technical efficiency, is crucial in order to determine the efficiency level [14]. The concept of efficiency can be seen not only in a department or organization [15] but also in traffic management systems. Since late 2015, the system has already been in development and tested, therefore, a run test that has been done along Persiaran Multimedia in Cyberjaya shows 65% reduction in waiting time at the traffic junctions, as seeing the usual travel duration of 32 min reduce to around 8–11 min [13]. However, the government may overlook the condition of the road where it also con-tributes to increased carbon if the infrastructure is not satisfactory. Hence, the improvement of road links and junctions is also important in reducing carbon emission rates.

In achieving low-carbon city standards, it is important to address issues affecting the rate of carbon emissions. The road link and junction inefficiency are also the cause of increasing carbon dioxide in the air. Since road infrastructure affects the rate of carbon emissions from vehicles, then improvement is needed in order to overcome this problem. If no upgrading takes place, then this problem will continue and the carbon rate will continue to increase.

An operation of national Traffic Management Center (TMC) as a hub has just started by Malaysia Highway Authority (MHA) to about 12 different highway control centers throughout Peninsular Malaysia. It was connected to and received the information from the 155 vehicle detection stations, closed circuit television with over 200 cameras, and 80 variable message signs (VMS). It can do a data, video, and voice-in-information extraction from the remote highway control center. It will collect, store, and process before being disseminated to the public via VMS, Web portal, SMS, and MMS for the latest traffic conditions and travel advise as being planned [16].

The expectations from the findings of this study can have a lot of impact not only on the authorities but also on potential researchers who may make studies that require the information contained in this study. It is hoped that the problems associated with increased carbon can be solved based on Malaysia's vision to become and make a low carbon city in the future.

2 Research Methodology

2.1 Selection of Research Methodology

An in-depth study on research methodology is considered essential in every research [17]. In order to satisfy the objectives of this study, a mixed method approach is selected. A mixed method approach for data collection is used as using both quantitative and qualitative research techniques that were adopted to achieve the research aim and objectives [18].

2.2 Data Analysis Technique

Data analysis is done by using SPSS data, which is to analyze the data from checklist. After the analysis is done, an outcome result is obtained and some recommendation is made based on the result. An analysis and findings of the study described sequentially according to the objectives and question of the research that want to achieve in the study [19].

2.3 Pilot Test

A pilot-test , which is a pre-test, is a mini version of full-scale re-search or experiments that conducted extensive research in preparation. It helps to improve the reliability of the questionnaire that was conducted by this study and the minimum value for inter-item respondent. Based on the study by [20], 30 participants suggested as a reasonable default value or starting point for pre-tests of questionnaire. Therefore, 30 residents are collected and used to carry out the pre-test. After pre-test, the questionnaire is distributed to the resident at the 11 chosen road in Batu Pahat. Table 1 shows the result for pilot test on pre-test and real test.

Table 1 Reliability statistics for pre- and actual test

	Cronbach's alpha	Cronbach's alpha based on standardized items	No. of items
Pre-test	0.697	0.677	12
Real test	0.691	0.696	12

3 Result and Discussion

3.1 Questionnaire

Causes of traffic congestions. Based on Table 2, the traffic light analysis shows 41.1% agree that the traffic congestion is caused by the traffic light. Next, 32.2% respondents also agree that junctions are the causes of the traffic congestion. Then, 30% respondents either agree or disagree that two ways are the causes of traffic congestion. Lastly, 50% of the respondents either agree or disagree that traffic congestion is caused by road structure.

Table 3 shows mean measurement for the causes of traffic congestion. Based on this table, it shows that traffic light mean is 3.81, which is the highest among other causes, followed by 3.3 for road structure. Then, mean for two way is 2.93, which is placed as the third highest. Mean for the junction is 2.87 while mean for roundabout is 2.67. Based on Table 3, scale for traffic light is high, while other causes are categorized as moderate. Its means that the main causes of the traffic congestion are traffic lights.

Steps Can Be Taken. Based on Table 4, 56.7% respondents strongly disagree that adding traffic lights is a step taken to reduce traffic congestion, meanwhile for lessening the traffic light, 31.5% respondents either agree or disagree to it.

Next, 36.7% of the respondents strongly agree that upgrading junctions can reduce the traffic congestion. 36.7% respondents disagree to adding two way, and at the same time 50% of the respondents strongly disagree that lessening two way steps can be taken to reduce the traffic congestion. For removing roundabout, 40% respondents

Table 2 Respondent percentages for causes of traffic congestion

Causes of traffic congestion	Strongly disagree	Disagree	Neutral	Agree	Strongly agree
Traffic light	3.0	17.4	6.7	41.1	31.9
Junction	20.7	21.9	16.3	32.2	8.9
Two way	10.0	26.7	30.0	26.7	6.7
Roundabout	33.3	6.7	30.0	20.0	10.0
Road structure	6.7	6.7	50.0	23.3	13.3

Table 3 Mean measurement for the causes of traffic congestion

	Mean	Scale
Traffic light	3.81	High
Junction	2.87	Moderate
Two way	2.93	Moderate
Roundabout	2.67	Moderate
Road structure	3.3	Moderate

Table 4 Respondent percentages for steps can be taken

Steps can be taken to reduce traffic congestion	Strongly disagree	Disagree	Neutral	Agree	Strongly agree
Adding traffic light	56.7	23.3	6.7	10	3.3
Lessening traffic light	3	14.1	31.5	30	21.5
Upgrading junction	27	12.2	6.7	17.4	36.7
Adding two way	13.3	36.7	23.3	26.7	0
Lessening two way	50	30	13.3	6.7	0
Removing roundabout	33.3	10	40	16.7	0
Upgrading road structure	0.4	2.2	13.7	34.8	48.9

Table 5 Mean measurement for steps can be taken

	Mean	Scale
Adding traffic light	1.8	Low
Lessening traffic light	3.53	High
Upgrading junction	3.24	Moderate
Adding two way	2.63	Moderate
Lessening two way	1.77	Low
Removing roundabout	2.4	Low
Upgrading road structure	4.3	High

either agree or disagree, meanwhile 48.9% respondents strongly agree that upgrading road structure can reduce traffic congestion.

Based on Table 5, it shows that upgrading road structure mean is 4.3 which is the highest among other steps, followed by lessening traffic light mean 3.53, which means the these two steps can be taken to reduce traffic congestion.

Based on Tables 4 and 5, it shows that the results are parallel with the causes which is the main causes are traffic light, and steps can be taken to reduce the traffic congestion are by lessening the traffic light in additional upgrading the road structure.

3.2 Observation

Based on Table 6, it shows that 8 out of 11 roads are in significant and considerable delays, whose means are high during traffic congestion. The highest duration is at Jalan Rahmat, which is 83 s per vehicle. The second highest is Jalan Tanjong Laboh where the average is 78 s per vehicle.

For Jalan Zabedah, the duration is 1.05 s per vehicle, Jalan Sultanah 69 s per vehicle, Jalan Mohd Akil 0.56 s per vehicle, and Jalan Pejabat 0.53 s per vehicle.

Table 6 Total motorcycle and vehicle per hour

	Total motorcycle	Total per hour	Total vehicle	Total per hour	Delayed per vehicle (second)	Technical descriptions
Jalan Rahmat	75	4	894	50	89	Significant delays
Jalan Tanjong Laboh	80	4	839	47	78	Significant delays
Jalan Sultanah	62	3	741	41	69	Significant delays
Jalan Zabedah	73	4	701	39	65	Significant delays
Jalan Mohd Akil	49	3	607	34	56	Considerable delays
Jalan Pejabat	54	3	567	32	53	Considerable delays
Jalan Syahbandar	45	3	545	30	50	Considerable delays
Jalan Mohd Salleh	65	4	542	30	50	Considerable delays
Jalan Ampuan	68	4	492	27	46	Minimal delays
Jalan Zaharah	54	3	417	23	39	Minimal delays
Jalan Masjid	20	3	154	9	14	Short delays

Jalan Mohd Salleh and Jalan Syahbandar have the same duration, which is 50 s per vehicle. Next is Jalan Ampuan where its duration is 46 s per vehicles. Jalan Zaharah's duration per vehicle is 39 s. Last is Jalan Masjid, the least duration among the roads, which is 14 s. This concludes that the traffic flow for two way stop intersections and intersections with traffic signal roads that delay is above 36 and 56 s for each type which is considered as not dynamic.

4 Conclusion

In order to reduce the traffic congestion, a strategic planning such as lessening the traffic light or upgrading the road structure can be taken by the local authorities. Lessening the traffic does not mean removing the current traffic light. Local authorities may lessen the waiting time, or use smart sensors, which will priorities the first waiting car. For example, the Intelligent Transportation System (ITS) technology has caught worldwide attention because of its great potential in solving road traffic congestion at a low cost [21].

Authorities are suggested to use Leaky Bucket Controller and Network Calculus, which a complicated element of traffic will incorporate into unified mathematical model called T-S Constrained Model that can help in deducing the flow assignment rate [22]. A system called Sydney Coordinated Adaptive Traffic System (SCATS), which used to calm the traffic flow at busy intersections across Sarawak. There is 12 installations of the system at three different are have costed RM4.5 million (RM 1.5 million for 4 systems) [23].

Local authorities also can upgrade the road structure. It is because bad road structures may cause the driver to slow down their car when using the road. When the driver slows down their car, it may cause traffic congestion during peak hours since the following car will also slow down. Therefore, in order to overcome this problem, local authorities may do some road restructure.

In conclusion, it can be stated that this study has achieved both objectives that were set out for this study, which are to identify dynamic traffic flow and to improve dynamic traffic flow. The data obtained from the questionnaire provided have been helpful in obtaining accurate analysis results to achieve the objectives of the study. The recommendations identified in advance are expected to assist future studies.

References

1. Treiber, M., Kesting, A.: Traffic flow dynamics. In: Traffic Flow Dynamics: data, Models and Simulation, Springer, Berlin, Heidelberg (2013)
2. Isa, N., Mohamed, A., Yusoff, M.: Implementation of dynamic traffic routing for traffic congestion: a review. In: International Conference on Soft Computing in Data Science, pp. 174–186. Springer, Singapore (2015)
3. Kuklicke, C., Demeritt, D.: Adaptive and risk-based approaches to climate change and the management of uncertainty and institutional risk: the case of future flooding in England. Glob. Environ. Chang. **37**, 56–68 (2016)
4. Carrao, H., Naumann, G., Barbosa, P.: Mapping global patterns of drought risk: an empirical framework based on sub-national estimates of hazard, exposure and vulnerability. Glob. Environ. Chang. **39**, 108–124 (2016)
5. Winn, M., Kirchgeorg, M., Griffiths, A., Linnenluecke, M.K., Günther, E.: Impacts from climate change on organizations: a conceptual foundation. Bus. Strateg. Environ. **20**(3), 157–173 (2011)
6. Slawinski, N., Pinkse, J., Busch, T., Banerjee, S.B.: The role of short-termism and uncertainty avoidance in organizational inaction on climate change: a multi-level framework. Bus. Soc. **56**(2), 253–282 (2017)
7. Chapman, L.: Transport and climate change: a review. J. Transp. Geogr. **15**(5), 354–367 (2007)
8. McKinnon, A.C., Piecyk, M.I.: Measurement of CO2 emissions from road freight transport: a review of UK experience. Energy Policy **37**(10), 3733–3742 (2009)
9. Senin, S.N., Fahmy-Abdullah, M., Masrom, M.A.N.: The implementation of green transportation towards low carbon city. In: IOP Conference Series: earth and Environmental Science, vol. 736, no. 1, p. 012063. IOP Publishing (2021)
10. Aziz, Z., Qasim, R.M., Wajdi, S.: Improving productivity of road surfacing operations using value stream mapping and discrete event simulation. Constr. Innov. **17**(3), 294–323 (2017)
11. Soebagyo, D., Fahmy-Abdullah, M.O.H.D., Sieng, L.W., Panjawa, J.L.: Income inequality and convergence in central java under regional autonomy. Int. J. Econ. Manag. **13**(1) (2019)
12. Trading Economics homepage.: (2021). https://tradingeconomics.com/malaysia/car-registrations

13. Wong, D.J.: You waste 158 h a year stuck in traffic-these 4 solutions could end that In: M'sia (2018). Retrieved from https://vulcanpost.com/633571/malaysia-traffic-jams-solutions/
14. Hamdan, H., Fahmy-Abdullah, M., Sieng, L.W.: Technical efficiency of malaysian furniture manufacturing industry: a stochastic frontier analysis approach. Int. J. Supply Chain Manag. **8**(6), 929–937 (2019)
15. Idris, A.I.M., Fahmy-Abdullah, M., Sieng, L.W.: Technical efficiency of soft drink manufacturing industry in Malaysia. Int. J. Supply Chain Manag. **8**, 908–916 (2019)
16. Hossain, M.: Integrated management of Malaysian road network operations through ITS initiatives: issues, potentials and challenges. In: International Seminar on Intelligent Transport Systems (ITS) in Road Network Operations, Kuala Lumpur, Malaysia (2006)
17. Norazlan, M.A E., Fahmy-Abdullah, M., Masrom, M.A.N.: Promoting carpooling and vanpooling program to reduce the use of private motorised transportation. In: IOP Conference Series: earth and Environmental Science, vol. 736, no. 1, p. 012050. IOP Publishing (2021)
18. Sabli, M.A.N., Fahmy-Abdullah, M., Sieng, L.W.: Application of two-stage data envelopment analysis (DEA) in identifying the technical efficiency and determinants in the plastic manufacturing industry in Malaysia. Int. J. Supply Chain Manag. **8**, 899–907 (2019)
19. Latif, M.S.A., Fahmy-Abdullah, M., Sieng, L.W.: Determinants factor of technical efficiency in machinery manufacturing industry in Malaysia. Int. J. Supply Chain Manag. **8**, 917–928 (2019)
20. Perneger, T.V., Courvoisier, D.S., Ange, P.M.H.: Sample size for pre-tests of questionnaires. Qual Life Res **24**, 147–151 (2015)
21. Necula, E.: Dynamic traffic flow prediction based on GPS data. In: 2014 IEEE 26th International Conference on Tools with Artificial Intelligence, pp. 922–929. IEEE (2014)
22. Peng, Y., Fan, W., Liu, J., Zhang, F.: The research of traffic flow assignment model based on the network calculus of computer networks. Inf. Technol. J. **11**(3), 307 (2012)
23. Ogilvy, G.: Traffic flow should be smoother (2017)

LCADP: a Low-Cost Accident Detection Prototype for a Vehicular Ad Hoc Network

Md. Julkar Nayeen Mahi, Sudipto Chaki, Shamim Ahmed, Iffat Tamanna, and Milon Biswas

Abstract Accident fatalities within vehicles are devastating incidents of today. In developing countries, vehicular accidents occur rapidly based on different use cases (i.e.: road problems or without following road safety policies). In our developed prototype, a cost-effective method is achieved to work within a single VANET controlled area. In this prototype, the Arduino UNO is equipped with an XBee module. The testbed is compiled with an FSR (i.e.: force-sensitive register) and successfully detects force or pressure value if bends completely. In terms of full bending, it gives an output of '1' and is regarded as true. If not, the output generates '0' as a value and shows a false alert. The true values justify the occupied (i.e.:each bend gets a resistance value, merges with location coordinates, and is configured within the prototype itself) accident spot. The prototype works in a specific pattern, for viewing accident (i.e.: location detection through resistance value and USB module-based LAN) occupied location. The accidental location is depicted by the GPS tracker and transfers data through a GSM network (i.e.: short-range, 20 meters) using text outputs. USB tracker devices carry user data and deliver towards the access repeaters customized roadside LANs or ERS (i.e.: 3–4, dedicated short-range console computer communication).

Keywords VANET · Low cost · GSM · Arduino · Accident detection

1 Introduction

Accidents that took place by the roads are a frequent incident of today and increasing rapidly. Recent statistics of newspapers and journals derive the number of people killed in road accidents. Particularly, the fundamental cause is heavy traffic

Md. J. N. Mahi (✉)
Computer Science and Engineering, City University, Birulia, Khagan 1216, Bangladesh

S. Chaki · S. Ahmed · I. Tamanna · M. Biswas
Computer Science and Engineering, Bangladesh University of Business and Technology, 1216 Dhaka, Bangladesh

M. S. Kaiser et al. (eds.), *Proceedings of the Third International Conference on Trends in Computational and Cognitive Engineering*, Lecture Notes in Networks and Systems 348, https://doi.org/10.1007/978-981-16-7597-3_32

congestion. In developing countries, child death and casualty percentages are estimated as 96 and 86% comparatively. But road accidental deaths have an increased percentage of 83%. In Bangladesh, every year, there occurred 3000 road fatalities, 3000 dreadful, and 3500 effortless injuries approximately, according to the police report of accidents. Other sources stated about the casualties and fatalities raised to 12000–20000 in a year. Hence, safety and protection problem is a must need for developing countries. In terms of Bangladesh, fatalities comparison to the international standards is at a rate of 60 to 150 per 10000 motor vehicles. Whereas this fatality result can be compared to 25, 16, 2, and 1.4 for India, Sri Lanka, USA, and the UK, respectively [1]. According to the mentioned statistics and referenced article, considering the number of deaths, it can be observed that ERS management is an essential need at the crucial time. On the other hand, proper vehicular management to avoid accidental problems is also needful for VANET communications, and also in this regard, Fig. 1. describes the uprising trends of the intelligent transportation market from the year 2012–2022.

Our developed prototype works as an alert module for delivering needful information (i.e.: accident detection spot) to the nearby ERS (i.e.: emergency response service). The developed module is considered to support a service module within a single VANET occupied area. VANET communication [2] acts as a secure wireless analogy to keep vehicles safe from various environmental (i.e.: accident, theft) problems. In our developed prototype, the Arduino UNO is merged with an XBee wireless module. The testbed hardware model is equipped with an FSR (i.e.: force-sensitive register) and fundamentally generates force or pressure value towards the

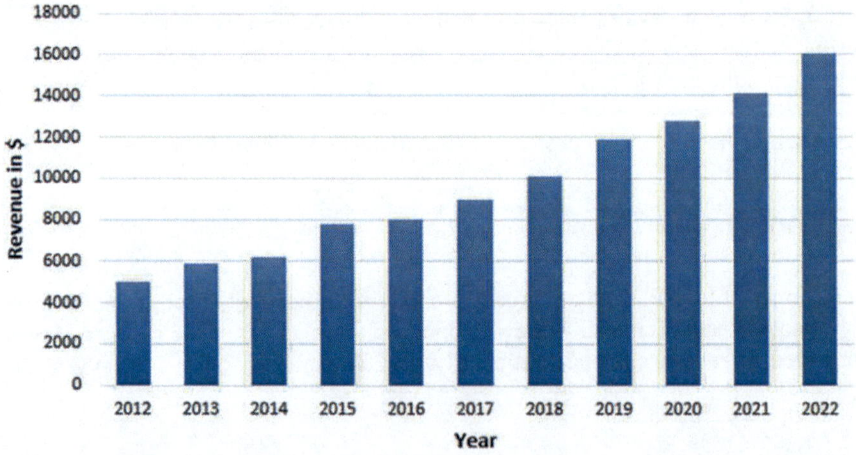

Fig. 1 In the future years, the ITS or the market for the smart transportation system is expected to expand considerably and reach revenues of $47.6 billion by 2022 [3]. The market of Asia-Pacific has the biggest development in the ITS industry (Based on traffic management, road safety security, freight management, coalition, detection and management, environment protection, automatic telemetries, and road user charging)

GSM network and ERS (i.e.: Local LAN-based 3–4 computer console communication) for outputs if bends completely. In terms of fully bending, it shows an output of '1' and is supported as true. If not, output prints '0' and determines as false. The true output validates (i.e.: each bend gets a resistance value, merges with location coordinates, and is configured within the prototype itself) occupied accident spot and transfer information to the ERS. It performs in a sequential pattern, for viewing accident (i.e.: location detection through resistance value and USB module-based LAN) occupied location [4]. GPS tracker views the accidental location while transferring data via a GSM network (i.e.: short-range, 20 meters) using text messages. The USB tracker module carries the needful data and transfers across the access points or repeaters of roadside-based customized LANs (i.e.: 3–4, dedicated short-range console computer communication or ERS).

The following are the contributions to this development prototype:

- We develop a prototype for a Single VANET control area to give an alert to accident occupied location and ERS management.
- The prototype development considers as a low-cost platform while comparing existing system managements.
- The configured prototype is beneficial for low-income countries.

The paper is organized as follows—Sect. 2 describes the Literature Review; Sect. 3 depicts the Decision Making Methodology and Prototype Development; Sect. 4 employs Hardware Configuration and Work in Progress; Sect. 5 shows Existing System Managements Versus. Developed Prototype; Sect. 6 represents Limitation and Future Works; and lastly, Sect. 7 illustrates the Conclusion through providing beneficiary results.

2 Literature Review

Reducing the frequency and severity of traffic accidents, detection modules served as an intermediary helper. Hence, proper detection module-based criteria need to be checked to acquire better data acquisition and aggregation. The author [5] proposed a system module based on the coordinates of the vehicles through simulated data validation within vehicular ad-hoc networks (VANETs). After data processing, the system sends alert messages to drivers. Moreover, it uses machine learning methods to detect accidents through ITS freeways. Isolation of common causes of accidents within traffic data is detected using supervised machine learning algorithms [6]. Zhou et al. described a collaborative identification structure for large data storage while ensuring high reliability of data transmission among vehicles. The platform also uses a reliable reputation-based cooperative communication method to minimize road traffic accidents [7]. Liang et al. proposed a novel ID-based wireless secured communication over VANETs [8]. A typical VANET-based accident management scenario is depicted in Fig. 2.

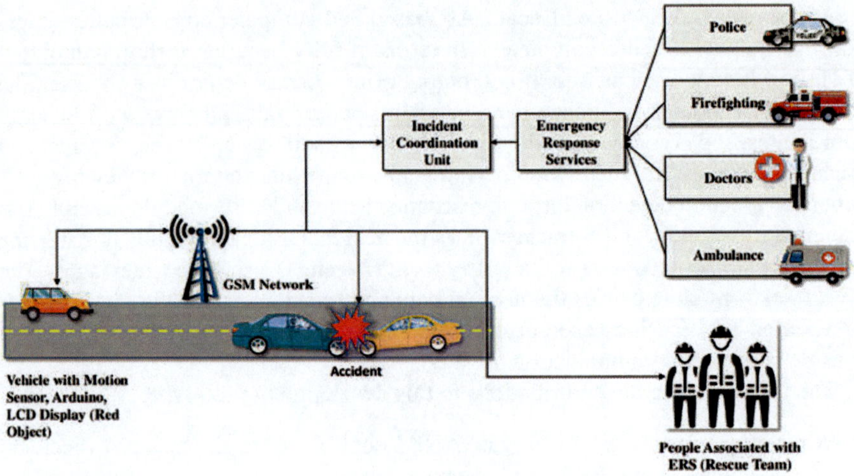

Fig. 2 Generic overview of accident detection management in a VANET

In particular, it has a feature extraction and classification-based algorithm. The algorithms classify Self-Organization Map (I-GHSOM) for ids within VANETs. The algorithm has two main features. The features detect the position of the traffic flow and distance [9]. The density of traffic, traffic jams, and accidents are increasing rapidly. In context with the developing countries, low-cost accident detection management is a needful project to work and develop on. In recent years, researchers are able to develop automatic accident detection systems by using artificial intelligence technologies. Normal and accident (i.e.: traffic flow) conditions are identified in those investigations. Time series, pattern recognition methods, fuzzy set filtering, and random forest classifier are some of the widely used techniques [10]. Since GPS data may not pinpoint the exact site of an accident, techniques for detecting locations can be devised without utilizing them. To detect unexpected traffic circumstances, machine learning algorithms provide beneficiary help to evaluate vehicle behaviors. Dense skyscrapers have hampered GPS-based localization as urbanization has evolved. Hence, finding the necessity of the development of high-reliability-based GPS-free localization. One of the major outcomes of the 5G network [11] is ultra-dense network activity.

The ultra-dense property functions simultaneously across several 'BST' coverage areas. This technology opens a whole new world of possibilities for vehicle localization. Furthermore, using a 5G antenna array to estimate direction-of-arrival (DOA) is a worthy option. A blended method of IoT-VANET technology is introduced in [13]. Different parameters used to evaluate the system's performance reflect the throughput of the IoT (i.e.: sensors) within VANET communications [2, 13–16]. If a passenger enters the car, they must manually activate their mobile NFC. The car number IDs and passenger IDs are supplied to the system and their memberships are checked throughout the system immediately after the NFC is switched on. The system then sends the passenger entering an alert message. Each occupant within the car is involved in this procedure. Low-energy Bluetooth (i.e.: BLE), which employs IoT protocol-based communications, may be used as a viable substitute for NFC.

3 Decision Making Methodology and Prototype Development

The developed system will work within a VANET controlled area. The fundamental purpose of VANET [9, 15, 16] is to make secure cellular communication to keep vehicles safe from different environmental (i.e.: accident, theft) consequences. In our developed prototype, the Arduino UNO is compiled with an XBee wireless module. The testbed module is equipped with an FSR (i.e.: force-sensitive resistor) and primarily detects force or pressure value if bends completely. For fully bending, it provides an output of ' 1 ' and is regarded as true. If not, output prints '0' and shows as false. The true value validates the occupied (i.e.:each bend gets a resistance value, merges with location coordinates, and is configured within the prototype itself) accident spot. It works in a specific pattern, for viewing accident (i.e.: location detection through resistance value and USB module-based LAN) occupied location [4]. The accidental location is shown by the GPS tracker and transfers data via a GSM network (i.e.: short-range, 20 m) using text messages (i.e.: bit exchange pattern with true or false). USB tracker modules carry the needful data and deliver across the access points or repeaters of roadside-based customized LANs or ERS (i.e.: 3–4, dedicated short-range console computer communication). We outline a systematic overview of our developed accident detection management system in Fig. 3. For an affirmative validation, processed data is delivered over the GSM network and distributed towards ERS provider (i.e.: USB module-based LAN-based computer) for taking necessary actions. Alternatively, if a negative outcome is addressed, concentrated data are transferred to the Arduino platform for efficient ERS validations.

4 Hardware Configuration and Work in Progress

In the developed system, Arduino UNO is added with XBee module. It uses a specific pattern for detecting the accident (i.e.: location detection through USB module-based LAN) [14].

The accidental spot is detected by the GPS tracker and transfers information towards the GSM network using a text message. The text message is carried by the USB module and access points of roadside-based customized LAN (i.e.: dedicated short-range). The whole scenario is illustrated in Fig. 3. If validation is affirmative, it sends further information to the GSM network and transfers towards ERS provider (i.e.: USB module-based LAN-based computer) for taking necessary actions. If a

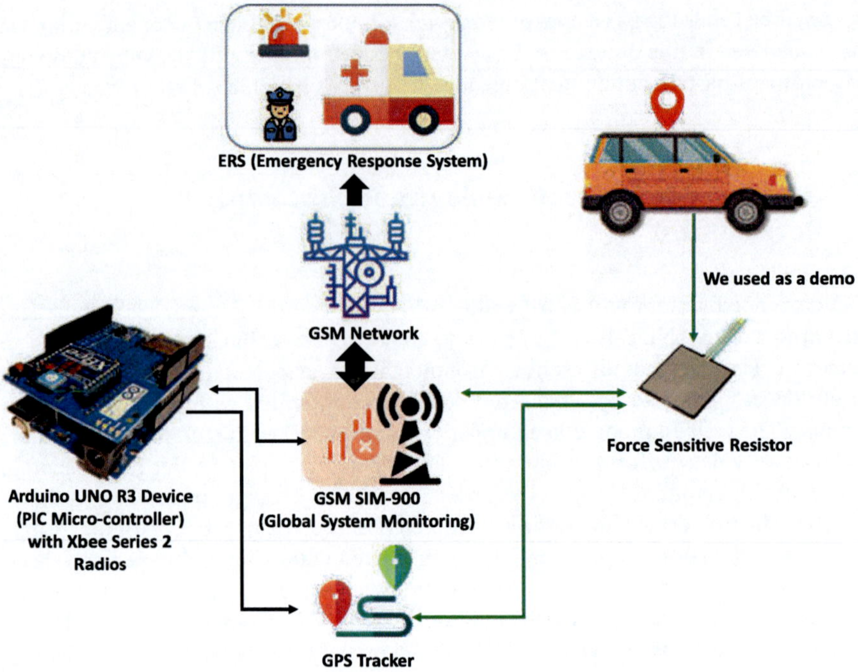

Fig. 3 LCADP prototype scenario of accident detection and management over a single VANET

negative outcome is addressed within the GSM network, the processed information is delivered to the Arduino UNO platform for the further validation process and ERS effectively.

The system can detect accidents by the use of GSM and roadside customized LAN (i.e.: dedicated configured short-range, 20 m) [4, 17]. The algorithmic scheme has both the criteria to check the accident occupied location by alarm notification and transfer emergency information to the road size configured LAN's. The workflow model of this proposed accident detection management is outlined in Fig. 4.

Algorithm 1: is based on two equation basics. They are:

- Accident occupied location.
- No accident occupied location.

Based on the use cases, three conditions were raised. They are stated below.

Condition 1: Firstly, the X attains an accident occupied spot and $x'(gsm)$, $x(gnet)$, and $i(usb)$ support data generation from GSM tracker, GSM network, and USB module-based computer communication (i.e.: exchange bit information value [either 0 or 1 for false and true, respectively]) [17].

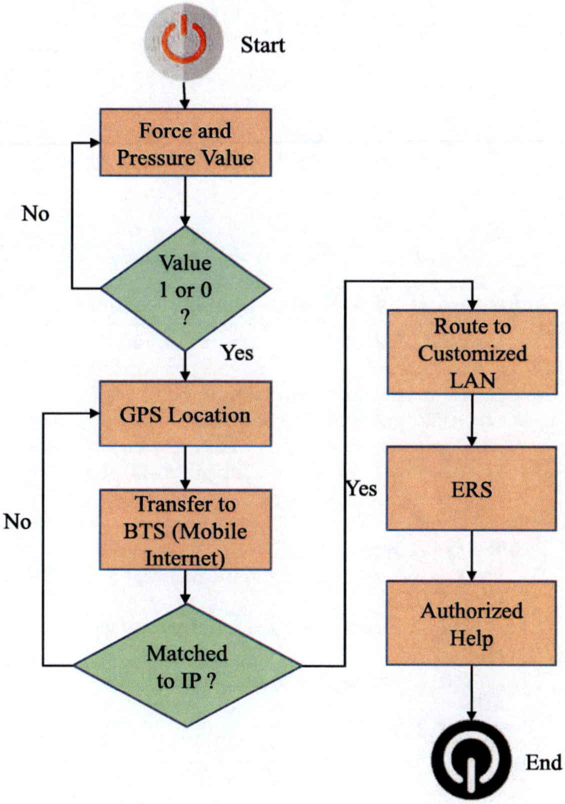

Fig. 4 Single VANET data processing (i.e.:workflow) within the developed prototype (LCADP)

Condition 2: Secondly, f supports accident occupied spot value if the force is greater than X. A certain amount of information will deliver to the USB communication-based LAN module. Thus, the information delivers an accident alert to the ERS. [4].

Condition 3: Thirdly, f supports no accident occupied spot value if the force is less than X. A certain amount of information will deliver via the GSM network to the USB LAN module. Thus, the information delivers no accident alert to the ERS authority.

The algorithm compares the value of X among y, z, $x'(gsm)$, $x(gnet)$, $i(USB)$, and f. The overall workflow of the algorithm detects the metrics by differentiating the above-mentioned parameters. For understanding the development, it is important to know GSM network activity, coordinates of accident detection spot, USB communication through roadside LAN's, and exchange data bit of communication (i.e.: 0 or 1). The software algorithmic scheme supports the following equations:

$$\sum(X) = \sum x'(gsm) + x(gnet) + i(usb) \tag{1}$$

$$\sum(z) = \sum(f) \geq \sum_{n->force<=0}^{n->force=1} (X) \tag{2}$$

$$\sum(y) = \sum(f) \leq \sum_{n->force<=0}^{n->0} (X) \tag{3}$$

Algorithm 1: LCADP algorithmic Procedure

```
/* Notation: i(usb) - usb tracker value update, x'(gsm) - gsm
   location coordinate value update, x(gnet) - gsm network
   communication exchange bits,                              */
/* X - Total communication value generated with force or
   without force, f - force resistance value, z - updated
   value for accident spot detection, y - updated value for
   non accidental spot detection                             */
initialization;
while x'(gsm), x(gnet), i(usb), y, z in X do
    find x'(gsm), x(gnet), i(usb) in X such that X=x'(gsm), x(gnet), i(usb);
    if f ≥ X then
        get() -accident take place and z is updated;
    else
        get() -i(usb) to transfer data towards ERS;
    end
    if f ≤ X then
        get() -accident not take place and y is updated;
    else
        get() -i(usb) to transfer data towards ERS;
    end
end
```

5 Existing System Managements Versus Developed Prototype

5.1 Airbag System

An Airbag system is a rubber (i.e.: silicone) made bag filled with air and compressed in a compact form upon the dashboard of a vehicle. The system itself is low cost and has higher accuracy [4]. Airbag only protects against frontal accidental issues. The fundamental demerit is it does not comply with both the left-right side and also with back-end sides of the vehicle. The frontal area is severely damaged if contacted accident.

5.2 *Accident Detection with Mobile*

A mobile-based detection system validates the location of an accidental area in no time [4]. A simple device integration acts as an information-sharing system while sending emergency queries to the nearest hospitals. It sends accurate and faster process queries to the responsible authorities (i.e.: hospitals). The overall system needs smartphone-based operating system support. On the other hand, this system requires specific area-based network coverage.

5.3 *Pervasive Fall Detection*

Against avoidable falls, various independent hardware have been used. The property uses the X, Y, and Z-axis for the accelerometer reading. The reading indicates whether a fall is occupied at an incident [4]. A threshold value is used for the effective detection of pervasive falling from different locations.

5.4 *Onstar Corporation Native System*

The United States-based vehicle declaration system. The system itself is equipped with various options like navigation by turn, theft vehicle identifier, auto accident response, and roadside unit checking. Overall system is confined with assistant shortage and redirects to expensive service inferiority's [4].

5.5 *Machine Learning in Accident Managements*

To detect accidents from traffic surveillance cameras, a machine learning and deep learning model based on the ideas of Clustering and Classification can be employed. A ResNet-50 architecture is used [6] to extract features from the acquired keyframes. K-Means clustering was used to create the Bag of Visual Words after getting the feature vectors of all movies (i.e.: BOVW). Finally, these visual terms are sent into a Support Vector Machine (i.e.: SVM) classifier, which determines whether or not a video contains an accident.

5.6 Vehicular Ad Hoc Network in Intelligent Transportation

VANET concepts can foresee a long-term solution for Intelligent Transportation. The VANET is a wireless network that uses moving vehicles as nodes. Each participating car is turned into a wireless router or node by the VANET network [5]. A vehicle enters or exits the network via a dynamic portable web. If a vehicle is outside the given signal range and leaves the network, additional cars will join and balance the portable web.

5.7 Internet of Things in Traffic Congestion Control

With the ongoing improvement of routing and mobility protocols, research on IoT and VANETs is progressing. As a result, in smart city management, a hybrid IoT-VANET architecture is employed [9] for accident detection. The simulation software Netsim is utilized. Wireless sensors, cable nodes, and a low-pan gateway [12] are used to depict an IoT scenario. The required situation is represented utilizing cars, Adhoc connections, and wireless links once the map has been exported from OSM to Netsim.

5.8 Developed Prototype (LCADP)

using the Vehicular Adhoc Network (VANET) and the Internet of Things, a prototype for automated accident detection is being developed (IoT). With the aid of mechanical and medical sensors installed in the car, the program may identify an accident and the severity of the emergency level. The developed prototype (based on LAN, GPS, GSM, Sensor, and ERS)uses only one particular mechanism for detecting the accident (i.e.: force-based location detection through USB module) [14]. The accidental spot is detected by the GSM tracker and sends information to the GSM network using a text message. The GSM information is carried by the USB repeaters and access points of roadside customized LAN. The developed system has criteria of one-time purchase by the owner itself having no other additional costs. In the event of an emergency, the message is delivered through VANET to a hospital, where our central server can determine the location of the accident and the nearest medical facility. After identifying basic information, it sends a message for an ambulance. The ambulance's client application sends out alert messages to clear the path on the approach to the accident site. A comparison is shown based on some attributes available in the market with our proposed system in Tables 1 and 2.

From Tables 1 to 2 data, we may say that our developed prototype (i.e.: LCADP) supports better in terms of the administrative option, data rates, and security privileges. Based on the development criteria, we can also say that the LCADP prototype

Table 1 Comparison of different attributes among available systems and proposed method

Refs.	Attributes	ML	IoT	VANET	Airbag	PF	Proposed
Zeadally et al. [2], Goh et al. [4], Agrawal et al. [6], Mohammadrezaei et al. [12]	Lower cost	Yes	No	No	Yes	No	Yes
Zeadally et al. [2], Goh et al. [4], Agrawal et al. [6], Hamdi et al. [15]	Larger service	Yes	Yes	Yes	Yes	No	Yes
Zeadally et al. [2], Goh et al. [4], Agrawal et al. [6], Hamdi et al. [15]	Vehicle dependent	Yes	Yes	Yes	Yes	Yes	Yes
Zeadally et al. [2], Goh et al. [4], Agrawal et al. [6], Hamdi et al. [15]	Simple imple-mentation	No	No	No	No	No	Yes
Zeadally et al. [2], Goh et al. [4], Agrawal et al. [6], Mohammadrezaei et al. [12]	Reliable perfor-mance	Yes	Yes	Yes	Yes	No	Yes
Goh et al. [4], Agrawal et al. [6], Mohammadrezaei et al. [12] Hamdi et al. [15]	Human supervi-sion	No	No	No	N/A	N/A	No

Notation: Ref.–Reference, ML–Machine Learning, IoT–Internet of Things, VANET–Vehicular Ad Hoc Network, PF–Pervasive Fall

Table 2 A comparative study among available systems in terms of budget and data rate

Refs.	Used technologies	Budget (USD)	Data rate	Security
Goh et al. [4]	Satellite GPS tracker	350	0.0458/s	Low
Balamurugan and Hariharan [5]	INET simulator	N/A	5.553 kbps	High
Agrawal et al. [6]	CCTV, GPU for ResNet-50	400–1000	N/A	High
Kaur et al. [9]	NetSim simulator, RFID sensor	N/A	0.00796 mbps	Low
Liu et al. [11]	5G	N/A	50/min	High
Proposed (LCADP)	Arduino, XBee, Customized LAN, USB Repeater, GPS Tracker, GSM-900, FSR Sensor	330	10 mbps	High

is useful for low-income countries; although many new technologies are available in these rapidly changing markets (i.e.: evolving technologies).

6 Limitation and Future Works

We have used a dedicated LAN network to communicate with ERS from the GSM network Internet (i.e.: BST). However, in this development, LAN connections need to be checked on time to establish a secured and potential communication among GSM and ERS. Short-distance paths could be traced back through a custom GSM network for avoiding traffic congestion to reach the ERS (i.e.: hospital or police) by time. Vacant ERS notifications can be viewed over smart mobile gateways (i.e.: SMS services). Vacant ERS could serve as a primary helper to the injured or wounded people by delivering proper scopes.

7 Conclusion

Increases in the number of people living in cities and the number of cars on the road contribute to increased traffic congestion and accidents. The absence of immediate medical attention in traffic accidents is the leading cause of mortality. In such a circumstance, automated accident detection can assist in preventing fatalities. This low-cost prototype can able to send an emergency notification towards the ERS in no time. The prototype detects accidental spots by generating an FSR resistance value if gets high pressure across its body plate. The developed model works on low battery power and generates effective alerts within a single VANET area. The configured prototype is useful for low-income-based countries. Local LAN-based ERS communication supports low bandwidth usage and low-cost (i.e.: money) delivery around area-based managements.

References

1. Forum.: http://archive.thedailystar.net/forum/2021/June/road.html
2. Zeadally, S., Hunt, R., Chen, Y.S., Irwin, A., Hassan, A.: Vehicular ad hoc networks (VANETS): status, results, and challenges. Telecommun. Syst. **50**(4), 217–241 (2012)
3. https://cagrvalue.com/asia-pacific-market-witness-highest-growth-intelligent-transport-system-industry-cagr-13-9-2015-2022/
4. Goh, K.N., Jaafar, J., Mustapha, E.E., Goh, E.T.E.: Automatic accident location detection system (AALDS). In: 2014 4th World Congress on Information and Communication Technologies (WICT 2014), pp. 63–69. IEEE (2014)
5. Balamurugan, R., Hariharan, M.M.: VANET based accident alerting system. In: 2021 5th International Conference on Trends in Electronics and Informatics (ICOEI), pp. 661–668. IEEE (2021)

6. Agrawal, A.K., Agarwal, K., Choudhary, J., Bhattacharya, A., Tangudu, S., Makhija, N., Rajitha, B.: Automatic traffic accident detection system using ResNet and SVM. In: 2020 Fifth International Conference on Research in Computational Intelligence and Communication Networks (ICRCICN), pp. 71–76. IEEE (2020)
7. Mohanty, A., Mohanty, S.K., Jena, B.: A real-time fuzzy logic based accident detection system in VANET environment. In: Cognitive Informatics and Soft Computing, pp. 685–696. Springer, Singapore (2021)
8. Liang, J., Chen, J., Zhu, Y., Yu, R.: A novel intrusion detection system for vehicular ad hoc networks (VANETs) based on differences of traffic flow and position. Appl. Soft Comput. **75**, 712–727 (2019)
9. Kaur, M., Malhotra, J., Kaur, P.D.: A VANET-IoT based accident detection and management system for the emergency rescue services in a smart city. In: 2020 8th International Conference on Reliability, Infocom Technologies and Optimization (Trends and Future Directions)(ICRITO), pp. 964–968. IEEE (2020)
10. Dogru, N., Subasi, A.: Traffic accident detection using random forest classifier. In: 2018 15th learning and Technology Conference (L&T), pp. 40–45. IEEE (2018)
11. Liu, J., Wan, J., Jia, D., Zeng, B., Li, D., Hsu, C.H., Chen, H.: High-efficiency urban traffic management in context-aware computing and 5G communication. IEEE Commun. Mag. **55**(1), 34–40 (2017)
12. Mohammadrezaei, M., Fard, H. S., Niaky, R.P., Dizaj, B.S.T.: IoT-Based Vehicular Accident Detection Systems (2020)
13. Barba, C.T., Mateos, M.A., Soto, P.R., Mezher, A.M., Igartua, M.A.: Smart city for VANETs using warning messages, traffic statistics and intelligent traffic lights. In: 2012 IEEE Intelligent Vehicles Symposium, pp. 902–907. IEEE (2012)
14. Khaliq, K.A., Qayyum, A., Pannek, J.: Prototype of automatic accident detection and management in vehicular environment using VANET and IoT. In: 2017 11th International Conference on Software, Knowledge, Information Management and Applications (SKIMA), pp. 1–7. IEEE (2017)
15. Hamdi, M.M., Audah, L., Rashid, S.A., Mohammed, A.H., Alani, S., Mustafa, A.S.: A review of applications, characteristics and challenges in vehicular ad hoc networks (VANETs). In: 2020 International Congress on Human-Computer Interaction, Optimization and Robotic Applications (HORA), pp. 1–7. IEEE (2020)
16. Mahi, M.J., Rahad, K.A., Biswas, M., Islam, R., Chowdhury, Z.I.: An accident detection system for a single VANET at low cost module. IEEE TechSym **4**, 44–45
17. Rahman, M.T., Mahi, M.J.N., Biswas, M., Kaiser, M.S., Al Mamun, S.: Performance evaluation of a portable PABX system through developing new bandwidth optimization technique. In: 2015 International Conference on Electrical Engineering and Information Communication Technology (ICEEICT), pp. 1–5. IEEE (2015)

Drop-Shaped Fractal Patch Antenna for THz Applications

Anita Garhwal, Muhammad Arif Jalil, Mufti Mahmud, M. Shamim Kaiser, Kanad Ray, Preecha Yupapin, P. Prabpal, Syed Zuhaib Haider Rizvi, Kavikumar Jacob, Anirban Bandyopadhyay, and Jalil Ali

Abstract In this paper, a drop-shaped fractal patch antenna is designed and simulated using Polyamide substrate. The designed antenna is simulated for 4.35–4.42 THz. The designed antenna resonates at 4.4 THz frequency. The maximum gain of 9.34 dBi is achieved. The designed antenna has applications in THz for communication, sensing, and 4.2, 4.3, and 4.4 THz frequency is used for quantum cascade laser. The proposed antenna is designed using CST software.

Keywords Drop shaped · Fractal · Patch · THz · Quantum cascade laser

A. Garhwal (✉) · J. Ali
Asia Metropolitan University, 6, Jalan Lembah, Bandar Baru Seri Alam, 81750 Masai, Johor, Malaysia
e-mail: anitagarhwal@amu.edu.my

M. A. Jalil
Departemnt of Physics, Faculty of Science, Unversiti Teknologi Malaysia, 81310 Skudai, Johor, Malaysia

M. Mahmud
Nottingham Trent University, Clifton Lane, Nottingham NG11, UK

M. S. Kaiser
Institute of Information Technology, Jahangirnagar University, Savar, Dhaka 1342, Bangladesh

K. Ray
Amity School of Applied Sciences, Amity University Rajasthan, Jaipur, India

P. Yupapin (✉) · P. Prabpal
Department of Electrical Technology, Faculty of Industrial Technology, Institute of Vocational Education Northeastern 2, Sakon Nakhon, Thailand
e-mail: preecha@techsakon.ac.th

S. Z. H. Rizvi · K. Jacob
Faculty of Applied Sciences and Technology, Universiti Tun Hussein Onn Malaysia, UTHM Kampus Pagoh, Jln Panchor, KM1 Muar, Johor, Malaysia

A. Bandyopadhyay
National Institute for Material Science, Tsukuba, Japan

M. S. Kaiser et al. (eds.), *Proceedings of the Third International Conference on Trends in Computational and Cognitive Engineering*, Lecture Notes in Networks and Systems 348, https://doi.org/10.1007/978-981-16-7597-3_33

1 Introduction

The optical antennas are in high demand which provides a higher data rate for a wireless network [1]. Fractal geometry is a new era of research. It helps in mathematical science, physics, biology, natural science, and computing. With the help of fractal geometry, we can easily understand the natural complex shapes. The important property of fractal geometry is self-similarity. "Fractal geometry" was proposed in 1976 by Mandelbrot. The meaning of fractal is "to break" into smaller components. Examples of fractals in nature are sea coastal lines, feathers, trees, etc. [2, 3]. The fractal shape also has an advantage as the antenna size is reduced. In fractal, a similar shape is repeated but at a different scale. Different types of patch antennas are designed for GHz applications as circle, hexagon, and octagon to a decagon. They all follow symmetry from circle to decagon [4–7]. A circular-shaped fractal patch antenna for THz application is also presented for 0.1–10 THz [8].

The THz frequency is in 10^{12} Hz range. It has various advantages as small wavelength, wider bandwidth, and higher speed in T-bits. A review on THz waves for communication and sensing is presented by Fitch et al. [9]. Where the development of sources, modulators, and detectors for THz communication is explained in detail. The THz radiation source at solid-state is first described in 2002 by Kohler et al. [10], which introduced the THz quantum cascade laser. The quantum cascade laser works around 4.4 THz frequency [11, 12]. A semiconductor-heterostructure laser for THz frequency is also presented which emits at a single mode frequency of 4.4 THz. The heterostructure was designed using GaAs/AlGa As, which shown good results for continuous wave and high-temperature operation in THz photonics [13]. Quantum cascade laser emitting at 4.2 THz is presented where repetition frequency locking is reported [14].

In this paper, a drop-shaped fractal patch antenna is proposed. The fractal concept is applied to reduce the size of the antenna. Along with this wider bandwidth and good gain also can be achieved by a fractal antenna.

2 Antenna Design

The basic shape of the antenna resembles to drop. The structure is designed by combining a circle and a triangle. The substrate material used is Polyamide of thickness 81.29 μm which has a relative permittivity of 4.3 and loss tangent of 0.004 [14, 15]. The patch and ground material is copper of thickness 7 μm and conductivity (σ) of 5.8×10^7 S/m.

First iteration-0 is designed which is a drop shape. Iteration-1 is achieved by extracting the same drop shape from iteration-0. Similarly, iteration-2 and -3 are designed. The partial ground is used to achieve better results. The designed antenna is simulated using CST (Computer Simulation Technology) software. Microstrip feed method is used. Iteration-3 is shown in Fig. 1.

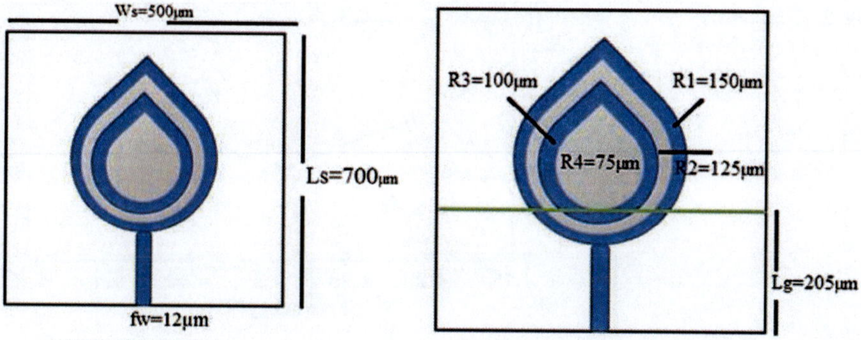

(a) Designed antenna
(b) antenna with partial ground

Fig. 1 Designed antenna and partial ground. The used antenna dimensions are given in Table 1

Table 1 Parameters used in antenna design

Sr. no	Parameter name	Dimension (μm)
1	L_s (Substrate length)	700
2	W_s (Substrate width)	500
3	L_g (Ground length)	200
4	f_w (Feed width)	12

Where L_s and W_s are substrate length and width while L_g and f_w are the length of ground and microstrip feed width, respectively.

3 Results and Discussion

The final antenna is the iteration-3 with the partial ground. The simulated reflection coefficient is shown in Fig. 2, where antenna resonates in the frequency band of 4.15–4.45 THz. The designed antenna has three resonating frequency bands at 4.21, 4.30, and 4.4 THz. The reflection coefficient is −34 dB, −32 dB, and −60 dB at 4.21 THz, 4.30 THz, and 4.4 THz frequencies, respectively. Antenna bandwidth is calculated the S11 below −20 dB. The band1 is in the range of 4.16T Hz–4.25 THz, band2 is

Fig. 2 Simulated S11 of iteration-3 antenna

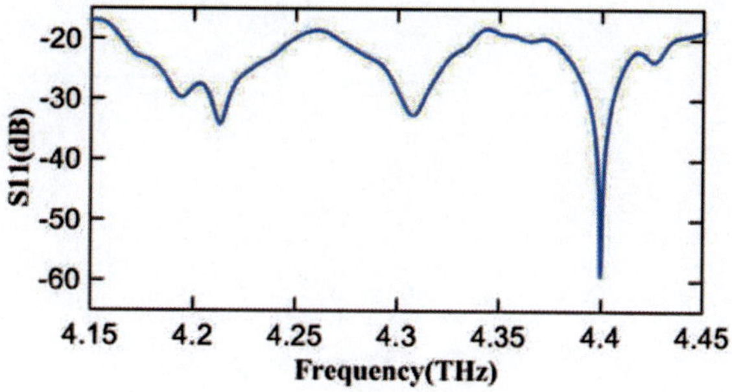

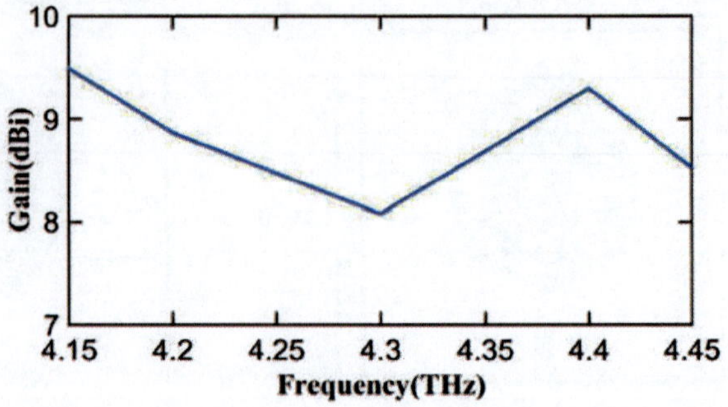

Fig. 3 Gain versus frequency plot. Maximum achieved gain is 9.34dBi at 4.4THz frequency

4.27 THz–4.33 THz, and band3 is 4.36–4.43 THz. The calculated bandwidths are 90 GHz, 60 GHz, and 70 GHz, respectively at band1, 2, and 3.

The antenna gain with frequency plot is depicted in Fig. 3. Here in the entire frequency range, the antenna gain is above 8dBi. The gain values at resonating frequencies are 8.87, 8.08, and 9.28 dBi. The gain is decreasing at 4.3 THz and then again increasing at 4.4 THz.

The antenna VSWR (voltage standing wave ratio) is less than <1.5 dB in the entire frequency range as shown in Fig. 4. The ideal value of VSWR is less than 2 dB.

In addition to reflection coefficient, VSWR, gain plots, radiation pattern, and radiation efficiency were also plotted. The radiation efficiency is above 0.87 in the entire frequency band as shown in Fig. 5. Both efficiencies are decreasing after 4.4THz.

The radiation pattern plot is shown in Fig. 6a, b for E and H fields, respectively. In Fig. 6a, the direction of the main lobs is at (i) 4°, (ii) 0°, and (iii) 153.0°. In Fig. 6b, the direction of main lobs are at (i) 57°, (ii) 40°, and, (iii) 41°.

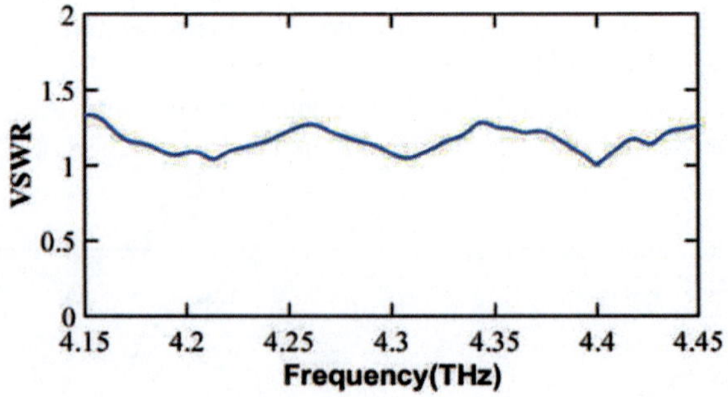

Fig. 4 VSWR of designed antenna

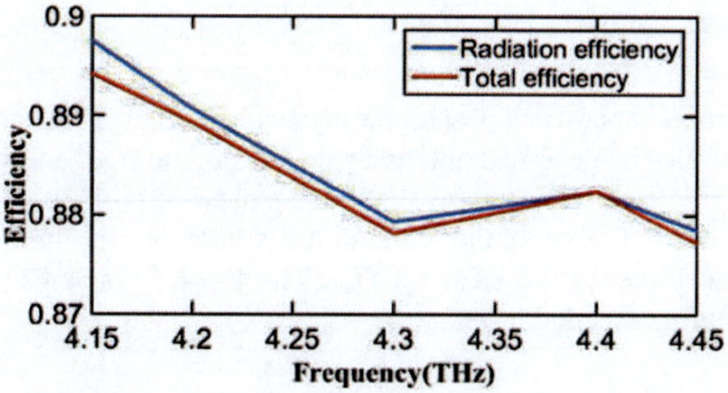

Fig. 5 Radiation and total efficiency of the proposed antenna

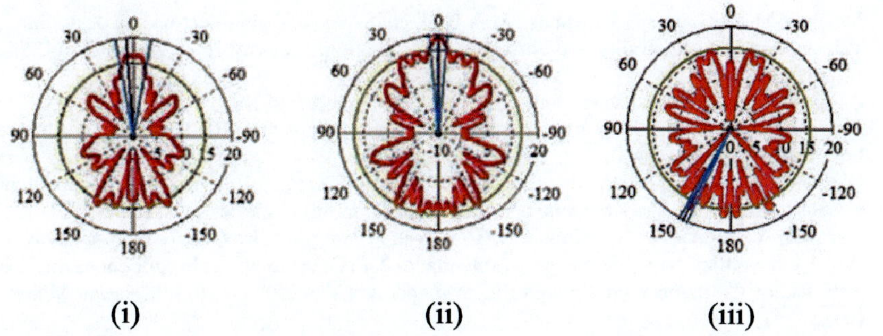

(a) E field Radiation pattern at (i) 4.2THz (ii) 4.3THz and (iii) 4.4THz

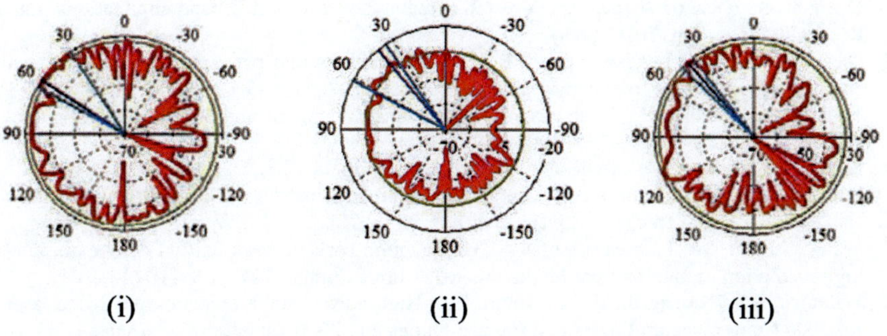

(b) H field radiation pattern at (i) 4.2THz (ii) 4.3THz and (iii) 4.4THz

Fig. 6 Radiation patterns of the designed antenna

4 Conclusion

The designed drop-shaped fractal antenna is of substrate size 700 and 500 μm of polyamide material. The designed antenna resonates in the frequency band of 4.15–4.45 THz. The simulated results show that good bandwidth of 620 GHz and gain of 9.2 dBi are achieved. The designed antenna has application in the THz band. The achieved antenna efficiency is 0.88 at 4.4 THz. The 4.2, 4.3, and 4.4 THz frequencies are used in quantum cascaded lasers.

References

1. Jamshed, M.A., Nauman, A., Abbasi, M.A.B., Kim, S.W.: Antenna selection and designing for THz applications: suitability and performance evaluation: a survey. IEEE Access **8**, 113246–113261 (2020)
2. Barnsley, M.F.: Fractals Everywhere, 2nd edn. Academic Press, San Diego (1993)
3. Jiang, B.: New paradigm in mapping: a critique on cartography and GIS. Cartogr.: Int. J. Geogr. Inf. Geovisualization **54**(3), 193–205 (2018)
4. Garhwal, A., Ahmad, M.R., Ahmad, B.H., Rawat, S., Singh, P., Ray, K.: A Bandyopadhyay, circular and elliptical shaped fractal patch antennas for multiple applications. IJEAT 8(5) (2019)
5. Garhwal, A., Ahmad, M.R., Ahmad, B.H., Rawat, S., Singh, P., Ray, K., Bandyopadhyay, A.: Wideband circular shaped fractal patch antenna for 2.45 GHz biomedical applications. In: IEEE Asia-Pacific Conference on Applied Electromagnetic (APACE) , pp. 1–5. Melacca, Malaysia (2019)
6. Garhwal, A., Ahmad, M.R., Ahmad, B.H., Rawat, S., Singh, P., Ray, K., Bandyopadhyay, A.: Mechanically reconfigurable hexagonal fractal patch antenna for ambient computing, Int. J. Innov. Technol. Explor. Eng. **8**(6) (2019)
7. Garhwal, A., Ahmad, M.R., Ahmad, B.H., Rawat, S., Singh, P., Ray, K., Bandyopadhyay, A.: Octagon and decagon shaped fractal patch antennas for S, C and X band applications. Int. J. Recent Technol. Eng. **7**(6) (2019)
8. Das, S., Sekhar, D.M., Chaudhuri, R.B.: Fracal loaded circular patch antenna for super wide band operation in THz frequency region, part 2. Optik **226** (2021)
9. Fitch, M.J., Osiander, R.: Terahertz Waves for Communications and Sensing. Johns Hopkins Apl Tech. Dig. **25**(4), 344–355 (2004)
10. Köhler, R., Tredicucci, A., Beltram, F., Beere, H.E., Linfield, E.H., Davies, A.G., Ritchie, D.A., Iotti, R.C., Rossi, F.: Terahertz semiconductor-heterostructure laser. Nature **417**, 156 (2002)
11. Jiang, T., Shen, C., Zhan, Z., Zou, R., Li, J., Fan, L., Xiao, T., Li, W., Deng, Q.H., Peng, L., Wang, X., Wu, W.: Fabrication of 4.4 THz quantum cascade laser and its demonstration in high-resolution digital holographic imaging. J. Alloys Compd. **771**, 106–110 (2019)
12. Williams, B.S., Kumar, S., Hu, Q., Reno, J.L.: High-power terahertz quantum cascade lasers. In: 2006 Conference on Lasers and Electro-Optics and 2006 Quantum Electronics and Laser Science Conference, pp. 1–2 (2006)
13. Köhler, R., Tredicucci, A., Beltram, F., Beere, H.E., Linfield, E.H., Davies, A.G., Ritchie, D.A., Iotti, R.C., Rossi, F.: Terahertz semiconductor-heterostructure laser, Macmillan Magazines Ltd, Lett. Nat. **417**, 156–159 9 (2002)
14. Guan, W., Li, Z., Zhou, K., Wan, W., Liao, X., Zhao, Y., Yang, S., Cao, J.C., Li, H.: Repetition frequency locking of a terahertz quantum cascade laser emitting at 4.2 THz. Terahertz Sci. Technol. **13**(1), 32–40 (2020)
15. Singhal, S.: Elliptical Ring Terahertz Fractal Antenna Optik (2019)
16. Keshwala, U., Ray, K., Rawat, S.: Ultra-wideband mushroom shaped half-sinusoidal antenna for THz applications, Optik 228 (2021)

Designing of Triple-Band, Quad-Band, and Super Wideband Microstrip Antennas for THz Application

S. K. Vijay, M. A. Jalil, B. H. Ahmad, Kanad Ray, Preecha Yupapin, Syed Zuhaib Haider Rizvi, K. K. Jacob, A. Bandyopadhyay, Kuryati Kipli, and Jalil Ali

Abstract In this paper microstrip ring antenna is designed for triple, quad, and super wideband terahertz application. Three antenna configurations are discussed here with application in upcoming 6G technology, Health monitoring systems, wireless sensor network, and the Internet of Things (IoT). A square ring resonator antenna designed on $710 \times 910 \times 10\ \mu m^3$ Rogers R03003 substrate. The first antenna is working on 374, 444, and 510 GHz with maximum bandwidth of 40 GHz. Further ground structure of ring antenna defected for super wide bandwidth of 192 GHz. Third antennas

S. K. Vijay
Department of Electronics & Communication Engineering, Amity University, Jaipur, Rajasthan, India

M. A. Jalil
Departemnt of Physics, Faculty of Science, Unversiti Teknologi Malaysia, 81310 Skudai, Johor, Malaysia

B. H. Ahmad
FECE, University Teknikal Malaysia Melaka, Melaka, Malaysia

K. Ray (✉)
Amity School of Applied Sciences, Amity University Rajasthan, Jaipur, India
e-mail: kray@jpr.amity.edu

P. Yupapin
Institute of Vocational Education Northeastern Region 2, Sakon Nakorn, Thailand

S. Z. H. Rizvi · K. K. Jacob
Faculty of Applied Sciences and Technology, Universiti Tun Hussein Onn Malaysia, UTHM Kampus Pagoh, Jln Panchor, 184600 Muar, Johor, KM, Malaysia

A. Bandyopadhyay
National Institute for Material Science, Tsukuba, Japan

K. Kipli
Department of Electrical and Electronics, Universiti, 94300 Kota Samarahan, Malaysia Sarawak, Malaysia

J. Ali
Centre for Research, Asia Metropolitan University, 6, Jalan Lembah, Bandar Baru Seri Alam, 81750 Masai, Johor, Malaysia

M. S. Kaiser et al. (eds.), *Proceedings of the Third International Conference on Trends in Computational and Cognitive Engineering*, Lecture Notes in Networks and Systems 348, https://doi.org/10.1007/978-981-16-7597-3_34

show quad-band performance with fractal rings. The result and performance indicate that the recommended antenna will be well-suited with compact future wireless devices. An soft High Frequency Structure Simulator (HFSS) is used for all simulation work.

Keywords Ring resonator · Terahertz antenna · Super wideband antenna · Triple-band antenna · Quad-band antenna · Fractal antenna

1 Introduction

Future wireless technologies require low profile, compact size and broadband planar antennas. Many researchers are attracted toward higher frequency due to the needs for higher data rates in current wireless devices. One of them is unlicensed Terahertz band ranging from 0.1 to 10 THz. The terahertz devices developing with wide bandwidth need miniaturization [1]. The need of higher band width can also be fulfilled with multiband antennas [2]. Several kinds of research on optical antenna application in triband [1], multiband [2], super wideband [3, 4], dual band [5], single band [6], wideband [7, 8] as well as ultra-wideband [9] frequency.

In recent years, terahertz antennas with various forms are realized. Few of them mentioned here like slotted bowtie antenna [1], slotted patch [2], Hexagonal fractal [3], T-shape parasitic loaded patch [5], Metasurface array [8], meander line [4], ring antenna [10] honeycomb shaped [11], sun flower shaped [12, 13], and elliptical fractal antenna [14].

Optical antenna requires size miniaturizations, compactness, and wide operating bandwidth with minimum losses. To address these issues a triple ring antenna on Rogers RO3003 substrate is proposed here. The overall antenna dimension is 710 $\times$ 910 μm^2. Three different antennas with triple, quad, and wide bandwidth are discussed with results and analysis. Wide bandwidth is achieved using defected ground structure. Finally ring antenna is modified by the insertion of square rings to make fractal geometry for quad-band frequency. Ansys electromagnetic software HFSSv15 is used for all the simulation work.

2 Triple-Band Ring Antenna Design and Analysis

Triple-band antenna is designed on Rogers RO 3003 substrate with dielectric constant of 3 and loss tangent of 0.0013. The lesser value of loss tangent provides better gain and efficiency with low power loss.

The size of substrate is chosen as 710 $\mu m \times 910$ μm with 10 μm of substrate thickness. 80 μm wide microstrip feed line is designed for 50 Ω impedance matching. To improve impedance matching 20 μm wide quarter wave transformer has been designed between feed line and patch. The 710 $\mu m \times 910$ $\mu m \times 1$ μm conductive ground plane of copper is designed on the back side of the substrate. Three square rings are inserted for triple-band antenna design. Three square rings with 20 μm width are used for designing triple-band antenna. Rings are placed at the gap of 20 μm. Copper is used here for designing of patch and feed line. The triple-band ring antenna is shown in Fig. 1. All the dimensions are listed in Table 1.

The antenna resonant characteristics can be observed from Figs. 2 to 3. The scattering characteristics show that antenna displays triple resonance on 374, 444, and 510 GHz. The first band covers 354–395 GHz frequency with 41 GHz bandwidth. Figure 2 shows that antenna shows good resonant characteristics 374, 444, and 510 GHz with a reflection coefficient of −25, −16.5, and −18 dB. The second and third band covers 433–458 and 501–519 GHz frequency. The recommended antenna has good VSWR characteristics. The value of VSWR at 374, 444, and 510 GHz is 1.1, 1.35, and 1.28. The low VSWR confirms good impedance matching.

The surface current effect on the triple-band ring antenna geometry has been presented in Fig. 4. The distribution of vector current is uniform and is concentrated on the rings equally at resonating frequencies of 374, 444, and 510 GHz. The radiation

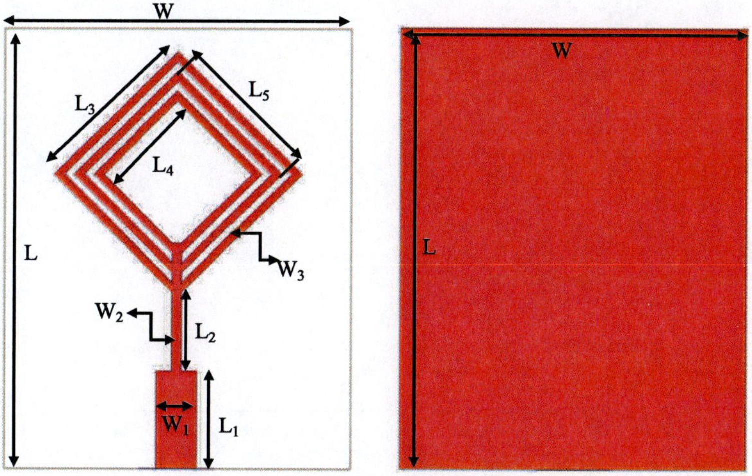

(a) Front View (b) Back View

Fig. 1 Triple-band ring antenna design

Table 1 Triple-band ring antenna optimized dimensions

Unit	Dimensions (μm)	Unit	Dimensions (μm)
L	910	W	710
L_1	200	W_1	80
L_2	159.5	W_2	20
L_3	353.5	W_3	20
L_4	297	L_5	212

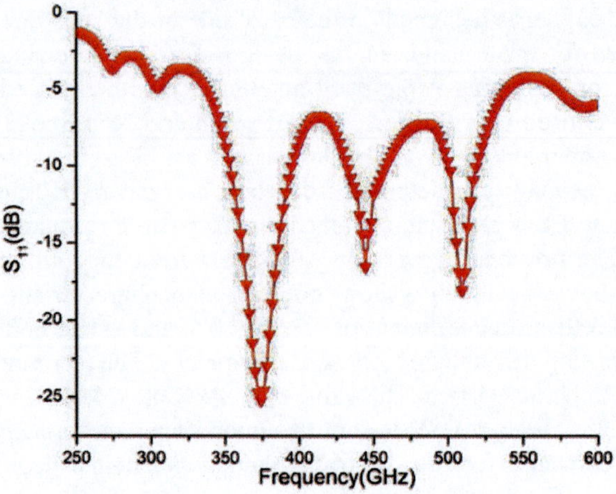

Fig. 2 S_{11}(dB) versus frequency for triple-band ring antenna

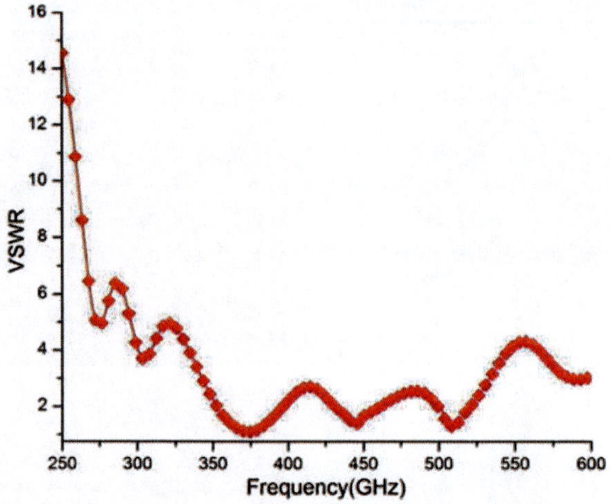

Fig. 3 Triple-band ring antenna VSWR versus frequency characteristics

characteristic of triple-band antenna has been presented in Fig. 5. The radiation pattern shows figure eight for the E-plane and omnidirectional for H-plane.

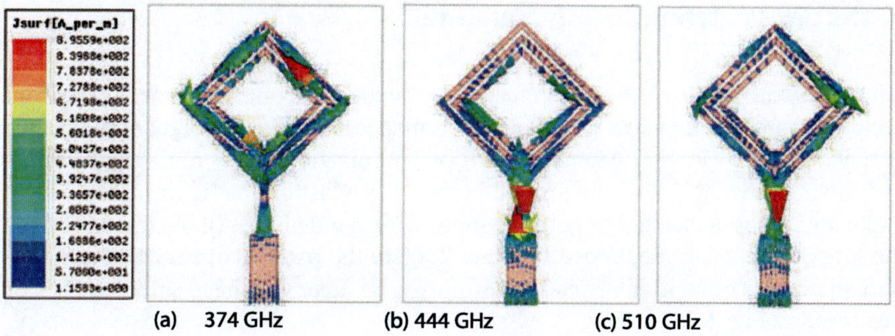

Fig. 4 Current distribution at various frequency on triple-band ring antenna

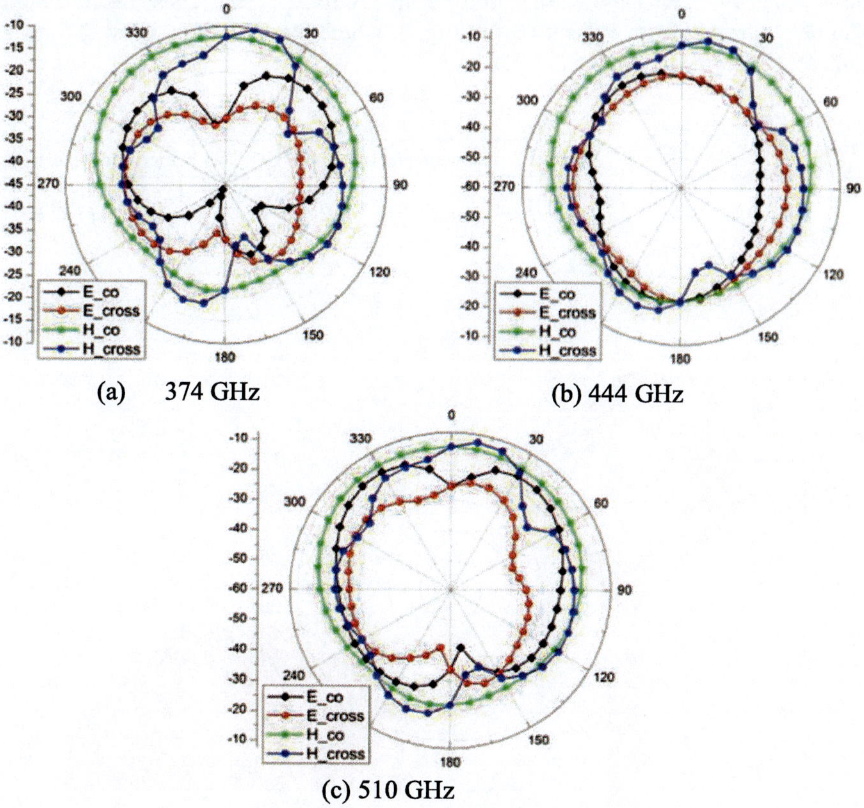

Fig. 5 Radiation characteristics of triple-band ring antenna

3 Super Wideband Ring Antenna

After analyzing the triple-band antenna, the ground of antenna mentioned in previous section is modified for super wideband performance. Initially the height of the ground is reduced to λ/4 length. After that various optimization stages are done to calculate the effective height of the ground plane. In the last ground is truncated on 45° angle and a slot is inserted to produce super wide bandwidth. All the dimensions of the ground plane are mentioned in Table 2. Here the patch geometry and substrate remain same as previous antenna. The prototype of super wideband antenna has been displayed in Fig. 6.

The reflection characteristics in terms of S_{11}(dB) have been displayed in Fig. 7. The recommended antenna shows a very high bandwidth. The graph shows that this antenna has 192 GHz bandwidth starting from 320 to 512 GHz. These results could also be confirmed from VSWR plot in Fig. 8, where the VSWR is below 2 for the complete band from 320 to 512 GHz.

Table 2 Super wideband antenna dimensions

Unit	Dimensions (μm)	Unit	Dimensions (μm)
L_6	110	W_5	35
L_7	320	W_6	105
L_8	56		

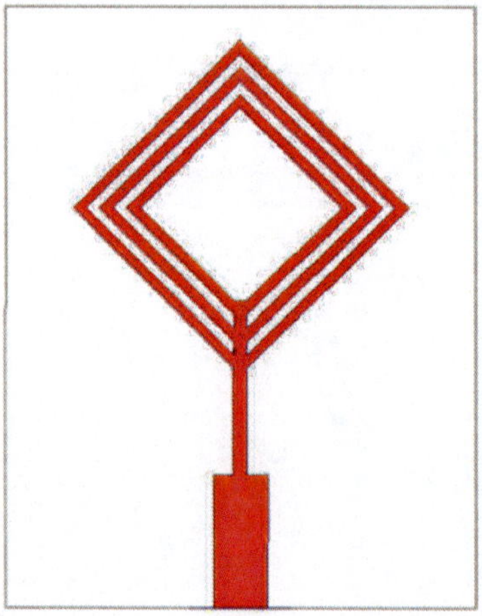
(a) Front view

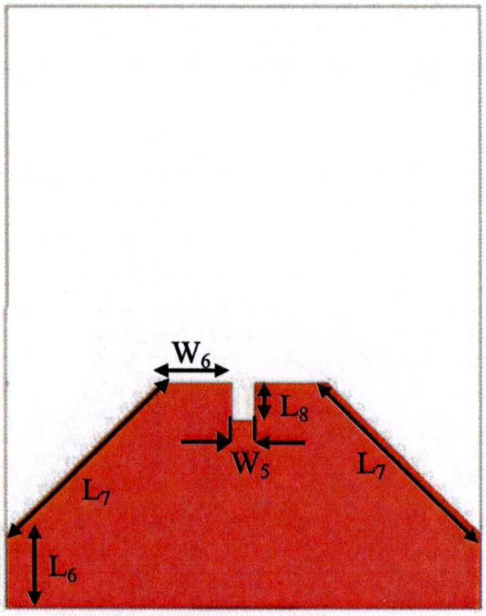

(b) Back view

Fig. 6 Super wideband ring antenna design

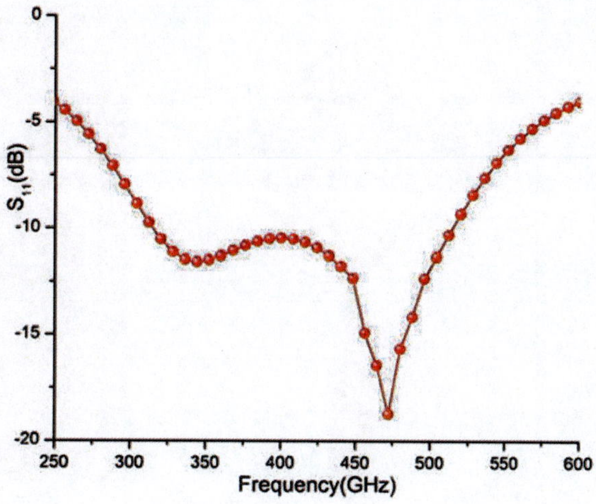

Fig. 7 S_{11}(dB) versus frequency of super wideband antenna

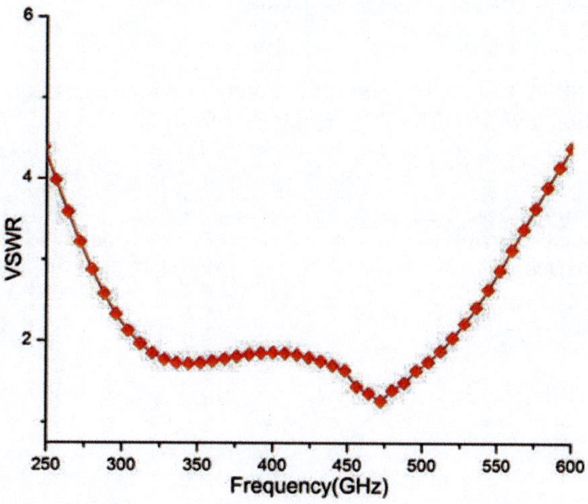

Fig. 8 VSWR characteristics of super wideband antenna

4 Quad-Band Ring Antenna

The antenna mentioned in Sect. 2 produces only triple-band characteristics. To achieve the quad-band performance various iterations are loaded to create fractal ring geometry. Initially slots are carved on the corners of ring, then multiple split ring resonators are loaded on rings. The loading of ring resonators increases the electrical length of the antenna and antenna. Due to an increment in equivalent electrical length the resonant frequency is shifted toward lower band. The quad-band antenna arrangement is shown in Fig. 9. The reflection coefficient of the proposed antenna is

(b) Front view (b) Back view

Fig. 9 Quad-band ring antenna design

plotted in Fig. 10. The reflection coefficient shows quad resonant band on 72, 347, 392, and 510 GHz. The result shows good quad-band pass characteristics.

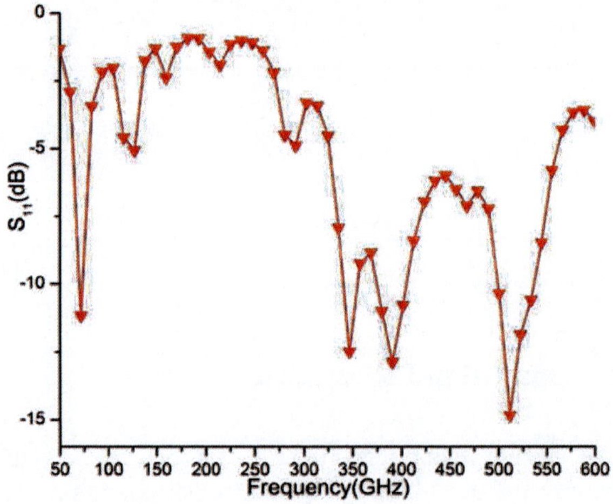

Fig. 10 Quad-band ring antenna S_{11}(dB) versus frequency

5 Conclusion

The proposed antenna covers triple-band, quad-band, and wideband resonance characteristics. Initially, a square ring microstrip is designed for optical application, and then the multiple ring-shaped geometry is inserted to make it triple-band antenna. After that bandwidth of the antenna is enhanced using defected ground structure. Finally, SRR structures are embedded in triple-band antenna to create fractal shape and provide quad-band resonance. The simple structure of recommended antenna with $710 \times 910\ \mu m^2$ size makes it a compact structure. The result and analysis indicate the proposed antenna operation in next generation wireless devices.

References

1. Bansal, G., Marwaha, A., Singh, A., Bala, R., Marwaha, S.: A triband slotted bow-tie wideband THz antenna design using graphene for wireless applications. Optik **185**, 1163–1171 (2019)
2. Shalini, M., Madhan, M.G.: Performance predictions of slotted graphene patch antenna for multi-band operation in terahertz regime. Optik—Int. J. Light Electr. Opt. **204**, 164223 (2020)
3. Singhal, S., Budania, J.: Hexagonal fractal antenna for super wideband terahertz applications. Optik 163615 (2019). https://doi.org/10.1016/j.ijleo.2019.163615
4. Ullah, S., Ruan, C., Sadiq, M.S., Haq, T.U., Fahad, A.K., He, W.: Super wide band, defected ground structure (DGS), and stepped meander line antenna for WLAN/ISM/WiMAX/UWB and other wireless communication applications. Sensors (Basel, Switzerland) **20**(6), 1735 (2020). https://doi.org/10.3390/s20061735
5. Kazemi, F.: Dual band compact fractal THz antenna based on CRLH-TL and graphene loads. Optik 164369 (2020).https://doi.org/10.1016/j.ijleo.2020.164369
6. Rubani, Q., Gupta, S.H., Pani, S., Kumar, A.: Design and analysis of a terahertz antenna for wireless body area networks. Optik (2018). https://doi.org/10.1016/j.ijleo.2018.10.202
7. Luk, K.M., Zhou, S.F., Li, Y.J., et al.: A microfabricated low-profile wideband antenna array for terahertz communications. Sci Rep **7**, 1268 (2017). https://doi.org/10.1038/s41598-017-01276-4
8. Hussain, N., Park, I.: Design of a wide-gain-bandwidth metasurface antenna at terahertz frequency. AIP Adv **7**(5), 055313 (2017). https://doi.org/10.1063/1.4984274
9. Nissanov, U., Singh, G.: MmWave/THz reconfigurable ultra-wideband (UWB) microstrip antenna. Progress Electromag. Res. C **111**, 207–224 (2021)
10. Malallah, R., Shaaban, R.M., Al-Tumah, W.A.G.: AEU—Int. J. Electron. Commun. **127**, 153473 (2020)
11. Keshwala, U., Rawat, S., Ray, K.: Honeycomb shaped fractal antenna with defected ground structure for UWB applications. In: 2019 6th International Conference on Signal Processing and Integrated Networks (SPIN), pp. 341–345 (2019). https://doi.org/10.1109/SPIN.2019.8711752
12. Singh, P., Ray, K., Rawat, S.: Analysis of sun flower shaped monopole antenna. Wirel. Person. Commun. **104**(3), 881–894
13. Garhwal, A., Ahmad, M.R., Ahmad, B.H., Rawat, S., Singh, P., Ray, K.: Circular and elliptical shaped fractal patch antennas for multiple applications. Int. J. Eng. Adv. Technol. (IJEAT) **8**(5) (2019)
14. Singh, P., Ray, K., Rawat, S.: Nature inspired sunflower shaped microstrip antenna for wideband performance. Int. J. Comput. Inf. Syst. Ind. Manag. Appl. **8**, 364–371 (2016)

A Smart Multi-user Wireless Nurse Calling System and E-Notice Board for Health Care Management

Liton Chandra Paul, Sayed Shifat Ahmed, and Kallol Krishna Karmakar

Abstract To provide smart health care service in a hospital, a prototype of a fingerprint recognition-based multi-user wireless nurse calling system has been designed and presented in this paper. The proposed system also has an attractive feature to circulate any common information or emergency message to the patients. This message can be sent by any authorized doctor or hospital management authority from any place within the LAN network. The use of fingerprint makes the system more robust and secure, i.e. there is no chance to access the system by any unauthorized person. The health care management system can work up to 700 m circular distance. The Arduino, ATmega328P, nRF24L01 + PA + LNA, fingerprint scanner, LED display, speaker, etc. hardware components have been used to build the prototype, and these hardware accessories are commercially available anywhere and comparatively low cost. Instead of a conventional wired setup, a wireless health care management system ensures easy maintenance and simpler and fast deployment. Flexibility and versatility are guaranteed by wireless tuning. Since the proposed system prototype shows satisfactory performance during practical trials, the system can be considered as a suitable candidate for the health care management system of a hospital or any health care center.

Keywords Fingerprint recognition · Calling system · Electronic notice board · Microcontroller · WiFi

L. C. Paul (✉) · S. S. Ahmed
Department of Electrical, Electronic and Communication Engineering, Pabna University of Science and Technology, Pabna, Bangladesh

K. K. Karmakar
Research Fellow, School of Information and Physical Sciences, University of Newcastle, Callaghan, Australia
e-mail: kallolkrishna.karmakar@newcastle.edu.au

M. S. Kaiser et al. (eds.), *Proceedings of the Third International Conference on Trends in Computational and Cognitive Engineering*, Lecture Notes in Networks and Systems 348, https://doi.org/10.1007/978-981-16-7597-3_35

1 Introduction

In the present era of the technological revolution, folks are searching for such a smart electronic system that can provide assurance of easy and smooth services with proper authenticity [1, 2]. In this article, we are talking about a fingerprint-based multi-terminal wireless nurse calling system with an electronic notice board which is designed to assist caregivers. A nurse calling service is simply a technical solution for serving the needs of individuals during treatment in a hospital or medical care center. A hospital authority chooses the appropriate system according to their needs. Wireless nurse calling systems are scalable from a few beds to so many [3]. Conventional nurse calling systems use push button switches placed at the patient's room. When it is pressed by any patient, it alerts the central nurse control room [4]. However, in the conventional system, anyone can use it and may deliberately create chaos, leaving a security threat, and improper use may result in haphazard use at the nurse station [5]. To mitigate such problems, there is a need to develop an authenticity verification feature in the system where only certain authorized users can access it. Biometrics are behavioral and physical human properties to be used to distinguish a person to grant access to a system [6]. A fingerprint scanner is a type of electronic component that identifies and authenticates the fingerprints of an individual to grant or deny access to a system [7]. Some monitoring systems are built on computer networks where there is a necessity for an Internet connection. However, our proposed calling system does not depend on Internet access. For an Internet access dependent system, if the Internet connection is interrupted, the whole system would be useless whereas our suggested model can still provide nurse calling service smoothly. In most traditional wired systems, the calling button is fixed at a position, so the patient must reach to dial the nurse; this challenge can be overcome by deploying the proposed wireless system. Each calling key in this design is compact and battery driven. As a result, if the electricity goes out, the device will operate on battery power. Whenever a person interacts with the system, the central nurse station is alerted by playing an audible sound and displaying the individual's name (patient or duty doctor), room number. There is also a provision for a real-time half-duplex voice communication system and this feature can access duty doctors only and not the patient. There is another part, the electronic notice board, which is integrated with the calling system. This electronic notice board can be placed anywhere within the hospital, which is useful to display common messages or emergency notifications. However, changing the message every day via analog procedure is a very tedious task, instead of using NodeMCU, the authority can change the message using the web browser remotely [8, 9]. The E-notice board can be used in the patient room as a medicine reminder or waiting room/corridor of the hospital to get the attention of visitors. The proposed system, however, contributes to reduce the transit time for doctors and nurses by providing the precise details of the patient who pressed the button that appears in the base station with an LED display module. During call setup, the same detailed information is also declared in the speaker. These features emphasize the key factors in emergency situations.

2 Concept Making and System Modeling

In today's rapidly developing technological environment, many smart systems are developed depending on the user's needs [10, 11]. Some are home security, some are IoT devices, some are plant monitoring, etc. In this work, we are intended to develop a fingerprint-based multi-terminal wireless nurse calling system and e-notice board for the health care management of a hospital. The concept of the health care management system is depicted in Fig. 1 as a block diagram. This system encourages patients and doctors to use them in a simple and convenient manner, seeking emergency help as soon as possible. The use of a fingerprint modulealso ensures security issues because biometrics are behavioral and physical human properties to be used to distinguish a person to grant access to systems. Instead of a wired network, the proposed system relies on wireless connectivity. Thus, the system's cost is minimized and the complexities of cables are eliminated. The terminal parts of the wireless calling system are placed in every room in the hospital, which are used by both authorized duty doctors and the patient(s) of that room. First, fingerprints of duty doctors and admitted patients who wish to call for assistance are collected by the hospital management authority during the admission of a patient and stored in the system's database. After admission, when a patient calls for assistance the module checks the input fingerprint within the database stored in the system and also categorizes the person between any one of two categories indicated as "Doctors" and "Patient". So when a doctor needs his assistant or a patient needs some help then just scan his/her finger then instantly the name of the doctor/patient and corresponding room number will be displayed to the control room and parallelly a call ringtone will be rung on the speaker. Therefore, the nurses from the control room can easily be notified about the location of the calling person. Moreover, sometimes a doctor

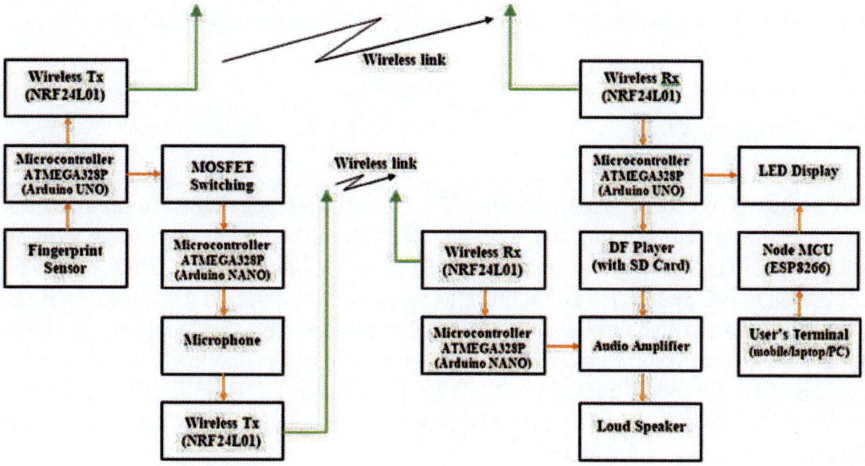

Fig. 1 Block diagram of the proposed calling system and e-notice board

may need to talk with central nurse control directly then it can be possible by activating a wireless transceiver module on this device, i.e. this device has a feature of two-way communication between doctor and assistant. The electronic notice board is another interesting feature of the proposed system. The E-notice board can control any authorized doctor/nurse from anywhere within the LAN network.

In the proposed system, nRF24L01 wireless transceiver module has been used which is controlled by an ATmega328P Arduino microcontroller. The fingerprint scanner collects the fingerprint of the calling person, verifies it, and communicates with the microcontroller via the universal asynchronous receiver transmitter module (nRF24L01). After receiving an authentication signal from the fingerprint module, the microcontroller determines whether or not to send a voice transfer as well as a command. The transmitting module (via nRF24L01) transmits that command referring to a specific fingerprint through the processing microcontroller (Arduino). The receiver part will receive the corresponding fingerprint signal via the receiving unit of nRF24L01 (Rx) and process through the ATmega328P Arduino microcontroller. According to transmitting corresponding fingerprint commands, it starts to process in the receiving section that was programmed previously. As per the signal, the microcontroller shows a preprogrammed text message on the LED screen and also plays a recorded ringtone stored in the SD card. Simultaneously, if a user needs to display real time, the user must first log in to the particular URL, fill the authentication process, and submit the text message. He can also modify the text anytime via a web browser.

3 Accessories of the Proposed System

In this section, the major components like Arduino Uno, fingerprint sensor, and radio transceiver module of the designed prototype have been discussed briefly.

3.1 Arduino Uno

The Arduino Uno board, which is a board based on the ATmega328P microcontroller, serves as the system's brain. The ATmega328P is an 8-bit microcontroller operating at 16 MHz [12]. Arduino Uno has only one SPI, one I2C, and several GPIO. The serial peripheral interface (SPI) is a synchronous serial communication interface specification used for short-distance communication, primarily in embedded systems. Using an SPI link, one master device (typically a microcontroller) controls the peripheral devices [13]. The board includes 32 KB of flash ROM, 2 KB of SRAM, and 1 KB of EEPROM. The Arduino board has a programming environment that allows C or C++ programming language.

3.2 Fingerprint Sensor Module

In any system, fingerprint authentication is the act of verifying an individual's identity. In our system, a fingerprint sensor module wired to the Arduino enrolls a new fingerprint and stores it directly within the flash memory with different IDs. After uploading the enrollment code, the serial monitor would prompt to insert credentials (fingerprint) to register. This module includes flash memory for storing fingerprints and can be used with any microcontroller or device that supports TTL serial [14]. It is able to store up to 127 different fingerprints. This module communicates with the microcontroller through software/hardware serial. Here, we used software serial by wiring the Tx pin of the fingerprint module to the digital pin-2 of the Arduino and the Rx pin of the fingerprint module to the digital pin-3 of the Arduino.

3.3 Radio Transceiver Module

The nRF24L01 + PA + LNA module is a radio transceiver module capable of communicating with one or more microcontrollers, i.e. it works as a multiceiver. The PA means a power amplifier which enhances the signal power transmitted from the nRF24L01+. The purpose of the low noise amplifier (LNA) is to receive incredibly weak signals from the antenna and extend it to a better standard. The radio transceiver's talk over distance is approximately 700 m; this distance can vary depending on the location and atmosphere. Any obstacles can limit the range of the transmitter and receiver. This gives us almost 125 channels with a 1 MHz spacing [16]. The nRF24L01+ transceiver module communicates over a 4-pin SPI with a maximum data rate of 10 Mbps. The channel fulfills a bandwidth of less than 1 MHz at 250 kbps and 1 Mbps air data rate. However, at 2 Mbps air data rate, 2 MHz bandwidth is used. It runs at 250 kHz, 1, and 2 MHz with a maximum bit rate of 2 Mbps and a transmit frequency of 1 mW in the 2.4 GHz unlicensed band using GFSK modulation. The MISO, MOSI, SCK, and CLK pins are used for communication. The NRF CE and SCN pins of nRF24L01 are connected to digital pins 7 and 8, respectively. The nRF24L01 module uses an IRQ pin for interrupting but it is not used in this scenario as the transmitter module does not need any power saving option. The nRF24L01 module needs 1.9–3.6 V to run. The module has programmable output capacity, which can be set to 0, −6, −12, or −18 dBm, and consumes just 12 mA during transmission at 0 dBm, which is less than a single LED [17]. The Arduino board includes a 3.3 V voltage regulator that can control an nRF24L01 module.

4 Project Design and Working Procedure

A schematic circuit diagram of the proposed system has been designed by using "Fritzing" software as shown in Fig. 2 which possesses two major parts: one is master presented in Fig. 2a and another is slave/child part depicted in Fig. 2b. The master part placed in the patient's room acts as the main system which can be used by doctors/patients while the slave part is placed at the central nurse control station.

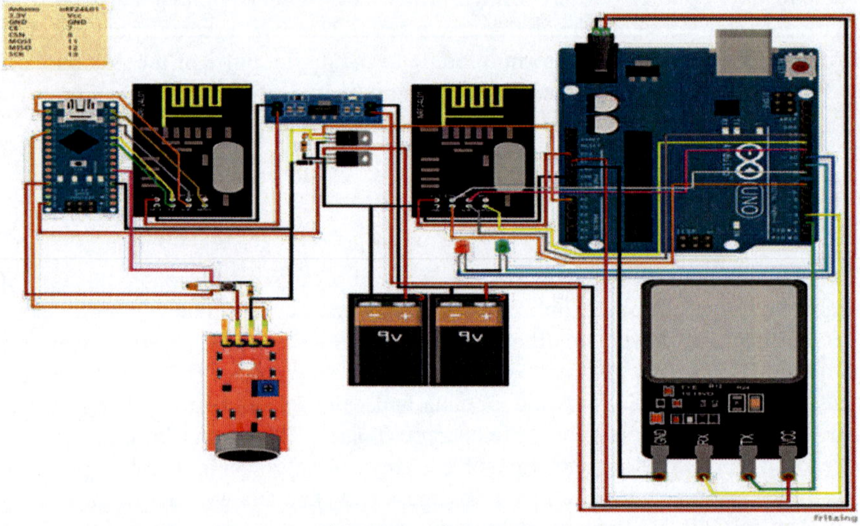

(a) Transmitter part

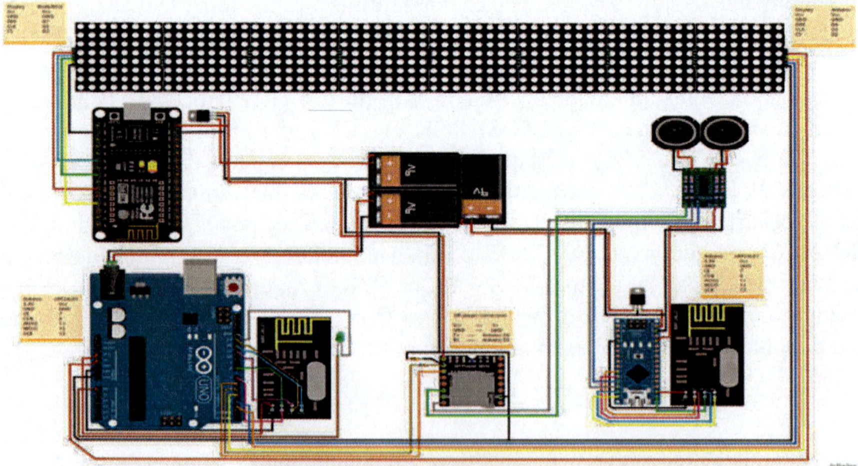

(b) Receiver part

Fig. 2 Circuit diagram: **a** Transmitter part. **b** Receiver part

The master part consists of the following components: a microcontroller (Arduino UNO) linked to the wireless module (NRF24L01), with two LEDs (green and red) that are used to indicate whether the user of the calling system (patient/doctor) is either authorized or not. Glowing the red LED suggests that the user's fingerprint is not in the database, i.e. he/she is not authorized to access the system, while a green LED indicates that the user is an authorized person. The prototype also has a voice communication feature that can use only duty doctors by again providing their fingerprints. When the patient/doctors pressed the fingerprint module, the controller located in the patient's room communicates using the wireless module nRF24L01 to the slave part that is located in the central nurse control room. The slave acknowledges the unique addresses of the master and opens a channel to establish communication between them. According to unique addresses sent by the master part, the LED screen displays the doctor or patient's name with the room number. There is no need to call a nurse on foot. By using such wireless electronic displays, a nurse from the control station can easily notify the patient about their medicine time if such a display is placed in the patient room.

Moreover, a centrally common e-notice board can be controlled by the hospital authority to announce important updates for their patient or visitors just by using a web browser for example: "Please leave the patient's relatives before 9 pm" as shown in Fig. 3. The entire signal flow and decision making scenario of the proposed system has been depicted as a flow chart in Fig. 4. The PCB layout and a designed

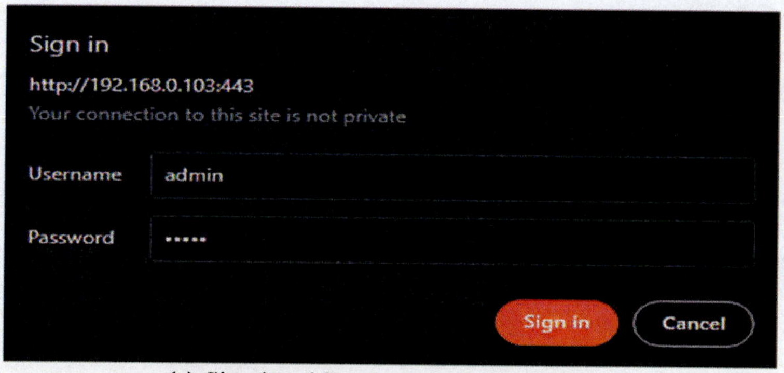

(a) Sign in with user name and password

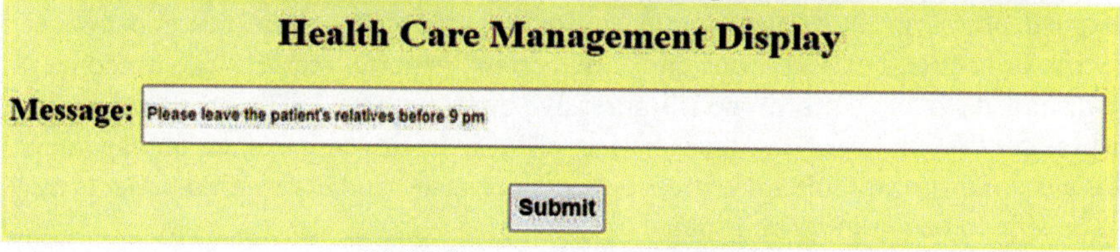

(b) Write a notice and submit

Fig. 3 E-notice board management: **a** Sign in with user name and password. **b** Write a notice and submit

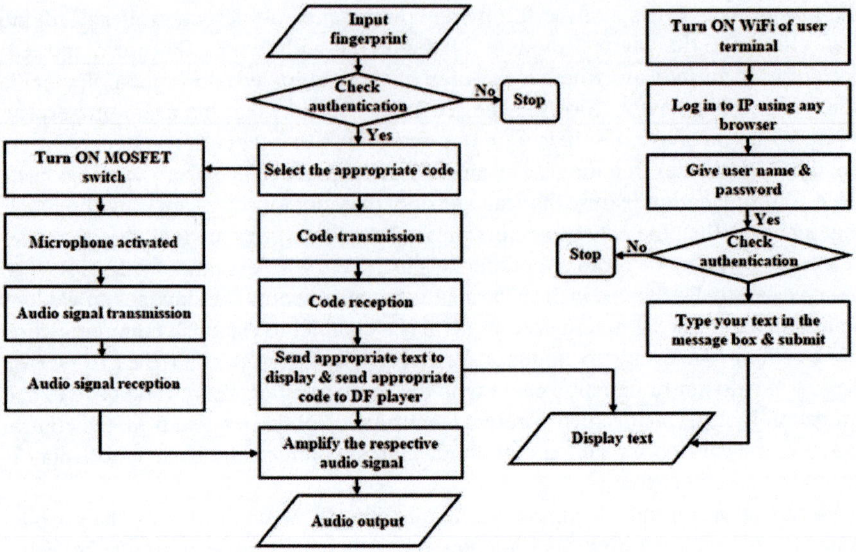

Fig. 4 Flow chart of the proposed calling system and e-notice board

prototype of the proposed calling system and e-notice board have been presented in Figs. 5 and 6, respectively.

5 Conclusion and Discussion

A prototype of a fingerprint recognition-based multi-terminal multi-user wireless smart calling system and electronic notice board for health care management of a hospital has been developed and proposed in this paper. The system has been implemented by using open source software tools Arduino IDE with C programming language, also used hardware interfaces ATmega328P, nRF24L01 + PA + LNA, fingerprint scanner, LED display, speaker, etc. For a hospital, a wireless health care calling system can play a vital role to provide smart management and service. The designed prototype is secure, stable, and economical. It can be tailored to meet specific desires and expectations and suited to any hospital budget. The outcome of the designed system's prototype is impressive during the trial. The proposed system ensures confidentiality and reduces human effort in health care system management. This system is not suitable for critical patients because critically injured patients may not be able to use fingerprint scanners.

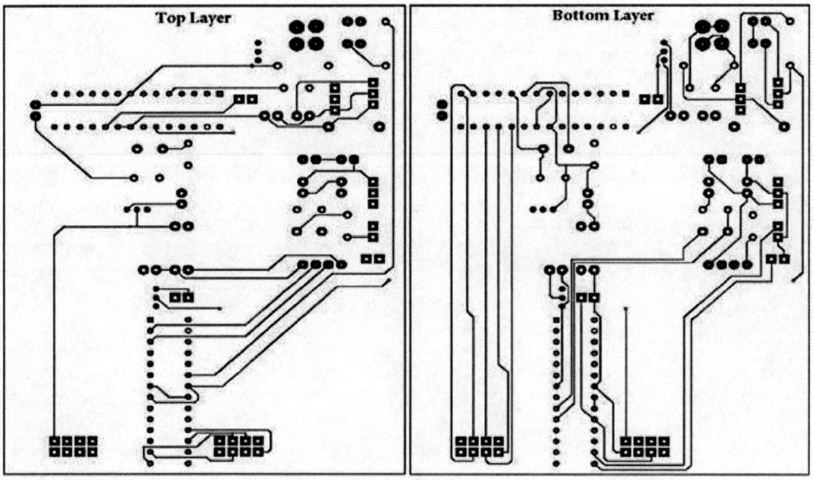

(a)Transmitter part

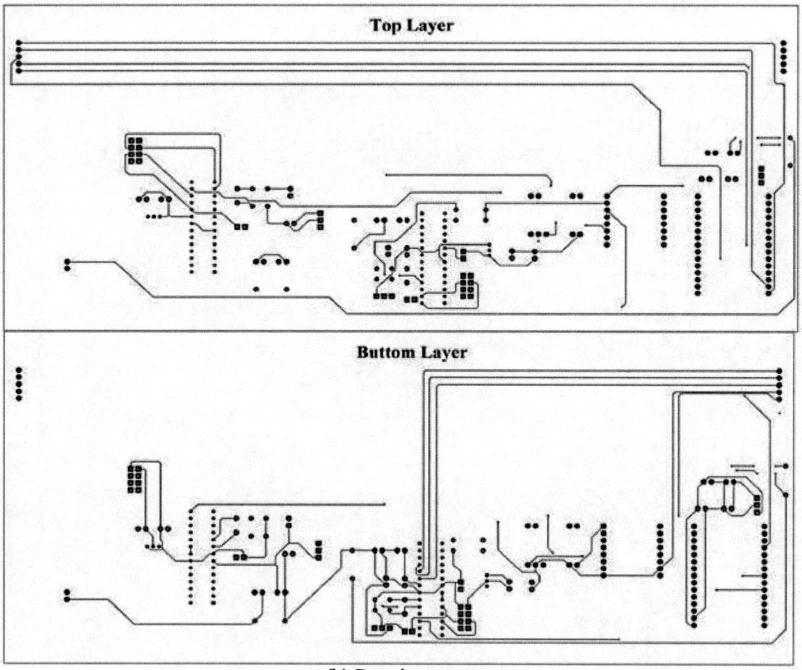

(b) Receiver part

Fig. 5 PCB layout: **a** Transmitter part. **b** Receiver part

(a) Transmitter part

(b) Receiver part

Fig. 6 Prototype of the proposed system: **a** Transmitter part. **b** Receiver part

References

1. Poongodi, M., Sharma, A., Hamdi, M., et al.: Smart healthcare in smart cities: wireless patient monitoring system using IoT. J. Supercomput. 1–26 (2021). https://doi.org/10.1007/s11227-021-03765-w
2. Latif, G., Shankar, A., Alghazo, J.M., et al.: I-CARES: advancing health diagnosis and medication through IoT. Wirel. Netw. **26**, 2375–2389 (2020). https://doi.org/10.1007/s11276-019-02165-6
3. https://www.fire-monitoring.com/what-is-a-nurse-call-system/. Accessed 30 Aug 2021
4. Aswin, S., Gopalakrishnan, N., Jeyender, S., Prasanna, R.G., Kumar, S.P.: Design development and implementation of wireless nurse call station. Annual IEEE India Conf. 1–6 (2011). https://doi.org/10.1109/INDCON.2011.6139633
5. Chen, Z., Hu, C., Liao, J., Liu, S.: Protocol architecture for wireless body area network based on nRF24L01. IEEE Int. Conf. Autom. Logist. 3050–3054 (2008). https://doi.org/10.1109/ICAL.2008.4636702
6. Hettikankanamage, N.D., Halgamuge, M.N.: Digital health or internet of things in tele-health: a survey of security issues, security attacks, sensors, algorithms, data storage, implementation platforms, and frameworks. In: IoT in Healthcare and Ambient Assisted Living, pp. 263–292 (2021). https://doi.org/10.1007/978-981-15-9897-5_13
7. https://www.techopedia.com/definition/29808/fingerprint-scanner. Accessed 30 Aug 2021
8. Abid, M.B., Rumon, M.R., Sraboni, T., Hossain, R., Ahmed, F., Uddin, J.: Design and implementation of an e-notice board using a NodeMCU. Int. Conf. Emerg. Technol. Comput. 288–295 (2020). https://doi.org/10.1007/978-3-030-60036-5_21
9. Macheso, P., Manda, T.D., Chisale, S., Dzupire, N., Mlatho, J., Mukanyiligira, D.: Design of ESP8266 smart home using MQTT and node-RED. In: International Conference on Artificial Intelligence and Smart Systems, pp. 502–505 (2021https://doi.org/10.1109/ICAIS50930.2021.9396027
10. Tripathi, G., Ahad, M.A., Paiva, S.: S2HS-A blockchain based approach for smart healthcare system. Healthcare **8**(1), 100391 (2020). https://doi.org/10.1016/j.hjdsi.2019.100391
11. Shi, Q., Dong, B., He, T., Sun, Z., Zhu, J., Zhang, Z., Lee, C.: Progress in wearable electronics/photonics—Moving toward the era of artificial intelligence and internet of things. InfoMat **2**(6), 1131–1162 (2020). https://doi.org/10.1002/inf2.12122
12. ATmega328P Datasheet
13. https://store.arduino.cc/usa/arduino-uno-rev3. Accessed 30 2021/08/30
14. https://randomnerdtutorials.com/fingerprint-sensor-module-with-arduino/. Accessed 30 Aug 2021
15. Ali, J., Ali, A.A.: Design and implementation of low-power hierarchical wireless sensors network. Annual Conf. New Trends Inf. Commun. Technol. Appl. 258–262 (2017). https://doi.org/10.1109/NTICT.2017.7976112
16. https://lastminuteengineers.com/nrf24l01-arduino-wireless-communication/. Accessed 30 Aug 2021

Evaluating Performances of VPN Tunneling Protocols Based on Application Service Requirements

Happy Akter, Sohely Jahan, Sajeeb Saha, Rahat Hossain Faisal, and Shariful Islam

Abstract A Virtual Private Network (VPN) creates a private communication path over the public network to ensure the security of the transmitted data. A plethora of VPN tunneling protocols has been developed for the site-to-site and remote access of the applications. A critical challenge for an organization is to select an appropriate VPN protocol considering the application requirements. In this paper, Internet Protocol Security (IPSec), Layer 2 Tunneling Protocol (L2TP), and Multiprotocol Label Switching (MPLS) protocols have been exploited to determine the application suitability of the VPN tunneling mechanism selected for remote access. End-to-end Throughput, Round Trip Time, and Jitter along heterogeneous hardware environments have been investigated to determine the performance of the VPN tunneling techniques. Experimental results depict that MPLS is suitable for real-time audio-video, web, e-mail, and file-sharing applications. IPSec and L2TP are more applicable to the web, e-mail, and file-sharing applications and not suitable for streaming applications due to high jitter and latency.

Keywords Virtual private network · MPLS · L2TP · IPSec

H. Akter · S. Jahan (✉) · R. H. Faisal
University of Barishal, Barishal, Bangladesh
e-mail: sojahan@bu.ac.bd

R. H. Faisal
e-mail: rhfaisal@bu.ac.bd

S. Saha
Jagannath University, Dhaka, Bangladesh

S. Islam
Daffodil International University, Dhaka, Bangladesh
e-mail: shariful.swe@diu.edu.bd

M. S. Kaiser et al. (eds.), *Proceedings of the Third International Conference on Trends in Computational and Cognitive Engineering*, Lecture Notes in Networks and Systems 348, https://doi.org/10.1007/978-981-16-7597-3_36

1 Introduction

A Virtual Private Network (VPN) is the collection of a private and public networks such as the Internet that performs secure transmission of data. A virtual private network can establish protected virtual links among different organizations or branch offices of the organization. A VPN transfers data over a public network between two systems in such a way that the transmitted data is visible to the other systems linked to the network. The application scope of VPN is increasing day by day as the organizations need a secure and private communication channel through the public Internet. More than 26% of all Internet users use VPN as it meets the four key factors of an enterprise that are compatibility, security, availability, and manageability [1].

Several applications use VPN protocols to provide security in the communication channel through the service requirements of these applications vary greatly from each other. Table 1 has showed application services requirements. For example, a few applications (Web, file transfer, and e-mail, etc.) have a flexible time requirement and are elastic to available bandwidth. On the other hand, a few applications (audio-video, real-time games, etc.) cannot tolerate more than a certain amount of delay and expects a minimum guaranteed bandwidth [11]. Hence, it is an imminent concern to determine an appropriate VPN protocol based on the service requirements of the application.

In state-of-the-art works, several approaches have been exploited to determine the performance of VPN protocols in different applications. In [10], Kotuliak et al. used AES, 3DES, and blowfish cipher to analyze the performance of OpenVPN and IPSec VPN. Due to its simple and first implementation, openVPN showed better performance in terms of throughput and response time. In [4], Thu et al. showed that security and reliability are the key parameters on the selection of VPN protocols in an organization. Faycal et al. studied the impact of the VPN tunnel layer on the performance of VOIP and showed that MPLS VPN can be a better option compared to IPSec VPN [6]. Lukas et al. in [12], analyzed the performance of IPSec, Open-VPN, and wireGuard with various cryptographic algorithms and the CPU core. The result depicted that the wireGuard can achieve a 30% better result whereas the result depends highly on the number of available CPU core. However, wireGuard VPN pro-

Table 1 Applications with their service requirements

Application	Bandwidth	Time sensitive
File transfer	Elastic	No
E-mail	Elastic	No
Web documents	Elastic	No
Internet telephony/video conferencing	Audio: few kbps to 1 Mbps, Video: 10 kbps to 5 Mbps	Yes: 100 s of msec
Streaming/stored audio/video	Audio: few kbps to 1 Mbps, Video: 10 kbps to 5 Mbps	Yes: few seconds

tocol supports only UDP protocol. In [13], Maximilian et al. analyzed the effect of the software architecture of different VPN gateways and proposed a new pipeline architecture. Zhipeng et al. analyzed the performance of different VPN technologies based on encryption speed, stability, availability, and security. The results showed that SSL VPN can be a better option, in general, compared to IPSec and MPLS VPN [14]. However, all the above works evaluated the performance of different VPN protocols considering various security parameters. None of them considered the service requirements of the running application on their selection of VPN protocol.

In this paper, we analyze the performance of three VPN tunneling protocols. We consider the service requirements of an application in the selection of the VPN protocol. The main contributions of our work are summarized as follows:

- We consider IPSec, L2TP with IPsec, and MPLS VPN protocols to analyze and measure the performance.
- Throughput, Round Trip Time, and Jitter parameter have been used to determine the sensitivity of the protocols.
- The result showed that all the stated protocols support elastic applications, whereas only MPLS is suitable for time-sensitive applications.

The remainder of the paper is organized as follows. An introduction of the target VPN protocols has been discussed in Sect. 2. Section 3 describes the formulation of the problem in the simulation environment. Section 4 discusses the results obtained from performance evaluation, and finally we conclude the paper in Sect. 5.

2 VPN Tunneling

A Virtual Private Network (VPN) is a feature of networking that creates a safe and secure virtual communication channel among different types of working groups. The data communication system should be more secure and reliable for transferring data through the network, need to ensure data authenticity, integrity, and confidentiality. In this paper, three widely used VPN protocols, namely, IPSec, L2TP with IPSec, and MPLS have been selected to analyze the performance considering various application requirements. A brief description of the abovementioned protocols is given below.

2.1 IPSec Tunneling Protocol

Based on established data transport protocols, Internet Protocol Security (IPSec) is one of the most complete, stable, and open standards VPN protocols that establishes a secure VPN tunnel over the public Internet. IPSec operates its functionality in the network layer. It includes protocols that help at the beginning of a session to establish mutual authentication between two communicating parties and to negotiate cryptographic keys to be used during the session. IPSec VPN is categorized into two groups: site-to-site access VPN and remote access VPN.

The IPSec is operated in two modes: transport mode and tunnel mode [7]. In transport mode, the encapsulation and decryption are done on the original IP data, whereas in tunnel mode encapsulation and decryption are done on the entire IP header. The Authentication Header (AH) and the Encapsulating Security Payload (ESP) are two main security protocols that are an essential part of the IPSec tunnel. The AH protocol mechanism provides integrity protection, data authentication, and anti-replay protection but not ensures privacy [8]. On the other hand, ESP mechanism provides data integrity, origin authentication, optional anti-replay, limited traffic flow confidentiality, and privacy. The utility service and security association (SA) is provided by the Internet Key Exchange (IKE) protocol. The Internet SA creates a secure channel between two peers. The Internet Security Association and Key Management Protocol (ISAKMP) offer the framework for the IKE exchange [8] that negotiates between the peer hosts for creating SA and cryptographic keys.

2.2 L2TP with IPSec Tunneling Protocol

L2TP stands for Layer 2 Tunneling protocol which is paired up with IPSec and provides a secure remote access VPN tunnel. L2TP does not provide data encryption by itself. IPSec is used for strong authentication, encryption, confidentiality, and integrity and hence is called L2TP with IPSec VPN tunneling technique. L2TP used Point-to-Point Protocol (PPP), challenge handshake authentication protocol (CHAP) authentication methods for authenticating the remote user. IPSec has obtained end-to-end security which encapsulates and authenticates each packet of data within a communication channel. L2TP with IPSec works in the client-server model for connecting remote users [8]. This tunnel has two communication ends points on the Internet L2TP access concentrator (LAC) and L2TP network server (LNS). The LAC was deployed by the customer network to the tunnel initiator, while the LNS was deployed by the network server to a tunnel concentrator waiting for tunnels to be built. LIN waits for a dial-up connection to establish a tunnel, if a tunnel is established the LAC and LNS can communicate bi-directionally. So the LAC and LNS can create more than one session at the same time. For connecting remote client, L2TP allowed the creation of a virtual private dialup network (VPDN) using a shared infrastructure. Two types of messages are required for L2TP: control messages and data messages. To build control L2TP tunnels, the control messages are used and have a versatile configuration mechanism called Attribute Value Pair (AVP) [9]. Data sessions that encapsulate PPP sessions use data messages. L2TP starts by building some L2TP tunnels and then uses data sessions inside the tunnels to communicate with peers.

2.3 MPLS Tunneling Protocol

The Multiprotocol Label Switching (MPLS) protocol is used on IP backbone networks. MPLS is based on version 4 of the Internet protocol which can be expanded by core MPLS technologies. Connection-oriented label swapping on connectionless IP networks is used in the MPLS VPN technique. MPLS VPN network is configured based on a set of equipment: Client Edge (CE), which is the client site's routers and it supports only IP. Provider (P) router is called the backbone of MPLS and it needs to support MPLS. Provider Edge (PE) is located between CE and P which must support MPLS and IP [5]. CE router is directly connected to PE router and CE cannot be established more than one peer with router PE. The actual goal of the MPLS is to ensure the security of the network in terms of confidentiality, integrity, authentication, and availability. MPLS VPN security requirements are as follows: VPN address and traffic separation, IP-spoofing protection model, MPLS core structure hiding, and attack robustness. For developers, the main protection criterion is to keep their traffic isolated from other users and the core side.

3 Problem Description

The common application scenarios of VPN technology are removing the physical barrier, communication with corporate office, communication for online education purposes, business conferencing, and accessibility of remote users. In the COVID-19 pandemic situation, many organizations moved their organizational activities on online platforms like online meetings, video conferencing, document sharing, etc. In a pandemic situation, the education system is dependent on the online platform, academic classes, seminars, assignment submissions, and also exams are held online. All of these application scenarios require to access e-mail, web documents, file transfer, store audio/video, streaming, etc. These application services should ensure high throughput and maximum bandwidth support with a minimum delay. It is essential to select the right protocol according to their application requirements. In this paper, the three most well-known VPN protocols have been exploited to identify their usage based on the versatile service demands of different applications.

3.1 Simulation of IPSec

Internet Security Protocol (IPSec) is one kind of site-to-site VPN tunneling technology. This tunneling protocol is implemented here with the GNS3 (version 2.2.12) simulation environment. The Site-to-Site IPSec VPN tunnel configuration can be divided into two steps as Phase 1 and Phase 2. In ISAKMP Phase 1, the Advanced Encryption Standard (AES 256) method, the authentication method (Pre-share), and

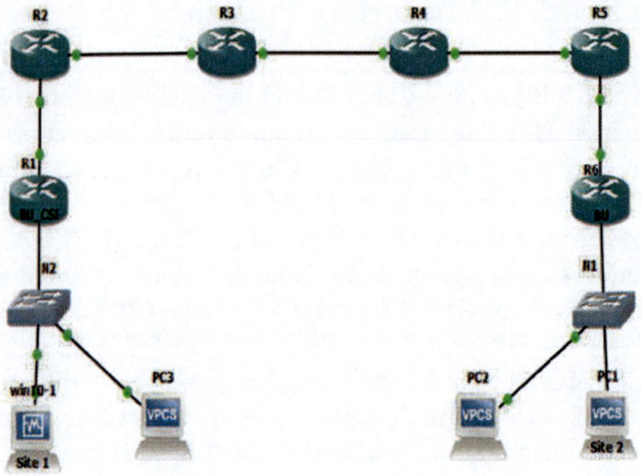

Fig. 1 Network topology for simulation

the hashing algorithm Massage Digest 5 (MD5) were used to create the first tunnel. In Phase 2, different types of operation such as creation extended Access control List (ACL), creation IPSec Transform, creation Crypto Map, and applying crypto map to the public interface occurred [8]. Open Shortest Path First (OSPF) routing protocol has been used in this simulation. In Fig. 1, only the N2 network of BU CSE and N1 network of BU uses IPSec tunnel for secure communication which means other network access is denied using the access control list.

3.2 *Simulation of Multi-protocol Label Switching (MPLS)*

The Multi-protocol Label Switching (MPLS) is a remote access VPN tunneling protocol that creates private virtual connections using the public networks. MPLS uses authentication protocol based on CE router. Microsoft Point-to-Point Encryption (MPPE) algorithm is used as an encryption algorithm. The same topology (Fig. 1) has been used to configure MPLS protocol, where routers R1 and R6 act as Customer Edge (CE) router; routers R2 and R5 act as a Provider Edge router; and routers R3, R4 act as Provider (P) router. Site1 is configured as an official network and site 2 is configured as a remote user. Cisco routers can be set up to act as MPLS servers, alternatively known as Virtual Private Dialup Network (VPDN) servers. The GNS3 and Virtual Box (Cross-platform Virtualization Software) are used as experimental tools. The products used for the implementation of MPLS are Cisco Router 7200, Cisco Switch, VPCS as a computer.

3.3 *Simulation of L2TP with IPSec VPN*

The Layer 2 Tunneling Protocol (L2TP) with IPSec remote access VPN tunneling protocol is usually implemented between a server and client similar to Fig. 1 topology, the server belonging to the enterprise network and the client being a remote workstation. L2TP with IPSec uses the Password Authentication Protocol and the Challenge Handshake Authentication Protocol (CHAP).

4 Performance Evaluation

In this section, we represent the simulation environment and the parameters used to evaluate the performance of the studied protocols.

4.1 *Simulation Environment*

To analyze the performance of the three tunneling protocols, we have created a topology for simulation with the help of Graphical Network Simulator (GNS3). The Oracle VM VirtualBox is used with GNS3 to create a separate operating system platform to communicate with the external environment. Simulation parameters are listed in Table 2.

4.2 *Throughput Analysis*

Throughput measures how much data packet transfer through a channel from source to destination per unit time. The throughput results of these protocols are depicted in Fig. 2 which shows that throughput is increasing linearly for all the studied protocols. The throughput increases gradually with the increasing data size where L2TP with

Table 2 Simulation parameters

Parameter	Value
Simulator	Graphical network simulator (GNS3)
Platform	Oracle VM VirtualBox
IOS router	Cisco IOS router 7200 edition
Routing protocols	RIP, OSPF
Performance parameters	Throughput, RTT, jitter, latency
Units of measurement	Millisecond

Fig. 2 Throughput

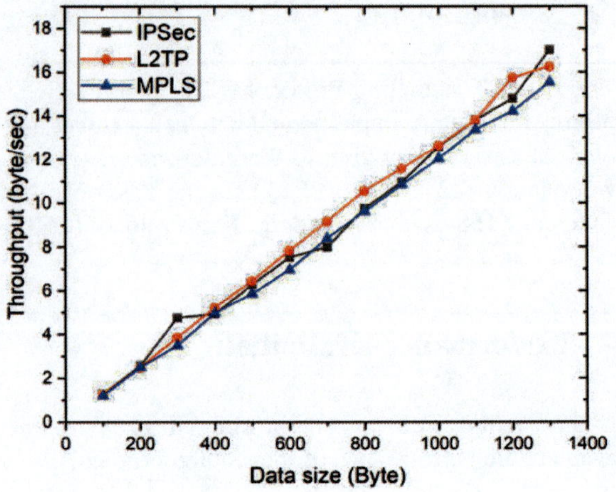

IPSec shows the highest rate and MPLS the lowest. The throughput of the IPSec protocol increases moderately with data size. Comparatively, the throughput of MPLS protocol is less than L2TP with IPSec, and the throughput of IPSec protocol fluctuates between the range of L2TP with IPSec and MPLS throughput range. Due to strong encryption, decryption method, and key exchange properties, IPSec protocol needs more time to reach the desired destination resulting in lower throughput. However, L2TP with IPSec provides the same security while providing reasonably higher throughput.

4.3 Latency Analysis

Latency (measured in milliseconds) defines the time delay between sending data from a sender to a receiver. Figure 3 depicts the latency graph of the studied protocols. The results show that the Multiprotocol Label Switching (MPLS) protocol has the highest latency among the three protocols. However, with the increasing data size, the rate of increase in the latency amount is very little. This is due to the fact that in MPLS protocol, data need to pass three routers, CE, PE, and P routers, which causes more time delay in the communication path. IPSec and L2TP with IPSec have lower latency compared to MPLS protocol. IPSec protocol used a more complex header and strong encryption mechanism, resulting in a higher fluctuation rate compared to the other two protocols. Although L2TP with IPSec uses IPSec encryption and authentication mechanism, it still performs better than IPSec.

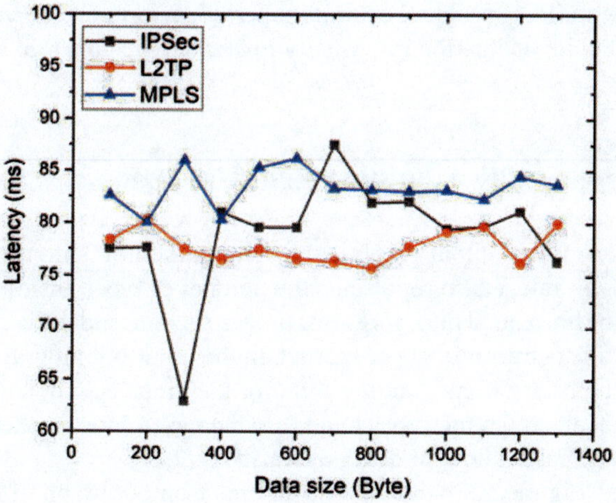

Fig. 3 Latency

4.4 *Jitter Analysis*

Jitter measures the variation of Round Trip Time caused by network congestion. The performance of an application heavily depends on the jitter amount. The jitter performance of the studied protocols is depicted in Fig. 4 which shows that colorgreen the IPSec protocol has large jitter and the fluctuation rate is very high compared to the other protocols. This gives an impression that the protocol is highly susceptible to time-sensitive and security-sensitive applications. This is due to the fact that the IPSec protocol uses a more complex header that requires more response time. L2TP

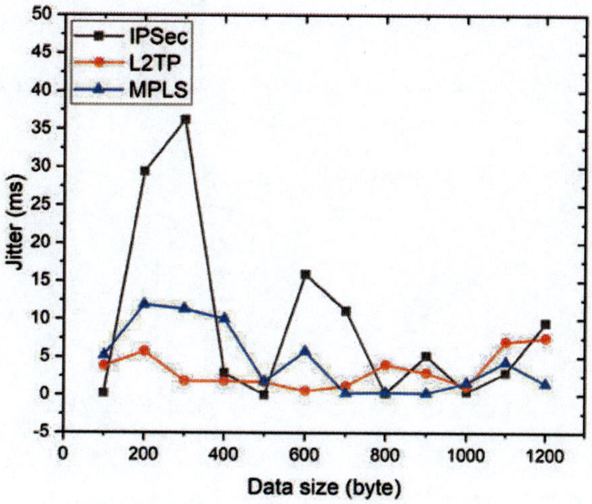

Fig. 4 Jitter

with IPSec protocol exhibits superior performance than IPSec and MPLS in this case and its fluctuation rate is very low when the data size is higher than 500 byte.

4.5 Input-Output Graph Analysis

The input-output graph is the combination of Throughput, Round Trip Time, Jitter, Loss rate, etc. It represents the number of bits transmitted at a particular time. Here, the horizontal line represents time in second, and the vertical line indicates the amount of data transmitted per second. In the input-output graph, a zero data traffic indicates the back-off and waiting time for the green acknowledgment of the data sent to the receiver. On the other hand, the increased inter-packet spacing is responsible for a decreasing flow of data transmission [2, 3].

Figures 5, 6, and 7 show the input-output graph of IPSec, L2TP with IPSec, and MPLS VPN tunneling techniques, respectively. The figures depicted that due to high inter-packet spacing and zero traffic, L2TP exhibits the lowest data rate whereas MPLS data rate is highest among the compared protocols. Moreover, MPLS shows a minimum fluctuation rate resulting in lower jitter in the transmission of data. On the other hand, the inter-packet spacing and zero traffic in both L2TP and IPSec make them unusable for time-sensitive applications.

Fig. 5 IPSec

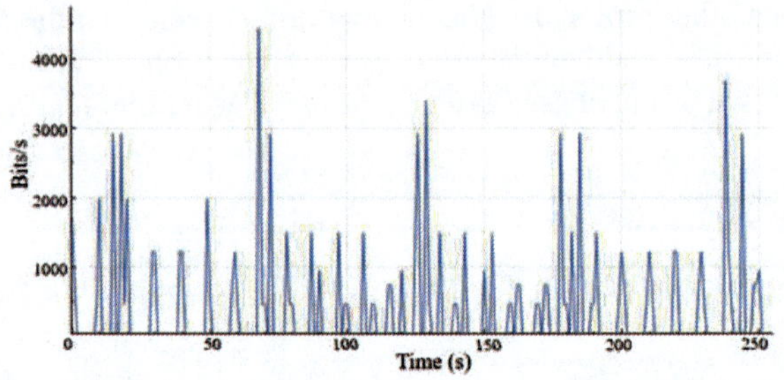

Fig. 6 L2TP with IPSec

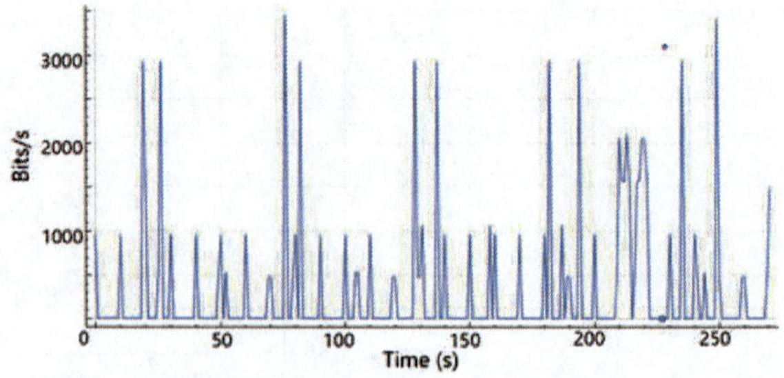

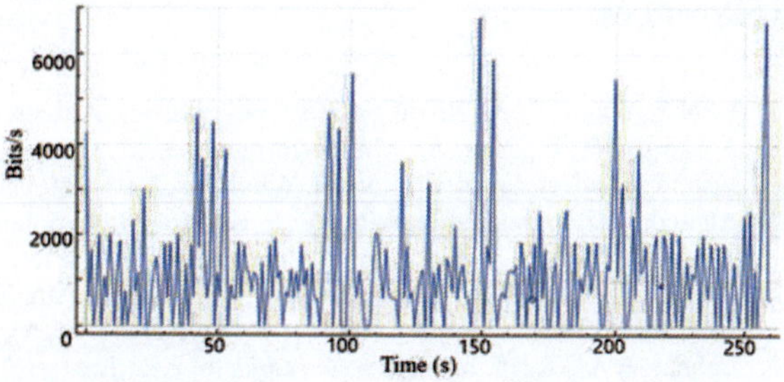

Fig. 7 MPLS

Table 3 Performance comparison of the studied protocols

IPSec protocol	L2TP with IPSec protocol	MPLS protocol
Applicable for elastic services	Applicable for elastic services	Applicable for elastic, bandwidth-sensitive services
Not applicable for time-sensitive services	Not applicable for time-sensitive services	Applicable for time-sensitive services
Applicable for document transfer, e-mail, file sharing	Applicable for document transfer, e-mail, file sharing	Applicable for live streaming, real-time audio/video as well as document transfer, e-mail, file sharing

Based on the simulation results and comparative analysis, Table 3 summarizes the performance and application area of IPSec, L2TP with IPSec, and MPLS VPN tunneling protocols.

5 Conclusion

Virtual Private Network (VPN) is an emerging technology in the networking area. VPN creates a secure private communication channel across the public network and provides an extra security layer that protects the original data packet. In this paper, three VPN protocols, namely, IPSec, MPLS, and L2TP with IPSec have been studied to determine the applicability of the protocols in different application service requirements. Our in-depth analysis showed that MPLS can perform better for both elastic and time-sensitive applications whereas IPSec and L2TP are suitable for elastic applications only. In the future, we plan to implement a real testbed to evaluate the performance while incorporating a few more congenial VPN protocols.

References

1. VPN Statistics. https://learn.g2.com/vpn-statistics. Accessed 06 Jun 2021
2. Agrawal, N., Tapaswi, S.: The performance analysis of honeypot based intrusion detection system for wireless network. Int. J. Wirel. Inf. Netw. **24**(1), 14–26 (2017)
3. Ahmed, M.Z., AbdallahHashim, A.H., Khalifa, O.O., Salami, M.J.: Border gateway protocol to provide failover in multihoming environment. Int. J. Inf. Technol. **9**(1), 33–39 (2017)
4. Aung, S.T., Thein, T.: Comparative analysis of site-to-site layer 2 virtual private networks. In: 2020 IEEE Conference on Computer Applications (ICCA), pp. 1–5. IEEE (2020)
5. Bahnasse, A., Talea, M., Badri, A., Louhab, F.E., Laafar, S.: Smart hybrid SDN approach for MPLS VPN management on digital environment. Telecommun. Syst. **73**(2), 155–169 (2020)
6. Bensalah, F., El Kamoun, N., Bahnasse, A.: Evaluation of tunnel layer impact on VOIP performances (IP-MPLS-MPLS VPN-MPLS VPN IPSec). Int. J. Comput. Sci. Netw. Secur. (IJCSNS) **17**(3), 87 (2017)
7. Forouzan, B.A., Mukhopadhyay, D.: Cryptography and Network Security. Mc Graw Hill Education (India) Private Limited, New York (2015)
8. Jahan, S., Rahman, M.S., Saha, S.: Application specific tunneling protocol selection for virtual private networks. In: 2017 International Conference on Networking, Systems and Security (NSysS), pp. 39–44. IEEE (2017)
9. Kara, A., Suzuki, T., Takahashi, K., Yoshikawa, M.: A DoS-vulnerability analysis of L2TP-VPN. In: 4th International Conference on Computer and Information Technology, 2004 (CIT'04), pp. 397–402. IEEE (2004)
10. Kotuliak, I., Rybár, P., Trúchly, P.: Performance comparison of IPsec and TLS based VPN technologies. In: 2011 9th International Conference on Emerging eLearning Technologies and Applications (ICETA), pp. 217–221. IEEE (2011)
11. Kurose, J.F., Ross, K.W.: Computer Networking: A Top Down Approach, 6th edn. Google Scholar Digital Library (2012)
12. Osswald, L., Haeberle, M., Menth, M.: Performance comparison of VPN solutions (2020)
13. Pudelko, M., Emmerich, P., Gallenmüller, S., Carle, G.: Performance analysis of VPN gateways. In: 2020 IFIP Networking Conference (Networking), pp. 325–333. IEEE (2020)
14. Zhipeng, Z., Chandel, S., Jingyao, S., Shilin, Y., Yunnan, Y., Jingji, Z.: VPN: a boon or trap?: a comparative study of MPLS, IPSec, and SSL virtual private networks. In: 2018 2nd International Conference on Computing Methodologies and Communication (ICCMC), pp. 510–515. IEEE (2018)

Tuning Wavelength of the Localized Mode Microcavity by Applying Different Oxygen Flows

María R. Jiménez-Vivanco, Godofredo García, Francisco Morales-Morales, Antonio Coyopol, Lizeth Martínez, Jocelyn Faubert, and J. E. Lugo

Abstract In this work, porous Si-SiO_2 UV microcavities were obtained by applying two stages of dry oxidation and different oxygen flows. In this way, we observed a maximum wavelength shift (134 nm) of the localized mode microcavity to higher energy when a low oxygen flow was used to obtain porous silicon microcavities. It also depicted a wavelength shift of 121 nm when a maximum oxygen flow was applied. We used an effective medium model to predict the refractive index for two media (silicon and air) and three media (silicon, silicon dioxide, and air) components. The result showed that UV microcavities obtained with higher oxygen flow absorb more UV light. Thus, there was less SiO_2 formation, and consequently the optical absorption increased. Besides, a decrease of the PBG bandwidth was achieved by incorporating SiO_2 within the porous silicon microcavities. This bandwidth decrease happened because there was less contrast between the high refractive index and the low index of porous Si-SiO_2 layers.

Keywords Oxygen flow · Microcavities · Dry oxidation

M. R. Jiménez-Vivanco · G. García · A. Coyopol
Semiconductor Devices Research Center, ICUAP, BUAP, Ciudad Universitaria, 72570 Puebla, Pue, Mexico

M. R. Jiménez-Vivanco
Institute of Physics, UNAM, Circuito de la Investigación Cientifica, Ciudad Universitaria, 04510 Mexico city, Mexico

F. Morales-Morales
Optics Research Center, A.C., Loma del Bosque 115, Col. Lomas del Campestre, 37150 León, Gto, Mexico

L. Martínez
Tepeji Graduate School, Industrial Engineering, Autonomous Hidalgo State University, Av. del Maestro No. 41, Col. Noxtongo 2ª Sección, 42855 Tepeji del Rio, Hidalgo, Mexico

J. Faubert · J. E. Lugo (✉)
Faubert Lab, School of Optometry, University de Montreal, Montreal, QC H3T 1P1, Canada

M. S. Kaiser et al. (eds.), *Proceedings of the Third International Conference on Trends in Computational and Cognitive Engineering*, Lecture Notes in Networks and Systems 348, https://doi.org/10.1007/978-981-16-7597-3_37

1 Introduction

Photonic Crystals (PC) are periodic structures, where the light can be tailored to propagate at specific frequencies and energies [1], among them we can find Bragg reflectors [2], Rugate filters [3], Fibonacci filters [4, 5], microcavities [6], and multiple coupled microcavities [7, 8]. In the effort to find the refractive index value of porous silicon (PS) and oxidized porous silicon (OPS), several authors have employed many methods [6, 9, 10], and they reported that if PS's morphology changes with porosity, the value of PS's refractive index is not easy to obtain [9]. In this way, different effective medium approximations had been proposed to calculate the theoretical values of PS's complex refractive index; Bruggeman [6, 11], Maxwell Garnet [12, 13], and Looyegan [6] approximations are the most used. On one hand, it is known that changing the applied current density during the PS fabrication modifies its oxidation state [2, 14]. On the other hand, PS has been used in several applications as vapor sensor [3], biosensor for insulin detection [15], ethanol sensors [16], photodetectors [17], photovoltaic devices [18], and antireflection coatings [19].

In order to fabricate PC in the UV range, PS microcavities were oxidized applying two states of dry oxidation and different oxygen flows during the oxidation process, and we calculated the theoretical complex refractive index of both PS and OPS applying the Maxwell Garnett's effective medium approximation [13], where it was demonstrated that after dry oxidation the complex refractive index of each PS layer that makes the microcavity up was decreased. Besides, we employed the matrix transfer method to predict the theoretical reflectance and transmittance spectrum of microcavities based in PS and OPS in the visible (Vis) and ultraviolet (UV) range and compared them with experimental results, where we observed an amplification of the transmission spectrum and a narrowing of the photonic bandgap (PBG) in the microcavity after dry oxidation. By using absorbance measurements we demonstrated that after applying a high oxygen flow, the microcavities showed strong UV light absorption [13], whereas employing a low oxygen flow the microcavities exhibited an increase in their transmittance spectrum in the UV band.

2 Materials and Methods

2.1 Porous Silicon and Oxidized Porous Silicon Microcavities

Firstly, PS microcavities were obtained by electrochemical etching on silicon (Si) wafer with 0.01–0.02 Ωcm resistivity, (100) orientation, before the electrochemical etching the Si wafer, with dimensions 1.5 cm × 1.5 cm, was immersed in a stirring solution of 20% HF for 5 min to eliminate native oxide; after that, it was rinsed with deionized water and dried with a dryer to remove the water droplets on the surface of the Si wafer to avoid any possible source of inhomogeneity on the photonic crystal surface during its fabrication.

The Si wafer was put in a Teflon cell with etching area of 1 cm^2; then an aqueous electrolyte of 40% HF and ethanol at 99.7% with a volume ratio of 1:1 was placed in the Teflon cell to carry out the electrochemical process. A power supply (Keithley 2460) controlled by a laptop was used to deliver the current profile for the microcavities by means of a ring-shaped tungsten electrode immersed in the electrolyte, which acted as cathode, and an aluminum plate that contacts the unpolished backside of the Si wafer was used as anode.

The current profile consisted in interchanging two different current pulses (5 mA/cm^2 for 4.1 s and 80 mA/cm^2 for 1.1 s). After each current pulse, a pause of 3 s was introduced to generate the flow of the electrolyte and prevent porosity gradients. Afterward, the microcavity lifting up was carried out using the same electrolyte that was mentioned before but a high current pulse of 450 mA/cm^2 for 2 s to the silicon substrate was applied, the electrolyte based in 40% HF and ethanol at 99.7% was removed completely adding ethanol to the solution. When the process finished, a free-standing microcavity was reached, which was immersed in an ethanol container and placed on a quartz substrate; finally, the sample was removed and dried in the environment.

All the microcavities were oxidized employing two stages of dry oxidation, it was oxidized at 350 °C for 30 min and then at 900 °C for 1 h, and during the oxidation process, the oxygen flow for each sample was changing from 1.15 to 4.52 SLPM (standard liter per minute) to modulate the optical characteristics of the microcavities.

The optical characterization of free-standing microcavities was carried out before and after oxidation with a Varian (Agilent) UV-Vis-NIR spectrophotometer at normal incidence and incident angle to 20° from 200 to 700 nm.

3 Results and Discussions

3.1 Unoxidized Porous Silicon Microcavity

The theoretical (red line) and experimental (black line) reflectance (solid line) and transmittance (dotted line) spectra from unoxidized PS microcavity are shown in Fig. 1. The microcavity was designed to exhibit a localized mode in the VIS range (490 nm). Transmittance and reflectance measurements were taken employing an UV-VIS-NIR spectrophotometer in the wavelength range from 200 to 700 nm. The reflectance spectrum was measured at 20° (Fig. 1a) and the transmittance spectrum was taken at normal incidence (Fig. 1b). The transfer matrix method was applied to obtain the theoretical transmittance and reflectance spectrum. To compare the experimental and theoretical results, the incidence angles were taken into account in the simulations.

In Fig. 1a, the theoretical and experimental reflectance spectra are displayed, and the theoretical results compared to the experimental results showed more reflection than the theoretical results predicting less losses than the experimental results,

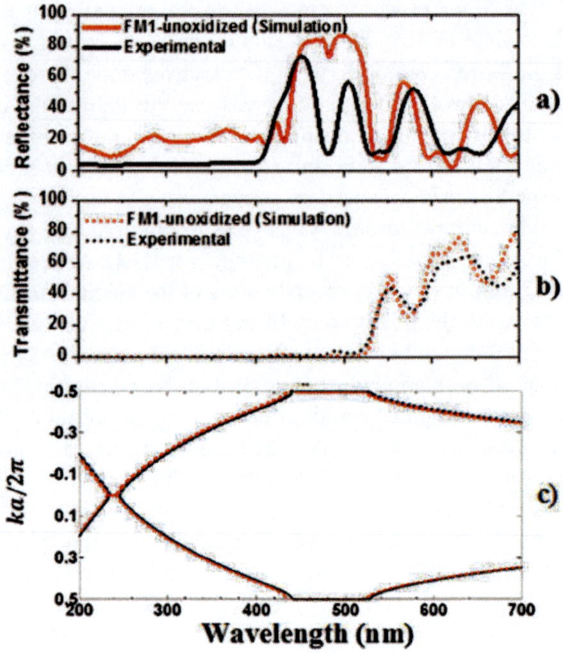

Fig. 1 The theoretical and experimental reflectance and transmittance spectra of an unoxidized microcavity and its photonic bandgap structure. **a** The theoretical and experimental reflection spectrum of an unoxidized microcavity on quartz substrate. **b** Theoretical and experimental transmittance spectrum of an unoxidized free-standing microcavity. **c** The photonic bandgap structure of the microcavity

whereas the theoretical and experimental transmittance spectra matched very well (see Fig. 1b). There is low transmission in the theoretical and experimental results from green to UV frequency range, showing a maximum transmission peak of 4% in the localized mode (490 nm), the low transmission is due to strong absorption of Si, while in the visible range from 555 to 700 nm there is high light transmission. This result is due to the complex refractive index values of Si, in particular, the extinction coefficient is very small in that wavelength interval; therefore, PS showed less optical losses by absorption. In spite of taking the reflectance and transmittance measurements of the same microcavity, its minimum reflection peak does not have the same position than the maximum transmission peak, moving to short wavelengths (483 nm), the reason is because reflectance measurements were taken at 20°, while the transmittance spectrum was taken at normal incidence. Figure 1c shows the PBG of an unoxidized microcavity, and their edges are 440 nm and 530 nm. The localized mode is displayed as a transmission maximum or a reflection minimum at the specific wavelength, which is found between the PBG edges. The PBG was obtained using the dispersion relation inside the first Brillouin zone, where measurement angles were considered.

Theoretical (red line) and experimental (black line) reflectance (solid line) and transmittance (dotted line) spectra from oxidized microcavities are presented in Fig. 2, both results showed a good fit.

The theoretical reflectance and transmittance spectra were obtained by the transfer matrix method taking into account both incidence angles as before. The microcavity

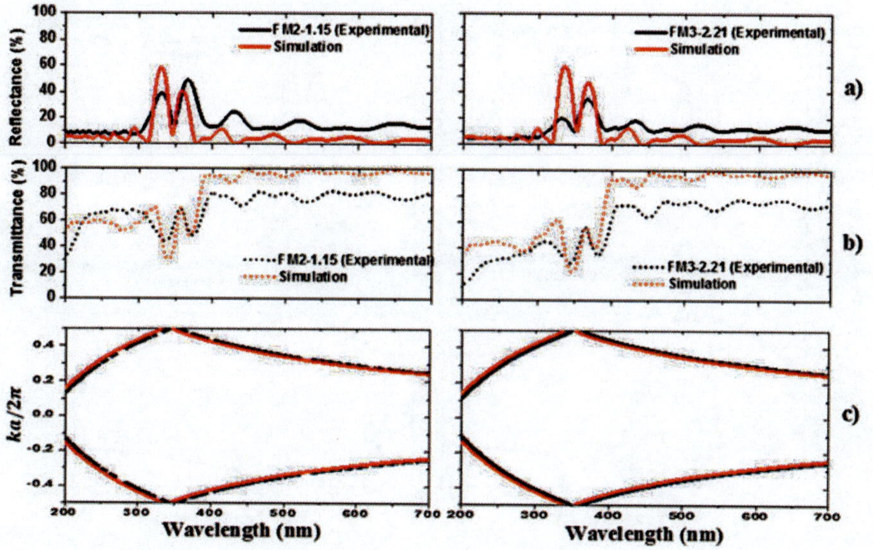

Fig. 2 The theoretical and experimental reflectance and transmittance spectra of UV microcavities and its photonic bandgap structure. **a** Theoretical and experimental reflectance spectrum of UV oxidized microcavities on quartz substrate. **b** Theoretical and experimental transmittance spectrum of UV free-standing microcavities. **c** The photonic bandgap structures of the microcavities

shown in left column in Fig. 2 was subjected to two stages of dry oxidation; firstly, it was oxidized at 350 °C for 30 min and then at 900 °C for 1 h, and the oxygen flow was 1.15 SLPM (standard liter per minute). In the right column, a second microcavity is shown, which was oxidized employing the same stages of dry oxidation mentioned above. In this case, an oxygen flow of 2.21 SLPM was used. The oxidized microcavity FM2 showed a transmittance maximum of 70% in its localized mode (356 nm), while the oxidized microcavity FM3 exhibited a maximum transmission of 56% in the localized mode position (364 nm). Consequently, the oxidized microcavity FM2 has less optical losses than the oxidized microcavity FM3. This was confirmed in a previous report, where Maxwell Garnett and J. E. Lugo approximations were used to calculate the complex refractive index of PS and OPS [6, 13], where it was demonstrated that after dry oxidation the complex refractive index components' values decreased as well as the absorption losses. On the other hand, the reflectance minimum location of both microcavities (FM2 and FM3) was located at 349 nm. The mismatch between both transmittance and reflectance peaks is again due to the difference between the incidence angles used to measure the spectra. The localized mode position of the maximum transmission and minimum reflection spectrum was found within the bandgap edges; this result is displayed in Fig. 2c that corresponds to the photonic bandgap of the microcavities FM2 and FM3.

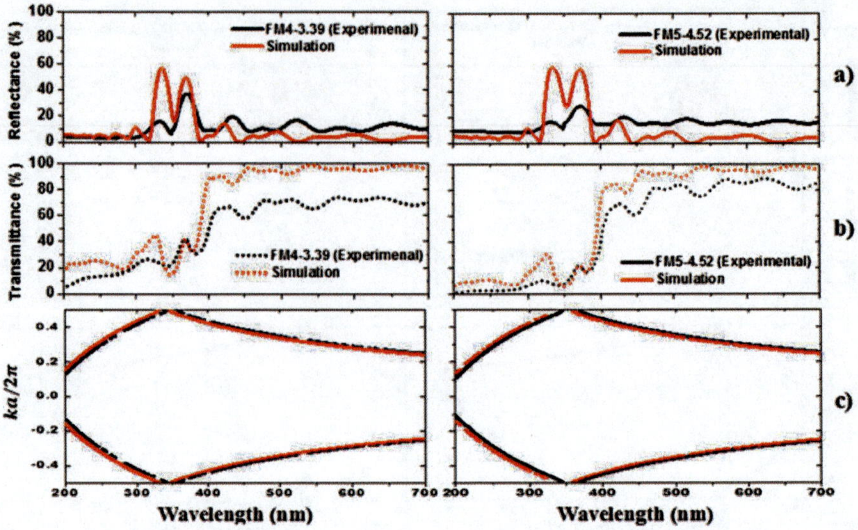

Fig. 3 Theoretical and experimental reflectance and transmittance spectra of oxidized microcavities and their photonic bandgap. **a** The theoretical and experimental reflectance spectra of oxidized microcavity obtained in the UV range. **b** Theoretical and experimental transmission spectra of free-standing UV microcavities. **c** The photonic bandgap structure of the microcavities

The theoretical (red line) and experimental (black line) reflectance (solid line) and transmittance results (dotted line) of two different oxidized microcavities (FM4 and FM5) are displayed in Fig. 3.

The oxidized microcavity to the left (FM4) was oxidized employing the same stages of dry oxidation mentioned above, and during the oxidation process an oxygen of 3.39 SLPM was applied. The microcavity FM4 has a maximum transmission peak of 42% at the localized mode position (368 nm). Meanwhile, the oxidized microcavity FM5 (shown in the right column) exhibited a maximum transmission peak of 21% in the UV range.

The oxygen flow employed for the sample FM5 was 4.52 SLPM. On the other hand, the theoretical and experimental reflectance spectra of the sample FM4 and FM5 presented a minimum reflection peak at 349 nm. The mismatch between the transmission and reflection peaks is again due to the different incidence angles used. Such difference was considered in the theoretical simulation as well. Furthermore, the PBG of both microcavities (FM4 and FM5) is displayed in Fig. 3c, and it can be observed that the localized mode position for samples FM4 and FM5 is found within the photonic bandgap edges. Besides, a decrease of the PBG bandwidth was observed (see Figs. 2b and 3b) in the oxidized microcavities, and this occurs because there was less contrast between the high refractive index and the low index of porous Si-SiO_2 layers.

The complex refractive index for each layer that made up the microcavity was calculated in a previous report, where it was demonstrated that the complex refractive

Fig. 4 Wavelength shift of the localized mode microcavity using different oxygen flows during the oxidation process

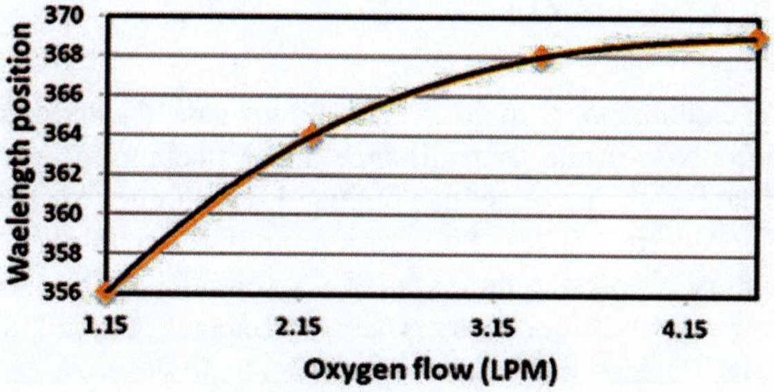

index components of PS are greater than the complex refractive index components of OPS [13].

Also, the maximum transmission peak of the localized mode wavelength in the samples FM3, FM4, and FM5 was reduced (see Figs. 2b and 3b) and shifted to low frequencies, and this occurred when the oxygen flow was increased gradually (see Fig. 4).

In addition, in a previous work, it was demonstrated that oxidized microcavities absorb more UV light, when high oxygen flows were employed [13].

Lastly, we observed a maximum wavelength shift (134 nm) of the localized mode position to higher energy when a low oxygen flow was used to oxidize the PS microcavity and a wavelength shift of 121 nm to high energy when a maximum oxygen flow was applied.

Several authors have reported the stabilization of the optical parameters of PS structures using different oxidation methods [20, 21]. In order to reach a wavelength blueshift in the reflectance and transmittance spectrum of PCs, the oxidation processes have been modified [20–22]. In the oxidation processes, the temperature and oxidation time variations on PS microcavities have been studied[6, 22], where it was reported that the localized mode microcavity moves to shorter wavelengths as the oxidation temperature increases [22] and the same occurs, when the time oxidation increased during the oxidation process in microcavities [6], that is, a greatest fraction of SiO_2 in the PS matrix leads to a more blueshift of the localized mode microcavity.

However, we reported in this work a high oxygen flow is used to obtain UV microcavities which causes a strong absorption in the UV range since the transmittance and reflectance spectra show a low reflectance and transmittance amplitude. The optimal conditions to obtain UV microcavities with good optical quality are attained employing two stages of dry oxidation and a low oxygen flow during oxidation process.

In addition, changing the oxygen flow and time oxidation during the oxidation process of PS microcavities is another way to manipulate its photonic bandgap.

4 Conclusions

In conclusion, a high oxygen flow made the oxidized microcavities exhibit less amplitude in the transmittance and reflectance spectra inside the UV light range. That is, UV microcavities obtained with higher oxygen flow absorb more UV light. Thus, there was less SiO_2 formation in the oxidized microcavity, and consequently optical absorption increased. Therefore, the optimal oxidation process to obtain UV microcavities found here is using two stages of dry oxidation along with a low oxygen flow. Besides, a decrease of the PBG bandwidth was achieved by incorporating SiO_2 within the porous silicon microcavities. This bandwidth decrease occurs because there was less contrast between the high refractive index and the low index of oxidized layers.

Appendix

MATLAB Program Code

This code is used to obtain the transmission and reflection spectra of microcavities in the Vis and UV range.

```
Clc
Clear all
For wavelength=200:800
θ1=sin((0*π)/180);
θH=abs(asin(n0/nH)*sin((0*π )/180))
θL=abs(asin(n0/nL)*sin((0*π )/180))
θc=abs(asin(n0/nc)*sin((0*π )/180))
θsi=abs(asin(n0/nsi)*sin((0*π )/180))
D0=[ 1 1;n0*cos(θ1) -n0*cos(θ1)]
D1=[ 1 1;nH*cos(θH) -nH*cos(θH)]
k1=((nH*2*π )/wavelength)*cos(θH)
x1=k1*dH
P1=[exp(i*x1) 0; 0 exp(-i*x1)]
D2=[1 1;nL*cos(θL) -nL*cos(θL)]
```

```
k2=((nL*2*π )/wavelength)*cos(θL)
x2=k2*dL
P2=[exp(i*x2) 0; 0 exp(-i*x2)]
Dc=[1 1;nc*cos(θc) -nc*cos(θc)]
Dsi=[1 1;nsi*cos(θsi) -nsi*cos(θsi)]
C=D1*P1*inv(D1)*D2*P2*inv(D2)
D=C^N1
E=C^N2
r=Inv(D0)*D*D2*P2*inv(D2)*E*Dc
t= Inv(D0)*D*D2*P2*inv(D2)*E*Dsi
m1=r(1:1)
m2=r(2:2)
m11=t(1:1)
T(wavlength)=abs((nc*cos(thetasi)/n0*cos(theta1))*(1/m11)
^2)
End
z=200:1:800
plot(z,T(z),z,R(z))
```

References

1. Joannopoulos, J.D., Johnson, S.G., Winn, J.N., Meade, R.D.: Photonic Crystals. Princeton University Press, Princeton (2011)
2. Ocier, C.R., Krueger, N.A., Zhou, W., Braun, P.V.: Tunable visibly transparent optics derived from porous silicon. ACS Photonics **4**(4), 909–914 (2017)
3. Kittle, J.D., Gofus, J.S., III., Abel, A.N., Evans, B.D.: Additive combination of spectra reflected from porous silicon and carbon/porous silicon rugate filters to improve vapor selectivity. ACS Omega **5**(31), 19820–19826 (2020)
4. Pérez, K.S., Estevez, J.O., Méndez-Blas, A., Arriaga, J., Palestino, G., Mora-Ramos, M.E.: Tunable resonance transmission modes in hybrid heterostructures based on porous silicon. Nanoscale Res. Lett. **7**(1), 1–8 (2012)
5. Vyunishev, A.M., Pankin, P.S., Svyakhovskiy, S.E., Timofeev, I.V., Vetrov, S.Y.: Quasiperiodic one-dimensional photonic crystals with adjustable multiple photonic bandgaps. Opt. Lett. **42**(18), 3602–3605 (2017)
6. Jimenéz-Vivanco, M.R., García, G., Carrillo, J., Agarwal, V., Díaz-Becerril, T., Doti, R., et al.: Porous Si-SiO 2 based UV microcavities. Sci. Rep. **10**(1), 1–21 (2020)
7. Ghulinyan, M., Oton, C., Bonetti, G., Gaburro, Z., Pavesi, L.: Free-standing porous silicon single and multiple optical cavities. J. Appl. Phys. **93**(12), 9724–9729 (2003)
8. Ghulinyan, M., Gelloz, B., Ohta, T., Pavesi, L., Lockwood, D., Koshida, N.: Stabilized porous silicon optical superlattices with controlled surface passivation. Appl. Phys. Lett **93**(6), 061113 (2008)
9. Estrada-Wiese, D., del Río, J.A.: Refractive index evaluation of porous silicon using Bragg reflectors. Rev. Mex. Física **64**(1), 72–81 (2018)
10. Martín-Palma, R., Ryan, J.V., Pantano, C.: Spectral behavior of the optical constants in the visible/ near infrared of GeSbSe chalcogenide thin films grown at glancing angle. J. Vac. Sci. Technol. A: Vac. Surf. Films **25**(3), 587–591 (2007)

11. Khardani, M., Bouaïcha, M., Bessaïs, B.: Bruggeman effective medium approach for modelling optical properties of porous silicon: comparison with experiment. Phys. Status Solidi C **4**(6), 1986–1990 (2007)
12. Lugo, J., Lopez, H., Chan, S., Fauchet, P.: Porous silicon multilayer structures: a photonic band gap analysis. J. Appl. Phys. **91**(8), 4966–4972 (2002)
13. Jimenéz-Vivanco, M.R., García, G., Carrillo, J., Morales-Morales, F., Coyopol, A., Gracia, M., et al.: Porous Si-SiO2 UV microcavities to modulate the responsivity of a broadband photodetector. Nanomaterials **10**(2), 222 (2020)
14. Rakhimov, R., Osipov, E., Dovzhenko, D., Martynov, I., Chistyakov, A.: Influence of electrochemical etching parameters on the reflectance spectra of porous silicon rugate filters. J. Phys.: Conf. Ser. (IOP Publishing) 012026 (2016)
15. Chhasatia, R., Sweetman, M.J., Prieto-Simon, B., Voelcker, N.H.: Performance optimisation of porous silicon rugate filter biosensor for the detection of insulin. Sens. Actuators B: Chem. **273**, 1313–1322 (2018)
16. Jiménez Vivanco, M.D.R., García, G., Doti, R., Faubert, J., Lugo Arce, J.E.: Time-resolved spectroscopy of ethanol evaporation on free-standing porous silicon photonic microcavities. Materials **11**(6), 894 (2018)
17. Rosli, N., Halim, M.M., Chahrour, K.M., Hashim, M.R.: Incorporation of zinc oxide on macroporous silicon enhanced the sensitivity of macroporous silicon MSM photodetector. ECS J. Solid State Sci. Technol. **9**(10), 105005 (2020)
18. Ramadan, R., Manso-Silván, M., Martín-Palma, R.J.: Hybrid porous silicon/silver nanostructures for the development of enhanced photovoltaic devices. J. Mater. Sci. **55**(13), 5458–5470 (2020)
19. Tregulov, V., Litvinov, V., Ermachikhin, A., Maslov, A.: Influence of deep level defects on photoelectrical processes in pn junction solar cells with porous silicon antireflection coating. In: 2020 ELEKTRO, pp. 1–3. IEEE (2020)
20. Morales, F., García, G., Luna, A., López, R., Rosendo, E., Diaz, T., et al.: UV distributed Bragg reflectors build from porous silicon multilayers. J. Eur. Opt. Soc.-Rapid Publ. **10** (2015)
21. Gelloz, B., Koshida, N.: Stabilization and operation of porous silicon photonic structures from near-ultraviolet to near-infrared using high-pressure water vapor annealing. Thin Solid Films **518**(12), 3276–3279 (2010)
22. El-Gamal, A., Ibrahim, S.M., Amin, M.: Impact of thermal oxidation on the structural and optical properties of porous silicon microcavity. Nanomater. Nanotechnol. **7** (2017). https://doi.org/10.1177/1847980417735702

Signal Processing, Computer Vision and Rhythm Engineering

An Approach to Detect Chronic Kidney Disease (CKD) by Removing Noisy and Inconsistent Values of UCI Dataset

Sabrina Jahan Maisha, Ety Biswangri, Mohammad Shahadat Hossain, and Karl Andersson

Abstract With the alarming rate of increase in chronic kidney disease (CKD) cases all over the world, researchers are trying to resolve it with state-of-the-art methods. It is evident that in a certain time period such disease gradually disrupts other organs functioning eventually causing death of patients. Early detection of CKD can diminish the chances of further damage to a great extent. Considering the UCI Machine Learning CKD dataset, this work attempts to present a more reliable approach, enabling handling of noisy data. However, CKD dataset contains noisy and inconsistent values, resulting in inaccurate prediction of CKD by using traditional machine learning algorithms. Therefore, this research presents an approach of handling noisy and inaccurate values of CKD dataset by employing a combination of deep neural network, statistical methods, Principal Component analysis (PCA), and "SMOTE". Consequently, the refined CKD dataset coming out of the mentioned pre-processed methods is used in various machine learning methods. Our results showed that RF outperformed with 98.5% accuracy among Support Vector Machine (SVM), K-Nearest Neighbor (KNN), Decision Tree (DT), Naive Bayes (NB), and Logistic Regression (LR) classifiers. Additionally, we found that the features such as serum-creatinine and blood urea exhibited their dominance in outcome prediction.

Keywords Chronic kidney disease · Machine learning · Principal component analysis · Deep learning imputation

S. J. Maisha · E. Biswangri
BGC Trust University Bangladesh, Biddyanagar, Chandanaish, Chattogram, Bangladesh
e-mail: sabrina@bgctub.ac.bd

M. S. Hossain (✉)
University of Chittagong, Chittagong, Bangladesh
e-mail: hossain_ms@cu.ac.bd

K. Andersson
Luleå University of Technology (LTU), Skellefteå, Sweden
e-mail: karl.andersson@ltu.se

M. S. Kaiser et al. (eds.), *Proceedings of the Third International Conference on Trends in Computational and Cognitive Engineering*, Lecture Notes in Networks and Systems 348, https://doi.org/10.1007/978-981-16-7597-3_38

1 Introduction

Chronic kidney disease (CKD) is an abnormal condition in the human body due to disorders in kidney functioning. It is a global public health concern remarked as a major non-contagious disease by World Health Organization (WHO) [19]. Clinically, CKD can be termed as the condition with a decreased rate of glomerular filtration than the normal category for consecutive 3 months. Besides these, getting a strain of a certain amount of albumin-to-creatinine ratio in the urine sample is also a way to identify CKD [18]. In simpler terms, CKD is the degradation stage of a human body losing its capability to conduct blood filtration and transformation of wastes in urine. The underlying causes for such conditions can occur due to having a family history of kidney problems, birth impairment, unhealthy lifestyles, hypertension, or being a diabetic patient. In a survey study of Global Burden Disease 2015, the researchers of the International Society of Nephrology found that globally since the year 2005 there has been an increase of mortality rate by 30% due to CKD. Moreover, annually 5–10 million people lose their lives due to kidney dysfunctionality [30]. However, the rate of patients suffering from the last stage renal disease (ESRD) requiring dialysis and kidney transplantation are also increasing rapidly over the span of years [28].

Basically, the deterioration of kidney performance in a human body takes place gradually over a certain period. It requires a persistent indication for at least 3 months to start treatment. The continuous degradation phenomenon remains mostly unidentified and undetected at the early stage of CKD. Although at the very beginning CKD does not exhibit any significant symptoms, a progressive change helps to identify the patient. Due to the delay in tracing the issue, it not only affects kidneys but also disrupts the proper functioning of the human nervous system or collapses the immune system. Additionally, the treatment of kidney disease imposes a huge financial burden on the patients even in developed countries [8].

Consequently, it has been observed that by detecting the kidney failure symptoms at early stages and performing tests on certain body features patients in a more vulnerable stage can be identified. So, in this attempt, we aim to find certain features that contribute greatly to the early detection of CKD.

Previously, many researches have been conducted on the UCI repository CKD dataset developed by a senior consultant nephrologist, Apollo Hospitals, India. Dealing with this dataset is a bit challenging because of inconsistency and missing values. In most of the previous works, various common statistical measures were undertaken to handle both categorical and numerical attributes of the dataset. Further, either applying two or three popular ML models such as SVM, NB, KNN, Artificial Neural Network (ANN), or DT the dataset had been analyzed. Health care being a severe issue advanced imputation techniques will help to create more dependable data to train our model. Considering this fact, in this proposed method an advanced deep neural network library will be used for data imputation. As this approach aims to detect the attributes responsible for CKD condition, principal component analysis will be implemented for feature extraction. Afterward class imbalance resolve technique, namely, "smote" technique will be used. Finally, a comparative analysis will be done on the behavior of the six most popular machine learning classifiers, i.e., SVM, KNN, DT, RF, LR, and NB to evaluate the dataset.

Therefore, the contribution of this research consists of: (a) imputation of missing values with deep neural network architecture-based advanced library, i.e., "datawig"; (b) feature extraction by principal component analysis; (c) resolving imbalance of class by "smote" technique; and (d) comparative analysis on the filtered dataset using six ML classifiers, i.e., SVM, KNN, DT, RF, LR, and NB. Eventually, this kind of strategy can be help to deduce comprehensive results with good accuracy in terms of real-world context.

In the following section the paper has been arranged in five sections. Section 2 reviews the previous works done in this field and the various approaches used by the researches. Section 3 discusses the framework that has been designed to conduct the proposed method. Section 4 demonstrates the output results obtained in the stated methodology in a detailed manner. Section 5 concludes the whole result of the work with future recommendations.

2 Literature Review

Over the years as the number of cases increased there has been a wide range of researches concerning CKD. Charlenon et al. [9] in 2016 had developed a predictive model on the UCI CKD dataset by comparing performances of LR, SVM, KNN, and DT. At the pre-processing step, they had transformed nominal attributes into binary. Applying BestFirst selection methods for feature selection, the SVM classifier outperformed with 98.3% accuracy. Comparison of classifiers in this work lacks proper handling of missing values during data processing.

In a study of 2017 by Basar et al. [6] that particular dataset had been evaluated in two approaches. Firstly, REPTree, BFTree, J48 Decision Tree, and SVM had been applied individually as well as boosting through ensemble method in the whole dataset. J48Tree classifier surpassed others with the highest accuracy of 100% accuracy. In the other approach, before going through a similar process 10 best features had been selected through the information gain attribute evaluator. BFTree classifier showed the maximum accuracy of 99.25% and 100%, respectively, without and with ensemble method. Though both BFTree and J48Tree classifiers manifested the highest possible accuracy, this may require a lot of computational time and as there was no scope of parameter tuning the accuracy somewhat indicates an overfitting issue.

In 2018, researcher Amirgaliyev et al. [5] had conducted a study to build up a model on the UCI CKD dataset. They had chosen the classifier that performs well in both classification and regression cases, i.e., SVM for this purpose. Prior to applying SVM, the missing values and the training phase had been handled using the sequential minimal optimization (SMO) method. Measuring its performance with 10-fold cross-validation criteria, an accuracy of 94.6% had been obtained. This work had just evaluated the impact of SVM in detecting the disease but cannot assure its reliability in terms of other methods.

On the same dataset, Marwa Almasoud and Tomas E Ward [4] in 2019 performed research for early detection of CKD patients using LR, SVM, RF, and gradient boosting algorithm. To handle the missing values and noisy data, the authors used multiple imputation (MI) technique. As a result of the classifiers integration by linear regression and logistic regression synthetic values were generated. Statistical measures such as Pearson's correlation, ANOVA test, and Cramer's V outlined the association of features with each other. These results further helped to select features through filter feature selection methods removing the redundancy. Although the gradient boosting technique showed higher accuracy by 99.1% accuracy, other classifiers exhibited nearby performance results. This work at the end deduced that the attribute is haemoglobin having stronger correlativity in detecting CKD.

While in another study in the same year 2019 by Almansour et al. [3], a comparative study had been done assessing through the neural network classifier ANN and SVM on the CKD dataset. During the pre-processing of the dataset, imputation of values had been accomplished using mean, median, standard deviation, and other popular statistical measures of the distribution. Selecting features using correlation co-efficient technique and with the help of several iterations, ANN yielded the highest accuracy of 99.8%.

UCI CKD dataset that has been considered contains imbalanced missing and noisy values. Going through the previous studies it has been observed that most of the findings either used mean, median, or other common statistical techniques to replace the missing values. On the contrary, feature selection techniques somewhat created biasness on the output results. Finding the solution to build up a good model to detect CKD for real-world application certain limitations of the prior works must be considered. In this respect, with the aim to address these issues, our proposed system will help to build up a more reliable model.

3 Methodology

In this section, the overall architecture of our system has been demonstrated. At the pre-processing stage, at first, with the combination of both deep neural network and mean categorical and numeric missing values will be imputed. Then the whole dataset needs to be normalized by MinMax Scaler to avoid biases in obtained results. With the help of the PCA technique, features will be extracted from the dataset. To handle oversampling, "SMOTE" will be applied to the features. Finally, the six ML classifiers will be implemented individually to analyze their behavior in terms of the CKD dataset. In Fig. 1, the structure of our proposed system has been shown.

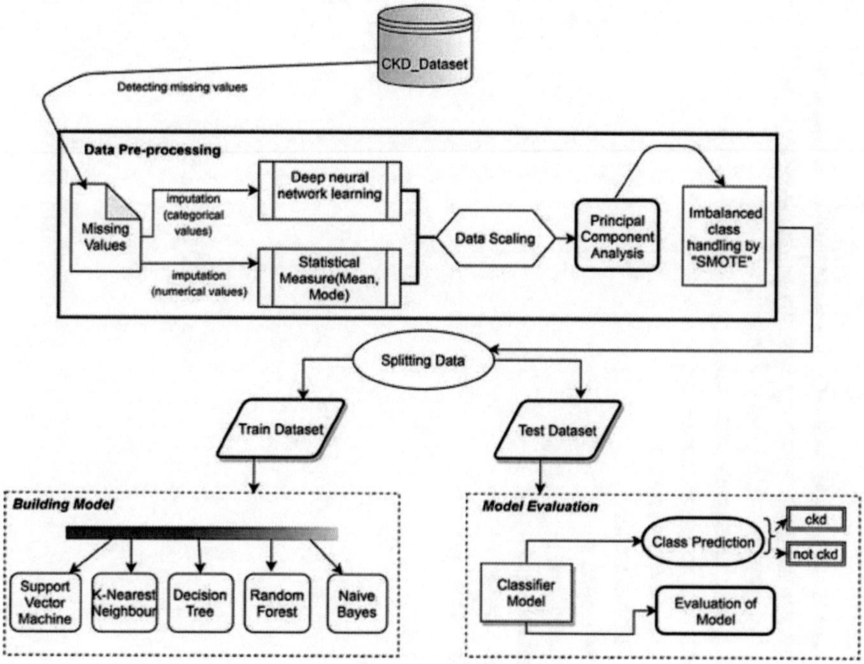

Fig. 1 Proposed methodology for prediction of chronic kidney disease

3.1 Dataset Description

To perform comparative study, CKD dataset accessible at the UCI repository is used in this system. This dataset was developed in 2015 over a period of 2 months by the Apollo Hospitals, India. In total, there are 400 cases, among them 250 of the records represents CKD patients and rest of the 150 records shows information of non-CKD patients. So the dataset is a binary class problem as there are two labels. Looking over the whole dataset it can be seen that there are 24 attributes to identify the output classes. However, all the attributes are not of the same type, i.e., 11 of them are numerical and remaining 13 are nominal. Basically, the dataset is a bit challenging to handle containing a lot of missing and imbalanced values. Prior to using the dataset, it is very crucial to understand the kind of values it exhibits which will help to handle the values more efficiently. Figure 2 shows the percentage of missing values in the dataset. It can be seen that the attribute red blood cell has more than 35% missing values. So it is important to handle the values in a more efficient manner.

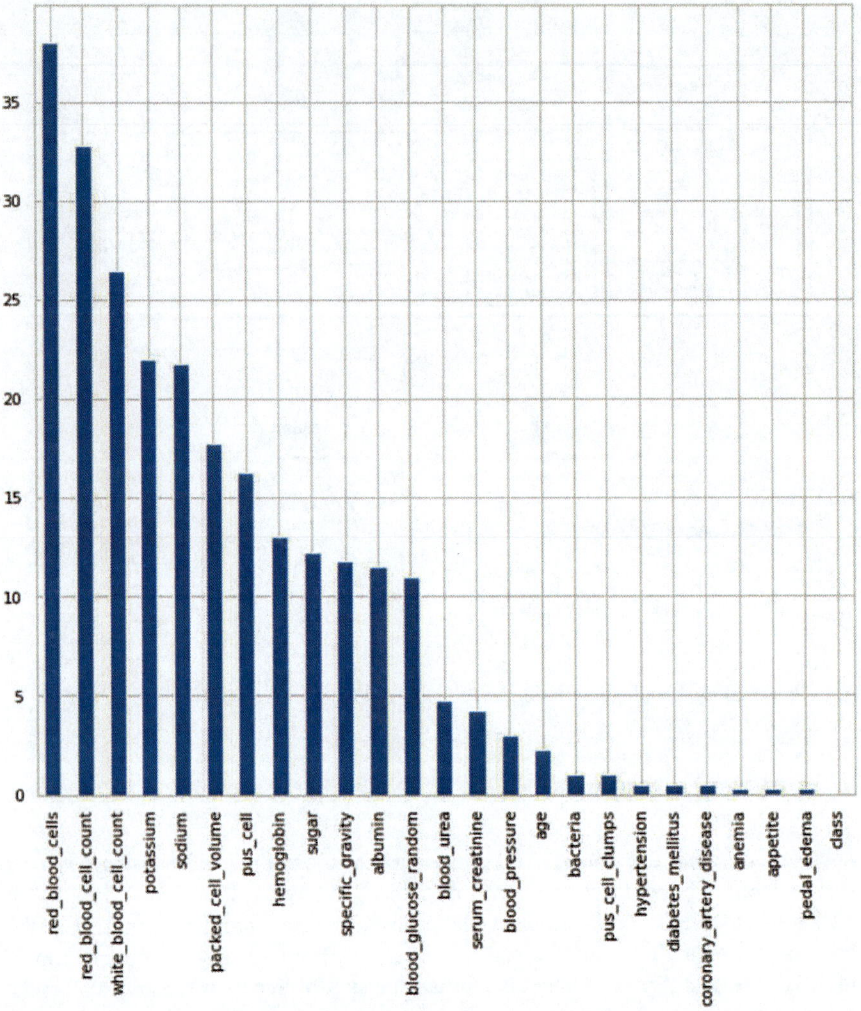

Fig. 2 Percentage of missing values in all attributes

3.2 Pre-processing of Dataset

In the real-world scenario, it is often common to have missing, inconsistent, noisy, and redundant values. Being a very sensitive arena, every value plays a very significant role in healthcare sector. These must be dealt within a sophisticated manner. Since machine learning classifiers work on by getting continuously trained up with various data, refined and polished data is a must to build up an accurate model.

Imputation of Missing Values: Dealing with datasets related to medical issues needs to be more finely tuned. In some of the studies, it can be observed that if certain columns have missing values it is discarded from the dataset. Eventually such kind of models, especially in the health industry, can create alarming conditions with misleading results. Few recent studies implemented the traditional techniques of common statistical measures (mean, median, and mode) to handle missing data of the CKD dataset. Although a good accuracy rate may be obtained with such tactics, it ignores the feature correlation and generates biased results. Thus reliability to use such models in real-life cases can be quite questionable. At first, the missing values are identified by finding the keyword "NaN" among the attribute values. The model then detects whether it is a categorical- or numeric-type feature. If numeric feature missing values are found, the mean of the distribution is imputed in the missing rows.

To handle the categorical values, we will use an advanced deep neural network architecture "datawig" which was developed by AWS labs in 2019 [7]. This library provides more robust and scalable features to impute data with dynamic hypertuning properties. In terms of heterogeneous data, this technique can show promising results. At the initial stage, when an attribute is sent to "datawig", it tries to figure out whether it is categorical or sequential (text data). Using the one-hot encoding, this approach transforms the bunch of available categorical values into numerical ones. To extract the features of the values, transformed values are then passed to the embedding layer. Finally, digging the features, components are merged into a logistic layer to accomplish the imputation process. In this way, a combination of deep learning module with neural architecture and successive optimization mechanism imputes categorical values. Thus, the dataset imputation process is completed at this stage.

Data Scaling: Normalization or scaling of data is a process to recompute the existing values in a certain boundary. Applying a range for the data values helps to remove biasness due to higher magnitude variables. In this system, MinMax Scaler technique is carried out to normalize the dataset. Generally, existing values of each feature are normalized by calibrating them from 0 to 1 ratio. To estimate each and every normalized, new values of that feature, the following equation is used:

$$x' = \frac{x - min(x)}{max(x) - min(x)} \tag{1}$$

So mapping the highest or lowest value in "x" variable 1 or 0 will be, respectively, the result of new scaled value x'. MixMax Scaler approach successively scales the values which later handles the sensitivity of ML classifiers.

Feature Extraction and Reduction using PCA: Among all the attributes of the dataset, some features play an important correlation in predicting the target result. In our work for early detection of CKD, it is required to pull out the dominant features from a large feature space and understand the correlation of those features with our expected output. To accomplish this task in our system, principal component analysis has been employed. This technique extracts and reduces features from a large set of feature space to the new set of independent features. However, due to use of PCA outlier's detection problem can also be resolved. Basically, at first, data are standardized in order to execute the technique in a productive way. Further calculating the values of eigenvectors an information regarding the pattern of the attributes is formed. Choosing the components on the basis of eigenvalues a feature vector is created. This phase basically reduces the redundant features and each time identifying the principal components a new feature set is created. Intuitively, PCA through this process presents an orthogonal projection of the relevant feature information responsible for characterizing expected output.

Imbalanced class handling by "SMOTE": Synthetic minority oversampling technique shorty known as "SMOTE" has shown satisfactory results in handling data imbalance problem. Substantially, "SMOTE" technique solves the overfitting problem of dataset by replacing the minority instances closer to that feature space. Selecting a random sample from the minority class it tries to pick some neighboring samples to synthesize new instances. To start the process a certain minority attribute is picked up at the very beginning stage. After choosing any random value of that attribute, its nearest neighbor values are calculated using the given Euclidean distance formula.

$$\sqrt{\sum_{i=1}^{n} X - X_i{}^2} \tag{2}$$

Using the k-nearest neighbor strategy, an intermediary set of features are created from existing ones with the help of random selection. For each minority values new instances are generated by comparing the initial value with intermediary ones and through some calculations. The following equation helps to generate those instances.

$$X + rand(0, 1) * |X - X_k| \tag{3}$$

Consequently, this process resolves the oversampling problem by increasing the percentage of minority classes replicating existing values.

3.3 Building Model

Preparing the dataset through various pre-processing stages, the dataset is split into two proportions to feed the machine learning classifiers. Here the most popular and traditional classifiers such as SVM, NB, KNN, RF, LR, and DT are used individually to build up a model.

SVM is a very popular classifier used mostly in classification problems [2]. It has shown satisfactory results in medical sector, pattern recognition, weather prediction, and many other significant areas of study. This computational algorithm performs well in higher dimensional space by creating an optimal hyperplane between the object classes separating them with maximum distance. As our classification problem is of binary type, linear kernel SVM can show better results.

The Bayesian classifier, NB, is an algorithm with less computational cost. It has been seen in many studies that this classifier has shown good results in cardiovascular and other diseases [21]. However, it can treat categorical variables more appropriately. Among the three types of NB classifier, Gaussian is selected due to having variation in the type of attributes.

Being a non-parametric classifier, KNN provides the facility to easily identify unknown instances from some neighboring data values using some distance metrics computations. In our analysis, Euclidean measure will be used to determine values of neighboring instances. Furthermore, in medical sector, this simple classifier is very often seen implemented in several studies [17].

RF classifier is basically an ensemble-approach-based technique machine learning model. To implement the ensemble learning technique, large number of decision trees are formed individually. Each of these trees performs a prediction of the target individually in multiple clusters. Thus being less susceptible to overfitting of data as well as works well in classification domain, RF is chosen in this method [22].

Another commonly used classifier for binary classification problems is logistic regression. It is commonly used in health analytic and has shown good results in this sector [20]. Unlike many other classifiers, it can handle both categorical and numerical values.

The tree-based classification approach following divide-and-conquer strategy is the basic idea of DT. As it extracts the underlying function persistent among the features and predicted outcome, implementation of such classifiers will help to understand the dominant features in classification cases.

3.4 Model Evaluation Criteria

Building the model evaluation is important as it helps to understand how better it can work actually. So for our system we have considered two approaches, i.e., splitting dataset and cross-validation.

Splitting process basically segments the dataset into training and testing ratio. Although the ratio is not fixed, in general, 70:30 is considered as the standard one [26]. Both the training and testing datasets are individually tested in terms of accuracy, sensitivity, specificity, and F1-score. Accuracy will measure how good the model can predict CKD in total number of observations. Sensitivity will show that the patients those who do not have CKD how much correctly they are predicted. Specificity will show the degree of wrong prediction in total actual negatives. On the other hand, F1-score is the average measure of specificity and sensitivity.

Another evaluation metric is cross-fold validation. In this evaluation process, randomly different ratios of training and testing datasets are taken. This evaluation strategy removes the biasness of the model. Most commonly 10-fold cross-validation is most suitable for the ML models [25].

In this way, all the six chosen ML classifiers are to be implemented individually in our proposed system. Then the results generated by each ML classifiers are to be compared through analyzing precision, sensitivity, F1-score, and accuracy. Finally, comparative study can vividly present the best performing classifier which is to be stated as the best suited model for CKD dataset.

4 Results

The experimental setup of our system has been implemented in Google Colab environment using scikit-learn and other necessary Python libraries. Before applying any strategy dataset must be thoroughly studied. So to understand the inconsistencies in dataset values, Fig. 3 can be observed which depicts that the attribute "while blood cell count" has a higher magnitude among all others in terms of inconsistency. In our

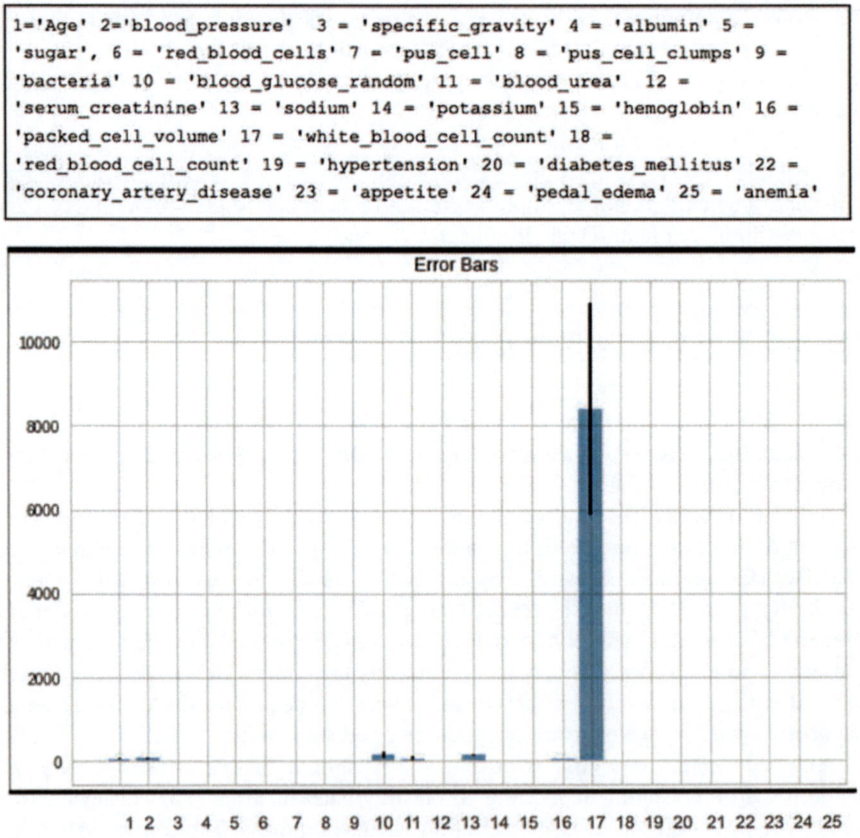

Fig. 3 Error bar chart showing inconsistencies in attributes of the dataset

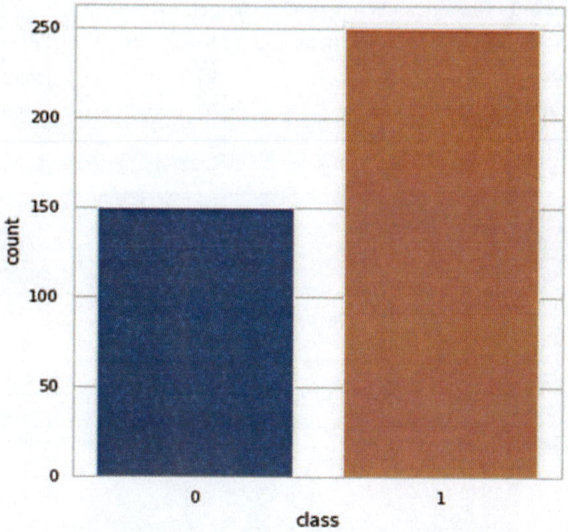

Fig. 4 Before oversampling class distribution

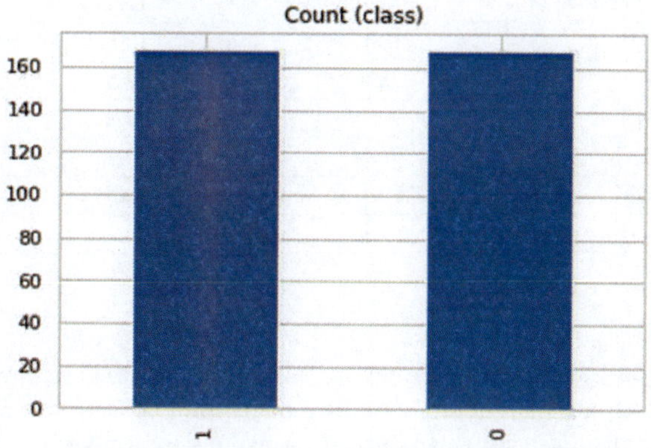

Fig. 5 After oversampling class distribution

proposed system, at first, we refined the dataset by imputing missing values with the combination of deep learning method and mean.

After data scaling through MinMax Scaler technique, the feature is extracted by PCA. As the data records had imbalance distribution of class "SMOTE" technique is employed. In Figs. 4 and 5, a comparative graphical representation shows the before and after conditions exhibiting the results of "smote" in effectively dealing with class imbalance issue.

After building a model it is required to evaluate the performance of our proposed system. At first, splitting evaluation is performed. The dataset is split into 67% training set and 33% testing set to assess the building models performance. Examining with different ratio numbers this proportion worked well in our case. The detailed classification report having sensitivity, precision, and F1-score for both CKD and

Table 1 Classification report of the ML classifiers

News type	Classifier name	Sensitivity (%)	Precision (%)	F1-score (%)
not_ckd	RF	97.8	98.2	98
ckd	RF	98	99.9	99
not_ckd	KNN	70	72	71
ckd	KNN	75.1	79	77
not_ckd	NB	96.5	99.8	98.1
ckd	NB	92.6	97.4	94.9
not_ckd	SVM	97.8	96.2	97
ckd	SVM	98.2	97.8	98
not_ckd	DT	91.8	94.3	93
ckd	DT	98.8	99.9	99.4
not_ckd	LR	98	92.2	95
ckd	LR	95.5	98.6	97

Table 2 Prediction results of classifiers using splitting and 10-fold cross-validation

Classifier	Training (%)	Testing (%)	Cross-validation (%)
RF	100	100	98.5
KNN	98.8	98.5	74
NB	99.1	99.2	96.5
SVM	98.5	97.7	97
DT	100	100	96.2
LR	97.9	96.2	97.5

non-CKD cases is shown in Table 1. In the next step, with the purpose to finely deal with the parameters avoiding overfitting of data samples, 10-fold cross-validation is also used. Both analytical procedures will give insight into whether our model is capable of exhibiting accurate results in a complex world. Table 2 shows the performance analysis in both splitting and 10-fold cross-validation procedures. Similarly in Fig. 6, the comparative results of the classifiers with their cross-fold validation, training, and testing accuracies are shown below.

Observing both the results in Tables 1, 2 and Fig. 6 it can be seen that RF and DT show highest performance of 100% accuracy in their training and test datasets. While in cross-validation procedure, RF classifier outperforms others with 98.5% accuracy. On the other hand, LR showed poor results having 97.9% and 96.2% accuracies, respectively, for training and testing datasets. Although KNN performed well in splitting phase, cross-validation evaluation indicated it to be a poor model with 74% accuracy. Going through the rank of performance, it can be observed that RF (98.5%), LR (97.5%), SVM (97%), NB (96.5%), DT (96.2%), and KNN (74%) which gives the descending order of classifiers.

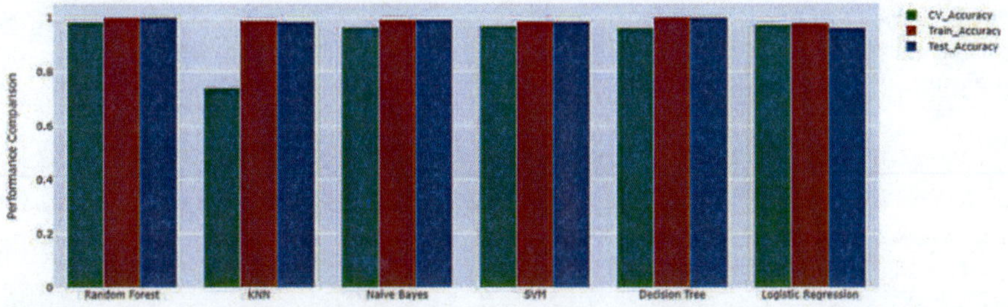

Fig. 6 Comparative analysis of the ML classifiers

Table 3 Comparison of proposed system and other existing studies on this CKD dataset

Author	Methodology	Handling technique for missing data	Accuracy
Charleonnan et al. 2016 [9]	LR, SVM, KNN, and DT	Best first search method for feature selection with no specific missing value handling technique	98.3% in case of SVM
Amirgaliyev et al. 2018 [5]	SVM	Sequential minimal optimization (SMO)	94.6% in case of SVM
Proposed method 2021	SVM, KNN, RF, DT, NB, and LR	Deep learning imputation method "Datawig"	98.5% in case of RF

RF approach analyzes dataset by forming various number of decision tree and each tree is independent in terms of performance from the other. Furthermore, this model created different sample tress utilizing various aspects of feature correlation. Obtaining the best result of the decision trees through voting, comparatively this classifier has been able to build a good model. Contrarily, having scaled dataset and reducing features cannot contribute to the learning of KNN. The reason behind such results lies in selecting random samples without considering other cases.

To realize the impact of our proposed system in real-world context, a comparative analysis of few studies with our proposed system using this UCI CKD dataset is shown in Table 3. The comparison analyzes how each of the study handled missing values along with accuracies. So through the inspection, our proposed system shows robustness in terms of better accuracy and well-developed data handling strategy. Accomplishing the comparative studies of the best classifier model, an important aspect of our study also needs to be analyzed. Feature correlation is the most crucial identifying factor in case of CKD early detection. In Fig. 7, feature correlation is shown. In this purpose, it can be observed that the two features such as serum-creatinine and blood urea dominate to detect the CKD patients. However, graphical analysis depicts that in certain cases serum-creatinine feature is more promising in predicting the outcome.

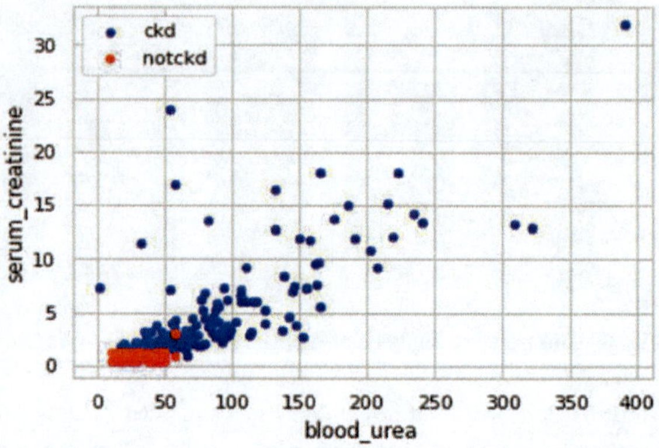

Fig. 7 Scatterplot of dominant features correlation

5 Conclusion

In this paper, an approach is outlined in early prediction of CKD patients. In this binary class problem, noisy and inconsistent dataset is handled with different techniques. Imputation of missing values using deep neural network learning tool, data normalization, higher dimensional feature space reduction by PCA technique, and resolving class imbalance via "smote" technique have been the prime focus of our work. Preparing our dataset, we have attempted to interpret the behavior of ML models. In this context, six ML classifiers, i.e., RF, SVM, DT, LR, NB, and KNN have been executed on the dataset. Out of all of them on the basis of standard evaluation metrics, RF showed promising results with 98.5% accuracy. Due to applying advanced approaches in data handling, our model can show comprehensive results in medical sector.

Further, our work can be extended by having comparative analysis with other CKD datasets consisting of similar attributes. Moreover, our existing dataset can be extended to get more data which will help to implement advanced deep learning models such as CNN, LSTM, and so on [1, 23, 24, 27]. This may boost the performance of our model in future. This work can be extended by applying belief rule-based expert system as it may yield good results in case of uncertainty [10–16, 29].

References

1. Afroze, T., Akther, S., Chowdhury, M.A., Hossain, E., Hossain, M.S., Andersson, K.: Glaucoma detection using inception convolutional neural network v3. In: International Conference on Applied Intelligence and Informatics, pp. 17–28. Springer (2021)
2. Aljawad, D.A., Alqahtani, E., Ghaidaa, A.K., Qamhan, N., Alghamdi, N., Alrashed, S., Alhiyafi, J., Olatunji, S.O.: Breast cancer surgery survivability prediction using Bayesian network and support vector machines. In: 2017 International Conference on Informatics, Health & Technology (ICIHT), pp. 1–6. IEEE (2017)

3. Almansour, N.A., Syed, H.F., Khayat, N.R., Altheeb, R.K., Juri, R.E., Alhiyafi, J., Alrashed, S., Olatunji, S.O.: Neural network and support vector machine for the prediction of chronic kidney disease: a comparative study. Comput. Biol. Med. **109**, 101–111 (2019)
4. Almasoud, M., Ward, T.E.: Detection of chronic kidney disease using machine learning algorithms with least number of predictors. Int. J. Soft Comput. Its Appl. **10**(8) (2019)
5. Amirgaliyev, Y., Shamiluulu, S., Serek, A.: Analysis of chronic kidney disease dataset by applying machine learning methods. In: 2018 IEEE 12th International Conference on Application of Information and Communication Technologies (AICT), pp. 1–4. IEEE (2018)
6. Basar, M.D., Akan, A.: Detection of chronic kidney disease by using ensemble classifiers. In: 2017 10th International Conference on Electrical and Electronics Engineering (ELECO), pp. 544–547. IEEE (2017)
7. Biessmann, F., Rukat, T., Schmidt, P., Naidu, P., Schelter, S., Taptunov, A., Lange, D., Salinas, D.: Datawig: missing value imputation for tables. J. Mach. Learn. Res. **20**(175), 1–6 (2019)
8. Chan, W.K., Huang, L., Gudikote, J.P., Chang, Y.F., Imam, J.S., MacLean, J.A., Wilkinson, M.F.: An alternative branch of the nonsense-mediated decay pathway. EMBO J. **26**(7), 1820–1830 (2007)
9. Charleonnan, A., Fufaung, T., Niyomwong, T., Chokchueypattanakit, W., Suwannawach, S., Ninchawee, N.: Predictive analytics for chronic kidney disease using machine learning techniques. In: 2016 Management and Innovation Technology International Conference (MITicon), pp. MIT–80. IEEE (2016)
10. Hossain, M.S., Andersson, K., Naznin, S.: A belief rule based expert system to diagnose measles under uncertainty. In: World Congress in Computer Science, Computer Engineering, and Applied Computing (WORLDCOMP'15): The 2015 International Conference on Health Informatics and Medical Systems 27/07/2015-30/07/2015, pp. 17–23. CSREA Press (2015)
11. Hossain, M.S., Monrat, A.A., Hasan, M., Karim, R., Bhuiyan, T.A., Khalid, M.S.: A belief rule-based expert system to assess mental disorder under uncertainty. In: 2016 5th International Conference on Informatics, Electronics and Vision (ICIEV), pp. 1089–1094. IEEE (2016)
12. Hossain, M.S., Rahaman, S., Mustafa, R., Andersson, K.: A belief rule-based expert system to assess suspicion of acute coronary syndrome (ACS) under uncertainty. Soft Comput. **22**(22), 7571–7586 (2018)
13. Hossain, M.S., Sultana, Z., Nahar, L., Andersson, K.: An intelligent system to diagnose chikungunya under uncertainty. J. Wirel. Mob. Netw. Ubiquitous Comput. Dependable Appl. **10**(2), 37–54 (2019)
14. Islam, R.U., Hossain, M.S., Andersson, K.: A novel anomaly detection algorithm for sensor data under uncertainty. Soft Comput. **22**(5), 1623–1639 (2018)
15. Islam, R.U., Hossain, M.S., Andersson, K.: A deep learning inspired belief rule-based expert system. IEEE Access **8**, 190637–190651 (2020)
16. Karim, R., Andersson, K., Hossain, M.S., Uddin, M.J., Meah, M.P.: A belief rule based expert system to assess clinical bronchopneumonia suspicion. In: 2016 Future Technologies Conference (FTC), pp. 655–660. IEEE (2016)
17. Khateeb, N., Usman, M.: Efficient heart disease prediction system using k-nearest neighbor classification technique. In: Proceedings of the International Conference on Big Data and Internet of Thing, pp. 21–26 (2017)
18. Levey, A.S., Eckardt, K.U., Tsukamoto, Y., Levin, A., Coresh, J., Rossert, J., Zeeuw, D.D., Hostetter, T.H., Lameire, N., Eknoyan, G.: Definition and classification of chronic kidney disease: a position statement from kidney disease: improving global outcomes (KDIGO). Kidney Int. **67**(6), 2089–2100 (2005)
19. Luyckx, V.A., Tonelli, M., Stanifer, J.W.: The global burden of kidney disease and the sustainable development goals. Bull. World Health Organ. **96**(6), 414 (2018)
20. Manogaran, G., Lopez, D.: Health data analytics using scalable logistic regression with stochastic gradient descent. Int. J. Adv. Intell. Parad. **10**(1–2), 118–132 (2018)
21. Miranda, E., Irwansyah, E., Amelga, A.Y., Maribondang, M.M., Salim, M.: Detection of cardiovascular disease risk's level for adults using naive Bayes classifier. Healthc. Inform. Res. **22**(3), 196 (2016)

22. Mohapatra, S.K., Mohanty, M.N.: Big data analysis and classification of biomedical signal using random forest algorithm. In: New Paradigm in Decision Science and Management, pp. 217–224. Springer, Berlin (2020)
23. Nahar, N., Ara, F., Neloy, M.A.I., Barua, V., Hossain, M.S., Andersson, K.: A comparative analysis of the ensemble method for liver disease prediction. In: 2019 2nd International Conference on Innovation in Engineering and Technology (ICIET), pp. 1–6. IEEE (2019)
24. Progga, N.I., Hossain, M.S., Andersson, K.: A deep transfer learning approach to diagnose covid-19 using x-ray images. In: 2020 IEEE International Women in Engineering (WIE) Conference on Electrical and Computer Engineering (WIECON-ECE), pp. 177–182. IEEE (2020)
25. Ramezan, C.A., Warner, T.A., Maxwell, A.E.: Evaluation of sampling and cross-validation tuning strategies for regional-scale machine learning classification. Remote Sens. **11**(2), 185 (2019)
26. Rácz, A., Bajusz, D., Héberger, K.: Effect of dataset size and train/test split ratios in QSAR/QSPR multiclass classification. Molecules **26**(4), 1111 (2021)
27. Rezaoana, N., Hossain, M.S., Andersson, K.: Detection and classification of skin cancer by using a parallel CNN model. In: 2020 IEEE International Women in Engineering (WIE) Conference on Electrical and Computer Engineering (WIECON-ECE), pp. 380–386. IEEE (2020)
28. Ruiz-Arenas, R., Sierra-Amor, R., Seccombe, D., Raymondo, S., Graziani, M.S., Panteghini, M., Adedeji, T.A., Kamatham, S.N., Biljak, V.R.: A summary of worldwide national activities in chronic kidney disease (CKD) testing. Ejifcc **28**(4), 302 (2017)
29. Sultana, Z., Nahar, L., Basnin, N., Hossain, M.S.: Inference and learning methodology of belief rule based expert system to assess chikungunya. In: International Conference on Applied Intelligence and Informatics, pp. 3–16. Springer (2021)
30. Wang, H., Naghavi, M., Allen, C., Barber, R.M., Bhutta, Z.A., Carter, A., Casey, D.C., Charlson, F.J., Chen, A.Z., Coates, M.M., et al.: Global, regional, and national life expectancy, all-cause mortality, and cause-specific mortality for 249 causes of death, 1980–2015: a systematic analysis for the global burden of disease study 2015. The Lancet **388**(10053), 1459–1544 (2016)

Deep Neural Networks for Brain Tumor Detection from MRI Images

Md. Kawsher Mahbub, Milon Biswas, Md. Abdul Mozid Miah, and M. Shamim Kaiser

Abstract Brain tumor patients have a significant mortality rate. If the tumors are misdiagnosed, it may result in ineffective medical treatment and reduce their life chances. As the risk of brain tumors increases with age and the world's population ages, there is an urgent need to develop low-cost, easy-to-use early detection technologies. MRI scans are commonly used to visualize a patient's brain. Artificial intelligence (AI), deep learning (DL), and its sub-domain have recently reduced the need for human judgment in detecting disorders. DL models are increasingly being employed in traditional supervised learning algorithms due to their inherent advantages of automatically obtaining the required features from images. The detection of brain tumors is one of the most challenging tasks in biomedical imaging. This study is intended to propose a deep neural network (DNN) based solution with a limited number of epochs and parameters. The experiment was conducted on two different datasets, and the proposed DNN obtained 99.22% accuracy, 98.94% sensitivity, 99.53% specificity, 99.57% precision, and 99.26% F1-Score for Dataset (D1) and 99.43% accuracy, 98.86% sensitivity, 100.0% specificity, 100.0% precision, and 99.43% F1-Score for dataset (D2). The results are comparable with the current state-of-the-art.

Keywords Brain tumor · MRI · Deep learning

Md. K. Mahbub (✉) · M. Biswas · Md. A. M. Miah
Computer Science and Engineering, Bangladesh University of Business and Technology, Dhaka 1216, Bangladesh
e-mail: milon@ieee.org

M. S. Kaiser
Institute of Information Technology, Jahangirnagar University, Savar, Dhaka 1342, Bangladesh
e-mail: mskaiser@juniv.edu

M. S. Kaiser et al. (eds.), *Proceedings of the Third International Conference on Trends in Computational and Cognitive Engineering*, Lecture Notes in Networks and Systems 348, https://doi.org/10.1007/978-981-16-7597-3_39

1 Introduction

A brain tumor is an abnormal cell mass that has grown out of control in the human brain. The number of people who die as a result of a brain tumor is highest in Asia [1]. Unlike other tumors in the human body, brain tumors have limited development areas due to the skull. As a result, a developing brain tumor may compress crucial brain areas that have been severely harmed by the endocrine system, resulting in catastrophic health issues. As a result, it is critical to correctly diagnose brain tumors at an early stage to provide treatments. For decades, data mining techniques have been used to extract meaningful information from large datasets. Classifying images or segments of images is one of the essential tasks in medical image analysis. Support vector machine (SVM), decision tree, random forest, fuzzy c-means, Bayesian classifiers, and ANNs [2–6] have all been used to identify or diagnose Covid-19 using chest x-rays [7] and brain tumor using MRI scans. In recent years, deep learning (DL) algorithm model models have been employed in place of traditional machine learning (ML) models because DL models do not require as many nodes as classic designs such as SVM or K-nearest neighbor [8] do. The brain tumor detection system Damodharan et al. [9] published using the feature-based segmentation approach, claiming the best accuracy of 80% classification. Rao et al. [10] developed a stacked CNN model-based prediction framework to the segment of the brain tumor, and 67% accuracy was achieved in the segmentation study. The models for categorization of brain cancers were basic CNN-based, given by Seetha et al. [11] and 97.5% precision was achieved. Ozyurt et al. [12] presented a segment-based algorithm based on neuroscopy and CNN (NS-CNN), which was then classified as a benign and malignant MRI brain picture. With a precision of 95.62%, and a sensitive 96.25%, NS-EMFSE + CNN + SVM model delivered the best performance metrics. There are several published article that attempts to determine the type of brain tumor. The authors used T1-weighted MRIs from a brain tumor data collection, created by Cheng et al. [13] in authors [14–18], for brain tumors to be classified as Meningioma or glioma. Sarhan et al. [14] suggested that the MRI images in the brain be decomposed with wavelets before they were sent to a CNN classifier. The approach suggested showed that the complexity of the CNN classifier was considerably reduced, and 99.3% for each classification accuracy was achieved. To classify brain types tumor from MRIs, Ismael et al. [16] used the modified ResNet50 [19] design. For training, total accuracy of 99% was applied with data increase for 3064 T1-weighted MRI images, with no increase at 95% for a three-class classification issue. Kaplan et al. [17] applied an extraction technique to build the classification process utilizing a kNN model employing the modified local binary pattern (nLBP). 95.56% was the most remarkable accuracy.

The three standard CNN networks, namely AlexNet, VGGNet, and GoogleNet, have been compared by Rehman et al. [18] to categorize brain tumor kinds. The training dataset utilized transfer learning techniques and data increase, which led to 98.69%, maximum accuracy in the VGG 16 architecture rating. Tahir et al. [20] evaluated the performance and accuracy of the classified statistical models using several popular preprocessing approaches. They showed 86% of their maximum accuracy compared to an SVM classification. A review article was published by authors [21, 22] to evaluate the use of DL, RL, and Deep RL algorithms in biological data mining. In addition, we examine the results of DL approaches when applied to diverse datasets in various application fields. Authors of [23] designed A 3D (CNN) architecture for feature extraction with a pre-trained CNN model to diagnose brain tumors and obtained 98.32% accuracy. For the classification of brain tumors, Rajinikanth et al. [24] modified a pre-trained VGG19 network and employed decision tree, KNN, and SVM as a classifier, achieving up to 99% accuracy. Researchers [25] looked at methodological research articles that proposed employing deep machine learning algorithms to detect neurodegenerative illnesses only using MRI data to detect neurodegenerative diseases. The findings demonstrate that DL based approaches can accurately determine the degree of chaos. This [26] study began with an introduction of Convolutional Neural Networks, the most frequently used DL approach, and its application to separate brain areas from Magnetic Resonance Imaging, aimed at novices. It then gave a quantitative analysis of the approaches under consideration and a thorough assessment of their effectiveness. Finally, a few outstanding difficulties were highlighted, and potential future work in medical image segmentation using DL.

Following are the most important contributions to this experiment:

- This paper proposes a DL framework for detecting brain malignancies from brain MRI images and explaining how a DL model works in the image processing of brain MRI data.
- The model is trained and tested on an open dataset of brain MRI images.
- To compare the model's performance with the current state of the art.

2 Proposed Methodology

Two phases of preprocessing (scale-up and convert image into grayscale) and network education demonstrate the flow of the suggested technique in Fig. 1. The following parts provide a detailed description.

2.1 Model Architecture

The model is presented in Figs. 1 and 2 and consists of an ensemble of blocks, each with multiple layers carrying out six fundamental activities, which are Conv2D, MaxPooling2D, Activation, Flatten, Dropout, Dense. The architecture of our model is demonstrated in Figs. 1 and 2.

The first layer of the architecture is the input layer, and the input shape is (224, 224, 1) and strides 2. The second layer is a convolution layer with 128 filters. The filter size of this layer is 1×1 followed by activation function relu and the 2×2 max

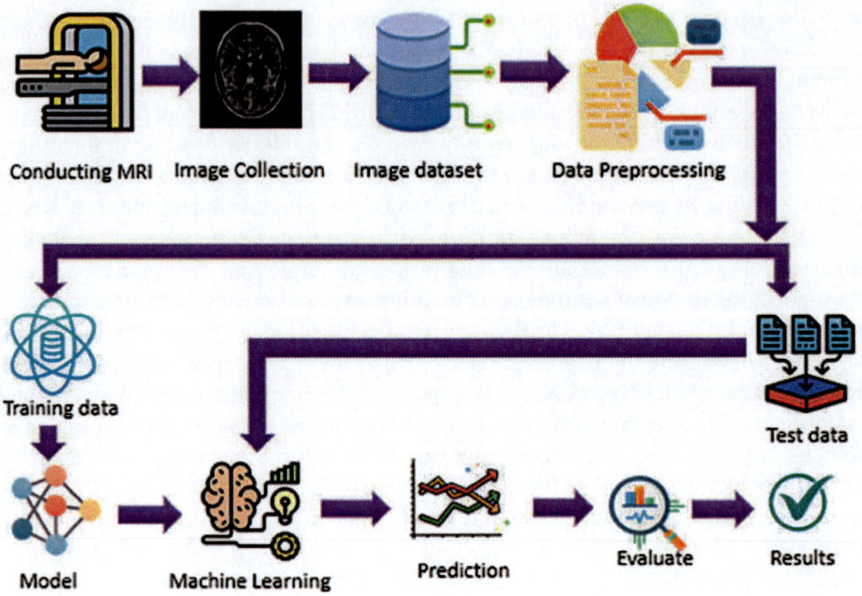

Fig. 1 Proposed deep learning flow for classification of brain tumor (source: using MS PowerPoint and internet)

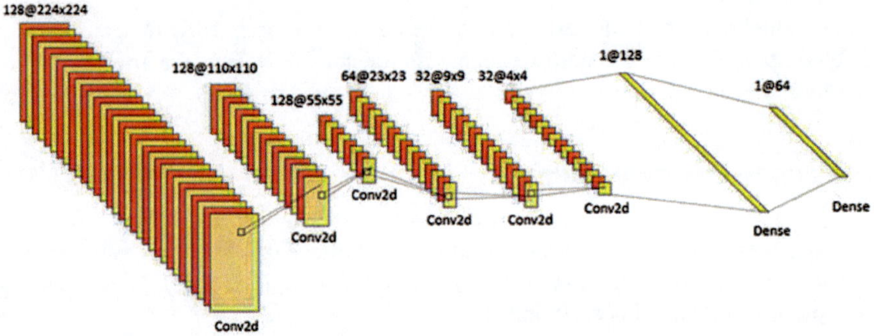

Fig. 2 Proposed deep learning architecture for classification of brain tumor

pooling layer. The next layer of the architecture is a convolution layer with 128 filters. The filter size of this layer is 3×3 followed by activation function relu and the 2, 2 maxpooling layer. The next layer of the architecture is a convolution layer with 64 filters. The filter size of this layer is 5×5 followed by activation function relu and the 2, 2 maxpooling layer. The next layer of the architecture is a convolution layer with 64 filters. The filter size of this layer is 3×3 followed by activation function relu and the 2, 2 maxpooling layer. The next layer of the architecture is a convolution layer with 32 filters. The filter size of this layer is 3×3 followed by activation function relu

Table 1 Number of learning parameter of our proposed architecture for 224 × 224 image

Layer (type)	Output shape	Param
conv2d (Conv2D)	(None, 224, 224, 128)	256
max_pooling2d (MaxPooling2)	(None, 112, 112, 128)	0
conv2d_1 (Conv2D)	(None, 110, 110, 128)	147584
max_pooling2d_1 (MaxPooling2)	(None, 55, 55, 128)	0
conv2d_2 (Conv2D)	(None, 51, 51, 64)	204864
max_pooling2d_2 (MaxPooling2)	(25, 25, 64)	0
conv2d_3 (Conv2D)	(None, 23, 23, 64)	36928
max_pooling2d_3 (MaxPooling2)	(None, 11, 11, 64)	0
conv2d_4 (Conv2D)	(None, 9, 9, 32)	18464
max_pooling2d_3 (MaxPooling2)	(None, 4, 4, 32)	0
conv2d_5 (Conv2D)	(None, 4, 4, 32)	1056
max_pooling2d_3 (MaxPooling2)	(None, 4, 4, 32)	0
flatten (Flatten)	(None, 512)	0
dense (Dense)	(None, 128)	65664
dropout_1 (Dropout)	(None, 128)	0
dense_1 (Dense)	(None, 64)	8256
dropout_2 (Dropout)	(None, 64)	0
dense_2 (Dense)	(None, 1)	65
Total parameters		483,137

and the 2, 2 maxpooling layer. The next layer of the architecture is a convolution layer with 32 filters. The filter size of this layer is 1 × 1 followed by activation function relu. The next layer of the architecture is a flatten or fully connected layer with 0.5 or 50% dropout. The next two layers are dense layers with 128 and 64 neurons with 0.5 or 50% dropout. The last layer is output layers with a sigmoid activation function. Proposed model specification as stated in Table 1.

3 Experiments and Results Discussion

3.1 Dataset

The dataset used in this work was obtained from Kaggle. The first dataset (D1) Br35H: Brain Tumor Detection 2020 (2 class of Images) [27] consists of 3060 of images of Benign Tumor, Malignant Tumor, Pituitary Tumor, and No tumor. The second dataset (D2) [28] Brain Tumor Classification (MRI) (4 classes of Images) consists of a total of 3264 images of Glioma Tumor, Malignant Tumor, Pituitary Tumor, and No tumor. From the second dataset, we only used Glioma Tumor, No

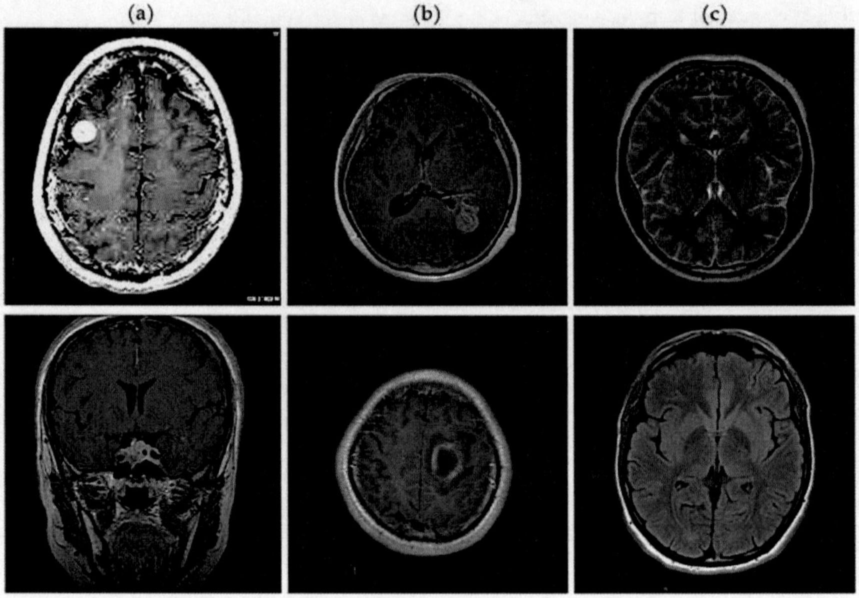

Fig. 3 Sample MRIs: **a** Pituitary Tumor, **b** Glioma Demented, **c** No Tumor; (source: Brain Tumor Classification (MRI) dataset, file name: y31.jpg, gg (10).jpg, no9.jpg, p (46).jpg, gg (24).jpg, image (18).jpg)

tumor images for our binary classification. There are 2107 images used for training the network and 903 images for testing for the first dataset (D1). There are 929 images used for training the network and 399 images for testing for the second dataset (D2). Data augmentation on images is done with rescale operation. Figure 3 shows some examples of images.

3.2 Data Preprocessing

We have preprocessed the MRI samples in our research. All of the images were in JPG format, 24-bit, 96 dpi, and 512×512 pixels, and we scaled down them to 224×224 pixels to match the input dimension of our proposed DNN. We turned all of the RGB photographs to greyscale images after scaling down. Figure 3 shows some examples of images.

3.3 Hyper Parameters

DL techniques encourage the solution of the problem to split the problem into distinct sections. The logic behind the outcome produced is not interpreted because of the unknown work of collective neurons underlying the dense network structure. The following parameters are therefore accessible for adjusting to improve the performance of the model.

- Loss function
- Optimizers
- Layers
- Drop out
- Epochs

The loss function is the field guide to notify the optimizer if it goes correctly or incorrectly by weighing the weights, optimizer shape and create the model in its most precise form. Dropout is utilized to prevent the network from being overridden by deliberately deactivating neurons; this is done with some likelihood and is generally desired to reduce the probability 0.5. The artificial increase in data increases the training size set to prevent network regularization. The accuracy of the model may be enhanced by adjusting the number of layers and periods.

3.4 Experimental Setup and Evaluation

The suggested model, which has a Tensorflow backend, is implemented using the Keras framework. The tests were conducted on an HP Intel Corei5 laptop with 4 GB of RAM. The model was trained using the Google Collaboratory GPU. Relu activation is applied for each CNN neuron. The output is separated into two groups: tumor patients and non-tumor patients. The loss function used is binary cross-entropy. The batch size employed in this experiment was 32. Adam is the optimizer, while sigmoid is the Dense activation function. The network is trained for 30 epochs.

The proposed DNN model's results are shown in Table 2 for both the training and validation sets, and performance is measured in terms of accuracy and loss. The loss provides the most accurate indication of the model's fit. Figure 4 shows the accuracy versus epoch and loss versus epoch graphs for the training and validation sets. For the training set, it can be seen that the accuracy reached 100.00%, and the loss was nearly zero. This indicates how far the model has progressed during its training phase. The validation set, on the other hand, provides a measure of the model's quality. Validation accuracy has reached 98.10%, which means that the model can predict new data detection with 98.10% accuracy.

Table 2 Performance of proposed DNN

Dataset	Optimizer	Training accuracy	Training loss	Validation accuracy	Validation loss
D1	Adam	1.0000	2.4627e-04	0.9810	0.2615
D2	Adam	1.0000	0.0022	1.0000	0.0039

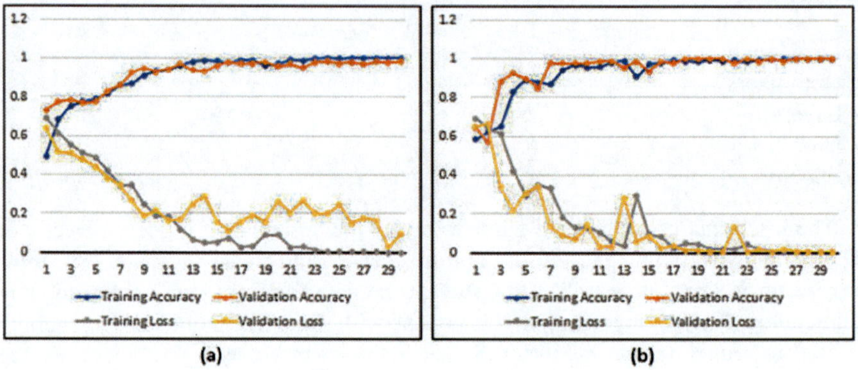

Fig. 4 Training accuracy versus validation accuracy and training loss versus validation loss: **a** For dataset D1 **b** For dataset D2

3.5 Validation Protocol and Evaluation Metrics

To validate our idea, we employed a 10-fold cross-validation technique for training and testing on our data set: D1 and D2. The following is how the 10-fold cross-validation process works: Assume there are 100 data samples in the total dataset. After that, samples 1–10 will be created in a sub-set, and designated fold-1 samples 11–20 will be labeled fold-2, etc. This results in ten different dataset sub-sets. The architecture to be assessed is then taught at 1–9 sub-sets before being tested at sub-set 10 (see Table 3).

Intention to provide more details about the performance confusion matrices in Fig. 5, a receiver operating characteristic curve (ROC) curve in Fig. 6 and classification reports in Table 4 are provided by our classifiers on test images. The confusion

Table 3 10-fold cross-validation accuracy rate (percentage) of each dataset

Dataset	K1	K2	K3	K4	K5	K6	K7	K8	K9	K10	Average
D1	99.67	99.34	99.00	98.67	99.67	99.34	100.0	99.34	98.67	99.67	99.34 (+/−0.42)
D2	100.0	99.25	99.25	100.0	100.0	100.0	98.50	100.0	99.24	99.24	99.55 (+/−0.50)

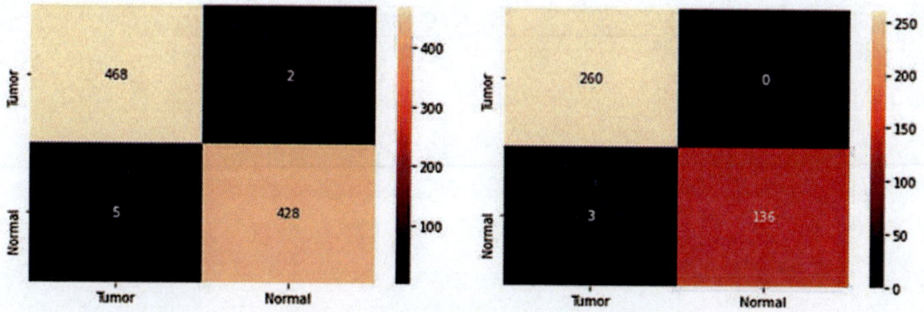

Fig. 5 Confusion matrix and of proposed deep learning model for dataset D1 & D2

Table 4 Results of applying the proposed algorithm on test set

Dataset	Accuracy (%)	AUC	Sensitivity (%)	Specificity (%)	Precision (%)	F1-score (%)
D1	99.22	1.00	98.94	99.53	99.57	99.26
D2	99.43	1.00	98.86	100.0	100.0	99.43

matrix depicts the distribution of correct and inaccurate predictions by class and circumstances. Our proposed DNN model correctly detected 468 of the 470 real tumor images, as shown in the confusion matrix of data sets D1 and D2 Fig. 5. At the same time, D1 was misdiagnosed in just 5 of 433 normal cases. Only three out of 139 normal patients from dataset D2 were misdiagnosed as tumors by the proposed DNN model, which correctly detected all 260 actual tumor images. As a result, our approach is highly accurate in detecting brain tumors.

A classification report is used to evaluate the predictive accuracy of a classification algorithm. The Classification report displays the essential classification metrics precision, recall, and f1-score for each class. On a per-class basis, the classification report of datasets D1 & D2 Table 4 shows the most crucial classification metrics precision, recall, and f1-score. For D2, we had a 100% recall rate, which means our model did not misclassify brain tumor patients as normal.

The performance of a model is depicted by a receiver operating characteristic curve (ROC) Fig. 6. AUC illustrates the degree or measure of separability, while ROC is a probability curve. It expresses how good the model is in distinguishing between classes. The AUC indicates how well the model predicts 0s as 0s and 1s as 1s. Our brain tumor detection model's AUC score is practical, indicating that it is capable of high-class separation. For datasets D1 and D2, the suggested classifier achieved the highest AUC score (1.00), indicating that our model is capable of high-class separation.

Now, we will provide a quick overview of a comparison of our experiment's results with several existing approaches presented in the literature review. Table 5 displays the findings of some of the most recent methodologies suggested for brain tumor detection. Sarhan et al. [14] proposed VGG-16-based architecture for brain tumor

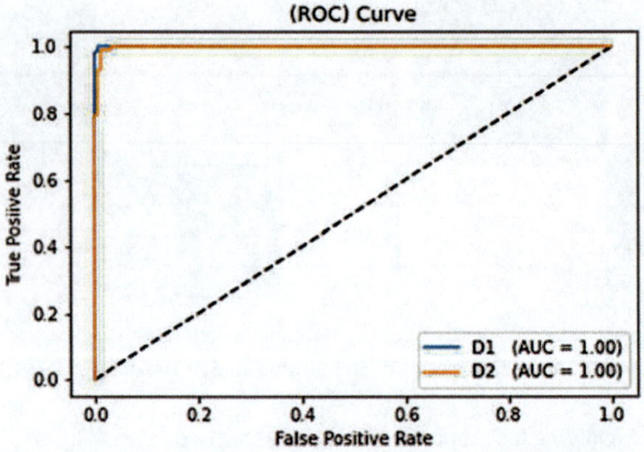

Fig. 6 ROC curve and of proposed deep learning model for dataset D1 & D2

Table 5 Performance comparison of proposed framework with other approaches

Authors	Year	Dataset	Purpose	Technique	Acc (%)
Sarhan et al. [14]	2020	Cheng's "brain tumor dataset"	Classification: menin-gioma/giloma/ pituitary tumor	WCNN	99.30
Bhanothu et al. [15]	2020			Faster-R-CNN	77.60
Ismael et al. [16]	2020			ResNet50	99.00
Kaplan et al. [17]	2020			KNN	95.56
Rehman et al. [18]	2020			VGG16	98.69
Tahir et al. [20]	2020			SVM	86.00
Sethy et al. [33]	2021			VGG19 + SVM	97.89
Gajula et al. [29]	2021	–		U-Net	96.90
Ahmadi et al. [30]	2021	–		CNN	96.00
Saffar et al. [31]	2021	–		MLP + SVM	91.02
Kaldera et al. [32]	2019	Cheng's "brain tumor dataset"		Faster-R-CNN	94.00
Proposed DNN		Kaggle		CNN	**99.43**

detection. Bhanothu et al. [16] used an enhanced DL approach based on residual networks for brain cancer MRI images classification; their presented models have more than 99% accuracy. On the other hand author of [15, 29–32] presented models have less accuracy than our model. So, we can conclude that our model gave a promising result for the detection of brain tumors.

4 Conclusion

Detecting brain tumors, this proposed customized DL model has an average 100% validation accuracy after 10-fold cross-validation. There is also an average accuracy of 99.22%, the sensitivity of 98.94%, specificity of 99.53%, and the accuracy of F1 of 99.26% for dataset D1 and an average accuracy of 99.43%, the sensitivity of 98.86%, specificity of 100%, and the accuracy of F1 of 99.43% for dataset D2. Further research will focus on performance metrics such as specificity, sensitivity, recovery, and F1 scores by upgrading the DNN model. The method employed in the suggested model may be expanded to include segmentation of the brain tumor in the following step, i.e., the model could locate the area impacted if a brain tumor is found. It would help the operators to decide the orientation and coverage area of the scan utilizing MRI equipment. This might help doctors discover and analyze brain cancers more effectively.

References

1. Khodamoradi, F., et al.: The incidence and mortality of brain and central nervous system cancer and their relationship with human development index in the world. World Cancer Res. J. **4**, e985 (2017)
2. Singh, A.: Detection of brain tumor in MRI images, using combination of fuzzy c-means and SVM. In: 2015 2nd International Conference on Signal Processing and Integrated Networks (SPIN), February, pp. 98–102. IEEE (2015)
3. Krishnakumar, S., Manivannan, K.: Effective segmentation and classification of brain tumor using rough K means algorithm and multi kernel SVM in MR images. J. Ambient. Intell. Hum. Comput. **12**, 6751–6760 (2021). https://doi.org/10.1007/s12652-020-02300-8
4. Khan, A.R., Khan, S., Harouni, M., Abbasi, R., Iqbal, S., Mehmood, Z.: Brain tumor segmentation using K-means clustering and deep learning with synthetic data augmentation for classification. Microsc. Res. Tech. (2021)
5. Maheswari, K., Balamurugan, A., Malathi, P., Ramkumar, S.: Hybrid clustering algorithm for an efficient brain tumor segmentation. Mater. Today: Proc. **37**, 3002–3006 (2021)
6. Gull, S., Akbar, S.: Artificial intelligence in brain tumor detection through MRI scans: advancements and challenges. Artif. Intell. Internet Things 241–276 (2021)
7. Mahbub, M., Biswas, M., Miah, A.M., Shahabaz, A., Kaiser, M.S.: COVID-19 detection using chest X-ray images with a RegNet structured deep learning model. In: International Conference on Applied Intelligence and Informatics, July, pp. 358–370. Springer, Cham (2021)
8. Mohsen, H., El-Dahshan, E.S.A., El-Horbaty, E.S.M., Salem, A.B.M.: Classification using deep learning neural networks for brain tumors. Futur. Comput. Inform. J. **3**(1), 68–71 (2018)

9. Damodharan, S., Raghavan, D.: Combining tissue segmentation and neural network for brain tumor detection. Int. Arab J. Inf. Technol. (IAJIT) **12**(1) (2015)
10. Rao, V., Sarabi, M.S., Jaiswal, A.: Brain tumor segmentation with deep learning. MICCAI Multimodal Brain Tumor Segm. Chall. (BraTS) **59** (2015)
11. Seetha, J., Raja, S.S.: Brain tumor classification using convolutional neural networks. Biomed. Pharmacol. J. **11**(3), 1457 (2018)
12. Ö-zyurt, F., Sert, E., Avci, E., Dogantekin, E.: Brain tumor detection based on convolutional neural network with neutrosophic expert maximum fuzzy sure entropy. Measurement **147**, 106830 (2019)
13. Cheng, L., et al.: Brain tumor dataset (2017). https://figshare.com/articles/brain_tumor_dataset/1512427/5. Accessed 13 Jul 2021
14. Sarhan, A.M.: Brain tumor classification in magnetic resonance images using deep learning and wavelet transform. J. Biomed. Sci. Eng. **13**(06), 102 (2020)
15. Bhanothu, Y., Kamalakannan, A., Rajamanickam, G.: Detection and classification of brain tumor in MRI images using deep convolutional network. In: 2020 6th International Conference on Advanced Computing and Communication Systems (ICACCS), March, pp. 248–252. IEEE (2020)
16. Ismael, S.A.A., Mohammed, A., Hefny, H.: An enhanced deep learning approach for brain cancer MRI images classification using residual networks. Artif. Intell. Med. **102**, 101779 (2020)
17. Kaplan, K., Kaya, Y., Kuncan, M., Ertunç, H.M.: Brain tumor classification using modified local binary patterns (LBP) feature extraction methods. Med. Hypotheses **139**, 109696 (2020)
18. Rehman, A., Naz, S., Razzak, M.I., Akram, F., Imran, M.: A deep learning-based framework for automatic brain tumors classification using transfer learning. Circuits Syst. Signal Process. **39**(2), 757–775 (2020)
19. He, K., Zhang, X., Ren, S., Sun, J.: Deep residual learning for image recognition. In: Proceedings of the IEEE Conference on Computer Vision and Pattern Recognition, pp. 770–778 (2016)
20. Tahir, B., Iqbal, S., Khan, Usman Ghani, M., Saba, T., Mehmood, Z., Anjum, A., Mahmood, T.: Feature enhancement framework for brain tumor segmentation and classification. Microsc. Res. Tech. **82**(6), 803–811 (2019)
21. Mahmud, M., Kaiser, M.S., Hussain, A., Vassanelli, S.: Applications of deep learning and reinforcement learning to biological data. IEEE Trans. Neural Netw. Learn. Syst. **29**(6), 2063–2079 (2018)
22. Mahmud, M., Kaiser, M.S., McGinnity, T.M., Hussain, A.: Deep learning in mining biological data. Cogn. Comput. **13**(1), 1–33 (2021)
23. Noor, M.B.T., Zenia, N.Z., Kaiser, M.S., Mahmud, M., Al Mamun, S.: Detecting neurodegenerative disease from MRI: a brief review on a deep learning perspective. In: International Conference on Brain Informatics, December, pp. 115–125. Springer, Cham (2019)
24. Rajinikanth, V., Joseph Raj, A.N., Thanaraj, K.P., Naik, G.R.: A customized VGG19 network with concatenation of deep and handcrafted features for brain tumor detection. Appl. Sci. **10**(10), 3429 (2020)
25. Rehman, A., Khan, M.A., Saba, T., Mehmood, Z., Tariq, U., Ayesha, N.: Microscopic brain tumor detection and classification using 3D CNN and feature selection architecture. Microsc. Res. Tech. **84**(1), 133–149 (2021)
26. Ali, H.M., Kaiser, M.S., Mahmud, M.: Application of convolutional neural network in segmenting brain regions from MRI data. In: International Conference on Brain Informatics, December, pp. 136–146. Springer, Cham (2019)
27. https://www.kaggle.com/ahmedhamada0/brain-tumor-detection
28. https://www.kaggle.com/sartajbhuvaji/brain-tumor-classification-mri
29. Gajula, S., Rajesh, V.: MRI brain image segmentation by fully convectional U-net. Rev. GEINTEC-Gest. Inov. Tecnol. **11**(1), 6035–6042 (2021)
30. Ahmadi, M., Sharifi, A., Jafarian Fard, M., Soleimani, N.: Detection of brain lesion location in MRI images using convolutional neural network and robust PCA. Int. J. Neurosci. 1–12 (2021)

31. Al-Saffar, Z.A., Yildirim, T.: A hybrid approach based on multiple eigenvalues selection (MES) for the automated grading of a brain tumor using MRI. Comput. Methods Programs Biomed. **201**, 105945 (2021)
32. Kaldera, H.N.T.K., Gunasekara, S.R., Dissanayake, M.B.: Brain tumor classification and segmentation using faster R-CNN. In: 2019 Advances in Science and Engineering Technology International Conferences (ASET), March, pp. 1–6. IEEE (2019)
33. Sethy, P.K., Behera, S.K.: A data constrained approach for brain tumour detection using fused deep features and SVM. Multimed. Tools Appl. (2021). https://doi.org/10.1007/s11042-021-11098-2

A Deep Learning Approach with Data Augmentation to Recognize Facial Expressions in Real Time

Tawsin Uddin Ahmed, Sazzad Hossain, Mohammad Shahadat Hossain, Raihan Ul Islam, and Karl Andersson

Abstract The enormous use of facial expression recognition in various sectors of computer science elevates the interest of researchers to research this topic. Computer vision coupled with deep learning approach formulates a way to solve several real-world problems. For instance, in robotics, to carry out as well as to strengthen the communication between expert systems and human or even between expert agents, it is one of the requirements to analyze information from visual content. Facial expression recognition is one of the trending topics in the area of computer vision. In our previous work, a facial expression recognition system is delivered which can classify an image into seven universal facial expressions—angry, disgust, fear, happy, neutral, sad, and surprise. This is the extension of our previous research in which a real-time facial expression recognition system is proposed that can recognize a total of ten facial expressions including the previous seven facial expressions and additional three facial expressions—mockery, think, and wink from video streaming data. After model training, the proposed model has been able to gain high validation accuracy on a combined facial expression dataset. Moreover, the real-time validation of the proposed model is also promising.

T. U. Ahmed (✉) · M. S. Hossain
Department of Computer Science and Engineering, University of Chittagong, Chittagong, Bangladesh

M. S. Hossain
e-mail: hossain_ms@cu.ac.bd

S. Hossain
Department of Computer Science and Engineering, University of Liberal Arts Bangladesh, Dhaka, Bangladesh
e-mail: sazzad.hossain@ulab.edu.bd

R. U. Islam · K. Andersson
Department of Computer Science, Electrical and Space Engineering, LuleÃě University of Technology, SkellefteÃě, Sweden
e-mail: raihan.ul.islam@ltu.se

K. Andersson
e-mail: karl.andersson@ltu.se

M. S. Kaiser et al. (eds.), *Proceedings of the Third International Conference on Trends in Computational and Cognitive Engineering*, Lecture Notes in Networks and Systems 348, https://doi.org/10.1007/978-981-16-7597-3_40

1 Introduction

Facial expression most of the time holds the information about the mental condition of an individual that is used as non-verbal communication. These facial expressions are the reflection of the inner feelings of an individual that are expressed intentionally or sometimes in the subconscious mind [1]. Humans have to undergo a variety of surrounding conditions that make them feeling joy, depressed, tensed, and so on. It is human nature to share and expose his feelings as well as intentions through behaviors. Most of the parts of the developed and even developing countries are under video surveillance for ensuring security. It would be so beneficial if the facial expression recognition system can be embedded in those systems in order to analyze the intention of a suspect by reading his expressions. So, facial expressions can be employed in this regard to establish a more efficient security surveillance system [2].

Sometimes it is not that comfortable to identify the expression accurately because of the variation in expressing emotions around the world. However, some universal expressions are taken into consideration because these universal expressions cover almost all basic emotions. One of the criteria based on what a facial expression recognition system can be formulated is identifying representative features. It can also be addressed as feature extraction. This feature extraction process can be conducted by adopting a manual or automated approach. In a manual approach, each of the facial expression features has to be identified by human interaction. That is why there may be a possibility of human error or being missed out on some important features which have major contributions to expression identification. If that happens, a negative impact must be taken place in the process of expression recognition. However, an approach that can be employed as the counter of this kind of situation is feature extraction automation [3]. It refers to the ability of a system to extract features from a dataset without the interference of human operators. In this context convolutional neural network is continuing its triumph for recent years.

In order to implement deep learning approach in computer vision problem, a convincing methodology which performs well with visual content is convolutional neural network (CNN) [4]. CNN differs itself from the other deep learning approach by its uniqueness in identifying features from images. CNN employs its convolution layers to learn the features which introduce the concept of feature extraction automation. Image classification is one of the computer vision problems, and CNN is continuing its triumph to solve this problem [5].

The objective of this research to add a contribution to computer vision area by delivering a facial expression recognition system that can detect faces from video streaming data and recognize the facial expression(s) that the face(s) of that videos contain by analyzing the frames. As the types of facial expressions, the research includes ten facial expressions—angry, disgust, fear, happy, neutral, sad, surprise, mockery, think, and wink.

Later in this article, we will discuss some related works regarding this topic, the proposed methodology, data collection and preprocessing, result and discussion, and so on.

2 Related Works

Mengyi Liu et al. proposed a method where the facial act is segmented into several units which are referred to as facial action units to recognize the facial expression [6]. The whole expression identification process is divided into two phases so that a robust facial expression recognition system can be delivered. In the first phase, convolution layers followed by max-pooling layer are occupied to extract the Micro-Action-Pattern. Basically, feature extraction is taken place in this phase. Then, the features are been grouped based on the correlated MAPs. At last, group wise sub-networks are formulated as a part of multi-layer learning process to construct the high-level representation. In order to evaluate the performance of the proposed model, three reputed facial expression datasets are been employed and the model's response is promising.

Bo-Kyeong Kim et al. formulate a facial expression recognition system using deep learning approach, to be specific, convolutional neural network [7]. Several deep convolutional neural networks are applied on multiple facial expression datasets which are SFEW2.0, FER-2013, TFD, and GENKI-4 k dataset and integrates the outputs of those CNNs. Experiments are conducted by altering network structure, normalization of inputs, and initializing the weights randomly. A hierarchical committee of deep convolutional neural networks is formulated to get a more robust classification of facial expressions—angry, disgust, fear, happy, sad, surprise, and neutral. This research is successful to achieve 61.6% of test accuracy of the EmotiW2015 dataset which consists of seven facial expressions.

Heechul Jung et al. focused on the techniques of feature extraction to avoid the complexity of manual feature detection from the face images [8]. Deep learning approach is considered as it can extract features automatically from raw data. Multiple deep networks are employed to extract several types of features. Temporal appearance features are considered to be the targeted type of features that deep networks are responsible for. Integration of these two techniques elevates the performance of facial expression recognition in CK + and oulu-CASIA datasets. This research is able to outperform the result of traditional state-of-the-art approaches like weighted summation and feature concatenation method.

Zhiding Yu et al. include a new combined face detection module which consists of joint cascade detection and alignment (JDA) detector, the Deep-CNN (DCNN) based detector, and MoT [9]. Input images are been passed through these three types of detectors in order to develop an efficient face detection process. This face detection is followed by the ensemble of convolutional neural networks. The proposed model is pre-trained on the facial expression dataset which is offered by the Facial Expression Recognition Challenge 2013. Minimization of log-likelihood and hinge loss are the structures to learn the network weight which are adopted to integrate several CNN models. This research is able to acquire 55.96 and 61.29% of validation and test accuracy in SFEW2.0 dataset beyond the challenge baselines which are 35.96 and 39.13%.

Nianyin Zeng et al. present an approach where the information of emotion is extracted from the facial features which are passed through the next phases of the proposed model [10]. As the types of features, the proposed model is considered with two types of features which are geometric features and appearance features. This feature extraction is followed by encoders which are referred to as deep sparse encoders (DSAE). DSAE is designed to execute the facial expression recognition task with higher accuracy than the state-of-the-art approaches. The proposed framework is effectual to achieve 95.79% of recognition accuracy on the extended Cohn-Kanade (CK +) dataset which consists of seven facial expressions. The proposed model includes another expression class (which is neutral) and maintains satisfactory recognition accuracy.

3 Convolutional Neural Network

Convolutional neural network [5] is an artificial neural network that is so far been most popularly used for analyzing images. Although image analysis has been the most widespread use of CNN's, it can also be used for other data analysis or classification problems as well. Basically, CNN is a kind of artificial neural network that has some type of specialization for being able to pick out or detect patterns and make sense of them. This pattern detection is what makes CNN so useful for image analysis. The convolution layer is the layer that makes CNN distinguish from the regular ANN. In this section, it will illustrate the CNN architecture of this integrated framework. Anyway, it should be noted that the reason for using convolutional neural network in this research because this research involves image dataset and CNN is well known to perform well for image classification task. However, CNN demands large image dataset for better classification accuracy. Also, the accuracy of the classification task is proportional to the size of the image dataset.

Figure 1 demonstrates the overall training workflow of our proposed real-time facial expression recognition system. Model architecture design and data preprocessing are discussed below.

According to Table 1, the organization of CNN started with two convolution layers with 64 filters each having a size of 3 × 3. All of these filters generate feature maps

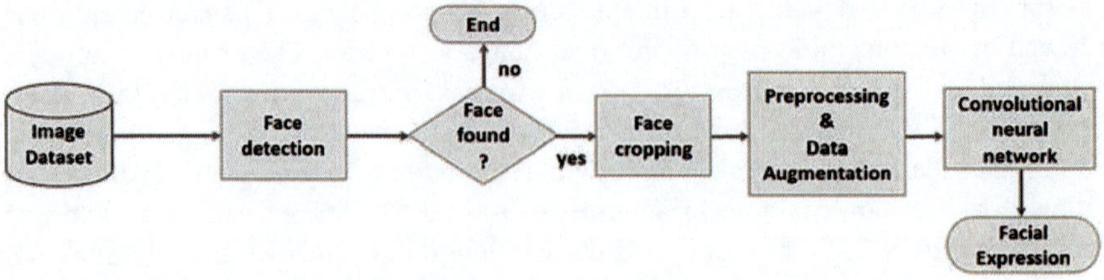

Fig. 1 System flow of the proposed model

Table 1 CNN architecture of the proposed model

Model content	Details
First and second convolution layer	64 filters of size 3 × 3, ReLU, input size 48 × 48
MaxPooling layer	Pooling size 2 × 2, Strides of (2,2)
Third and forth convolution layer	128 filters of size 3 × 3, ReLU
MaxPooling layer	Pooling size 2 × 2, Strides of (2,2)
Flatten layer	Convert 2D matrix into 1D vector
First hidden layer	1024 nodes, ReLU
Dropout layer	Deactivates 40% nodes randomly
Second hidden layer	1024 nodes, ReLU
Dropout layer	Deactivates 40% nodes randomly
Output layer	10 nodes for 10 classes, SoftMax
Optimization function	AdaBound
Learning rate	0.001
Callbacks	EarlyStopping, ReduceLROnPlateau, ModelCheckpoint, TensorBoard

to extract different features from the images. Then, to fetch the maximum features from the feature maps MaxPooling layer with pool size 2 × 2 and strides of (2,2) are included. Following the previous first two layers, second and third convolution layers are constructed by 128 filters of size 3 × 3 each. Like the previous pooling layer, this time MaxPooling with pool size 2 × 2 is inserted. Since hidden layers are not capable of dealing with 2-dimensional structure as input, a flatten layer is embedded just after the last pooling layer. As a part of fully connected layer, 2 hidden layers with 1024 nodes per layer are formulated. Hence, each hidden layer is followed by a dropout layer which randomly deactivates 40% of the active nodes so that the model cannot be biased to the training data. In other words, it helps the model to avoid overfitting [11]. The activation function that is applied in all convolution as well as hidden layers is ReLU. At last, CNN is concluded with an output layer which consists of 10 nodes for eight classes and softmax is used as activation function is this layer. To optimize the learning process AdaBound optimizer with 0.001 learning rate is applied [12]. Since overfitting is one of the concerned factors in CNN training, EarlyStopping is considered as one of the callbacks. That means the CNN part of the proposed model stops its training when there is no improvement in learning.

4 Data Collection and Preprocessing

The proposed convolutional neural network is trained on a combined dataset collected from the different sources of standard datasets. The origins of this combined dataset are illustrated below.

1. Ck CK + : Initial release which is known as CK contains 486 sequences from 97 posers started with neutral expression gradually proceeds to the peak expression. CK + includes both posed and non-posed face images where the number of sequences swells up with 22% and the number of subjects with 27% [13].
2. Fer2013: This dataset is one of the most popular data in facial expression recognition. It is a collection of 35,887 grayscale 48 × 48 images which is a combination of seven expressions—angry, disgust, fear, happy, sad, surprise, and neutral [14].
3. The MUG facial expression dataset: In this dataset 86 subjects including 35 women and 51 men performing six facial expressions which are anger, disgust, fear, happiness, sadness, and surprise [15].
4. KDEF AKDEF: 4900 pictures of KDEF has been used more than 1500 research publications. It contains seven facial expressions—anger, disgust, fear, happiness, sadness, and surprise.

In addition to seven regular facial expressions, three facial expressions, which are mockery, think, and wink, are added by downloading quality images from Shutterstock and adobe stock (Fig. 2). Actually, in this research, a new dataset is going to be proposed for mockery, think, and wink that might help other researchers in this domain.

Fig. 2 Dataset sample of seven basic expressions [16]

Fig. 3 Additional three facial expressions

Before the training phase, the image dataset needs some modification to pave the way of getting a better result out of it (Fig. 3). In order words, it can be defined as data preprocessing. In this phase, the dataset of the proposed system has to undergo some steps. Initially, face detection is executed to extract the region of interest with the help of Haar Cascade classifier. The detected face region is cropped and converted to grayscale to reduce the color complexity of the image. In order to diminish low-intensity problem, the images of the dataset are normalized. After the image normalization process, images are fed into a convolutional neural network input for training.

Variation in dataset helps deep learning models to learn an object from various points of view. It contributes to efficient model learning (Fig. 4). Hence, it makes a model not to get biased to a single angle of view. A model can deeply understand the features that are involved to define the object more precisely. In fact, if image dataset does not have diversity; the model might not able to recognize the unseen images which include, for example, position variants.

In order to avoid this kind of situation as well as extend the size of the training dataset, data or image augmentation approach can be encountered. Image augmentation is a method or process that is applied on a dataset which results in data extension by applying some image transformation parameters, such as zoom, flip, shear, shift, and so on. It should be noted that the extended images are identical to each other. Basically, in data augmentation process additional image dataset is generated from the existing dataset such a way that the transformed images involve diversity which allows model not to get biased on some specific images. As a result, overfitting cannot be an issue in this case [17].

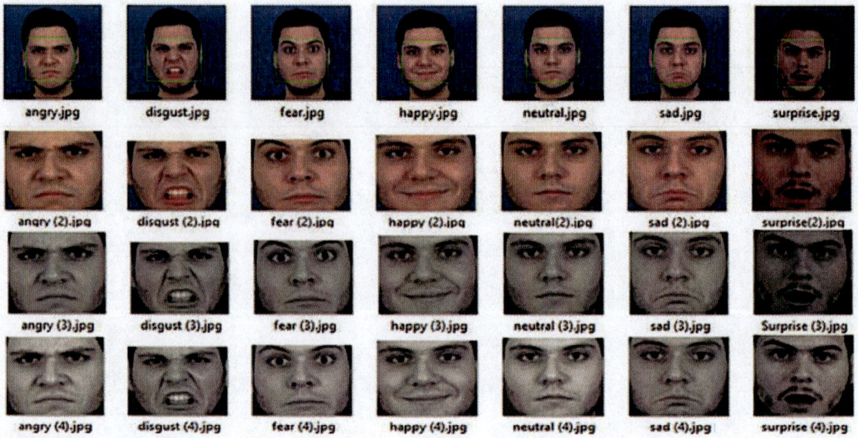

Fig. 4 Data preprocessing [16]

5 Data Augmentation

In a data-driven approach, the performance or recognition rate of a trained model depends on some significant factors—system framework, dataset size, proper data preprocessing, and so on. Image variation in dataset should be maintained in order to increase the model's efficiency by avoiding the risk of model overfitting. If the training phase is considered, some monitorings are executed to make the training process more effective. Callbacks like ModelCheckpoint, ReduceLROnPlateau, and EarlyStopping offered by keras are employed to monitor the learning process of the proposed model [18]. Experiments consist of multiple phases of training (Fig. 5). In the earlier training phase, the model is trained on 13,100 images where each of the ten classes includes 1310 images. Since the model's accuracy is proportional to the size of dataset, the number of images per class is needed to be increased.

Fig. 5 Data augmentation [16]

Then data augmentation method designed by keras comes into the scenario in order to elevate the size of dataset avoiding the hustle of collecting dataset from the field level. As data augmentation parameters zoom, shear, rotation, and flip are considered. After data augmentation, the dataset size has become reached to 39,300 where each class consists of 3930 images. Data augmentation generates variation in data so that overfitting is not a concern in this case [17]. Anyway, each input image has a dimension of 48 × 48. During the learning phase, 80% of the images have been selected for model training and the remaining 20% is for system validation.

6 System Implementation

The system proposed to identify facial expression from video streaming data. It allows users to select videos from local directory as well as enables webcam to capture live video. A python graphical user interface tkinter is embedded so that users can upload any video from the local directory [19]. In addition to that, users can also use their webcams with the help of OpenCV [20]. The video, provided into the system, is segmented into multiple frames, and each frame is considered to be the subject of expression classification. Haar cascade classifier, a classifier offered by OpenCV to detect faces, is employed in order to extract the region of interest [21]. Then the detected region of each frame is preprocessed exactly the same as the model got trained on. The trained model is embedded in the system as 'model.h5'. With the help of this file, the resultant frames are classified into ten classes. From Fig. 6 it can be observed that the proposed system is successful to identify the expressions in the unseen real-time video frames provided by the user.

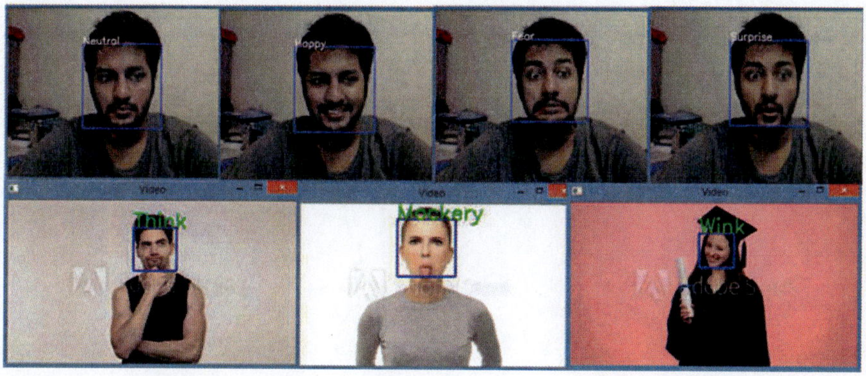

Fig. 6 Real time validation of the proposed model

7 Results and Discussion

Due to hardware limitations of local CPU, the proposed model was trained on google colaboratory [22]. It is a cloud server for machine learning model training. This cloud server includes a K80 GPU and a jupyter notebook environment. Google colab allows a model to get trained for up to 12 h unless the browser is closed before this time limit. As google colab offers high gpu to train model, the model training of the proposed model was fast. In fact, it takes only 2 h and 25 min to complete its training where each epoch allocates around 270 s. It can be noticed that our model takes only 30 epochs to reach the training accuracy of 98.28%. Moreover, it is early stopped with the help of the callback named EarlyStopping so that model cannot get overfitted. However, its validation accuracy is also promising which is 90.93%. Low difference between training and test accuracy indicates that the model has good prediction ability on unseen images also. According to Fig. 7, the model was learning fast as well as effectively in the period between 1 and 10 epochs. After that improvement in learning become slower. At that moment the learning rate of model training is reduced to resolve the oscillation problem. In order to depict the recognition rate for each class, a confusion matrix is delivered in Fig. 8. So, it can be observed that the proposed model can classify each of the classes of facial expression with high accuracy. As multiple experiments are conducted in this research with and without data augmentation, a comparison between these two experiments is plotted in Fig. 9 in terms of model training. The figure shows how data-augmentation boosts the prediction accuracy of the model. The reason lies in the extended size of dataset that the augmentation process offers. The learning curve of the experiment with augmentation remained higher than the other experiment right from the very beginning of model training. Moreover, experiment with data augmentation ended its training with the validation

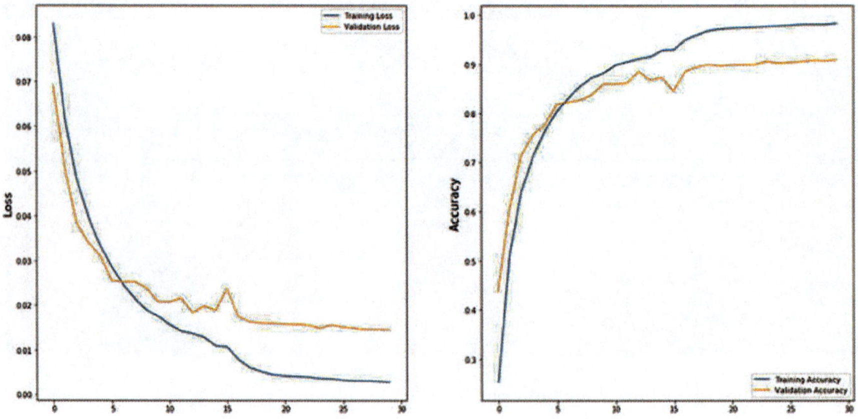

Fig. 7 Loss and accuracy of the proposed model

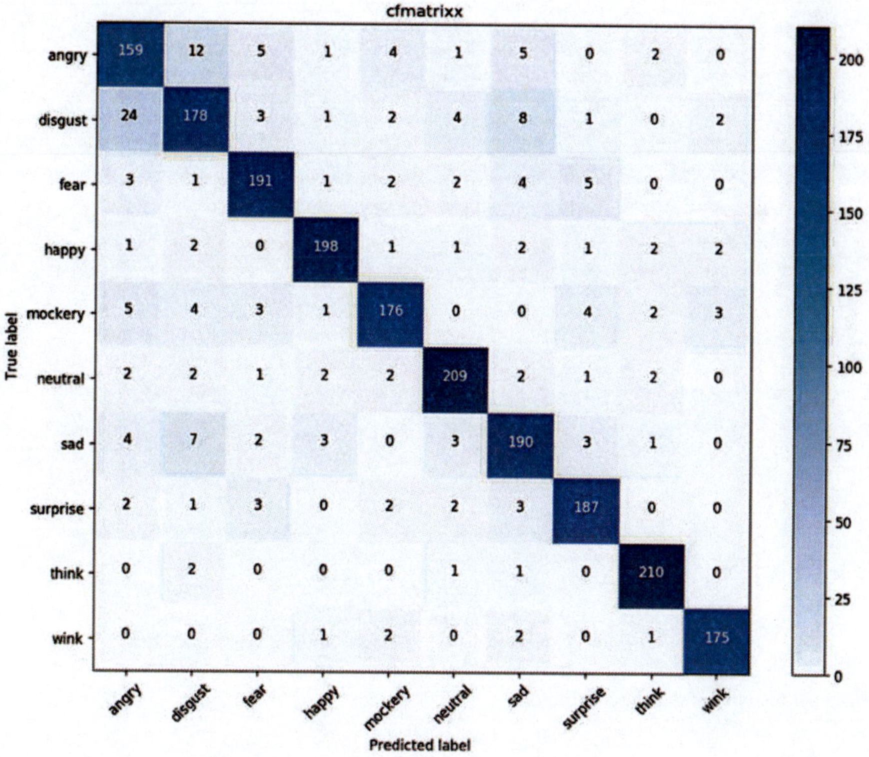

Fig. 8 Confusion matrix of the proposed model

accuracy of 90.93% whereas experiment without data augmentation completed its training with the validation accuracy of 86.20%.

Table 2 presents a comparative study of the models' system results. When HOG features are being used as learning resources instead of actual frames, the SVM's accuracy rate improves marginally from 0.66 to 0.68. The random forest model, on the other hand, has a better predictive performance of 0.72 by using raw pictures instead of HOG features as learning parameter. In terms of all performance assessment measures, however, our model beats the other machine learning models by a significant margin: validation accuracy (90.93%) and f1 score (0.90). Tensorboard callback offered by keras is employed to visualize the model training. Confusion matrix will be included to have an overview of the recognition rate for each class.

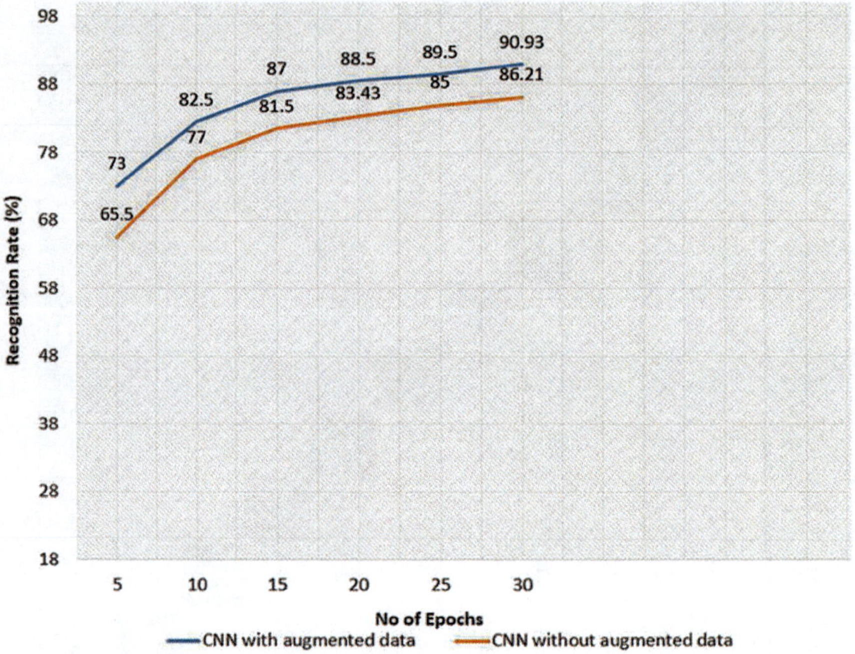

Fig. 9 Effectivity of data augmentation

Table 2 Performance comparison table between several models

Model	Val. accuracy (%)	F1 score
Support vector machine	66.34	0.67
Support vector machine with HOG features	68.11	0.68
Random forest	72.04	0.72
Our model	90.93	0.90

8 Conclusion and Future Work

This research aims to focus on various other types of facial expressions along with the regular basic seven facial expressions (angry, disgust, fear, happy, neutral, sad, and surprise). Objectives include delivery of a facial expression recognition system that has the ability to identify ten facial expressions that are angry, disgust, fear, happy, mockery, neutral, sad, surprise, think, and wink from video streaming data. Moreover, maintaining a proper and nearly equal recognition rate are taken into account while constructing the proposed model. The proposed system can be embedded in modern security system where visual content processing is required i.e. CCTV footage analysis. Since facial expressions are the most common way of emotion exposure, a security system analyzer can able to identify suspects based on the model output. Although the model's recognition rates of the classes are quite prominent, improvement needs to be gained in recognizing disgust class. Validation accuracy can also be raised by gathering more data as cnn is a data-driven approach. These improvements are considered as future works. Hence, another research that can be included in the future work list is to integrate Belief Rule Based Expert System (BRBES) with convolutional neural network in order to assess the overall mental state of an individual analyzing facial expressions under uncertainty [23, 23].

References

1. Ekman, P.: Facial expression and emotion. Am. Psychol. **48**(4), 384 (1993)
2. Al-modwahi, AAM, Sebetela O, Batleng LN, Parhizkar B, Lashkari AH. Facial expression recognition intelligent security system for real time surveillance. In: Proceedings of the International Conference on Computer Graphics and Virtual Reality (CGVR). The Steering Committee of The World Congress in Computer Science, Computer âĂę; 2012. p. 1.
3. Li, H., Chutatape, O.: Automated feature extraction in color retinal images by a model based approach. IEEE Trans. Biomed. Eng. **51**(2), 246–254 (2004)
4. Srinivas, S., Sarvadevabhatla, R.K., Mopuri, K.R., Prabhu, N., Kruthiventi, S.S., Babu, R.V.: A taxonomy of deep convolutional neural nets for computer vision. Front. Robot. AI. **2**, 36 (2016)
5. Krizhevsky, A., Sutskever, I., Hinton, G.E.: Imagenet classification with deep convolutional neural networks. In: Advances in neural information processing systems. pp. 1097–1105 (2012)
6. Liu, M., Li, S., Shan, S., Chen, X.: Au-inspired deep networks for facial expression feature learning. Neurocomputing **159**, 126–136 (2015)
7. Kim, B.K., Roh, J., Dong, S.Y., Lee, S.Y.: Hierarchical committee of deep convo- lutional neural networks for robust facial expression recognition. J. Multimodal User Interfaces. **10**(2), 173–189 (2016)
8. Jung, H., Lee, S., Yim, J., Park, S., Kim, J.: Joint fine-tuning in deep neural networks for facial expression recognition. In: Proceedings of the IEEE International Conference on Computer Vision, pp. 2983–2991 (2015)
9. Yu, Z., Zhang, C.: Image based static facial expression recognition with multiple deep network learning. In: Proceedings of the 2015 ACM on International Conference on Multimodal Interaction ACM, pp. 435–442 (2015)
10. Zeng, N., Zhang, H., Song, B., Liu, W., Li, Y., Dobaie, A.M.: Facial expression recogni- tion via learning deep sparse autoencoders. Neurocomputing **273**, 643–649 (2018)
11. Srivastava, N., Hinton, G., Krizhevsky, A., Sutskever, I., Salakhutdinov, R.: Dropout: a simple way to prevent neural networks from overfitting. J. Mach. Learn. Res. **15**(1), 1929–1958 (2014)
12. Luo, L., Xiong, Y., Liu, Y., Sun, X.: Adaptive gradient methods with dynamic bound of learning rate. In: Proceedings of the 7th International Conference on Learning Representations. New Orleans, Louisiana (2019)
13. Lucey, P., Cohn, J.F., Kanade, T., Saragih, J., Ambadar, Z., Matthews, I.: The extended cohn-kanade dataset (ck+): a complete dataset for action unit and emotion-specified expression. In: 2010 IEEE Computer Society Conference on Computer Vision and Pattern Recognition-Workshops, IEEE, pp. 94–101 (2010)
14. Ng, H.W., Nguyen, V.D., Vonikakis, V., Winkler, S.: Deep learning for emotion recognition on small datasets using transfer learning. In: Proceedings of the 2015 ACM on International Conference on Multimodal Interaction ACM, pp. 443–449 (2015)

15. Koestinger, M., Wohlhart, P., Roth, P.M., Bischof, H.: Annotated facial landmarks in the wild: a large-scale, real-world database for facial landmark localization. In: IEEE International Conference on Computer Vision Workshops (ICCV Workshops), IEEE, pp. 2144–2151 (2011)
16. Ahmed, T.U., Hossain, S., Hossain, M.S., ul Islam, R., Andersson, K.: Facial expression recognition using convolutional neural network with data augmentation. In: 2019 Joint 8th International Conference on Informatics, Electronics Vision (ICIEV) and 2019 3rd International Conference on Imaging, Vision Pattern Recognition (icIVPR), pp. 336–341 (2019)
17. Perez, L., Wang, J.: The effectiveness of data augmentation in image classification using deep learning (2017). arXiv preprint arXiv:171204621
18. Gulli, A., Pal, S.: Deep learning with Keras. Packt Publishing Ltd (2017)
19. Shipman, J.W.: Tkinter 8.4 reference: a GUI for Python. New Mexico Tech Computer Center (2013)
20. Bradski, G., Kaehler, A.: Learning OpenCV: computer vision with the OpenCV library. " O'Reilly Media, Inc. (2008)
21. Wilson, P.I., Fernandez, J.: Facial feature detection using Haar classifiers. J. Comput. Sci. Coll. **21**(4), 127–133 (2006)
22. Bisong, E.: Google colaboratory. In: Building Machine Learning and Deep Learning Models on Google Cloud Platform, Springer, pp. 59–64 (2019)
23. Hossain, M.S., Rahaman, S., Kor, A.L., Andersson, K., Pattinson, C.: A belief rule based expert system for datacenter pue prediction under uncertainty. IEEE Trans. Sustain. Comput. **2**(2), 140–153 (2017)
24. Ul Islam, R., Andersson, K., Hossain, M.S.: A web based belief rule based expert system to predict flood. In: Proceedings of the 17th International Conference on Information Integration and Web-based Applications & Services, ACM, p. 3 (2015)

Automatic Localization of the Left Ventricle from Short-Axis MR Images Using Circular Hough Transform

Zakarya Farea Shaaf, Muhammad Mahadi Abdul Jamil, and Radzi Ambar

Abstract The localization of the left ventricle (LV) automatically from cardiac magnetic resonance images (CMRI) is an important initial step during segmentation to quantify global and regional volumetric functions of LV such as volume, mass, and ejection fraction. The exact quantification of these indicators results in an accurate system for cardiac disease detection. Conventionally, LV localization in magnetic resonance (MR) images is implemented manually by experts, which is a tedious task and time-consuming. This paper proposes an automated localization method for LV from the cardiac short-axis MR images. The technique applies a circular Hough transform (CHT) to localize the LV with optimized performance. The results revealed that the accuracy of the proposed technique is 89.5% based on the range radius of the LV circle.

Keywords Left Ventricle (LV) · LV localization · Cardiac Magnetic Resonance Images (CMRI) · Circular Hough Transform (CHT)

1 Introduction

American Heart Association (AHA) indicates that human life expectancy can be extended by 10 years if cardiovascular diseases (CVDs) can be efficiently prevented. On the other hand, if all types of cancer can be cured, human life expectancy can be extended to only 3 years. Currently, cutting-edge medical imaging tools provide great options for diagnosing CVDs at early stages, such as computed tomography (CT) [1] and cardiac magnetic resonance imaging (CMRI) [2, 3] technologies. CMRI is a non-invasive modality that produces high-resolution images, which is the most used among different imaging modalities [4].

Early diagnosis is a significant task, and prior to diagnosis, complicated methods are required to track the region of diagnosis and localize the region of interest (ROI).

Z. F. Shaaf · M. M. A. Jamil (✉) · R. Ambar
Department of Electronic Engineering, Faculty of Electrical and Electronic Engineering, Universiti Tun Hussein Onn Malaysia, 86400 Parit Raja, Batu Pahat, Johor, Malaysia
e-mail: mahadi@uthm.edu.my

M. S. Kaiser et al. (eds.), *Proceedings of the Third International Conference on Trends in Computational and Cognitive Engineering*, Lecture Notes in Networks and Systems 348, https://doi.org/10.1007/978-981-16-7597-3_41

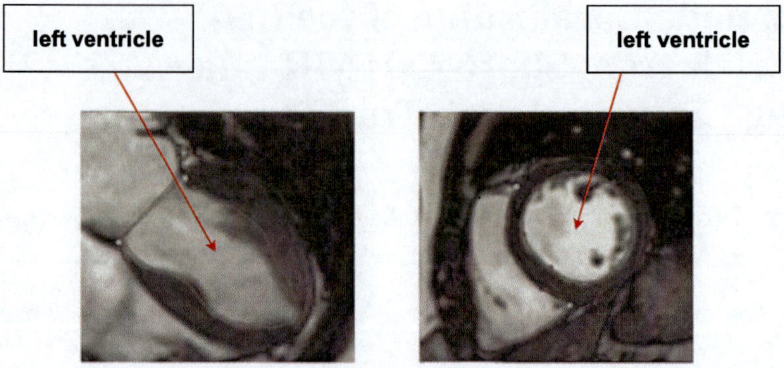

Fig. 1 Long-axis cardiac MRI (left) and short-axis cardiac MRI (right)

The localization of the heart's left ventricle (LV) morphological structures plays a perfect role for segmentation to evaluate essential indicators, namely left ventricle mass (LVM), left ventricle volume (LVV), and ejection fraction (EF) that result in the detection of cardiovascular diseases such as ischemic heart disease, heart failure and others [5]. Short-axis MR images have been considered as a standard technique for evaluating such indicators. LV has an elliptical shape in the long-axis cardiac MRI while it has a circular shape in the short-axis cardiac MRI, as illustrated in Fig. 1.

Previous studies have proposed several methods for LV localization and detection in cardiac MRI [3, 6–8]. Table 1 summarizes recent studies of the LV localization from short-axis MR images. Helwan and Uzun [9] designed a sliding window to detect LV in MRI images based on a machine learning system, and Wang et al. [10] proposed a landmark localization for LV in the cardiac MRI by utilizing deep distance metric learning. The automatic detection of LV has a crucial role in reducing user interaction. Furthermore, the detection of LV is an essential step for the segmentation by the deep learning algorithms, which reduces the computational cost and time. Therefore, CHT is applied to approximate the location of LV [3, 11], and He et al. [12] currently use CHT to detect the location of LV from MRI before the segmentation by using the capsule network. The detection of LV in cardiac MR images is still a challenging task due to the non-homogenous intensity distribution within the myocardium.

Table 1 Recent studies for the LV detection from MR images

Reference	Dataset	Method
[9]	MRI	BPNN
[6]	MRI	Fourier transform + thresholding + score object
[11]	MRI	CHT
[13]	MRI	Sum of absolute difference (SAD)
[12]	MRI	Fourier transform + CHT
[14]	MRI	Level set + CHT

To overcome the issue, this study proposes a localization technique to detect LV from short-axis MR images based on the circular Hough transform (CHT). The robustness of this method compared with previous methods is based on the direct detection of LV from MR images without image pre-processing. An image sequence for four cases is used in this method. The detection generated by CHT was examined in two stages, detection using image binarization as preprocessing for CHT and detection without the preprocessing step. The structure of this paper is organized as follows. Section 2 presents the methodology and outlines datasets and the parameters of the experiment. Section 3 explains the results and discussion. Section 4 describes the conclusion.

2 Methodology

This approach consists of major processes to locate LV from DICOM (digital imaging and communication in medicine) short-axis MR images, as shown in Fig. 2. First, DICOM MR images were stored in the image datastore. Then, the images were read and cropped to get only the region of interest (ROI) area, which is the LV. After that, the CHT algorithm was applied to locate the LV. These processes are explained in the next subsections. MATLAB R2020b software was utilized in the proposed system.

2.1 Datasets

The dataset used in this paper is provided by the challenge of medical image computing and computer-assisted intervention (MICCAI 2009) [15]. The dataset is the DICOM short-axis MR images, with a size of 256 × 256 pixels, obtained at 10–15 s breath-hold with a slice depth of 8–10 mm, and a temporal resolution of 20 slices for the cardiac cycle. The dataset was acquired from several cardiac planes of 45 patients with four classes of morphologies: healthy (N), left ventricular hypertrophy (HYP), heart failure with infarction (HF-I), and heart failure without infarction (HF-NI). Thus, only one case was selected from each class, and the dataset for this method is described in Table 2.

2.2 Circular Hough Transform (CHT) Algorithm

Circular Hough transform is a robust feature extraction tool in digital image processing to extract circular shapes from images. The advantage of this method is its capability to detect the circular object in the presence of noise. Several steps such as the edge detection, canny, and image binarization are required to locate a circle in an image using CHT. In this method, LV is detected in the image utilizing

Fig. 2 Flowchart of LV detection process using CHT

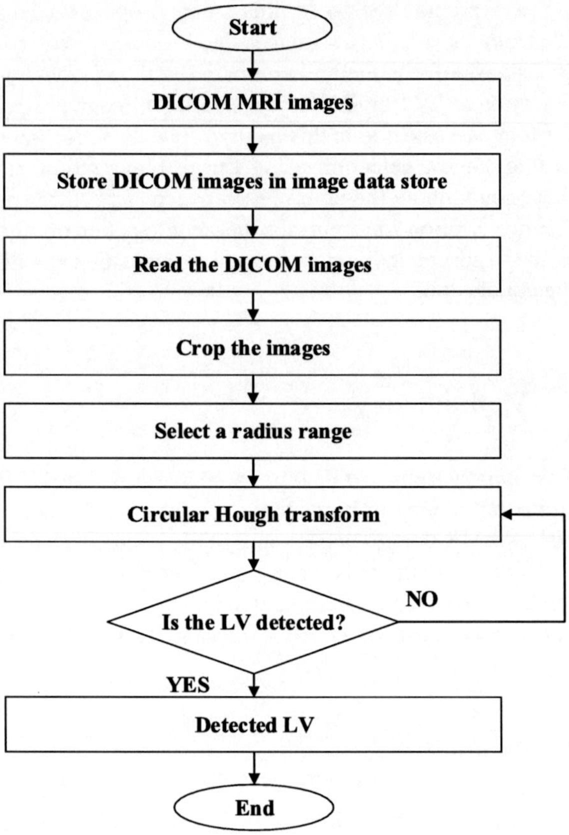

Table 2 Details of the datasets of the four cases from MICCAI 2009

Case	Gender	Age	Pathology
Case 1	Male	53	HF-I
Case 2	Female	63	HF-NI
Case 3	Male	42	HYP
Case 4	Female	53	N

image binarization, and the original image excluded the image binarization step. The circle in the image is described in Eq. (1) by Abdelazeem [16], where $\boldsymbol{r}$ is the radius of a circle, (a, b) are the coordinates of its center in the $\boldsymbol{x}$ and $\boldsymbol{y}$ directions, respectively. The parametric form of the circle in the analytical plane is depicted in Eqs. (2) and (3), where $\boldsymbol{\theta}$ is a changeable angle in the range of $0° < \boldsymbol{\theta} < 360°$ to determine if there are points on the circle with a certain radius.

$$(x - a)^2 + (y - b)^2 = r^2 \quad (1)$$

$$x = a + r\sin\theta \tag{2}$$

$$y = b + r\cos\theta \tag{3}$$

2.3 LV Localization

This approach consists of three main processes, namely, the extraction of ROI, applying the circular Hough transform (CHT) algorithm, and the final step is the localization of LV, as shown in Fig. 3. The extraction of ROI is 80 × 80 pixels images, achieved by creating a rectangular center-cropping window. Then, the images are used as input for the CHT algorithm to detect the LV area. In the short-axis MR images, LV appears as a bright circle with different radius ranges. Thus, the minimum and maximum radiuses were selected as 17 and 23, respectively, based on the extensive measurements that result in an optimized radius range for LV detection at the end-diastolic and end-systolic phases. The performance of this system is explained explicitly in the results and discussion section.

3 Results and Analysis

Figure 4 depicts the outcomes of the proposed algorithm for the LV detection at the end-diastolic (ED) and end-systolic (ES) phases for four different cases. It is necessary to extract the ROI in the image to increase the proposed system's performance. Hence, a rectangular cropping window with a size of 80 × 80 pixels is applied to extract LV in all images. The advantage of this proposed system is that the preprocessing step is optional. Furthermore, this algorithm can locate LV in the original image (grayscale) better than in the preprocessing image (binary image), which is clearly seen in Fig. 4 for Case 1 and Case 2 at the end-systolic phases.

The accuracy of the system is measured based on the exact detected LV images compared with all used images as in Eq. (4). The accuracies of the cases are depicted in Table 3 with the overall average accuracy of the model.

$$Accuracy = \left(\frac{detected\ images\ by\ CHT}{All\ images}\right) \times 100\% \tag{4}$$

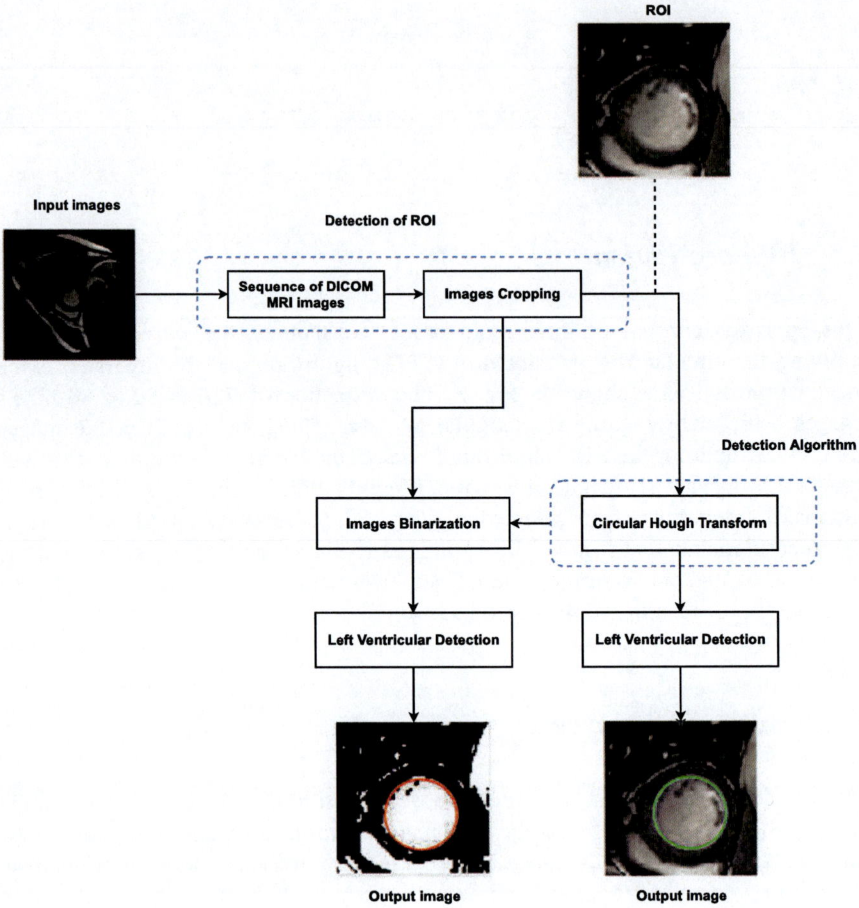

Fig. 3 The block diagram of the proposed system

4 Conclusions

Object detection in images is crucial to localize only the ROI and exclude unwanted features. Moreover, this is an essential step for segmentation by deep learning algorithms, which reduces the computational cost and time. In this study, a simple and effective method utilizing circular Hough transform (CHT) has been proposed for the left ventricle (LV) detection from the short-axis MRI. It produced efficient results for the middle and basal slices, while for the apical slices, the detection accuracy was much lower. Even without further preprocessing such as thresholding and image binarization, the detection produced by CHT has shown robust performance. The low accuracy of the LV detection in apical slices was due to the out-of-range radius that is the suggestion for improvement in future work.

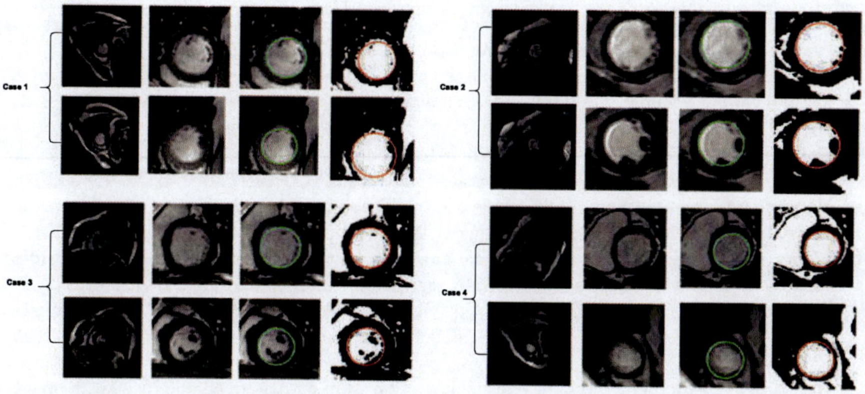

Fig. 4 Results of CHT model for the LV detection in four cases (upper row represents ED images, while the bottom row represents ES images, for each case)

Table 3 The proposed system's performance based on datasets cases

Case	Image sequences	Percentage of localization accuracy (%)
1	50	92
2	50	87
3	50	89
4	50	90
Overall average accuracy		89.5

Acknowledgements This research was supported by the Ministry of Higher Education (MOHE) through Fundamental Research Grant Scheme (FRGS) (FRGS/1/2020/TK0/UTHM/02/16) and Universiti Tun Hussein Onn Malaysia (UTHM) through FRGS Research Grant (Vot K304).

References

1. Baskaran, L. et al.: Automatic segmentation of multiple cardiovascular structures from cardiac computed tomography angiography images using deep learning. PLoS One 15, e0232573 (2020)
2. Albà, X., et al.: Automatic initialization and quality control of large-scale cardiac MRI segmentations. Med. Image Anal. **43**, 129–141 (2018)
3. Zhong, L., Zhang, J.-M., Zhao, X., Tan, R.S., Wan, M.: Automatic localization of the left ventricle from cardiac cine magnetic resonance imaging: a new spectrum-based computer-aided tool. PLoS One 9, e92382 (2014)
4. Sharif, M., Arfan Jaffar, M., Tariq Mahmood, M.: Optimal composite morphological supervised filter for image denoising using genetic programming: application to magnetic resonance images. Eng. Appl. Artif. Intell. **31**, 78–89 (2014)
5. Wu, B., Fang, Y., Lai, X.: Left ventricle automatic segmentation in cardiac MRI using a combined CNN and U-net approach. Comput. Med. Imaging Graph. **82**, 101719 (2020)

6. Tan, L.K., et al.: Automatic localization of the left ventricular blood pool centroid in short axis cardiac cine MR images. Med. Biol. Eng. Comput. **56**, 1053–1062 (2018)
7. Abdeltawab, H. et al.: A deep learning-based approach for automatic segmentation and quantification of the left ventricle from cardiac cine MR images. Comput. Med. Imaging Graph. 81, 101717 (2020)
8. Hellwig, S., et al.: Evaluation of left ventricular function in patients with acute ischaemic stroke using cine cardiovascular magnetic resonance imaging. ESC Hear. Fail. **7**, 2572–2580 (2020)
9. Helwan, A., Uzun Ozsahin, D.: Sliding window based machine learning system for the left ventricle localization in MR cardiac images. Appl. Comput. Intell. Soft Comput. 1–9 (2017)
10. Wang, X., Zhai, S., Niu, Y.: Left ventricle landmark localization and identification in cardiac MRI by deep metric learning-assisted CNN regression. Neurocomputing **399**, 153–170 (2020)
11. Kurzendorfer, T., Brost, A., Forman, C., Maier, A.: AUTOMATED LEFT VENTRICLE SEGMENTATION IN 2-D LGE-MRI. IEEE 14th Int. Symp. Biomed. Imaging (ISBI 2017) 831–834 (2017)
12. He, Y., et al.: Automatic left ventricle segmentation from cardiac magnetic resonance images using a capsule network. J. Xray. Sci. Technol. **28**, 541–553 (2020)
13. Irshad, M., Muhammad, N., Sharif, M., Yasmeen, M.: Automatic segmentation of the left ventricle in a cardiac MR short axis image using blind morphological operation. Eur. Phys. J. Plus **133**, 148 (2018)
14. Lu, J. et al.: Segmentation of the cardiac ventricle using two layer level sets with prior shape constraint. Biomed. Signal Process. Control **68**, 102671 (2021)
15. Radau P, Lu Y, Connelly K, Paul G, Dick A.J.W.G.: Evaluation framework for algorithms segmenting short axis cardiac MRI. MIDAS J. (2009)
16. Abdelazeem, S.: Micro-aneurysm detection using vessels removal and circular Hough transform. In: Proceedings of the Nineteenth National Radio Science Conference, Alexandria University, pp. 421–426 (2002). https://doi.org/10.1109/NRSC.2002.1022650

Facial Detection for Neonatal Infant Pain Using Facial Geometry Features and LBP

Jarin Tasnim Ritu, Md. Shahadat Hossen Shakil, Md. Nahian Imtiaz Hasan, Shamim Al Mamun, M. Shamim Kaiser, and Mufti Mahmud

Abstract Neonatal pain assessment is essential for infants concerning their health issues. There have been several studies to assess the pain of infants using image processing in the field of computer vision. In this paper, we propose a different approach to detect pain in infants that outperforms previous research in this field. We merged a face area-based feature collection method with a local binary pattern (LBP). Moreover, three different machine learning algorithms have been executed to find the best parameter to get a decent accuracy on the iCOPE dataset. The proposed method uses the SVM classifier to achieve 86% of testing accuracy compared to other methods.

Keywords Facialgeometry · LBP · iCOPE · Pain detection · Neonates

J. T. Ritu (✉) · Md. S. H. Shakil · Md. N. I. Hasan · S. Al Mamun · M. S. Kaiser
Institute of Information Technology, Jahangirnagar University, Savar, Dhaka 1342, Bangladesh
e-mail: shamim@juniv.edu

M. S. Kaiser
e-mail: mskaiser@juniv.edu

S. Al Mamun · M. S. Kaiser
Applied Intelligence and Informatics Lab (AII-Lab), Jahangirnagar University, Savar, Dhaka 1342, Bangladesh

M. Mahmud
Department of Computer Science, Nottingham Trent University, Clifton Lane, Nottingham NG118NS, UK
e-mail: muftimahmud@gmail.com; mufti.mahmud@ntu.ac.uk

Medical Technologies Innovation Facility, Nottingham Trent University, Clifton Lane, Nottingham NG118NS, UK

Computing and Informatics Research Centre, Nottingham Trent University, Clifton Lane, Nottingham NG118NS, UK

M. S. Kaiser et al. (eds.), *Proceedings of the Third International Conference on Trends in Computational and Cognitive Engineering*, Lecture Notes in Networks and Systems 348, https://doi.org/10.1007/978-981-16-7597-3_42

1 Introduction

Face recognition technology aims to identify a person based on their unique facial landmarks and skin texture [1]. Computer Vision is an affordable and popular technology for health care and diagnostics today. Cheap and non-invasive means of diagnosis by looking at the face will lead our futuristic society. There has been much work on this topic at the moment, such as measuring heart rate [2], diagnosing brain disorders [3, 4], measuring the distance to an object, measuring social distance [5, 6], and diagnosing non-communal diseases. Also, robots and computer vision [7, 8] are essential applications that are being created in the health sector. Avowedly, this technology will take the health sector to a new level. Pain is a significant source of suffering that can have both short- and long-term effects on the human body. It is fundamental to detect pain appropriately to properly treat it. In comparison to adults, detecting the source and cause of pain in neonates is more challenging because they cannot verbalize their pain or their pain experience. In addition to being associated with cognitive disorders, newborn infants are subjected to painful experiences that may significantly raise their short- and long-term morbidity and mortality. There can be many possible causes of neonatal pain. Anand proposed an updated uniform taxonomy in 2017 based on temporal features, pain characteristics, and secondary effects. Based on these three characteristics, he proposed the following to classify neonatal pain: Acute recurrent, Prolonged, Persistent, and Chronic [9]. Acute procedural pain is usually triggered by a brief dull ache and ends when the source of pain is taken away. The acute prolonged pain is provoked by a target movement and has a well-defined starting and estimated endpoint; the intensity of this type of pain decreases as the time since the painful stimulus occurs increases [10]. If a baby is not treated when it is needed, it can lead to serious consequences and permanently changes the brain structure and functions [11]. Several long-term effects of repeated pain exposure in childhood have been reported in pediatric studies. For example, repeated painful experience in newborns has been linked to transformations in the cerebral white matter and subcortical gray matter, as well as delayed cortico-spinal growth. These neurodevelopmental changes can result in a wide range of behavioral, developmental, and learning disabilities [12]. Other long-term effects of pain exposure observed in school-aged children include delayed visual perceptual development, lower IQs, and internalizing behavior [13]. The annual cost of treatment for unfavorable neurodevelopmental outcomes in neonates alone is estimated to be more than 7 billion [14]. So it is very much essential to detect pain and the origin of pain to prevent further complications. The accuracy of pain assessment is vital because it helps parents of a newborn baby to understand the gravity of the situation and take action accurately for appropriate treatments. Even though it is difficult to tell whether an infant is in pain or not because of their lack of ability [15] to be misinterpreted, unappreciated, or interact, many advanced techniques for detecting neonatal pain are being developed these days. The face is a distinct part of the human body, and for infants, the best way to determine whether or not they are in pain could be through facial expression as facial expression study presents accurate, sensitive, and precise information about the nature and severity of pain.

Some work has been done on facial expressions. But we proposed a new approach which is LBP and triangle distance between some key points of the face in our study. The following is the rest of the paper: In Sect. 2, the feature collection technique and experimental design are briefly covered. In Sect. 3, we describe the dataset we worked with, which is the Infant COPE (iCOPE) dataset. We go over our classification methods and algorithms in Sect. 4. In Sect. 5, we covered our technique, including how we gathered and classified our data and features. We discussed the outcomes of our tests in Sect. 4, and we draw conclusions and make recommendations for further study in Sect. 5.

Our contribution in this paper is as follows:

(i) to review about the spatial feature of the face using deep learning algorithm or feature-based learning.
(ii) to detect neonatal pain detection using machine learning approach.

The rest of the paper is organized as follows. In Sect. 2, we discussed some literature reviews of existing researches and disease identification methods in Sect. 3. In Sect. 4, i-Health Monitoring System in Society 5.0 is discussed. In Sect. 5, report studies concentrate on opportunities and challenges. Finally, conclusion is discussed in Sect. 6.

2 Feature Collection

The original photos were taken from the iCOPE database and are 3008*2000 pixels. To begin, we rotated all of the landscape photos.

The color image is then transformed into a grayscale intensity image. The brightness is retained in the grayscale image, but the color and saturation information is removed. Following that, we used the dlib package to locate 68 points on a face, from which we identified our target points and put them in separate arrays (Fig. 1).

Five triangular sections were created on the face for the features. One for the left eye, one for the right eye, one for the nose, and two for the mouth are the options. Each triangle's area was determined as follows:

$$y = m * x + c \tag{1}$$

3 Dataset

In all the classification studies mentioned in this article, we utilized the Infant COPE (Classification Of Pain Expressions) database. This collection includes 204 photos, from 18 h to 3 days old, of a group of 26 Caucasian newborns (13 children and 13

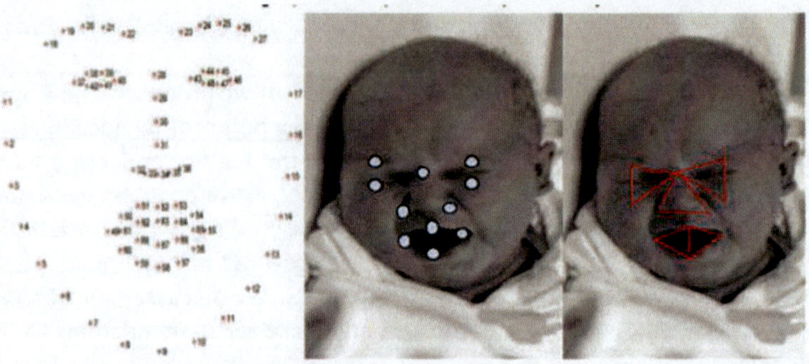

Fig. 1 Feature facial landmark points and area coverage

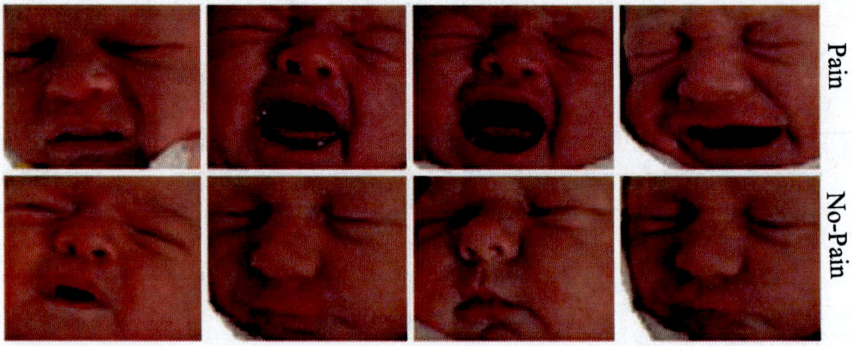

Fig. 2 Sample photos from iCOPE dataset

girls) [16, 17]. In the primary rest period and experiencing many noxious stimuli, photographs were taken of the toddlers, including physical disturbance, air stimulation on the nose, friction on the outer side of the heel, and heel discomfort. Among the 204 photos, 60 were in the pain category and 144 were in the non-pain category. We have divided the data in a manner which keeps 80% of the data for training and 20% for testing (Fig. 2).

4 Classification Algorithm

SVM: The vector support machine is a non-linear, supervised technique with various applications for biomedical or image processing.

SVM is typically utilized for double-grade issues, and the supply of kernel functions extends to multi-class applications of SVM. SVM distinguishes between two classes by generating a classification hyperplane in a high-dimensional feature space. The equation for a hyperplane is

$$w * x + b = 0 \tag{2}$$

w is the weight vector, which is orthogonal to the hyperplane. The parameter b stands for bias or threshold.

We used a radial base function as a kernel for the categorization of emotions in our study. This classifier performance depends on the cost (C) and kernel (μ) values. This study uses the grid-search technique to discover the best value for C from $2 - 14$ to $2 + 15$ and α from $2 - 12$ to $2 + 8$ to reach a higher rating. RF: Classification, regression, and pattern-identification applications of the random forest classifier are the ensemble learning approach employed. The main idea of this classifying system is based on the features of the data and on the characteristics of the test data, which fits training. The judgment is based on training. The classification performance depends on the number of trees utilized. This study shows a heuristic range from 20 to 500 of the trees in the RF method.

LR: Linear regression tries to represent the connection of two variables using a linear equation of data observed. One variable is called the explanatory variable (X), and the other is the dependent variable (Y).

So, for example, a designer could want to link people's weights in a linear regression model with their heights. A modeling model should first identify whether or not a connection exists between the variables of interest before applying a linear model to the data seen. Dispersion can be a valuable tool to determine the strength of the link between two variables. If the suggested explanatory and dependent variables do not appear to be associated (that is to say, the dispersion plot does not reveal any increase or decrease in trends), then the usable model is likely not to fit a linear regression model with the data. The correlation coefficient value between -1 and 1 reflecting the strength of the combination of the data seen for the two variables is a meaningful numerical measure of the association of two variables. A linear line of regression has a

$$y = c + m * x \tag{3}$$

equation in which, x is an explanatory and dependent variable. The slope of the line is m, and c is the intercept which is the value of y when $x = 0$.

5 Methodology

The method proposed in this paper to detect infant pain is divided into three stages: A. Data collection. B. Feature extraction. C. Classification using extracted features illustrated in Fig. 3. The following section contains a detailed description of these stages.

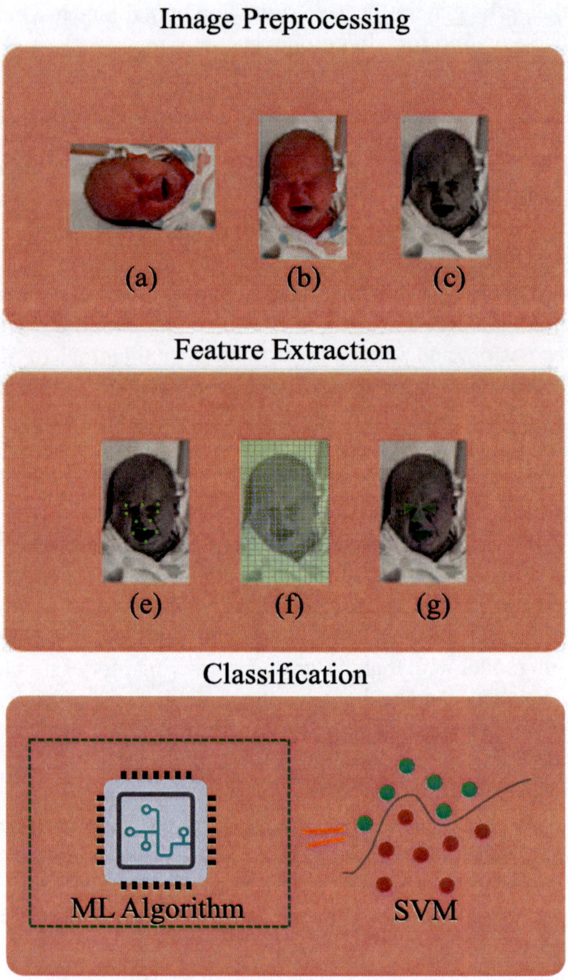

Fig. 3 The main iCOPE dataset has raw images with vertical orientation. In the image prepossessing part, the image is rotated and converted into grayscale images to get raw features from the image. Subsection **a**, **b**, and **c** illustrated the process. In the feature extraction part, the proposed method used three feature collection methodology, **e** 11 facial landmark coordinates **f** applying the LBP method to get the textural features from the train images and **g** uses geometric area features from the facial parts. Finally, for the classification, 5-fold cross-validation ML algorithm is applied and SVM for the classification has been used with the best parameter searching

5.1 Data Collection

The original images were taken from the Classification of Pain Expressions (COPE) database. All the images are categorized into two types: pain and non-pain. There are 60 images of babies which are of pain. The other types are rest, cry, air stimulus, and friction. These types are considered as non-pain types in this paper. For testing and training, an 80:20 ratio was maintained.

5.2 Features

Features were extracted from the images of the COPE dataset using facial geometry, i.e. triangular area of a specific location of the face and Local Binary Pattern (LBP).

5.2.1 Triangular Area

We consider 68 points of facial landmarks from which some specific points are taken as features. Five triangular sections were created on the face for the features. One for the left eye, one for the right eye, one for the nose, and two for the mouth are the options. Table 1 shows the points taken as features.

We use the following formula using the landmark points to compute and store the distance between the specific points of eye, nose, and mouth [18].

$$d = \sqrt{((x_2 - x_1)^2 + (y_2 - y_1)^2)} \tag{4}$$

Then with the help of Heron's formula [19], we calculate the triangle area of some particular region [shown in Fig. 2]

$$\frac{s * (s - a) * (s - b) * (s - c)}{2} \tag{5}$$

5.2.2 Local Binary Pattern

A Local Binary Pattern (LBP) [20] is a local texture operator with powerful discrimination, low computational complexity, and less sensitivity to changes in illumination. The original LBP operator was introduced by Ojala et al. [21]. The LBP works in a block size of 3 × 3, in which the center pixel is used as a threshold for the neighboring pixel, and the LBP code of a center pixel is generated by encoding the computed threshold value into a decimal value. The mathematical expression of LBP is given as

Table 1 Calculated facial landmarks number

Area	Point number on landmark
Left eye	[36–40]
Right eye	[42–46]
Nose	[27–36]
Mouth	[48–60]

Table 2 Result summary

Pain category	Training image	Testing image	Correctly detected images	Accuracy
Pain, no-pain	80%	20%	95%	87%

$$LBP = \sum_{i=0}^{P-1} S\,(n_i - G_C)\,2^i \tag{6}$$

$$S(x) = \begin{cases} 1, & if\ x > 0 \\ 0, & if\ x \le 0 \end{cases} \tag{7}$$

5.3 *Classification*

In our paper for categorization of pain and non-pain faces, we employed Euclidean distance, triangular distance, and support vector machine (SVM). SVMs have proven to be beneficial in a variety of pattern recognition and classification, including image and facial action detection. They are highly adapted to the task of pain versus no-pain classification since SVM is a supervised learning-based binary classifier and performs better than other classifiers. SVM distinguishes between two classes by generating a classification hyperplane in high-dimensional feature space (Table 2).

6 Discussion

In the paper, the proposed methodology's performance has been evaluated in terms of accuracy.

The database was split into two sections. One portion has just pain images and is classified as Pain, whereas the other contains rest, sucrose, cry, friction, and air images and is classified as No Pain. Within this two-part data collection, the Training and Testing dataset was segregated, where some corrupted and blur images were avoided during the process. Various methodologies were used to compare the outcomes collected in different steps during the experimental inquiry. Facial Geometry and LBP were integrated, and each image's feature was retrieved. Three methods were used to extract features: The dlib frontal face detector was used to detect 11 facial coordinate points in the first method. Following that, we calculated facial geometry on the observed points and saved the findings as the method's final features. In the third method, we extracted 100 feature points from each image using LBP, and finally, we merged all the feature points together and classified them using multiple classification algorithms.

Table 3 Different classification model accuracy

Model	Accuracy	Best params
SVM	0.826667	'C': 1, 'kernel': 'linear'
Random forest	0.815556	'*n* estimators': 10
Logistic regression	0.783334	'C': 1
Proposed method	0.866333	'C': 1, 'kernel': 'linear'

Table 4 Average accuracy over 26 subjects

Features	Accuracy (%)
LBP	77.52
HOG	81.75
HOG + LBP	82.80
VGG face	82.42
Proposed method	86.63

The table shows the classification accuracy for each of the algorithms (Tables 3 and 4).

7 Conclusion

In newborn care, pain evaluation is critical. Inadequate pain therapy in babies causes unnecessary suffering and can have long-term consequences. In this research, we offer a system for detecting newborn pain that uses face geometries and Local Binary Pattern algorithms, as well as data-learned characteristics, to encode pain suggestive signals. The features were then loaded into a support vector machine, which sorted them into one of two predefined classes: pain or no pain. With the proposed method, we got 86.63% of test accuracy.

References

1. Al Mamun, S., Daud, M.E., Mahmud, M., Kaiser, M.S., Rossi, A.L.D.: ALO: AI for least observed people. In: International Conference on Applied Intelligence and Informatics, pp. 306–317. Springer (2021)
2. Rahman, M.M., Al Mamun, S., Kaiser, M.S., Islam, M.S., Rahman, M.A.: Cascade classification of face liveliness detection using heart beat measurement. In: Proceedings of International Conference on Trends in Computational and Cognitive Engineering, pp. 581–590. Springer (2021)
3. Mahmud, M., Kaiser, M.S., Rahman, M.M., Rahman, M.A., Shabut, A., Al-Mamun, S., Hussain, A.: A brain-inspired trust management model to assure security in a cloud based IoT framework for neuroscience applications. Cogn. Comput. **10**(5), 864–873 (2018)

4. Noor, M.B.T., Zenia, N.Z., Kaiser, M.S., Al Mamun, S., Mahmud, M.: Application of deep learning in detecting neurological disorders from magnetic resonance images: a survey on the detection of Alzheimer's disease, Parkinson's disease and schizophrenia. Brain Inform. **7**(1), 1–21 (2020)
5. Al Mamun, S., Lam, A., Kobayashi, Y., Kuno, Y.: Single laser bidirectional sensing for robotic wheelchair step detection and measurement. In: International Conference on Intelligent Computing, pp. 37–47. Springer (2017)
6. Kaiser, M.S., Mahmud, M., Noor, M.B.T., Zenia, N.Z., Al Mamun, S., Mahmud, K.M.A., Azad, S., Aradhya, V.N.M., Stephan, P., Stephan, T., et al.: iWorksafe: towards healthy workplaces during covid-19 with an intelligent Phealth app for industrial settings. IEEE Access **9**, 13814–13828 (2021)
7. Kaiser, M.S., Al Mamun, S., Mahmud, M., Tania, M.H.: Healthcare robots to combat covid-19. In: COVID-19: Prediction, Decision-Making, and Its Impacts, pp. 83–97. Springer, Berlin (2021)
8. Kaiser, M.S., Zenia, N., Tabassum, F., Al Mamun, S., Rahman, M.A., Islam, M.S., Mahmud, M.: 6G access network for intelligent internet of healthcare things: opportunity, challenges, and research directions. In: Proceedings of International Conference on Trends in Computational and Cognitive Engineering, pp. 317–328. Springer (2021)
9. Guinsburg, R., de Araújo Peres, C., de Almeida, M.F.B., Balda, R.D.C.X., Berenguel, R.C., Tonelotto, J., Kopelman, B.I.: Differences in pain expression between male and female newborn infants. Pain **85**(1–2), 127–133 (2000)
10. Anand, K.J.S.: Defining pain in newborns: need for a uniform taxonomy? Acta Pædiatrica **106**(9), 1438–1444 (2017)
11. Zamzmi, G., Goldgof, D., Kasturi, R., Sun, Y.: Neonatal pain expression recognition using transfer learning (2018). arXiv preprint arXiv:1807.01631
12. Cignacco, E., Mueller, R., Hamers, J.P.H., Gessler, P.: Pain assessment in the neonate using the Bernese pain scale for neonates. Early Hum. Dev. **78**(2), 125–131 (2004)
13. Field, T.: Preterm newborn pain research review. Infant Behav. Dev. **49**, 141–150 (2017)
14. Behrman, R.E., Butler, A.S., et al.: Preterm Birth: Causes, Consequences, and Prevention. The National Academies Press of Science, Engineering and Medicine, Washington, D.C. (2007)
15. Buchholz, M., Karl, H.W., Pomietto, M., Lynn, A.: Pain scores in infants: a modified infant pain scale versus visual analogue. J. Pain Symptom Manag. **15**(2), 117–124 (1998)
16. Brahnam, S., Nanni, L., Sexton, R.S.: Neonatal facial pain detection using NNSOA and LSVM. In: IPCV, pp. 352–357 (2008)
17. Brahnam, S., Nanni, L., McMurtrey, S., Lumini, A., Brattin, R., Slack, M., Barrier, T.: Neonatal pain detection in videos using the iCOPEvid dataset and an ensemble of descriptors extracted from gaussian of local descriptors. Appl. Comput. Inform. (2020)
18. Schiavenato, M., Byers, J.F., Scovanner, P., McMahon, J.M., Xia, Y., Lu, N., He, H.: Neonatal pain facial expression: evaluating the primal face of pain. Pain **138**(2), 460–471 (2008)
19. Keith, K.: Is a 2000-year-old formula still keeping some secrets?. Amer. Math. Monthly. **107**, 402–415. (2000) https://doi.org/10.2307/26952956
20. Nanni, L., Lumini, A., Brahnam, S.: Local binary patterns variants as texture descriptors for medical image analysis. Artif. Intell. Med. **49**(2), 117–125 (2010)
21. Ojala, T., Pietikainen, M., Maenpaa, T.: Multiresolution gray-scale and rotation invariant texture classification with local binary patterns. IEEE Trans. Pattern Anal. Mach. Intell. **24**(7), 971–987 (2002)

A Deep Learning-Based Ophthalmologic Approach for Retinal Fundus Image Analysis to Detect Glaucoma

Lutfun Nahar, Mohammad Shahadat Hossain, Promi Das, Tanzeem Alam, and Karl Andersson

Abstract Glaucoma is a retinal condition caused by increasing intraocular pressure inside the eye. It is called a silent killer disease because the damage caused by this is un-rectifiable, and the consequence is inevitable blindness if it is left untreated. Ophthalmologists diagnose glaucoma through a number of retinal tests, which include ophthalmoscopy, tonometry, perimetry, gonioscopy and pachymetry. Each of these procedures is time-consuming, expensive and sometimes not available. Early detection and treatment of these diseases are quite important to avoid ultimate blindness and sight loss. By reducing the computational complexity, a computer-aided diagnostic system may advantageous in the early detection and screening of glaucoma. In this paper, we proposed two state-of-the-art glaucoma detection techniques using deep learning model, VGG16 and ResNet152. Here, the extraction of structural characteristics, namely, cup-to-disk ratios, disc-liability-scale damage and superior nasal temporal lower areas are considered to diagnose glaucoma. Moreover, the segmentation method was evaluated using the Fundus Image dataset.

Keywords VGG16 · ResNet152 · Glaucoma · Fundus Image · Deep Learning

1 Introduction

Glaucoma, which is known as the "silent thief of sight", is a chronic eye disease that gradually degrades the patient's vision while the patient is not known that he or she is losing sight. Glaucoma is a dangerous and ultimately blinding disorder. Early detection and treatment of the illness have been shown to avoid ultimate blindness and sight loss. Glaucoma has no obvious symptoms in the early stages of the illness,

L. Nahar (✉) · P. Das · T. Alam
International Islamic University Chittagong, Chittagong, Bangladesh

M. S. Hossain
University of Chittagong, Chittagong, Bangladesh

K. Andersson
Lulea University of Technology, Skelleftea, Sweden

M. S. Kaiser et al. (eds.), *Proceedings of the Third International Conference on Trends in Computational and Cognitive Engineering*, Lecture Notes in Networks and Systems 348, https://doi.org/10.1007/978-981-16-7597-3_43

therefore, the only method to diagnose is via regular eye examinations. As we said it is a degenerative ailment that desolates the eye and obliterating all vision. Through the exertion, it has been effectual in generating a program that complies the diagnosis of glaucoma utilizing the detection of the properties of the glass and tough to assess whether or not there is damage to the optic nerve. There has been some work done related to this topic such as CDR (cup to disk ratio) where an automated cup to disk ratio assessment system based on a two-dimensional fundus picture to address the glaucoma problem [1]. Another algorithm uses image processing methods [2, 3, 5] to automatically look up to the cup and the optical disc of glaucoma. The optical disc is varied by its orange color in pictures of fundus of the eye, whereas the brightest region introduces the eye cup. Deep learning is an approach [20, 29], which has the capabilities to learn and represent information. The research presented in this paper is an attempt to develop a method that can detect glaucoma in an early stage.

Thus, this paper focuses on building an operative tool for the early detection of glaucoma by introducing using deep learning model VGG16 and ResNet152. The VGG16 model is considered because it can handle a large amount of raw data with comparatively less pre-processing and less convolution layer for computation. In addition instead of having a large number of hyper-parameter, they focused on having convolution layers of 3×3 filter with a stride 1 and always used same padding and maxpool layer of 2×2 filter of stride 2 where ResNet152 architecture allows multiple layer switching on weight basis and hence reduces unnecessary computational costs.

Therefore, ResNet152 plays an important role to achieve better model accuracy. This research is further enhanced by implementing more image preprocessing techniques and accesses better feature extraction.

The remaining of the paper is structured as follows. Section 2 analyses the different works related to glaucoma detection. Section 3 outlines the methodology undertaken in order to develop the model. Section 4 presents the results; while Section 5 concludes the paper with a discussion on future work.

2 Literature Review

There has been lots of work related to glaucoma detection using Machine learning. In [15–17], they represent an automated glaucoma screening method based on a hybrid cloud platform. In the automated glaucoma screening module, an innovative SDC technique is assigned segmenting and rebuilding the optic disc, scantiness and diversity restraints are taken into account.

In Juneja et al. [18], they proposed a cloud-based architecture for glaucoma screening and monitoring services that operates a technique for cup to disk ratio estimation based on two-dimensional retinal fundus pictures. The cloud-based platform merges patients, ophthalmologists, glaucoma screening algorithms and diagnostic devices into a unified system.

In Xu et al. [4], author focused on structural features such as CDR. Images were split into red, green and blue channels where accuracy was 0.77, which is an important drawback factor. Some systems are implemented to support tools for the presumptive diagnosis of glaucoma through the identification and processing of medical images of the human eye.

In Eziama et al. [6], they proposed an algorithm, which automatically looks up to the cup and the optical disc. To better distinguish the cup and optical disc, the original picture of the fundus of the eye is resolved to grayscale. In Mohr et al. [7], the creation of this method is based on the detection of probable glaucoma diagnoses using pictures of the fundus of the eye as a foundation.

In [8, 9], the digital processing of images of the human eye fundus is required, which is a more complex process to analyze. Moreover, in [32, 33, 37–42], they proposed some image preprocessing techniques and CNN method to access better feature extraction.

In [10–14], supervised machine learning method is used with the selected feature. In another paper, author proposed [27, 28] a convolutional neural network based on the six-layer architecture was used to perform the classification of patterns observed in the patient's eye pictures for glaucoma detection. The ORIGA and SCES datasets are both used; their respective values are 0.822 and 0.882. It can be observed that none of the studies as discussed above considered the VGG16 and Resnet152 to extract disc-liability-scale damage and superior nasal temporal lower areas to detect glaucoma.

Therefore, in this research, deep learning model has been proposed to detect glaucoma, which will be described in the following section.

3 Methodology

3.1 Data Classification

Here, around 5000 images of retinal fundus are used to detect glaucoma and have five classes. These images are divided into two sets:

- Training Set: Most algorithms can be trained to learn from and make predictions on data. Such algorithms use data to arrive at a prediction or judgment. They use data to develop a mathematical model, and then test that model against new data. The model uses 4000 images for training, which is 80% of retinal image dataset.
- Test Set: The dataset used for validation is called the test dataset and is essential to producing an impartial assessment of the final model fit on the training dataset. 1000 images are used for testing from the dataset (Fig. 1).

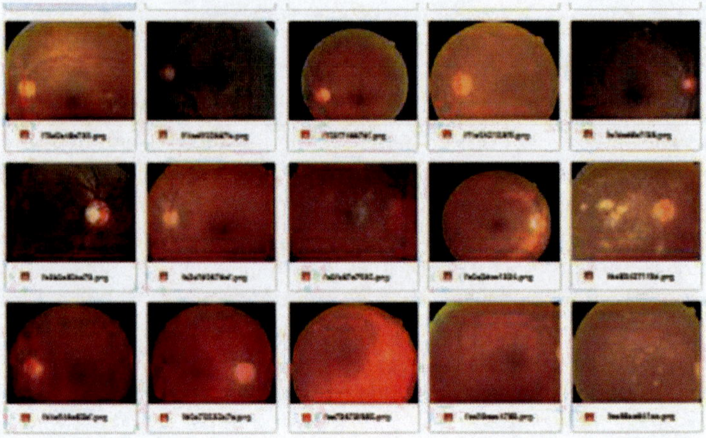

Fig. 1 Sample dataset

3.2 Algorithmic Analysis

A glaucoma analysis system has been created by using the VGG16 and ResNet152 architecture. Finally, apart from the general issues that the framework explores, such as the distinction between dataset types and classifications, and making data analysis and interface easier, the analysis system was also concerned with obtaining more data that deals with spatial and temporal patterns, and offering it to customers.

Our proposed system is based on supervised machine learning approach [19, 24, 25]. This approach is usable with the labeled data. Below is the description of two deep learning models considered in our research.

- VGG16: VGG16 or Visual Geometric Group is a supervised learning model that is used for classifying and regression analysis, with related learning methods. A VGG16 algorithm constructs a model that associates fresh samples with one of two categories, making it a non-probabilistic binary linear classifier
- ResNet152: ResNet152 short for Residual Networks is a traditional neural network that serves as the foundation for many computer vision applications. It enabled to effectively train incredibly deep neural networks with 150 + layers. Prior to ResNet152, training extremely deep neural networks was challenging owing to the issue of vanishing gradients.

Figure 2 illustrates the pre-processing steps and Fig. 3 describes the main components of our proposed system.

Data preprocessing is a data mining [21–23, 26] approach that applies a standard transformation on raw data to provide the user with a digestible and intelligible format. It is frequently the case that the data we have access to in the real world is partial, inconsistent and deficient in specific behaviors or patterns, and as a result,

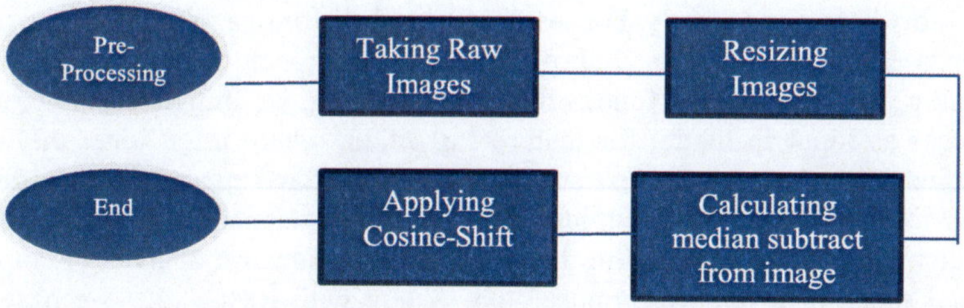

Fig. 2 Data pre-processing framework

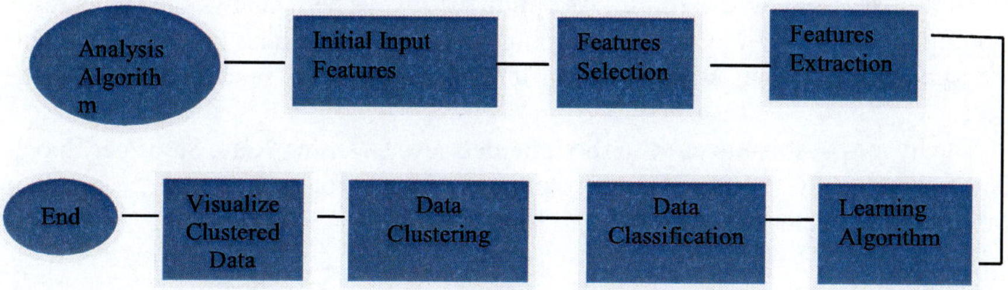

Fig. 3 System framework

will include numerous inaccuracies. That's why we have applied some pre-processing techniques through the process of work.

Figure 2 illustrates the pre-processing steps. This is a kind of computer software designed to analyze incoming data and provide predicted output values that are within an acceptable range.

Here, we tuned our model parameters individually for both VGG16 and ResNet152, which we have applied as algorithm in the process. At first, we have applied some preprocessing techniques in VGG16 and ResNet152 for training fundus images of glaucoma. In initial stage, they were raw dataset. Techniques used in VGG16 are given below:

- Resize image: Sometimes algorithm faces some trouble to understand the features in dataset if we don't keep constant the image size for all images in the dataset. So it's necessary so that we can feed our images into our model algorithm smoothly.
- Subtract median: we have chosen this so that we can detect the changes of features between the images correctly, if we say more clearly, to detect the edge line from where we will only take the weighted features. Less weighted features will be removed.
- Radius reduction: we have chosen this so that we can only take our region of interest (ROI). Other areas will be reduced, this technique we have applied on median subtract.
- Cosine shift: ultimately we have applied cosine shift, so that in ROI where feature density will be visible by color code. By selecting this, we can represent our feature

visibility better and clearly. For resnet152, we have only applied image resize as preprocessing that explained above and got our expected result. After preprocessing was done for the fundus images in training, we initialized the prepared images as input in model for feature extraction, where after done, the tuning parameter vgg16 got very less accuracy. Then we have approached our method with resnet152 where after tuning the hyper-parameters we got our expected results, which we explained in the result section. Because of its high computational speed of layer, we proposed our system with Resnet152. We have only worked with input layer of these models and added dense layer for multiclass glaucoma.

The optimizer used here is Adam. As optimizer always help change the attributes of training model such as weights or learning rate so that it reduces the losses and help to reach faster minima. Adam is best as it helps the model to reach minima in faster pace.

The hyper-parameters used for both models are: Learning Rate, Steps per Epoch, Batch size, Seed Comparative hyper-parameter tuning for each model are shown in Tables 1 and 2 below.

Table 1 Vgg16 parameters tuning

SL	VGG16 Tuning parameter stages	
	Hyper-parameter	Accuracy
1	Learning rate: 0.1	0.59
	Epochs: 20	
	Batch size: 32	
	Seed: 42	
2	Learning rate: 0.1	1 (model over fitted)
	Epochs: 60	
	Batch size: 32	
	Seed: 42	
3	Learning rate: 0.001	0.62
	Epochs: 30	
	Batch Size: 32	
	Seed: 42	
4	Learning rate: 0.01	0.69
	Epochs: 8	
	Batch size: 32	
	Seed: 42	
5	Learning rate: 0.01	0.72 (final)
	Epochs: 10	
	Batch size: 32	
	Seed: 42	

Table 2 Resnet152 parameter tuning

	ResNet152 Tuning parameter stages	
	Hyper-parameter	Accuracy
1	Learning rate: 0.1	1 (model over fitted)
	Epochs: 20	
	Batch size: 32	
	Seed: 42	
2	Learning rate: 0.01	1 (model over fitted)
	Epochs: 10	
	Batch size: 32	
	Seed: 42	
3	Learning rate: 0.00001	0.41
	Epochs: 5	
	Batch size: 32	
	Seed: 42	
4	Learning rate: 0.00001	0.77
	Epochs: 15	
	Batch size: 32	
	Seed: 42	
5	Learning rate: 0.00001	0.87 (final)
	Epochs: 20	
	Batch size: 32	
	Seed: 42	

As we can see from the table, after overall tuning for both models, we got our ultimate result where ResNet152 achieved 87% accuracy and VGG16 is 72%. So ResNet152 performs better significantly over VGG16.

4 Result and Discussion

The train data are classified into five different categories in the dataset. Therefore, the overall expansion of these data with respect to their volume has been plotted using histogram. At first, we plotted histogram to show the balance of our dataset that shown in Fig. 4.

After plotting the data histogram, five images from each type have been printed on the notebook for getting an idea of how each image of each category looks like, shown in Fig. 5. The next step is image preprocessing where a series of process has been done. These processes are described below according to their code cell. First, all the raw images are called as shown in Fig. 6.

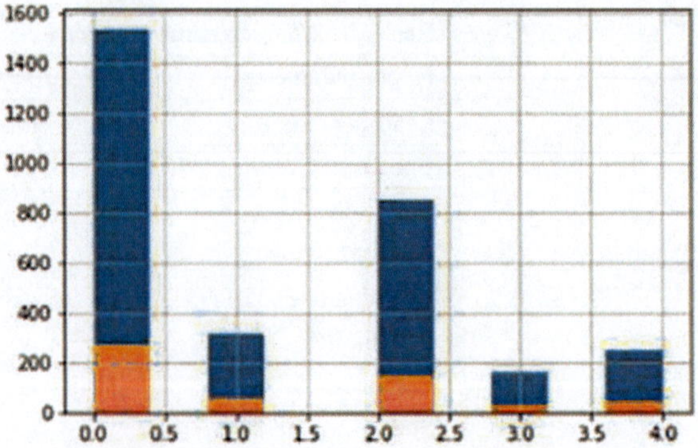

Fig. 4 Plotting histogram of data concentration

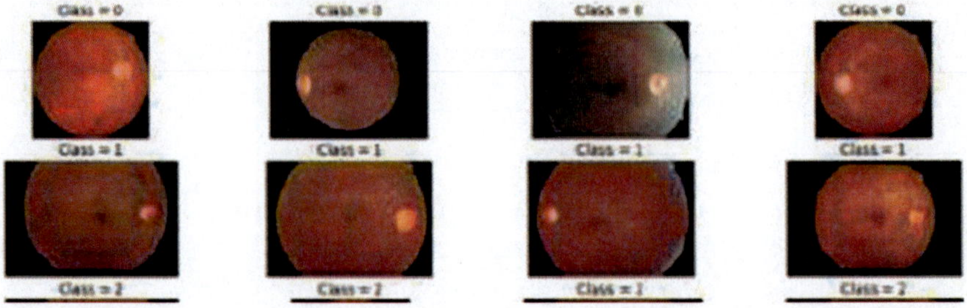

Fig. 5 Each class of dataset

Fig. 6 Initial image after applying the pre-processing technique

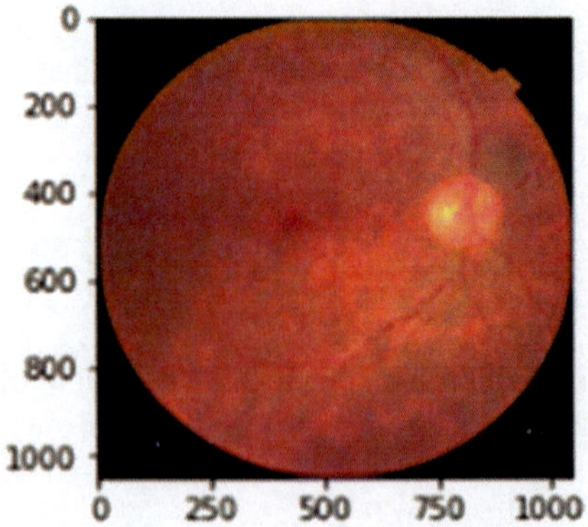

Then image center and radius are calculated using the following figure for detecting the circumferential area, and after that, the image resizing function has initialized. The function for removing irregularities has called afterward. The outputs from image resize are shown in Fig. 7.

Finally, the output from Radius Reduction is shown in Figs. 8 and 9.

To estimate each method used in this research, precision, recall and F1-score are used to show the performance matrices. TP (true positive) and FP (false positive) are used as parameters to calculate the performance metrics. Precision focuses on decreasing the false-positive values, so that it induces a precise model. Recall reduces the false-negative value. F1-score reduces the false-positive and false-negative values, in order to increase F1-score for the model. Accuracy focuses on true positive and true negative values to identify the classes accurately. Table 3 provides the classification results of ResNet152 and VGG16 model for standard dataset.

Table 3 illustrates the comparison between the ResNet152, VGG9 and VGG16 and CNN. It can be observed from the values of the evaluation metrics that ResNet152

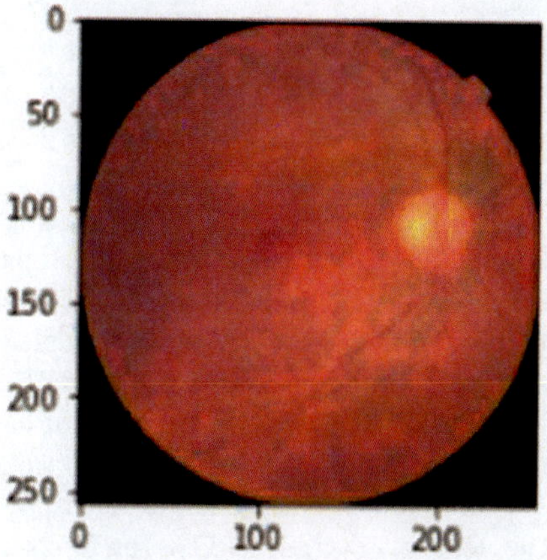

Fig. 7 Output from image resize for both vGG16 and ResNet152

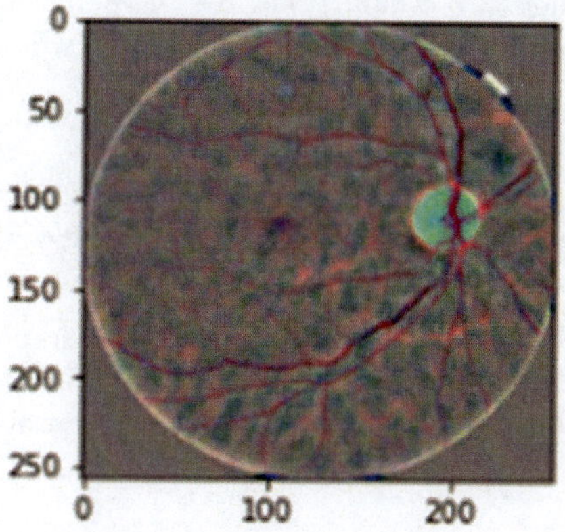

Fig. 8 Output from subtract median function

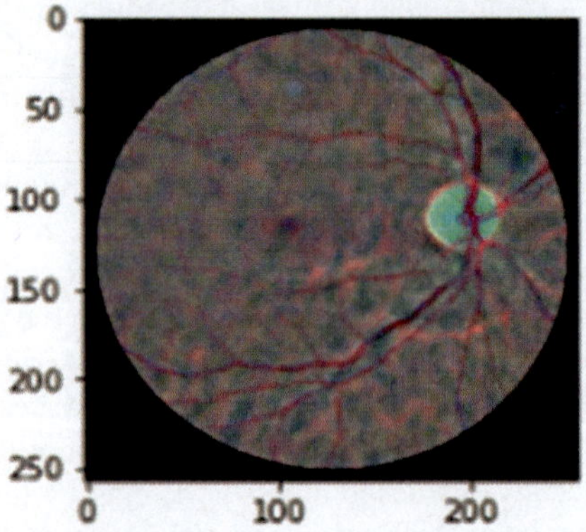

Fig. 9 Output from radius reduction function

Table 3 Comparison between CNN, VGG16, VGG9 and ResNet152

Model	Precision	Recall	F1-Score	Training accuracy	Testing accuracy
CNN	0.5	0.3	0.37	0.75	0.70
VGG9	0.9	0.9	0.9	0.62	0.62
VGG16	0.7	0.7	0.7	0.73	0.75
ResNet152	0.8	0.8	0.8	0.78	0.87

demonstrates high performance in comparison to other CNN models. In contrast with VGG9 and VGG16 model, it can be observed that ResNet152 performs better than the other model as a well-fitted model. However, the training accuracy of VGG9 is only 62%. Although VGG16 is a well model, its accuracy is 73%, which is lower than ResNet152 model. Therefore, it can be concluded that the ResNet152 model performs well for Fundus Image dataset.

From the test data generator class, we have cross-checked the multiclass output. The accuracy we got is 87% for ResNet152 where it is 73% for VGG16 shown in Fig. 10. We have found that ResNet152 architecture outperforms VGG16. Figure 10 shows the accuracy and loss curve. It can be seen from both the curves that an initial end at 20 epochs has been achieved where the learning rate is 0.0001 and batch size is 32.

5 Conclusion and Future Work

The contribution of this system is to implement a system that propounds an assorted image processing technique for glaucoma datasets. The image dataset can be embodied competently in preparation for feeding the data to the network model. We have carved the dataset (around 5000 images) into 80% of train set and 20% of

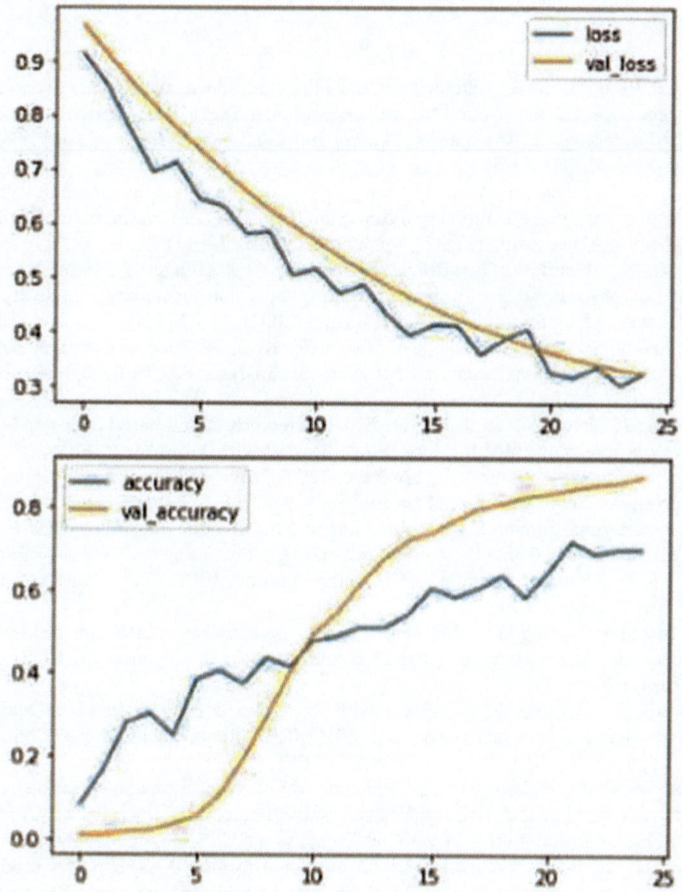

Fig. 10 Loss and accuracy curve from ResNet152

test set. While we train a machine learning model, one of the chief reasons that one tried to avoid is overfitting. This happens when the model fits the training data well.

Some future work can be implemented later by further research such as to amend the accuracy more for better compliance of the work, we will consider the Belief Rule Base Expert system (BRBES) [30, 31, 34–36, 43]. It is a knowledge-driven approach used to capture uncertain knowledge and accomplish evidential reasoning under uncertainty.

References

1. Andreas, J., Purnomo, M.H., Hariadi, M.: Controlling the hidden layers output to optimizing the training process in the Deep Neural Network algorithm. In :2015 IEEE International Conference on Cyber Technology in Automation, Control and Intelligent Systems, IEEE-CYBER 2015, pp. 1028–1032. IEEE (2015)
2. Quan, Y., Cheng, J., Lee, B., Yow, A., Wong, D.: Automatic glaucoma screening hybrid cloud system with pattern classification algorithms. In: 2017 IEEE 2nd International Conference on Signal and Image Processing (ICSIP), pp. 219–222. IEEE (2017)
3. Abra'moff, M., Alward, W., Greenlee, E., Shuba, L., Kim, C., Fingert, J., Kwon, Y.: Automated segmentation of the optic disc from stereo color photographs using physiologically plausible features. Invest. Opthalmology Vis. Sci. **48**, 1665 (2007)
4. Xu, J., Chutatape, O., Sung, E., Zheng, C., Chew Tec Kuan, P.: Optic disk feature extraction via modified deformable model technique for glaucoma analysis. Pattern Recogn. **40**, 2063–2076 (2007)
5. Xu, Y., Lin, S., Wong, D., Liu, J., Xu, D.: Efficient reconstruction-based optic cup localization for glaucoma screening. Adv. Inf. Syst. Eng. 445–452 (2013)
6. Eziama, E., Jaimes, L., James, A., Nwizege, K., Balador, A., Tepe, K.: Machine learning-based recommendation trust model for machine-to-machine communication. In: 2018 IEEE International Symposium on Signal Processing and Information Technology (ISSPIT) (2018)
7. Mohr, F., Wever, M., Tornede, A., Hullermeier, E.: Predicting machine learning pipeline runtimes in the context of automated machine learning. IEEE Trans. Pattern Anal. Mach. Intell. **43**, 3055–3066 (2021)
8. Ma, J.: Machine learning in predicting diabetes in the early stage. 2020. In: 2nd International Conference on Machine Learning, Big Data and Business Intelligence (MLBDBI), pp. 167–172. IEEE (2020)
9. Gouripeddi, P., Gouripeddi, R., Gouripeddi, S.: Toward machine learning and big data approaches for learning analytics. In : 2019 IEEE Tenth International Conference on Technology for Education (T4E), pp. 256–257. IEEE (2019)
10. Taher, K., Mohammed Yasin Jisan, B., Rahman, M.: Network intrusion detection using supervised machine learning technique with feature selection. In: 2019 International Conference on Robotics, Electrical and Signal Processing Techniques (ICREST), pp. 643–646. IEEE (2019)
11. Qassim, H., Verma, A., Feinzimer, D. :Compressed residual-VGG16 CNN model for big displaces image recognition. In: 2018 IEEE 8th Annual Computing and Communication Workshop and Conference (CCWC), pp. 169–175. IEEE (2018)
12. Pinos-Velez, E., Flores-Rivera, M., Ipanque-Alama, W., Herrera-Alvarez, D., Chacon, C., Serpa-Andrade, L.: Implementation of support tools for the presumptive diagnosis of Glaucoma through identification and processing of medical images of the human eye. In: 2018 IEEE International Systems Engineering Symposium (ISSE) (2018)
13. Hu, Z., Abràmoff, M., Kwon, Y., Lee, K., Garvin, M.: Automated segmentation of neural canal opening and optic cup in 3D spectral optical coherence tomography volumes of the optic nerve head. Invest. Opthalmology Vis. Sci. **51**, 5708 (2010)
14. Abràmoff, M., Lee, K., Niemeijer, M., Alward, W., Greenlee, E., Garvin, M., Sonka, M., Kwon, Y.: Automated segmentation of the cup and rim from spectral domain OCT of the optic nerve head. Invest. Opthalmology Vis. Sci. **50**, 5778 (2009)
15. Palakvangsa-Na-Ayudhya, S., Sapthamrong, T., Sunthornwutthikrai, K., Sakiyalak, D.: GlaucoVIZ: assisting system for early glaucoma detection using mask R-CNN. In: 2020 17th International Conference on Electrical Engineering/Electronics, Computer, Telecommunications and Information Technology (ECTI-CON), pp. 364–367. IEEE (2020)
16. Sun, X., Xu, Y., Tan, M., Fu, H., Zhao, W., You, T., Liu, J.: Localizing optic disc and cup for glaucoma screening via deep object detection networks. Comput. Pathol. Ophthalmic Med. Image Anal. 236–244 (2018)

17. Sevastopolsky, A.: Optic disc and cup segmentation methods for glaucoma detection with modification of U-Net convolutional neural network. Pattern Recognit. Image Anal. **27**, 618–624 (2017)
18. Juneja, M., Thakur, S., Wani, A., Uniyal, A., Thakur, N., and Jindal, A : DC-Gnet for detection of glaucoma in retinal fundus imaging. In: Machine Vision and Applications, 31, pp. 1–14 (2020)
19. Al-Bander, B., Williams, B. M., Al-Nuaimy, W., Al-Taee, M.A., Pratt, H and Zheng, Y.: Dense fully convolutional segmentation of the optic disc and cup in color fundus for glaucoma diagnosis. In: Symmetry (Basel), 10 (2018)
20. Saxena, A., Vyas, A., Parashar, L and Singh, U.:A Glaucoma Detection using Convolutional Neural Network.In: Proceedings of the International Conference on Electronics and Sustainable Communication Systems (ICESC), pp. 815–820. IEEE (2020)
21. Zhang, S.F., Zhai, J.H., Xie, B.J., Zhan, Y., Wang, X.: Multimodal representation learning: advances, trends and challenges. In: Proceedings—International Conference on Machine Learning and Cybernetics (2019)
22. Ray, S.: A quick review of machine learning algorithms. In: Proceedings of the International Conference on Machine Learning, Big Data, Cloud and Parallel Computing: Trends, Prespectives and Prospects (COMITCon), pp. 35–39. IEEE (2019)
23. Horio, Y.: Chaotic neural network reservoir. In: Proceedings of the International joint Conference on Neural Networks (2019)
24. Chowdhury, N., Kashem, M.A.: Comparative analysis of feed-forward neural network & recurrent neural network to detect intrusion. In: Proceedings of ICECE 2008—5th International Conference on Electrical and Computer Engineering, pp. 488–492. IEEE (2008)
25. Karayiannis, N.B., Xiong, Y.: Training reformulated radial basis function neural networks capable of identifying uncertainty in data classification. In: IEEE Transactions Neural Networks, 17, pp. 1222–1233 (2006)
26. Song, Q., Wu, Y., Soh, Y.C.: Robust adaptive gradient-descent training algorithm for recurrent neural networks in discrete time domain. In: IEEE Transactions Neural Networks 19, pp. 1841–1853 (2008)
27. Uçkun, F.A., Özer, H., Nurbaş, E., Onat, E.: Direction finding using convolutional neural networks and convolutional recurrent neural networks. In: 28th Signal Processing and Communications Applications Conference (SIU), pp. 1–4. IEEE (2020)
28. Hu, Z., Li, Y., Yang, Z.: Improving convolutional neural network using pseudo derivative ReLU. In: 2018 5th International Conference on Systems and Informatics (ICSAI), pp. 283–287. IEEE (2018)
29. Chen, C.H., Lin, P.H., Hsieh, G., Cheng, S.L., Eng, H.: Robust multi-class classification using linearly scored categorical cross-entropy. In: Proceedings of the 3rd IEEE International Conference on Knowledge Innovation and Invention (ICKII), pp. 200–203. IEEE (2020)
30. Hossain, M.S., Andersson, K., Naznin, S.: A belief rule based expert system to diagnose measles under uncertainty. In: World Congress in Computer Science, Computer Engineering, and Applied Computing (WORLDCOMP'15), pp. 17–23. (2015)
31. Andersson, K., Hossain, M.S.: Smart risk assessment systems using belief-rule-based DSS and WSN technologies. In: 2014 4th International Conference on Wireless Communications, Vehicular Technology, Information Theory and Aerospace & Electronic Systems (VITAE), pp. 1–5. IEEE (2014)
32. Ahmed, T.U., Hossain, M.S., Alam, M.J., Andersson, K.: December. An integrated cnn-rnn framework to assess road crack. In: 2019 22nd International Conference on Computer and Information Technology (ICCIT). pp. 1–6. IEEE (2019)
33. Chowdhury, R.R., Hossain, M.S., Hossain, S., Andersson, K.: Analyzing sentiment of movie reviews in bangla by applying machine learning techniques. In: 2019 International Conference on Bangla Speech and Language Processing (ICBSLP), pp. 1–6. IEEE (2019)
34. Ahmed, T.U., Jamil, M.N., Hossain, M.S., Andersson, K., Hossain, M.S.: An integrated real-time deep learning and belief rule base intelligent system to assess facial expression under uncertainty. In: 2020 Joint 9th International Conference on Informatics, Electronics & Vision

(ICIEV) and 2020 4th International Conference on Imaging, Vision & Pattern Recognition (icIVPR), pp. 1–6. IEEE (2020)
35. Hossain, M.S., Monrat, A.A., Hasan, M., Karim, R., Bhuiyan, T.A., Khalid, M.S.: A belief rule-based expert system to assess mental disorder under uncertainty. In: 5th International Conference on Informatics, Electronics and Vision (ICIEV), pp. 1089–1094. IEEE (2016)
36. Islam, R.U., Hossain, M.S., Andersson, K.: A deep learning inspired belief rule-based expert system. IEEE Access **8**, 190637–190651 (2020)
37. Abedin, M.Z., Akther, S., Hossain, M.S.: An artificial neural network model for epilepsy seizure detection. In 5th International Conference on Advances in Electrical Engineering (ICAEE), pp. 860–865. IEEE (2019)
38. Basnin, N., Nahar, L., Hossain, M.S.: An integrated CNN-LSTM model for micro hand gesture recognition. In: International Conference on Intelligent Computing & Optimization. pp. 379–392. Springer, Cham (2020)
39. Basnin, N., Nahar, L., Hossain, M.S.: An integrated cnn-lstm model for Bangla lexical sign language recognition. In: Proceedings of International Conference on Trends in Computational and Cognitive Engineering, pp. 695–707. Springer, Singapore (2021)
40. Rezaoana, N., Hossain, M.S., Andersson, K.: Detection and classification of skin cancer by using a parallel CNN model. In: IEEE International Women in Engineering (WIE) Conference on Electrical and Computer Engineering (WIECON-ECE), pp. 380–386. IEEE (2020)
41. Nahar, N., Hossain, M.S. and Andersson, K.: A machine learning based fall detection for elderly people with neurodegenerative disorders. In: International Conference on Brain Informatics, pp. 194–203. Springer, Cham (2020)
42. Afroze, T., Akther, S., Chowdhury, M.A., Hossain, E., Hossain, M.S., Andersson, K.: Glaucoma detection using inception convolutional neural network V3. In: International Conference on Applied Intelligence and Informatics, pp. 17–28. Springer, Cham (2021)
43. Sultana, Z., Nahar, L., Basnin, N. and Hossain, M.S.: Inference and learning methodology of belief rule based expert system to assess Chikungunya. In: International Conference on Applied Intelligence and Informatics, pp. 3–16. Springer, Cham (2021)

Classification of ECG Ventricular Beats Assisted by Gaussian Parameters' Dictionary

Sh Hussain Salleh, Fuad Noman, Hadri Hussain, Chee-Ming Ting, Syed Rasul bin G. Syed Hamid, Hadrina Sh-Hussain, M. A. Jalil, A. L. Ahmad Zubaidi, Syed Zuhaib Haider Rizvi, Kuryati Kipli, Kavikumar Jacob, Kanad Ray, M. Shamim Kaiser, Mufti Mahmud, and Jalil Ali

Abstract Automatic processing and diagnosis of electrocardiogram (ECG) signals remain a very challenging problem, especially with the growth of advanced monitoring technologies. A particular task in ECG processing that has received tremendous attention is to detect and identify pathological heartbeats, e.g., those caused by premature ventricular contraction (PVC). This paper aims to build on the existing

S. H. Salleh · H. Hussain (✉) · H. Sh-Hussain
HealUltra PLT, Taman Pulai Utama, No-20 Jalan Pulai 18, 81300 Skudai, Johore, Malaysia

S. H. Salleh
e-mail: sh.hussain@tutanota.com

F. Noman · H. Hussain · C.-M. Ting
School of Information Technology, Monash University Malaysia, 47500 Bandar Sunway, Selangor, Malaysia
e-mail: fuad.noman@monash.edu

C.-M. Ting
e-mail: ting.cheeming@monash.edu

S. R. G. S. Hamid
Department of Cardiothoracic Surgery, Hospital Sultanah Aminah, Johor Bahru, 80100 Johor, Malaysia

M. A. Jalil
Department of Physics, Faculty of Science, Unviersiti Teknologi Malaysia, 81310 Skudai, Johor, Malaysia
e-mail: arifjalil@utm.my

A. L. A. Zubaidi
Faculty of Medicine, Medical Campus, Universiti Sultan Zainal Abidin, 20400 Kuala Terengganu, Terengganu, Malaysia

S. Z. H. Rizvi · K. Jacob
Faculty of Applied Sciences and Technology, Universiti Tun Hussein Onn Malaysia, Parit Raja, Malaysia
e-mail: syedzuhaib@uthm.edu.my

K. Jacob
e-mail: kavi@uthm.edu.my

M. S. Kaiser et al. (eds.), *Proceedings of the Third International Conference on Trends in Computational and Cognitive Engineering*, Lecture Notes in Networks and Systems 348, https://doi.org/10.1007/978-981-16-7597-3_44

methods of heartbeat classification and introduce a new approach to detect ventricular beats using a dictionary of Gaussian-based parameters that model ECG signals. The proposed approach relies on new techniques to segment the stream of ECG signals and automatically cluster the beats for each patient. Two benchmark datasets have been used to evaluate the classification performance, namely, the QTDB and MIT-BIH Arrhythmia databases, based on a single lead short ECG segment. Using the QTDB database, the method achieved the average accuracies of 99.3% ± 0.7 and 99.4% ± 0.6% for lead-1 and lead-2, respectively. On the other hand, identifying ventricular beats in the MIT-BIH Arrhythmia dataset resulted in a sensitivity of 82.8%, a positive predictivity of 62.0%, and F1 score of 70.9%. For non-ventricular beats, the method achieved a sensitivity of 96.0%, a positive predictivity of 98.6%, and F1 score of 97.3%. The proposed technique represents an improvement in the field of ventricular beat classification compared with the conventional methods.

Keywords ECG · Classification · Gaussian kernels · Segmentation · Template extraction

1 Introduction

Electrocardiogram (ECG) is a non-invasive tool used to record and monitor cardiac electrical activity. An ECG cycle (heartbeat) is obtained as a sequence of P, QRS, and T waves, representing atrial-ventricular depolarization and repolarization sequences. The advanced computerized modeling and analysis techniques of the ECG are nowadays well-established practices, leading to significant improvements to assess cardiologists for long-term ECG recordings interpretation [1–3]. Premature ventricular contraction (PVC) is a premature heartbeat originating from the early contraction of the ventricles. As a result, the ECG morphology of a PVC beat differs from that

K. Kipli
Faculty of Engineering, Universiti Malaysia Sarawak (UNIMAS), Sarawak, Malaysia
e-mail: kkuryati@unimas.my

K. Ray
Amity School of Applied Sciences, Amity University, Jaipur, India

M. S. Kaiser
Institute of Information Technology, Jahangirnagar University, Savar, Dhaka, Bangladesh
e-mail: mskaiser@juniv.edu

M. Mahmud
Department of Computer Science, Nottingham Trent University, Clifton Lane, Nottingham NG11 8NS, UK
e-mail: mufti.mahmud@ntu.ac.uk

J. Ali
Asia Metropolitan University, Masai, Malaysia
e-mail: jalilali@amu.edu.my

of a regular heartbeat [4]. The main purpose of automatic analysis is to classify and diagnose cardiovascular diseases.

The development of automated cardiovascular disease diagnostic techniques has been a challenge for many decades. Many algorithms for ECG heartbeats classification were reported using a variety of features and several classification methods [2–4], including support vector machine (SVM), artificial neural network (ANN), linear discriminant, hidden Markov model (HMM), Gaussian Markov model (GMM), and deep-learning methods. Most of these methods classify beats without any local expert (unsupervised machine learning) by automatically detecting the ECG fiducial point and clustering the ECG segments [3, 5–7]. In contrast, those requiring cardiologist expertise (supervised machine learning) are clustering the heartbeats referring to the cardiologists' annotations. This approach takes advantage of a cardiologists' labeling to train a static classifier and improve the classification performance [5, 6, 8, 9]. In addition, model-based ECG methods have been investigated and applied in different ECG analysis frameworks. For example, the underlying dynamics of the ECG signals make it suitable to be modeled by a sum of Gaussian functions. This approach was first proposed by [12], to filter, compress, and classify the ECG signals, ECG synthesis and denoising [11–13], ECG segmentation [16], ECG fiducial point extraction [17], and ECG Classification [4, 18, 19].

Derya Übeyli [20] used a recurrent neural network (RNN) for ECG heartbeat classification. The authors used a partially connected RNN, which is simply a feed-forward neural network (NN) with a carefully selected set of feedback connections. The Elman network was a four-layer NN where the neurons in one layer are fully connected with neurons in the next layer. Martis et al. [21] developed an ECG heartbeat automatic screening method based on an error backpropagation neural network (ERBNN). A simple three-layer ERBNN with 13 neurons in the input layer, six nodes in the hidden layer, and two neurons in the output layer to represent the binary classes of heartbeats. Chen and Yu [22] suggested a feedforward backpropagation neural network (FFBNN) to classify seven different heartbeat classes. FFBNN is a three-layer network with a hidden layer that has 60 nodes. Martis et al. [23] implemented a four-layer ERBNN for heartbeat classification. The network consists of an input layer, two hidden layers, and the output layer. The gradient descent optimization method was used to update the network weights during the learning phase. The output layer produced five outputs for the heartbeat classes. Prasad et al. [24] also used a three-layer fully connected feedforward NN, which consisted of 10 nodes for the input layer, six neurons for the hidden layer, and three neurons for the output layer corresponding to the three heartbeat classes. Martis, Acharya, and Min [25] followed the same NN structure as in [24] with some adjustments in the number of neurons, 12 nodes for input layer, experimentally chosen 10 neurons for the hidden layer, and 5 neurons for the heartbeat classes (normal ectopic beats, supraventricular ectopic beats, ventricular ectopic beats, fusion beats, and unknown beats). Elhaj et al. [26] recently used a NN with 28 input nodes, a hidden layer of 40 neurons, and 5 nodes for the output layer to distinguish between the different five heartbeat classes. A method for heartbeat classification was developed in [5, 61] using a probabilistic neural network (PNN). Javadi et al. [29] developed an ECG classifier based on NN ensembles. Since the proposed method used more than individual NN, it would have a higher computational cost and a large number of hyperparameters to set prior to the training and testing phase.

Oster et al. [4] proposed a switching Kalman filter (SKF) for ventricular beat detection in electrocardiogram (ECG) signals. The ECG signal is first segmented, then clustered, and each cluster is then assigned to one of three switching modes with different state-space formalisms. Gaussian parameters are estimated for each relevant mode (class) by fitting seven Gaussian kernels over the set of all cycles in that class. However, the proposed segmentation method covers the complete heartbeat cycle, where the ventricular beat can be detected by only using the QRS complex. In this paper, we address this limitation by suggesting various segmentation methods and a weighted-segment classification approach. Oster et al. [4, 30] also estimated the Gaussian parameters from the whole set of ECG cycles in each class. In real-world ECG signals, most of the cycles that belong to the same subjects are similar. Therefore, we propose to reduce the amount of data required for training by processing only the clusters' centroids of each recording.

ECG signals may hold unique dynamics despite that they belong to the same class. Therefore, the complicated morphological changes of ECG waveforms make the clustering process more difficult. In this paper, for a given ECG recording and the cardiologists' annotations, the data are first segmented into heartbeat segments using three different approaches. Then, the heartbeat morphologies are clustered and labeled by creating a template (typical heartbeat) or more for each record. The templates (i.e., the clusters' centroids) are automatically fitted to a Gaussian model to build a dictionary of Gaussian parameters. We compare the classification performance of two different databases, QT database (QTDB) and MIT-BIH Arrhythmia database (MIT-BIH AR). The QTDB database is used for patient-specific classification where the training dataset is the template extracted from each patient recording using the proposed clustering method, and the testing dataset is the segmented heartbeats. The MIT-BIH AR database is used for ventricular beat classification, where half of the available recordings are used for training and the other unseen half is used for testing. The assignment to a cluster consequently classifies the MIT-BIH AR training heartbeats into normal, ventricular, and fusion; and classifies the QTDB heartbeats. The proposed technique creates a dictionary from the modeled ECG segments. This approach learns the underlying dynamics for most heartbeats, which is then used to improve the heartbeats' clustering and classification.

2 Methodology

Figure 1 illustrates the primary steps for ECG classification using the Gaussian parameters dictionary (GPD) approach. In peak detection, three fiducial points are mainly required for the proposed processing methods, namely, Q-wave onset, the R-peak, and the T-wave offset. The segmentation stage will reassemble the ECG data into a multivariate stack of cardiac cycles followed by duration normalization

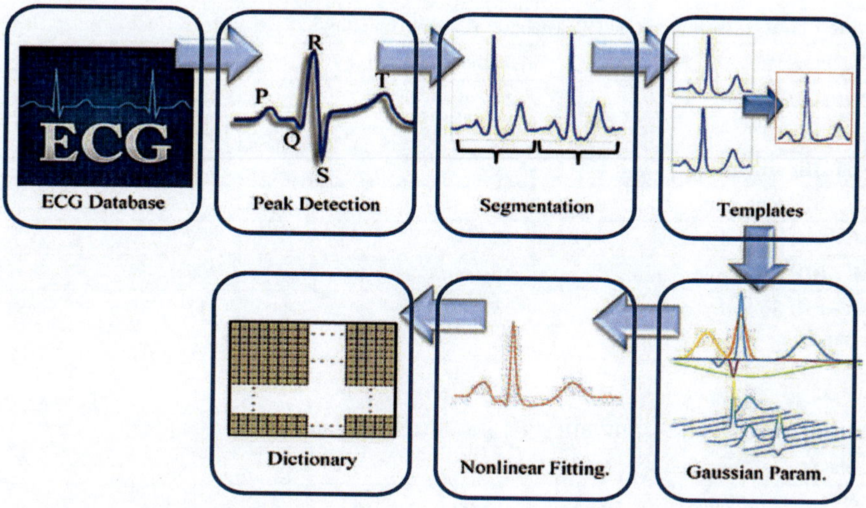

Fig. 1 Block diagram describing the proposed GPD approach for ventricular beat classification. The arbitrary length ECG signal undergoes several processing steps (i.e., peak detection, segmentation, and templates extraction), as such a set of Gaussian parameters are extracted from each template by means of nonlinear fitting. A class-specific dictionary of Gaussian parameters is built using multiple ECG recordings from the training set

and automatic clustering to extract the best representative ECG templates. The core method in this classification approach is the modeling of the cardiac beats to form a new class of parametric functions (seven Gaussian kernels), which were specifically designed to model each cardiac cycle with a sum of seven Gaussian kernels (e.g., two for P-wave, three for QRS and two kernels for T-wave). In the proposed framework of the machine learning classifier, the Gaussian mixtures were extracted from the templates to represent and further discriminate abnormal beats (ABs) from normal beats (NBs). For each class (ABs or NBs), each heartbeat in the training set was classified into its particular centroid as the template. The unknown test data will be used to evaluate the classification performance.

2.1 ECG Database

The experiments performed in this paper were carried out using two different databases available on Physionet web store [31]. The relevant details of QTDB [32] and MIT-BIH AR database are summarized in Tables 1 and 2, respectively.

QTDB Database. The QTDB database, as shown in Table 1, contains 105 two-lead recordings of 3622 heartbeats. This database comprises ECG signals that were chosen to represent a wide variety of QRS and ST-T morphologies. Additionally, the database includes some records from the MIT-BIH database (Arrhythmia, ST

Table 1 Data distribution (QTDB database) (Patient-Specific)

Database name	Number of records	Total cycles
BIH sudden death	24	744
European STT	33	1041
MIT-BIH arrhythmia	15	673
MIT-BIH long term	4	141
MIT-BIH normal	10	300
MIT-BIH ST change	6	206
MIT-BIH SV arrhythmia	13	517
Total	105	3622

Table 2 Merged Classes (MIT-BIH AR database—extracted beats)

Dataset	Purpose	N	S	V	F	Q	#Rec
DS1	Train	45,847	944	3788	414	8	22
DS2	Test	44,239	1837	3220	388	7	22
Total		90,086	2781	7008	802	15	44

N: Normal, V: PVC, F: Fusion (ventricular and normal), Q: Unclassified, S: Superventricular premature (ectopic), F: Fusion (paced and normal)

Change, Supraventricular Arrhythmia, Normal Sinus Rhythm, Sudden Death, and Long Term). All the records have been sampled at 250 Hz. In this paper, we use the QTDB database to show the patient-specific classification performance using the proposed classification method.

MIT-BIH Arrhythmia Database. The MIT-BIH AR database contains two-lead 48 recordings of approximately 30 min long. Several major life-threatening arrhythmias, as well as regular sinus rhythm heartbeats, are included in the data, which total roughly 100,000 annotated beats. All the recordings are sampled at 360 Hz.

In this paper, the MIT-BIH AR database recommendations and methodologies of the state-of-the-art methods as described in [2, 4] were followed. The MIT-BIH AR database was divided into training (DS1) and unknown testing (DS2) sets as in Table 2. Differently, the QTDB was used for patient-specific classification, where the extracted templates for each of the available seven classes are grouped together to represent the training dataset, and the unclustered heartbeats for each class are considered as the testing dataset. In some QTDB database recordings, the tail beats (start and end of the annotated interval) were excluded from fulfilling the segmentation method carried out in this paper, details can be found in Sect. 2.3.

2.2 Pre-Processing

The data were first smoothed using a bandpass filter—cascaded FIR high-pass and low-pass filters with order 24 and cut-off frequencies of 5 Hz and 15 Hz; most of the power is expected to be in the 5–15 Hz band for a clean ECG [31]. The baseline drift was then extracted using a single-stage median filter with a window length of $Fs/2, (Fs = 1sec)$, followed by subtraction from the noisy recording [32]. Some of the recordings were contaminated by powerline noise; an FIR notch filter with a cut-off frequency of 60 Hz was used. The ECG recordings were then smoothened again using the previously mentioned FIR bandpass filter. Finally, the data undergo standard preprocessing steps, which were standardized by removing the mean and dividing by the standard deviation for the proposed GPD classifier.

2.3 Segmentation

The heart rate is usually not consistent through the whole ECG recording of the same patient, the segmentation of the heartbeats may result in varied segment lengths and R-peak location. The segment length variation can be solved by using any anti-aliasing interpolation method; however, the R-peak locations will remain unsynchronized. The experiments performed in this paper follow three different segmentation approaches detailed as follows:

Method 1. With reference to the cardiologists' annotations, the ECG was segmented into heartbeat cycles. Each segment starts from the beginning of P-wave and ends at the end of T-wave (see Fig. 2(a)). For that, only the important interval of atrial and ventricular activity was considered in this segmentation.

Method 2. We used the previously suggested phase-based segmentation approach to extract the heartbeat segments allowing the synchronization of heartbeats durations and QRS localizations [12, 20]. In order to calculate the cyclic phase of a specific heartbeat, the preceding, current, and consequent QRS-peak locations have to be

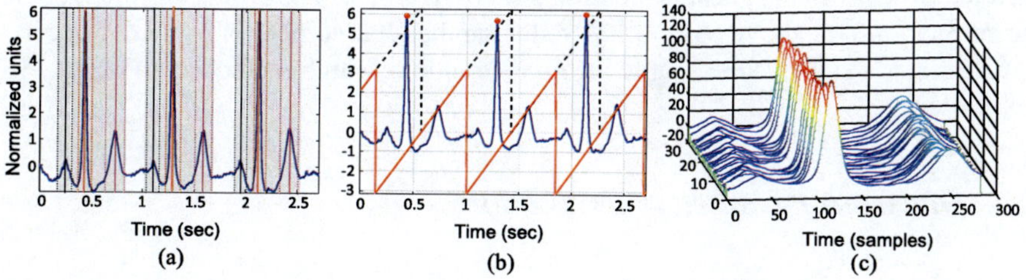

Fig. 2 **a** ECG signal segmentation using reference annotations. **b** ECG signal segmentation by phase calculation. **c** Stack of ECG heartbeats segmented around R-peak

available. Hence, we proposed a method to delineate the missing annotations. The proposed delineation method is tested on the QTDB database. For that, the ECG signal was segmented around the QRS from $-\pi$ to π (see Fig. 2(b)), creating a cyclic phase of the RR intervals. In order to include the whole atrial and ventricular activity, the R-peak location is set to $-\pi/3$ radians, as suggested in [4].

Method 3. As in Method 2, the ECG signal was segmented, with each segment is centered around the R-peak (see Fig. 2(c)). Each segment takes a ratio of the sampling frequency (Fs) to calculate the target intervals around R-peak (e.g., 25% and 45% of Fs to the left and right of R-peak, respectively).

For QTDB database, the P, QRS, and T wave reference annotations are only available for some selected heartbeats. We used the R-peak detection method to locate the required R-peak locations so that the three segmentation approaches were used to segment the QTDB database. For MIT-BIH AR database, only the R-peak locations are available, which allows to select the phase-related segmentation approach.

2.4 *Template Extraction*

The QTDB and MIT-BIH AR are rich datasets of ECG morphologies. The data consist of two-channel ECG Holter recordings. Each channel of each record was analyzed separately. Hence, a template is extracted from each ECG channel. In some cases, a single channel might contain more than one template.

The segmentation stage will reassemble the data into a multivariate stack of heartbeat cycles. The average of this stack is considered as the template/mean cycle for the current record. The R-peak alignment was performed if necessary, based on the correlation lag. First, each heartbeat within a given channel was compared with the centroid/template heartbeat using the similarity measure of cross-correlation. If the correlation was over a given threshold ($t_1 = 0.92$), then the heartbeat was assigned to the main class.

The class centroids are updated sequentially, and the cross-correlations are repeated till all the heart cycles are clustered. Finally, the proposed clustering approach will check the inter-similarity between the created clusters. A second cross-correlation between the clusters' centroids was performed. The clusters are merged if the correlation was above ($t_2 = 0.93$), reducing the total number of created clusters. The thresholds values were adapted from the random search method in [4].

2.5 *Gaussian Parameters Estimation*

In this experiment, for QTDB database, we considered each source of the seven available datasets as a unique class of ECG recordings. The Gaussian parameters are estimated for each relevant class by fitting eight Gaussian functions nonlinearly across the set of all template cycles present in that class. While for MIT-BIH AR database, we empirically maintain to fit only seven Gaussian kernels on each denoised heart cycle. Each Gaussian parameter requires some initial values for nonlinear fitting. The quality of the fit will be determined by those parameters, particularly the initial position of each Gaussian center. Since the locations of P and T waves are unknown, $N - 2$ (N is the number of Gaussians), Gaussians were assigned to the first half of the ECG cycle and 2 Gaussians for the other half. These values were chosen at random over the course of $\{-\pi, \pi\}$, and the procedure was repeated iteratively if the normalized mean square error (NMSE) between the Gaussian-based function and the template was greater than 5%.

Each ECG heartbeat segment was modeled as a mixture of independent Gaussian functions in the dynamical ECG model introduced by [33] and extended in [34]. Assuming the baseline drift is filtered, the heartbeat is modeled as follows,

$$z = \sum_{i=1}^{N} a_i e^{-(\Delta\theta_i^2/2b_i^2)}, \tag{1}$$

where N is the number of Gaussian functions that represent the ECG waveforms $\{P^+, P^-, Q, R, R', S, T^+, T^-\}$, a_i is the non-normalized magnitude, $\Delta\theta_i$ is the time-shift in radians $\{-\pi \sim \pi\}$ from the beginning of the modeled ECG heartbeat, and b_i is the width of the Gaussian function. For example, to model the biphasic P-wave$\left\{P^+, P^-\right\}$, two Gaussian functions are used as follows,

$$P_{wave} = a_{P+} e^{-\left(\Delta\theta_{P+}^2/2b_{P+}^2\right)} + a_{P-}\, e^{-\left(\Delta\theta_{P-}^2/2b_{P-}^2\right)} \tag{2}$$

The model coefficients were computed by solving the nonlinear least-squares problem [34] and minimize the model fitting error as follows,

$$\min_{a_i,\theta_i,b_i} \|\varepsilon_r\|_2^2 = \min_{a_i,\theta_i,b_i} \sum_{i=1}^{m} \varepsilon_r^2, \tag{3}$$

where ε_r is the model fitting error, m is the heartbeat length. The fitting error or the model residual is the error between the original ECG heartbeat $s(t)$ and the estimated heartbeat $z(t)$ in Eq. (1), the model residual is,

$$\varepsilon_r = s(t) - z(t) \tag{4}$$

2.6 QRS Detection

The QRS detection procedure proposed in [35] was adopted in this study. The method employs a windowed tangent calculation on the ECG recordings, which involves the use of adaptive thresholds due to the presence of the unstable QRS morphologies available in the abnormal database. Furthermore, the algorithm also applies the sampled threshold to avoid fluctuations in the QRS morphology, particularly when a sudden change of the heart occurs (e.g., the tachycardia and bradycardia cases).

2.7 Classification (Experiment Setup)

Figure 3 shows the block diagram of the proposed ventricular beat classification system.

Patient-specific Classifier (QTDB Database). The QTDB contains ECG recording from seven different sources:

1. MIT-BIH Arrhythmia Database (AR).
2. MIT-BIH ST Change Database (STT).
3. MIT-BIH Supraventricular Arrhythmia Database (SV).
4. MIT-BIH Normal Sinus Rhythm Database (NSR).
5. European ST-T Database (EST).
6. "Sudden Death" patients from BIH database (SD).
7. MIT-BIH Long-Term ECG Database (LT).

Each of the seven QTDB sub-datasets is considered a unique class. The templates were extracted from each recording and combined together to represent that particular class. As shown in Fig. 1 and explained previously, the 24 Gaussian parameters are

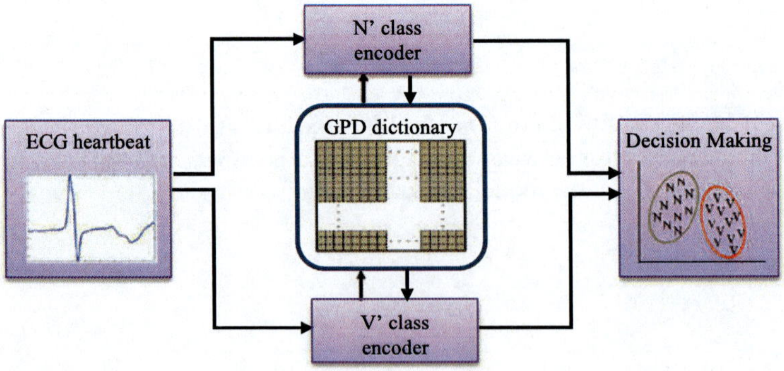

Fig. 3 The proposed classifier block diagram. The input ECG heartbeat is encoded (i.e., extract Gaussian parameters) then tested by means of Frobenius norm to check the similarity with the pre-build GPD dictionary

vectorized and concatenated to form a labeled dictionary for the processed templates. In the testing stage, the Frobenius norm between the true test data and the estimates is calculated, and the test heartbeat is assigned to the dictionary label, which gives the highest likelihood (lowest normalized mean square error).

Ventricular Beat Classification (MIT-BIH AR. Database). The MIT-BIH AR database is available from Physionet along with more than 175 000 experts' annotations. The annotations categorize the beats into 15 different classes, which merged to form only five dominant classes, normal (N), supraventricular (S), ventricular (V), fusion (F), and unclassified (Q) [3, 23]. In this study, only two beat classes were considered. The N and S classes are merged and denoted by N', the F and V classes are merged and denoted as V' [4], the Q class is discarded.

The data were split into 22 train dataset (DS1) and 22 datatest (DS2) as per [2, 4] and illustrated in Table 2. For each class (N' or V'), each heartbeat in the training dataset was automatically clustered. The clusters' centroids are considered as the templates. The Gaussian parameters are then extracted from all templates to represent the class those templates belong to. Finally, the testing dataset, which contains unseen heartbeats, is used to evaluate the classification performance. Frobenius norm between the true test data and the GPD dictionaries. The Frobenius norm estimation is calculated within three overlapping intervals to predict the label of the test ECG.

3 Results

3.1 QTDB Peak-Detection

Since the QTDB database is available with the fundamental ECG waves experts' annotations. Unfortunately, these annotations are only for a certain segment of the whole recording interval. The phase-based segmentation approach (see Sect. 2.c) requires all the R-to-R intervals, including the non-annotated beats at the start and end of that particular segment. Hence, the aim of the R-peak detection method is to locate the non-annotated R-waves. Table 3 summarizes the performance evaluation of the suggested R-peak delineation algorithm. The performance measure followed was the mean $\pm$ standard deviation ($m \pm std$), where, m and std are the average and standard deviations of the time difference between the experts' annotations and the results of this algorithm. The average value indicates the similarity in the results of the proposed algorithm to that of the manual results, whereas the standard deviation explains the accuracy of the results. The method achieved slightly lower QRS delineation errors compared with Martinez et al. [37] and Laguna et al. [38].

Table 3 QTDB QRS delineation performance

Method	Parameter	QRS_{on}	$QRS_{off'}$
(This work)	# annotations	3623	3623
	Se (%)	99.95	99.95
	P + (%)	100	100
	m ± std (ms)	−0.08 ± 4.32	−4.68 ± 7.28
Martinze et al. [37]	# annotations	3623	3623
	Se (%)	99.97	99.97
	P + (%)	N/A	N/A
	m ± std (ms)	4.6 ± 7.7	0.8 ± 8.7
Laguna et al. [38]	Se (%)	99.92	99.92
	P + (%)	N/A	N/A
	m ± s (ms)	−3.6 ± 8.6	−1.1 ± 8.3

3.2 *Patient-Specific Classifier (QTDB Database)*

Following the block diagram presented in Fig. 3, the QTDB database was considered as a test dataset. The dictionary is constructed from the Gaussian parameters fitted on the templates. This experiment is performed to show that each sub-database in QTDB is holding different underlying dynamics. In this study, a single lead heartbeat classifier is developed, in which the reported results in this section show the performance of each lead separately. In addition, the three different segmentation approaches are also considered.

Figure 4 shows that the method achieved approximately an average detection rate of 96% ± 4% for recordings in lead-1, and 97% ± 3% for lead-2. The segmentation approach (2) achieved the highest accuracies, 99.3% ± 0.7, 99.4% ± 0.6% for lead-1 and lead-2, respectively. The reason for this high accuracy stems from the

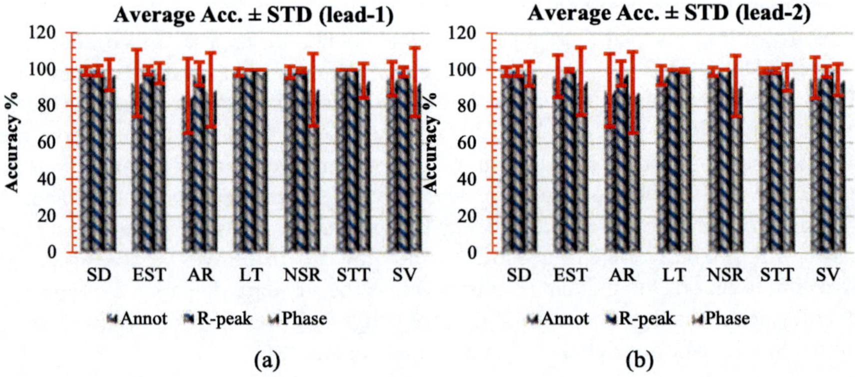

Fig. 4 Classification performance, **a** QTDB—Lead 1, **b** QTDB—Lead 2. The bars show the average accuracies of each class for the three segmentation methods (Annot, Phase, and R-peak). The error bars are the standard deviation of the averaged accuracies. Refer to Sect. 2.7 for datasets abbreviations

Table 4 Confusion matrix (MIT-BIH AR DS2) without F class

		Algorithm		
		n'	v'	Total
Truth	N'	44,245	1831	46,076
	V'	619	2989	3608
	Total	44,864	4820	49,684

fact that, during the interval between the end of T-wave to the start of P-wave, the heart electrically is inactive. For that, this approach of segmentation focuses on the clinically important intervals in the ECG heartbeat. Despite the fact that the R-peak segmentation obtains superior scores, the other segmentation approaches maintain to obtain accuracies greater than 94% on average.

3.3 Ventricular Beat Classification (MIT-BIH AR. Database)

The MIT-BIH AR test dataset confusion matrix is presented in Tables 4 and 6 with and without the fusion class, wherein Table 4, the fusion (F) class is merged with the ventricular (V) class and denoted as (V').

Table 6 compares the scores obtained in this study with some of the previously published methods. The method Oster et al. [4] achieved the highest performance with 91.8% F1 score for V' beat classification. The authors used a statistically advanced method involving the dynamic switching systems to learn the underlying dynamics of each class inherently. However, compared with the proposed method in this paper, the switching Kalman filter method is more computationally expensive and needs more parameters to tune in the initial stage. In addition, the authors also performed a signal quality index approach to remove the cardiac cycles that have a low quality (noisy), in which the high performance comes with the price of removing almost 5000 cardiac cycles from the whole MIT-BIH AR dataset and using a semi-automatic approach, which requires human intervention in the middle of the processing stage. Comparing with Llamedo and Martinez [6], where conventional feature extraction and classification methods were used, the proposed method obtained a higher F1 score of 70.9%. When including the Fusion class as depicted in Table 7, Llamedo and Martinez [6] method achieved a higher F1 score; however, these results are for the balanced classes. The balanced classes are generated by multiplying the V and F classes in the confusion matrix of Table 5 by some weights in order to achieve similar total beats as in N class (46,076 beats). The proposed method shows a higher F1 score than that reported in [6] in balanced classes, as shown in Table 7.

Table 5 Confusion matrix (MIT-BIH AR DS2) with F class

		Algorithm		
		n'	v	Total
Truth	N'	44,245	1831	46,076
	V	279	2941	3220
	F	340	48	388
	Total	44,864	4820	49,684

Table 6 Performance on MIT-BIH AR DS2 without F class. The numbers in parentheses represent the classifier performance after applying the weight cost for imbalanced classes

	N'			V'		
Classifier	*Se*	*P+*	*F1*	*Se*	*P+*	*F1*
This work	96.0	98.6	97.3	82.8 (82.8)	62.0 (85.4)	70.9 (88.7)
Oster et al. [4]	99.7	99.0	99.4	87.6	96.4	91.8
Llamedo and Martinez [6]	81.7	99.4	89.3	93.3	28.6	43.7

Table 7 Performance on (MIT-BIH AR DS2) with F class. The numbers in parentheses represent the classifier performance after applying the weight cost for imbalanced classes

	N'			V		
Classifier	Se	P+	F1	Se	P+	F1
This work	96.0	99.4	97.8	91.3 (91.3)	61.6 (95.8)	75.4 (93.5)
Oster et al. [4]	99.7	99.5	99.6	92.7	96.2	94.5
Llamedo and Martinez [6]	–	–	–	94.6	88.1	91.2

4 Conclusion

This study introduced a classification method for ECG signals based on new segmentation approaches and a proposed automatic clustering method. Two different ECG databases were used to measure the performance of the proposed method in different experimental setups (known testing dataset for QTDB data and unknown testing dataset for MIT-BIH AR data). The implementation of the classification technique involved single lead heartbeat data. The method performance assessment for ventricular beats on MIT-BIH DS2 resulted in a sensitivity of 82.8%, a positive predictivity of 62.0%, and F1 score of 70.9%, while for the non-ventricular beat, the method achieved a sensitivity of 96.0%, a positive predictivity of 98.6%, and F1 score of 97.3%. The proposed GPD approach for ECG automatic analysis used the Gaussian parameters as features to build the classifier. The approach showed a good performance for patient-specific QTDB classification with the novelty of using the templates only as a training set rather than splitting the cardiac cycles into train and test. However, this approach was estimated to achieve high performance compared with the subject-specific of MIT-BIH AR database. In the subject-specific, the MIT-BIH AR database was split into two train-test datasets, with each one consists of ECG recordings from a different patient.

Acknowledgements This work was supported by the Ministry of Higher Education under Fundamental Research Grant Scheme with grant number FRGS/1/2019/STG06/UTM/02/14 (UTM vote no: R.J130000.7851.5F157) and associated facilities.

References

1. Llamedo, M., Martinez, J.P.: An automatic patient-adapted ECG heartbeat classifier allowing expert assistance. IEEE Trans. Biomed. Eng. **59**(8), 2312–2320 (2012)
2. Jortveit, J., Lislevand, T.H., Rysstad, L., Dahlslett, T., Sjøli, B.: Long-term ECG recording: findings and implications. Tidsskr. Den Nor. legeforening (2020)
3. Altay, Y.A., Kremlev, A.S.: Analysis and systematization of noise arising by long-term recording of ECG signal. In: IEEE Conference of Russian Young Researchers in Electrical and Electronic Engineering (EIConRus) **2018**, 1053–1057 (2018)
4. Oster, J., Behar, J., Sayadi, O., Nemati, S., Johnson, A.E.W., Clifford, G.D.: Semisupervised ECG ventricular beat classification with novelty detection based on switching Kalman filters. IEEE Trans. Biomed. Eng. **62**(9), 2125–2134 (2015)
5. Philip de, C., Dwyer, M.O., Reilly, R.B.: Automatic classification of heartbeats using ECG morphology and heartbeat interval features. IEEE Trans. Biomed. Eng. **51**(7), 1196–1206 (2004)
6. Llamedo, M., Martinez, J.P.: Heartbeat classification using feature selection driven by database generalization criteria. IEEE Trans. Biomed. Eng. **58**(3), 616–625 (2011)
7. Hu, Y.H., Palreddy, S., Tompkins, W.J.: A patient-adaptable ECG beat classifier using a mixture of experts approach. IEEE Trans. Biomed. Eng. **44**(9), 891–900 (1997)
8. Soria, M.L., Martinez, J.P.: An ECG classification model based on multilead wavelet transform features. Comput. Cardiol. **2007**, 105–108 (2007)
9. Park, K.S., et al.: Hierarchical support vector machine based heartbeat classification using higher order statistics and hermite basis function. Comput. Cardiol. **2008**, 229–232 (2008)
10. Lagerholm, M., Peterson, C., Braccini, G., Edenbrandt, L., Sornmo, L.: Clustering ECG complexes using Hermite functions and self-organizing maps. IEEE Trans. Biomed. Eng. **47**(7), 838–848 (2000)
11. de Chazal, P., Reilly, R.B.: A patient-adapting heartbeat classifier using ECG morphology and heartbeat interval features. IEEE Trans. Biomed. Eng. **53**(12), 2535–2543 (2006)
12. Clifford, G.D., Shoeb, A., McSharry, P.E., Janz, B.A.: Model-based filtering, compression and classification of the ECG. Int. J. Bioelectromagn. **7**(1), 158–161 (2005)
13. Sayadi, O., Shamsollahi, M.B., Clifford, G.D.: Synthetic ECG generation and Bayesian filtering using a Gaussian wave-based dynamical model. Physiol. Meas. **31**(10), 1309 (2010)
14. Sameni, R., Shamsollahi, M.B., Jutten, C., Clifford, G.D.: A nonlinear Bayesian filtering framework for ECG denoising. IEEE Trans. Biomed. Eng. **54**(12), 2172–2185 (2007)
15. Clifford, G.D., Nemati, S., Sameni, R.: An artificial vector model for generating abnormal electrocardiographic rhythms. Physiol. Meas. **31**(5), 595 (2010)
16. Sayadi, O., Shamsollahi, M.B.: A model-based Bayesian framework for ECG beat segmentation. Physiol. Meas. **30**(3), 335 (2009)
17. Sayadi, O., Shamsollahi, M.B.: Model-based ECG fiducial points extraction using a modified extended Kalman filter structure. In: First International Symposium on Applied Sciences on Biomedical and Communication Technologies, 1–5 (2008)

18. Hua, X., et al.: A novel method for ECG signal classification via one-dimensional convolutional neural network. Multimed. Syst. 1–13 (2020)
19. Wang, J., Li, R., Li, R., Fu, B.: A knowledge-based deep learning method for ECG signal delineation. Futur. Gener. Comput. Syst. **109**, 56–66 (2020)
20. Derya Übeyli, E.: Recurrent neural networks employing Lyapunov exponents for analysis of ECG signals. Expert Syst. Appl., **37**(2), 1192–1199 (2010)
21. Martis, R.J., et al.: Automated screening of arrhythmia using wavelet based machine learning techniques. J. Med. Syst. **36**(2), 677–688 (2012)
22. Chen, Y.-H., Yu, S.-N.: Selection of effective features for ECG beat recognition based on nonlinear correlations. Artif. Intell. Med. **54**(1), 43–52 (2012)
23. Martis, R.J., Acharya, U.R., Mandana, K.M., Ray, A.K., Chakraborty, C.: Cardiac decision making using higher order spectra. Biomed. Signal Process. Control **8**(2), 193–203 (2013)
24. Prasad, H., Martis, R.J., Acharya, U.R., Min, L.C., Suri, J.S.: Application of higher order spectra for accurate delineation of atrial arrhythmia. In: Engineering in Medicine and Biology Society (EMBC), 2013 35th Annual International Conference of the IEEE, pp. 57–60 (2013)
25. Martis, R.J., Acharya, U.R., Min, L.C.: ECG beat classification using PCA, LDA, ICA and Discrete Wavelet Transform. Biomed. Signal Process. Control **8**(5), 437–448 (2013)
26. Elhaj, F.A., Salim, N., Harris, A.R., Swee, T.T., Ahmed, T.: Arrhythmia recognition and classification using combined linear and nonlinear features of ECG signals. Comput. Methods Programs Biomed. **127**, 52–63 (2016)
27. Wang, J.-S., Chiang, W.-C., Hsu, Y.-L., Yang, Y.-T.C.: ECG arrhythmia classification using a probabilistic neural network with a feature reduction method. Neurocomputing **116**, 38–45 (2013)
28. Gutiérrez-Gnecchi, J.A., et al.: DSP-based arrhythmia classification using wavelet transform and probabilistic neural network. Biomed. Signal Process. Control **32** (2017)
29. Javadi, M., Arani, S.A.A.A., Sajedin, A., Ebrahimpour, R.: Classification of ECG arrhythmia by a modular neural network based on mixture of experts and negatively correlated learning. Biomed. Signal Process. Control **8**(3), 289–296 (2013)
30. Rao, A., Gupta, P., Ghosh, P.K.: P-and T-wave delineation in ECG signals using parametric mixture Gaussian and dynamic programming. Biomed. Signal Process. Control **51**, 328–337 (2019)
31. Behar, J.: Extraction of clinical information from the non-invasive fetal electrocardiogram (2016)
32. Sameni, R.: Open Source ECG Toolbox (OSET) (2006)
33. McSharry, P.E., Clifford, G.D., Tarassenko, L., Smith, L.A.: A dynamical model for generating synthetic electrocardiogram signals. IEEE Trans. Biomed. Eng. **50**(3), 289–294 (2003)
34. Clifford, G.D.: A novel framework for signal representation and source separation: applications to filtering and segmentation of biosignals. J. Biol. Syst. **14**(02), 169–183 (2006)
35. Izan, N.F., et al.: Clinical interpretations of the effectiveness of changes in body position during aerobic fitness after neurologic injury. J. Integr. Neurosci. **19**(3), 479–487 (2020)
36. ECAR; AAMI, Recommended practice for testing and reporting performance results of ventricular arrhythmia detection algorithms. Assoc. Adv. Med. Instrum. (1987)
37. Martinez, J.P., Almeida, R., Olmos, S., Rocha, A.P., Laguna, P.: A wavelet-based ECG delineator: evaluation on standard databases. IEEE Trans. Biomed. Eng. **51**(4), 570–581 (2004)
38. Laguna, P., Jané, R., Caminal, P.: Automatic detection of wave boundaries in multilead ECG signals: validation with the CSE database. Comput. Biomed. Res. **27**(1), 45–60 (1994)

Oil Palm Tree Detection and Counting for Precision Farming Using Deep Learning CNN

Kuryati Kipli, Paul Lee Jaw Bin, Sam Huai En, Annie Joseph, Hushairi Zen, Brandon Gan Yong Kien, M. A. Jalil, Kanad Ray, M. Shamim Kaiser, and Mufti Mahmud

Abstract Oil palm tree is a very important crop in Malaysia and other tropical areas. The number of oil palm trees in a plantation area is crucial as it could help to estimate the potential yield of palm oil, monitoring the growing situation of palm trees after plantation such as the age and the survival rate and also the amount of fertilizer and pesticides needed. In this paper, a deep learning-based oil palm tree detection and counting method is proposed and designed into a functioning app. Images of oil palm plantation are collected by using drones then they are pre-processed. The pre-processed images are used to train and optimize the convolutional neural network (CNN). After the CNN model is trained, it is used to predict the label for all the samples in an image dataset collected through the sliding window technique. Its performance is tested. The performance of the classifier is tested on three different tree conditions, from small number of properly separated trees to big number of crowded trees. Based on the result, accuracy ranging from 83.5% to 100% is obtained. Finally, the method is built into an application for a better user experience.

Keywords Oil palm tree · Detection · Deep learning · Convolutional neural network (CNN)

K. Kipli (✉) · P. L. J. Bin · S. H. En · A. Joseph · H. Zen · B. G. Y. Kien
Department of Electrical and Electronics, Faculty of Engineering, Universiti Malaysia Sarawak, 94300 Kota Samarahan, Malaysia
e-mail: kkuryati@unimas.my

M. A. Jalil
Department of Physics, Faculty of Science, Universiti Teknologi Malaysia, 81310 Skudai, Johor, Malaysia

K. Ray
Amity School of Applied Sciences, Amity University, Rajasthan 303001, India

M. S. Kaiser
Institute of Information Technology, Jahangirnagar University, Savar Dhaka-1342, Bangladesh

M. Mahmud
Nottingham Trent University, Clifton Lane, Nottingham NG11 8NS, UK

M. S. Kaiser et al. (eds.), *Proceedings of the Third International Conference on Trends in Computational and Cognitive Engineering*, Lecture Notes in Networks and Systems 348, https://doi.org/10.1007/978-981-16-7597-3_45

1 Introduction

Oil palm tree is a very important crop in Malaysia due to its high economic value. Not only it can produce palm oil but also used to produce various useful products such as furniture, paper, plywood, etc. Detecting and counting oil palm trees are crucial as the data about the number and the location of oil palm trees in a plantation can be useful in many ways. First, it can be used to estimate the yield of palm oil. Second, it gives important information about the age, the survival rate and the health of the palm trees. Besides, it can help to improve productivity. Thus, a good and efficient method to achieve detection and counting is very important.

There are various methods of oil palm trees, detection and counting are reviewed along with the problems that come with them. The objectives are then proposed to help overcoming the problems as well as suggesting a better way to detect and count oil palm trees.

The traditional method of manually counting trees is labor-intensive and posts the potential problem of lack of accuracy [1]. Detecting and counting the number of oil palm trees require many human workers because the oil palm plantation is usually very large, which is unmanageable by just a few workers. Regarding the counting accuracy, counting mistakes and errors can happen from time to time because it is very hard for human force to deal with crowded and randomly aligned trees. This method may work just fine when dealing with small-scale plantations, but it could be a big challenge if the plantation is very large. Many detections of trees can be missed out and govern low counting accuracy, which, in turn, results in obtaining incorrect information. Therefore, a more efficient and accurate way is required for oil palm tree detection and counting.

Furthermore, the performance of some existing technique-based methods could decline when dealing with complicated regions. For example, very young and small trees were unable to be detected by the local maximum filter-based method [2] because the shape of the top of young palm trees could not be located by the local maximum of filter. Moreover, the template matching approach did not work well when detecting crowded palm trees especially when their crowns overlap [3]. These methods suffer from lower performance when detecting trees in crowded tree regions so a better way to do that is needed.

The main objectives of this work are:

1. To propose a deep learning CNN method for oil palm tree detection and counting.
2. To optimize the CNN model for it to perform with high accuracy.
3. To build the whole system into an app for user application.

2 Literature Review

2.1 Deep Learning

Deep learning is a subfield of machine learning (ML), which, in turn, is a subset of artificial intelligence (AI) (see Fig. 1). Deep learning is a type of machine learning method that has its algorithms inspired by the structure and function of the human brain called artificial neural networks. It works like what human brain does by teaching computers to do what humans are capable of and that is learning by example.

In deep learning, computer learns to carry out classification through neural network that learns features from data in form of images, text or sound without the need for manual feature extraction. Therefore, it usually requires a large amount of data to achieve great classification accuracy or even exceed human-level performance. Here, it should be highlighted that the main feature of deep learning is the ability of network itself to self-generate the elements and to predict their hierarchical value [4].

Deep learning works by applying neural network architecture that consists of input layer, one or multiple hidden layers and output layer as shown in Fig. 2 on input data and returns output data after computation. Neuron, the core entity of neural network is where the information processing takes place.

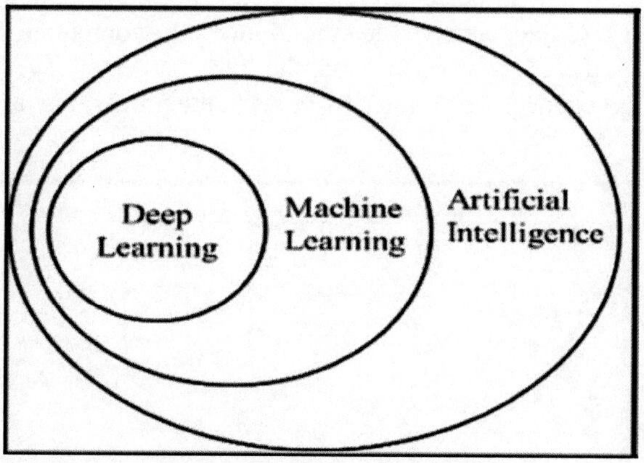

Fig. 1 Venn diagram as suggested at [4]

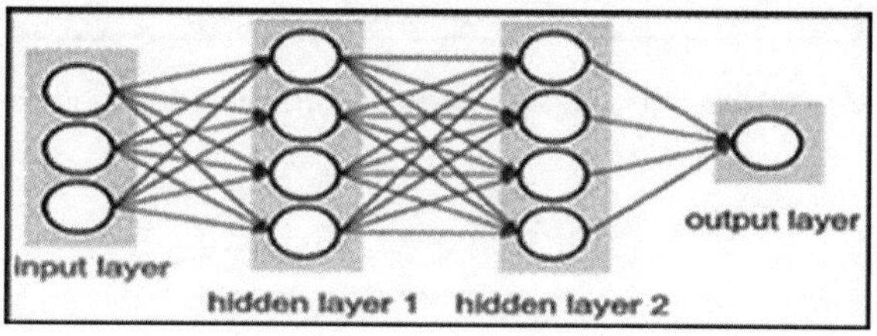

Fig. 2 Neural network

Each data, for example, each pixel of an image is fed to a neuron in the first layer of the neural network. This forms the input layer, which receives input data. The information is then passed from the input layer to the hidden layer where mathematical computations are carried out on the inputs. Deciding the number of hidden layers and the number of neurons for each layer can be challenging. Traditional neural networks only have around two hidden layers, while deep networks can have its number way more than that [5]. The depth of the network is determined by the number of hidden layers [4]. The output layer returns the output data.

2.2 Convolutional Neural Network

One of the most popular deep neural networks is the convolutional neural network (CNN). CNN is one of the best-known deep learning architectures [4]. The CNN was introduced in 1998 by Yann Le Cun. It is popularly used for analyzing images as well as other data analysis or classification.

CNN has three main layers, including convolutional layer, pooling layer and fully connected layer as shown in Fig. 3. CNN has the ability to detect various features of an image using multiple convolutional layers and the complexity of the learned image features increases for every layer it has. For instance, the first convolutional layer could learn how to detect shapes, edges or textures while the last one could learn how to detect more complex shapes. For pooling layer, it reduces the spatial size of the convolved feature to lower the computational power required to process the data. On the other hand, the fully connected layer is acted as a classifier. It has full connections to all activations in the previous layer.

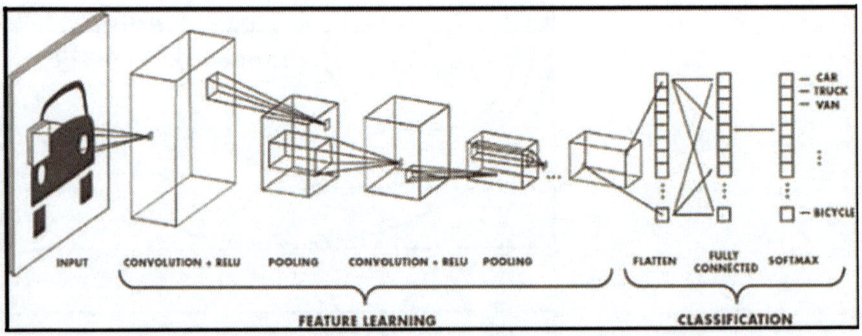

Fig. 3 Main layers of convolutional neural network as suggested at [5]

2.3 CNN Application

CNN can be used in medical field as a better way to analyze medical image. James et al. [6] had proposed a CNN-based method for the detection of diseased heart using CT images, and the result was promising as the accuracy was able to reach around 78.9%. Such automated method to analyze medical image to distinguish between healthy and unhealthy heart could greatly shorten the diagnostic time and improve accuracy. CNN can also be used for prediction. Chen and He [7] had proposed a stock movement prediction method of Chinese stock market by applying CNN approach, and it was done by fetching important stock data into the architecture and performing model training, turned out, the result suggested that it was a bit reliable to use such method. Prediction can be made easy with such automated method.

2.4 Related Work

There are a number of related works that have been done on oil palm tree detection and counting. There are two main ways to acquire images and they are satellite imagery and unmanned aerial vehicle (UAV) [8–10]. Satellite imagery is image of Earth collected by imaging satellites operated governments and businesses around the world. The majority of the researchers chose to use satellite imagery such as QuickBird, WorldView-2 and WorldView-3 while the rest used images captured by using UAV. This is probably due to the fact that satellite image is much easier to obtain and it covers larger area as compared with UAV. UAV, however, does have its own advantages over satellite imagery, and they are low cost and high image clarity.

For image resolution, it is interesting to see that the researcher's approaches are different as they employed different image resolutions ranged from as big as 20,782 × 14,420 pixels to as low as 250 × 250 pixels. This can be explained that they were using different methods to make use of the data therefore only a certain resolution of image suits their needs. As for image spatial resolution, those who used satellite imagery have their image spatial resolution ranged from 0.3 m per pixel to 2.4 m per pixel.

Furthermore, the methods of processing data and making it useful are different from one to another. CNN takes the majority of the places followed by the other methods like machine learning, support vector machine (SVM), Haar-based method, etc. Every method used yields different performance or accuracy.

3 Methodology

This study is conducted based on the flowchart as shown in Fig. 4. It has the high-resolution remote sensing images as the inputs. The inputs are then preprocessed and further categorized into two dataset subgroups, which are training set and testing set. Image datasets are randomly selected from the initial images. Training set and testing set are used to train the CNN model. During this process, parameter optimization is

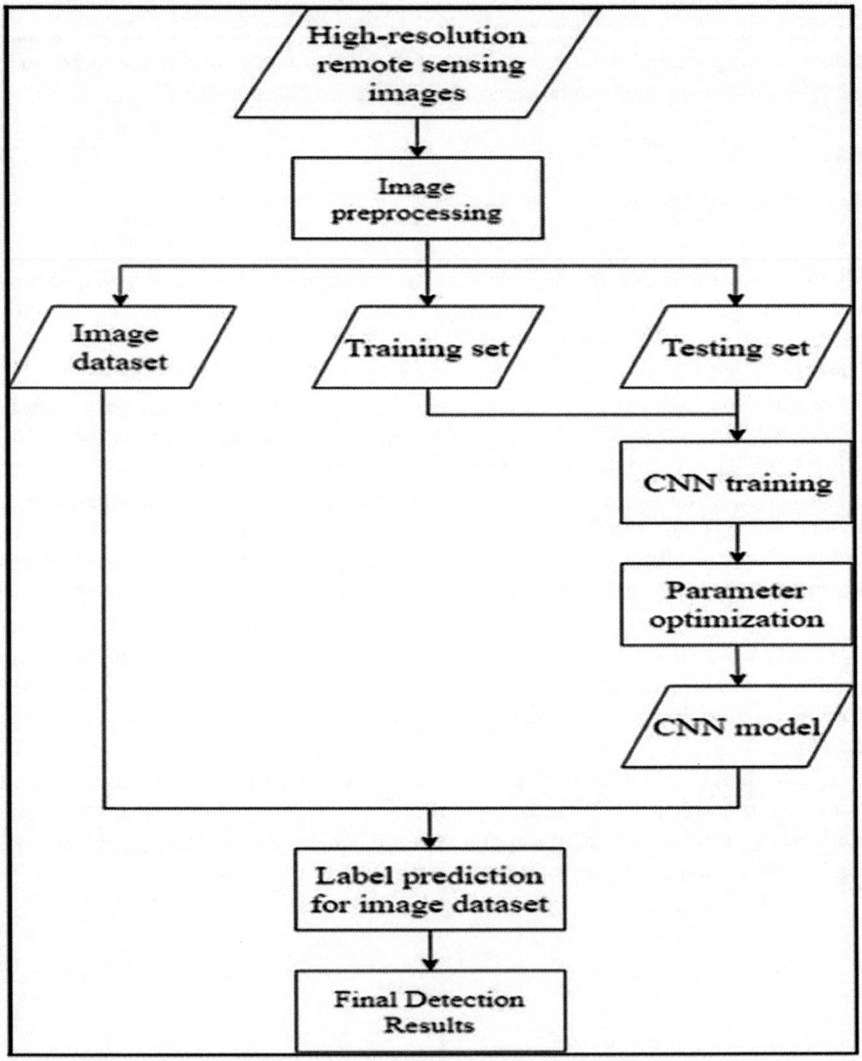

Fig. 4 Research flowchart

carried out. After a CNN model is trained, label prediction is performed in which the image dataset is tested with the obtained model. Finally, final detection results are shown, and the performance of the model is evaluated.

In this research, the preparation of dataset is conducted at an oil palm tree plantation located at Jalan Chin Ping, 94,300, Kota Samarahan, Sarawak, Malaysia. The area of the plantation is approximately 85,000 m^2, which is large enough to have a sufficient number of oil palm trees to be the study region.

Dataset acquisition is first performed by flying the drone over the study region and capturing the images of oil palm plantation. The images should have high resolution, preferably 5472 × 3078 pixels or above and every single tree in the images should hold at least 227 × 227 pixels [11, 12]. After that, image preprocessing is carried out to crop out every single oil palm tree with a fixed size of 227 × 227 pixels from the images as well as the background for training purposes as illustrated in Fig. 5. 1000 oil palm tree samples and background samples are needed. Next, these oil palm tree samples are divided into two subgroups, which are training set and testing set.

After that, CNN training is carried out and parameter optimization is constantly performed to get the best model performance. Here, transfer learning is applied by modifying AlexNet. After the model is trained, label prediction on the image dataset is then carried out by using it with the sliding window technique.

The sliding window technique [13] as illustrated in Fig. 6 is a technique that runs the CNN classification over the entire image. The part of the image framed by the red box will be classified by the model and labeled. The red box will then slide

Fig. 5 Tree cropping

Fig. 6 Sliding window

horizontally across the rest of the image to perform the next classification. The loop keeps on going until the whole image is completely classified. When oil palm tree is detected, a red dot will be left at its coordinate.

During label prediction, the classifier might detect the same tree multiple times, leaving more than one coordinate, which will affect the counting. To avoid this situation, an algorithm of grouping ups all the nearby coordinates is used. The mean coordinates of those coordinates are obtained as the only coordinate of that tree.

The performance of the model is evaluated with three measurement metrics. There is precision, recall and overall accuracy. The precision is the probability that an oil palm tree is detected correctly. The precision is defined as the number of true positives (TP) over the number of true positives plus the number of false positives (FP). The recall is the probability that an oil palm tree in ground truth is detected. The recall is defined as the number of true positives (TP) over the number of true positives plus the number of false negatives (FN). The overall accuracy is the average of precision and recall.

4 Results

For testing, three images with different number of trees, height and complicity are chosen and tested with the system. To obtain the respective accuracy, the test result is compared with the ground truth and carried out the calculation.

For test 1, image with small number of trees is selected. The oil palm trees in the image also properly separate from each other, forming a simple scene. On the left is the image after the testing while the original image is on the right (see Fig. 7). For test 2, image with increased number of trees is selected. This time, the trees do not separate from each other but with tree branches overlapping each other (see Fig. 8). For test 3, image with more number of trees is selected. Not only the tree branches are overlapping each other but also it is rather crowded (see Fig. 9).

Fig. 7 Test 1 result (left), original image (right)

Fig. 8 Test 2 result (left), original image (right)

Fig. 9 Test 3 result (left), original image (right)

Table 1 Detection and counting result (%)

Test Number	Precision	Recall	Accuracy
1	100	100	100
2	77.2	100	88.6
3	87.0	80.0	83.5

As shown in Table 1, testing 1 achieved 100% of precision and recall. In testing 2, the precision obtained is the least, which is only 77.2% but it is able to obtain 100% of recall, in testing, 3, 87.0% for precision and 80.0% recall. For overall accuracy, it can be observed that the overall accuracy is 100% in testing 1, 88.6% in testing 2, while achieving 83.5% in testing 3.

Figure 10 shows the oil palm tree detection and counting app. After inserting oil palm trees image, user gets to choose the value of resize, window size and sliding step size. After running the test, the app will tell user the number of oil palm tree detected in the image.

After getting the best model, it is then built into an app using App Designer that allows user to carry out classification on the oil palm trees.

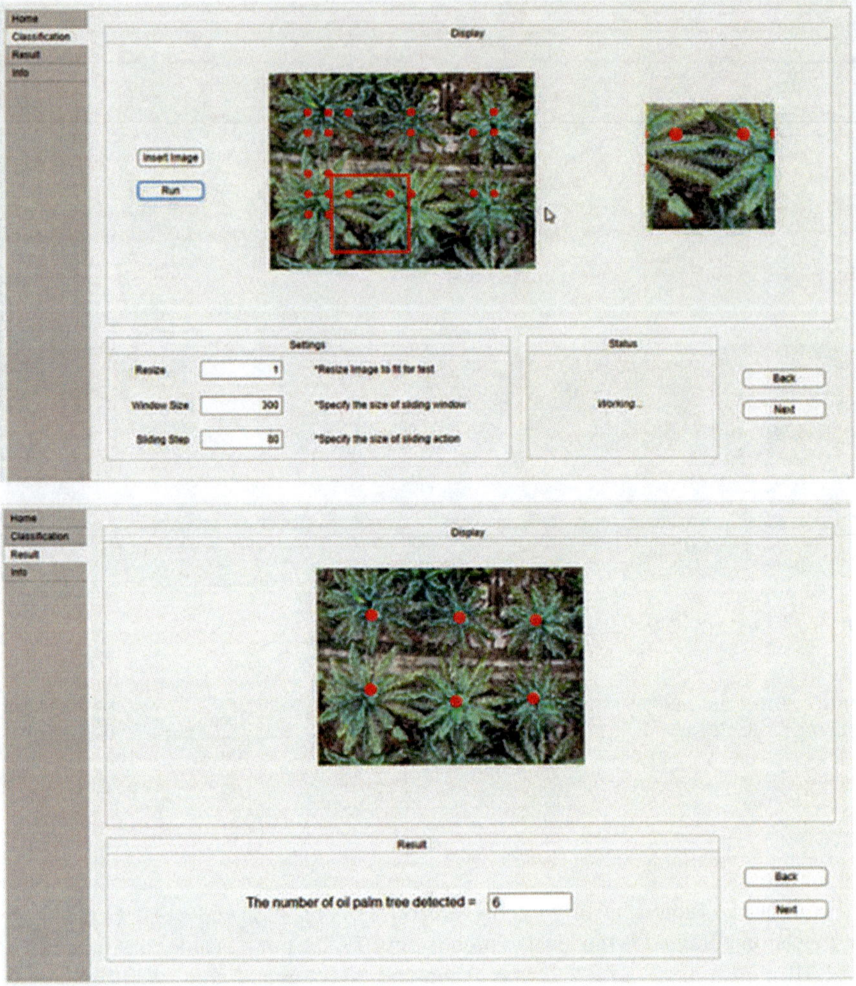

Fig. 10 Oil palm tree detection and counting app

5 Discussion

In term of accuracy, the system can actually be able to achieve 83.5%–100%. There are several factors that affect the performance include image complicity, window size and sliding step.

For complicated image, when two or more tree branches are overlapping each other, the classifier would detect and count it as one oil palm tree. When this happens, it causes the difference between detected number of oil palm trees and actual number of oil palm tree. Besides, in crowded region where small and bigger trees stick to each other, precision issue could happen thus lowering the performance. It can be said that the higher the tree crowdedness, the lower the classifier accuracy. The classifier performs the best when dealing with a neatly arranged tree. In this case, using it on detecting and counting young oil palm tree in a plantation could be better instead of on crowded tree region.

Moreover, window size and sliding step could heavily affect the overall accuracy. Window size needs to be adjusted properly because if it is too big, it could perform classification on multiple trees at once, but it is undesired as it is the best that the window size has same size as a single tree and only classifies one tree at once [10]. Sliding step is the moving distance of the sliding window in each step. Sliding step must be carefully selected too to avoid missing out the detection of oil palm trees especially when it is too big.

The advantage of our system is that it automates the counting work using CNN method and it shows high accuracy as well. The disadvantages include lower performance on certain image conditions with high complicity. Besides, the user might have to adjust window size and sliding step multiple times before getting the best result. The significance of this project is that it proves that the deep learning method can be applied in the agriculture field.

6 Conclusion

In conclusion, a deep learning-based method for oil palm tree detection and counting was achieved. The trained CNN model carried out detection and counting of oil palm tree by applying sliding window technique and it is proved to have fairly good performance of classification. A simple app was also developed for a better user experience.

From this study, it is shown that deep learning could be very helpful in precision farming. Not only the number of trees but it could also be designed to perform various tasks that reduce human efforts such as monitoring the health of trees and many others. It is recommended that the system could be working in real time. For example, instead of using still images, real-time detection and counting can be performed on the video taken from the drone.

Acknowledgements The authors would like to thank University Malaysia Sarawak (UNIMAS) for supporting this research.

References

1. Santoso, H., Tani, H., Wang, X.: A simple method for detection and counting of oil palm trees using high-resolution multispectral satellite imagery. Int. J. Remote Sens. **37**(21), 5122–5134 (2016)

2. Pouliot, D., King, D., Bell, F., Pitt, D.: Automated tree crown detection and delineation in high-resolution digital camera imagery of coniferous forest regeneration. Remote Sens. Environ. **82**(2–3), 322–334 (2002)
3. Shafri, H., Hamdan, N., Saripan, M.: Semi-automatic detection and counting of oil palm trees from high spatial resolution airborne imagery. Int. J. Remote Sens. **32**(8), 2095–2115 (2011)
4. Zakharova, M.: Automated coconut tree detection in aerial imagery using deep learning (2019). https://iiw.kuleuven.be/onderzoek/eavise/mastertheses/mariia-zakharova.pdf. Convolutional Neural Network, *Mathworks.com*, 2019
5. Dormer, J., Halicek, M., Ma, L., Reilly, C., Fei, B., Schreibmann, E.: Convolutional neural networks for the detection of diseased hearts using CT images and left atrium patches, Medical Imaging 2018: Computer-Aided Diagnosis (2018)
6. Chen, S., He, H.: Stock prediction using convolutional neural network. IOP Conference Series: Materials Science and Engineering, Vol 435, p. 012026 (2018)
7. Monthly Export of Oil Palm Products 2019, Bepi.mpob.gov.my, 2019. http://bepi.mpob.gov.my/index.php/en/statistics/export/371-export-2019/925-monthly-export-of-oil-palm-products-2019.html. Accessed 19 Dec 2019
8. Li, W., Fu, H., Yu, L., Cracknell, A.: Deep learning based oil palm tree detection and counting for high-resolution remote sensing images. Remote Sens. **9**(1), 22 (2016)
9. Malek, S., Bazi, Y., Alajlan, N., AlHichri, H., Melgani, F.: Efficient framework for palm tree detection in UAV images. IEEE J. Selected Topics in Appl. Earth Observ. Remote Sens. **7**(12), 4692–4703 (2014)
10. Manandhar, A., Hoegner, L., Stilla, U.: Palm tree detection using circular autocorrelation of polar shape matrix, in ISPRS Annals of Photogrammetry, Remote Sensing and Spatial Information Sciences, Vol. 3, pp. 465–472 (2016)
11. Coskun, M., Ucar, A., Yildirim, O., Demir, Y.: Face recognition based on convolutional neural network, in 2017 International Conference on Modern Electrical and Energy Systems (MEES) (2017)
12. Mubin, N., Nadarajoo, E., Shafri, H., Hamedianfar, A.: Young and mature oil palm tree detection and counting using convolutional neural network deep learning method. Int. J. Remote Sens. **40**(19), 7500–7515 (2019)
13. Daliman, S., Abu-Bakar, S., Azam, S.: Development of young oil palm tree recognition using Haar- based rectangular windows. IOP Conference Series: Earth and Environmental Science, vol 37, p. 012041 (2016)

Correction to: Cognitive Engineering for AI: An Octave Drawing Test for Building a Mathematical Structure of a Subconscious Mind

Anindya Pattanayak, Tanusree Dutta, Piyush Pranjal, Pushpendra Singh, Pathik Sahoo, Soumya Sarkar, and Anirban Bandyopadhyay

Correction to:
Chapter "Cognitive Engineering for AI: An Octave Drawing Test for Building a Mathematical Structure of a Subconscious Mind" in: M. S. Kaiser et al. (eds.), *Proceedings of the Third International Conference on Trends in Computational and Cognitive Engineering*, Lecture Notes in Networks and Systems 348,
https://doi.org/10.1007/978-981-16-7597-3_11

In the original version of the book, the following affiliation correction have been incorporated in Chapter "Cognitive Engineering for AI: An Octave Drawing Test for Building a Mathematical Structure of a Subconscious Mind".

The affiliation of author "Piyush Pranjal" affiliation has been corrected from "Management Department, Jindal Global University, Sonipat, O.P 131001, India" to "Management Department, O. P. Jindal Global University, Sonipat 131001, India".

The chapter and book have been updated with the changes.

The updated version of this chapter can be found at
https://doi.org/10.1007/978-981-16-7597-3_11

M. S. Kaiser et al. (eds.), *Proceedings of the Third International Conference on Trends in Computational and Cognitive Engineering*, Lecture Notes in Networks and Systems 348, https://doi.org/10.1007/978-981-16-7597-3_46

Author Index

M. S. Kaiser et al. (eds.), *Proceedings of the Third International Conference on Trends in Computational and Cognitive Engineering*, Lecture Notes in Networks and Systems 348, https://doi.org/10.1007/978-981-16-7597-3